2nd International Seminar on Ocean Sciences and Sustainability (ISOSS 2024)

BIO Web of Conferences Volume 168 (2025)

Bogor, Indonesia
5 July 2024

ISBN: 979-8-3313-1630-3

Printed from e-media with permission by:

Curran Associates, Inc.
57 Morehouse Lane
Red Hook, NY 12571

Some format issues inherent in the e-media version may also appear in this print version.

For additional information, please contact EDP Sciences – Web of Conferences at the address below.

EDP Sciences – Web of Conferences
17, Avenue du Hoggar
Parc d'Activité de Courtabœuf
BP 112
F-91944 Les Ulis Cedex A
France

Phone: +33 (0) 1 69 18 75 75
Fax: +33 (0) 1 69 28 84 91

contact-edps@webofconferences.org

Additional copies of this publication are available from:

Curran Associates, Inc.
57 Morehouse Lane
Red Hook, NY 12571 USA
Phone: 845-758-0400
Fax: 845-758-2633
Email: curran@proceedings.com
Web: www.proceedings.com

TABLE OF CONTENTS

Departemen ITK
Fakultas Perikanan dan Ilmu Kelautan
IPB University
ISOSS
Kampus Merdeka
gef
lpdp
WORLD UNIVERSITY RANKINGS
NGI
HYBRID CONFERENCE
THE 2nd INTERNATIONAL SEMINAR ON
OCEAN SCIENCES AND SUSTAINABILITY (ISOSS) 2024
BOGOR, JULY 8 - 9, 2024
AMBON, OCT 9 - 10, 2024
Mapping the Future : Hydrography, Ocean Sciences, and Sustainability
BIO Web of Conferences
Selected papers will be published in International reputable index Scopus Proceeding (BIO web of conf)
IMPORTANT DATES
01 CALL FOR ABSTRACT
02 DEADLINE FOR ABSTRACT SUBMISSION
03 ANNOUNCEMENT OF ABSTRACT APPROVAL
04 FULL PAPER SUBMISSION
05 EVENT
06 PAPER PUBLISHED
MORE INFORMATION :
+62 822 1004 4968
(Ratna Suhita)
Web : https://ipb.university/isoss2024
Email : isoss@apps.ipb.ac.id
OPENING SPEECH
Prof. Arif Satria
Rector of IPB University
Indonesia
WELCOME SPEECH
Prof. Fredinan Yulianda
Dean of Faculty of Fisheries and Marine Sciences, IPB University
Indonesia
KEYNOTE SPEAKER
Vice Admiral TNI Dr. Budi Purwanto, ST, MM
Indonesian Chief Hydrographer
Indonesia
INVITED SPEAKERS
Prof. Neviaty P Zamani
IPB University
Indonesia
Prof. Henry M Manik
IPB University
Indonesia
Dr. Ong Meng Chuan
Universiti Malaysia Terengganu
Malaysia
Dr. Gonzalo Perez-Rosales
Hong Kong University of Science and Technology
Hongkong
TOPICS :
Climate change and ocean threats
Marine bioprospecting and food security
Renewable blue energy
Sustainable blue economies
Sustainable ocean technology
This event will be held in Bogor & Ambon, there will be an excursion to Banda Neira On October 2024
CALL FOR PAPERS
Registration & Upload your Abstract on
May 1 - July 6, 2024
itk.ipb.ac.id
isoss@apps.ipb.ac.id
itk.ipb

edp sciences

Web of Conferences

Statement of Peer review

In submitting conference proceedings to *Web of Conferences*, the editors of the proceedings certify to the Publisher that

1. They adhere to its **Policy on Publishing Integrity** in order to safeguard good scientific practice in publishing.
2. All articles have been subjected to peer review administered by the proceedings editors.
3. Reviews have been conducted by expert referees, who have been requested to provide unbiased and constructive comments aimed, whenever possible, at improving the work.
4. Proceedings editors have taken all reasonable steps to ensure the quality of the materials they publish and their decision to accept or reject a paper for publication has been based only on the merits of the work and the relevance to the journal.

Title, date and place of the conference The 2nd International Seminar on Ocean Sciences and Sustainability Bogor, July 5th 2024 and Ambon, November 4-5th

Proceedings editor(s):

March 13, 2025

Prof/ neviatyt P. Zamani MSc

Date and editor's signature

Statement of Peer review

In submitting conference proceedings to *Web of Conferences*, the editors of the proceedings certify to the Publisher that

1. They adhere to its **Policy on Publishing Integrity** in order to safeguard good scientific practice in publishing.
2. All articles have been subjected to peer review administered by the proceedings editors.
3. Reviews have been conducted by expert referees, who have been requested to provide unbiased and constructive comments aimed, whenever possible, at improving the work.
4. Proceedings editors have taken all reasonable steps to ensure the quality of the materials they publish and their decision to accept or reject a paper for publication has been based only on the merits of the work and the relevance to the journal.

Title, date and place of the conference: The 2nd International Seminar on Ocean Sciences and Sustainability Bogor, July 5th 2024 and Ambon, November 4-5th 2024

Proceedings editor(s): Prof. Henry M. Manik, Ph.D.

Date and editor's signature

edp sciences

Web of Conferences

Statement of Peer review

In submitting conference proceedings to *Web of Conferences*, the editors of the proceedings certify to the Publisher that

1. They adhere to its **Policy on Publishing Integrity** in order to safeguard good scientific practice in publishing.
2. All articles have been subjected to peer review administered by the proceedings editors.
3. Reviews have been conducted by expert referees, who have been requested to provide unbiased and constructive comments aimed, whenever possible, at improving the work.
4. Proceedings editors have taken all reasonable steps to ensure the quality of the materials they publish and their decision to accept or reject a paper for publication has been based only on the merits of the work and the relevance to the journal.

Title, date and place of the conference: The 2nd International Seminar on Ocean Sciences and Sustainability Bogor, July 5th 2024 and Ambon, November 4-5th

Proceedings editor(s): Prof. Dr. Jonson Lumban Gaol

Date and editor's signature: 10 March 2025

Biosorption of copper by *Alcaligenes faecalis* and *Delftia tsuruhatensis* isolated from wastewater

Yumna Husna Nisaa[1,2*], *Anto Budihardjo*[2], *Sri Pujiyanto*[2]

[1]Biolology Graduate Program, Faculty of Sains and Mathematics, Diponegoro University
[2]Department of Biology, Faculty of Sains and Mathematics, Diponegoro University

Abstract: Biosorption techniques can remove Cu. Copper-resistant bacteria isolated from Cu-contaminated environments. This study test the ability of *Alcaligenes faecalis* and *Delftia tsuruhatensis* to absorb Cu. The concentrations of copper used for the biosorption tests were 50, 100, and 150 mg/L in the nutrient broth media. The concentration of Cu in the nutrient broth media after the biosorption test was measured using an atomic absorption spectrophotometer. *A Scanning Electron Microscope* (SEM) was used to confirm metal biosorption on bacterial cell surfaces. The ability of *Alcaligenes faecalis* to remove copper was 34%, and *Delftia tsuruhatensis* was 45% at the highest concentration. The results showed that *Delftia tsuruhatensis* has the highest capacity for copper biosorption. The results of the condition of *Delftia tsuruhatensis*, with copper metal (Cu), showed that the bacterial cells had an irregular cell shape, and there was a build-up of heavy metal molecules.

1 Introduction

Water pollution can cause public health problems, reduce agricultural productivity, and damage natural biology [1]. The Indonesian Central Statistics Agency reported that water pollution will affected 6.160 villages in 2021. Heavy metals are hazardous pollutants in the environment because they can be accumulate in organisms that consume contaminated food in the food chain [2]. Copper is an example of toxic heavy metal at high concentrations in living organisms at high concentrations [3]. It can cause liver and kidney damage, anemia, and irritation of the digestive tract [4]. Bioremediation could be a useful bacterial technique for reducing or recovering heavy metal to minimize their harmful forms [5].

The advantages of using the bioremediation process to reduce heavy metal waste are cost-effectiveness and environmental friendliness. This bacterial utilization technique is likely able to degrade pollutants and convert them into harmless properties through a redox process

* Corresponding author: yumnahusnanisaa06@gmail.com

[5]. Bacteria are biosorbents that can help improve polluted environments because these cells can grow under controlled conditions and survive in polluted environments [6]. The mechanism of heavy metal removal occurs through metabolic or physicochemical processes by the biomass of active or inactive bacterial cells [7], [8]. It occurs in bacterial cell walls [9]. Bacterial cells must adapt to changes in physical, chemical, and bioreactor configurations during metal remediation [10].

The bacteria used in the bioremediation process were isolated from an environment polluted with heavy metals. In this study, the bacteria were isolated from Sayung Demak ditch surrounded by industrial surroundings, including electronics, pet food, and textile companies. Recent research has shown that the copper content in the Gonjol River is 0.05 mg/L - 0.42 mg/L [11] ,which exceeds the water quality threshold 0.02 mg/L. A previous study showed that nine bacteria isolated from domestic liquid waste were resistant to Cu at a concentration of 25 mg/L [12]. Bacteria isolated from polluted environments have been used for the biosorption of copper metal, such as bacteria from the Cisadane River, namely *Pantoea agglomerans* (1.42 mg/g), *Klebsiella penumoniae* (1.55 mg/g), and *Shigella flexneri* (0.92 mg/g) [13]. Therefore, this research to show the effectiveness *Alcaligenes faecalis* and *Delftia tsuruhatensis* of the Sayung Demak ditch as biosorption agents for copper metals.

2 Materials and methods

2.1 Isolation and identification of bacteria

Bacteria were isolated from the wastewater samples using a serial dilution technique (dilution 10^{-1} to 10^{-4}) [14]. Next, 0.1 ml of each of these dilutions was inoculated using the spread plate technique on the nutrient agar media, which had been supplemented with 5 mg/L of Cu metals and incubated at 37 °C for 2 days. Bacterial growth was purified and then characterized based on elevation, edges, color, shape of bacterial colonies, and optical appearances, such as cell walls and shapes of bacterial cells [15]. Purified bacterial isolates were inoculated using the continuous streak plate method on NA medium, supplemented with Cu. The Cu concentrations used were 50, 100, and 150 mg/L. The culture were then incubated for 24 h at room temperature. Colony growth indicated that resistance to Cu metals was isolated. Isolates grow at a concentration of 150 mg/L were selected for further research. Isolation of bacterial DNA using the Quick-DNA Fungal/Bacteria Minirep kit. Bacterial DNA amplification was carried out using 16S rRNA primer, 27F and 1492R [16]. DNA sequencing was performed using Sanger Sequencing. The results of the bacterial 16S rRNA DNA sequencing were analyzed using the Basic Local Alignment Search Tool (BLAST) from NCBI (https://www.ncbi.nlm.nih.gov/). Phylogenetic tree construction was made using software called (MEGA) version 10.0 [17].

2.2 Test the ability of bacteria to remove copper metals

20 ml bacterial isolates were inoculated on to NB (80 ml incubated at 150 rpm for 24 h). A bacterial starter culture (100 ml was inoculated aseptically into 100 ml of nutrient broth liquid medium and incubated for up to 12 h. The starter culture on NB contained 50, 100, and 150) mg/L Cu up to a volume of 25 ml. Incubation was performed on a rotary shaker (150 rpm) for 24 h. The culture was centrifuged at 4500 rpm for 5 min to separate the cells from the filtrate that was collected and put into a test tube by; adding HNO_3. Cu metal was analyzed

using an atomic absorption spectrophotometer [18]. The percentage of heavy metals absorbed by the bacteria was calculated as follows [19]:

$$Q\,\% = \frac{Ci - Cf}{Ci}\ x\ 100\%$$

Notes:
Ci = initial concentration
Cf = final concentration of metal ions.

2.3 Scanning electron microscopy analysis

The surface morphology of the bacterial cells after the biosorption test for heavy metals was analyzed using scanning electron microscopy (SEM). Samples for observation using a scanning electron microscope were bacterial cells grown in nutrient broth media and bacterial cells grown in nutrient broth media containing copper metals. The bacterial isolates analyzed exhibited the highest ability to reduce copper.

3 Results and discussion

3.1 Morphology and identification of bacterial resistant to copper (Cu)

The isolation results from the Sayung Demak ditch showed two bacterial isolates that grew on Nutrient Agar media containing 150 mg/L of copper metal (Cu) YL18 and YL315. These bacterial colonies exhibited macroscopic and microscopic morphological differences. Microscopic observations showed that the isolates had an bacilli shape. Gram-negative bacteria were detected by Gram staining (Table 1). The Gram-negative bacteria in this study were able to survive in media with a concentration of 150 mg/L of Cu metal. Gram-negative bacteria contain lipopolysaccharides (LPS), proteins, and phospholipids. Lipopolysaccharides and phospholipids in the cell walls of gram-negative bacteria are composed of phosphate and carboxyl groups, respectively, which are compounds capable of binding metal ions [20].

Table 1. Morphological Characteristics of Bacteria

Isolate Code	Bacterial Colony Morphology						
	Colony Type					Microscopic	
	Color	Margins	Elevation	Texture	Form	Grams	Cell Shape
YL18	White	Lobalate	Flat	Moist	Irregular	Negative	Bacilli
YL315	Beige	Entire	Concex	Mucoid	Circular	Negative	Bacilli

The results of the YL18 and YL315 isolates were analyzed using the blast program on the NCBI (National Center of Biotechnology Information) website to align the DNA sequences of YL18 and YL315 isolates with the bacterial DNA sequence database available on the NCBI (National Center of Biotechnology Information) website. The results of this process provided information in the form of species that were closely related to isolates YL18 and YL315 (Table 2). Species derived from alignment with the NCBI database and bacterial isolates YL18 and YL315 were reconstructed into a phylogenetic tree, as shown in Fig 1.

Table 2. Blast Results

Isolat e Code	Result Blast					
	Species Name	Ma x Score	Quer y Cover	E - Value	Identit y	Accession
YL18	*Alcaligene s faecalis*	177 9	100%	0	98,51%	NR_11360. 1
YL31 5	*Delftia tsuruhatensis*	256 2	100%	0	100%	MT374626. 1

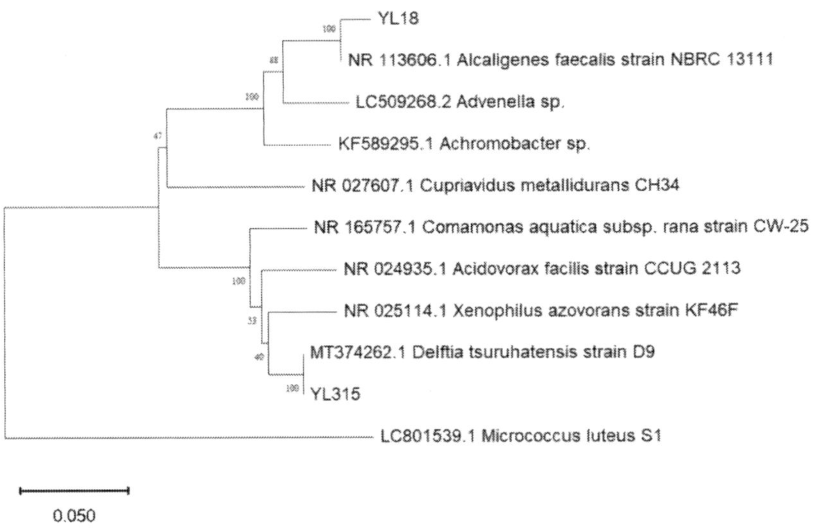

Fig 1. Phylogenetic tree reconstruction of bacterial isolates YL18 and YL315

The 16S rRNA gene isolate YL18 is the *species Alcaligenes faecalis*. The YL18 isolate derived from wastewater has characteristics of bacilli-shaped cells and gram-negative bacteria. The characteristics of *Alcaligenes faecalis* are rod-shaped (bacilli), gram-negative, aerobic, and polyvitric. Optimal growth of *Alcaligenes faecalis* bacteria isolated from waste water and soil at temperatures ranging from 20 to 37 °C [21]. The *Alcaligenes* genus can grow in environments containing Ni, Cd, Cu, Zn, Pb, and Cr. *Alcaligenes* contain bacterioferritin, porin, ABC transporters, and ATPase, which form a metal homeostasis system so that *Alcaligenes* survive in polluted environments [22]. The bacterial isolate, YL315, was identified as *Delftia tsuruhatensis*. *Delftia tsuruhatensis* is a rod-shaped gram-negative bacterium. *Delftia tsuruhatensis* was isolated from domestic wastewater in Japan [23]. *Delftia tsuruhatensis* has the genes copA, copB, copC, copF, copG, 2copR, copS, and cusA, which can live in an environment containing copper metal [24].

3.2 The ability of bacteria to remove copper

Biosorption testing of copper metals showed a decrease in metal concentrations in nutrient broth media without bacteria (Table 3). This decrease can be attributed to the presence of peptone in the nutrient broth medium, which is composed of amino acids containing reactive groups, such as carboxylic and amine groups [25]. The negatively charged carboxylate and amine groups can bind positively charged metals such as zinc and copper [26]. This causes the Nutrient Broth media to experience a decrease in the concentrations of zinc and copper metals. A decrease in metals in nutrient broth media without bacteria also occurred in a previous study [27] by as much as 24% at a concentration of 5 ppm.

Table 3. Concentration Media NB Without Bacteria

	Initial Concentration (mg/L)	**Final Concentration (mg/L)**
Copper (Cu)	50	1,024 (K1)
	100	1,027 (K2)
	150	1,454 (K3)

The test results for Cu metal removal showed that increasing the Cu metal concentration would decreased the absorption capacity (Table 4). The concentration of Cu without bacteria (50 mg/L) decreased to 1,024 mg/L, while in *Alcaligenes faecalis* bacteria it decreased to 1,1995 mg/L (Table 4). This condition shows that the final concentration of copper metal without bacteria is better than the addition of *Alcaligenes faecalis*, resulting in a negative copper metal reduction efficiency value (Table 4). The test results of *Alcaligenes fecalis* in absorbing copper metal show an upward and downward graph (Fig 2). The up and down values of metal absorption occurs because of the desorption ability of bacteria as a form of self-defense in environmental conditions that are not favorable for bacteria. Desorption is the process of releasing ions bound to the active groups on the adsorbent. In addition, it is also influenced by the surface area of the cell wall, aeration, and the specific gravity of the metal [28] .

Table 4. Ability of bacterial biosoption copper

Bacteria	**Initial Concentration (mg/L)**	**Final Concentration (mg/L)**	**Bioremoval Efficiency (%)**
Alcaligenes faecalis	1,024 (K1)	1,1995	-17%
	1,027 (K2)	0,625	39%
	1,454 (K3)	0,965	34%
Delftia tsuruhatensis	1,024 (K1)	0,349	66%
	1,027 (K2)	0,4635	55%
	1,454 (K3)	0,8015	45%

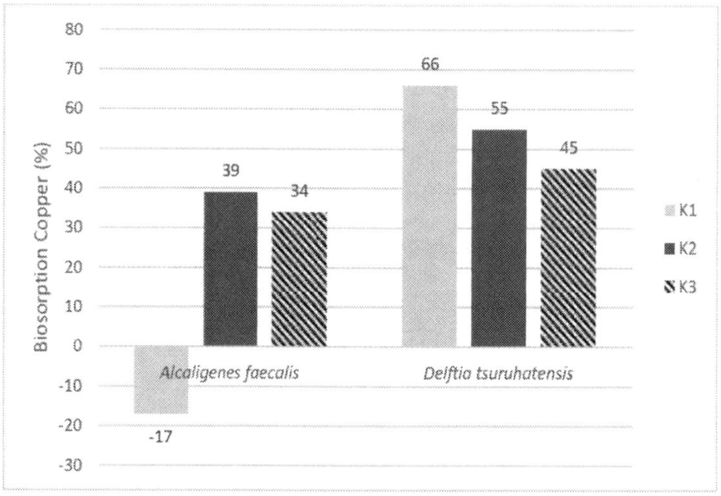

Fig 2. Bioremoval Efficiency of Copper by Bacterial Isolates

Delftia tsuruhatensis exhibited the highest copper reduction ability, with K3 showing the greatest reduction value of 45% (Fig 2). *Rhodococcus erythropolis* exhibited an absorption ability of 40% in copper media at a concentration of 0.5 mg/L. An increase in the concentration of copper metal causes a decrease in the percentage of copper removal [29]. The process of absorbing heavy metals by bacteria is carried out by ion exchange on cell surfaces, precipitation on cell and extracellular surfaces, physical adsorption on cell surfaces, complexation on cell surfaces, and transport of heavy metal ions [8].

3.3 Scanning electron microscopy analysis

The condition of *Delftia tsuruhatensis* cells before the addition of copper was observed using Scanning Electron Microscopy (Fig 3A and 3B). The results of the condition of *Delftia tsuruhatensis*, with copper metal (Cu), showed that the bacterial cells had an irregular cell shape, and there was a build-up of heavy metal molecules (Fig 3). The addition of metals to bacterial media can induce changes in the surface structure of the bacterial cells. Changes in the structure of bacterial cells in the presence of heavy metals were probably caused by the toxicity of metal ions on bacterial cell walls and the process of secretion of extracellular polymeric substances during metal biosorption [30]. In addition, a characteristic indicating the process of metal absorption is that the surface of the bacterial cells is covered with heavy metal molecules. This was possibly due to the process of secretion of extracellular polymeric substances or metabolism through metal binding to the cell surface, ion exchange, or precipitation [31].

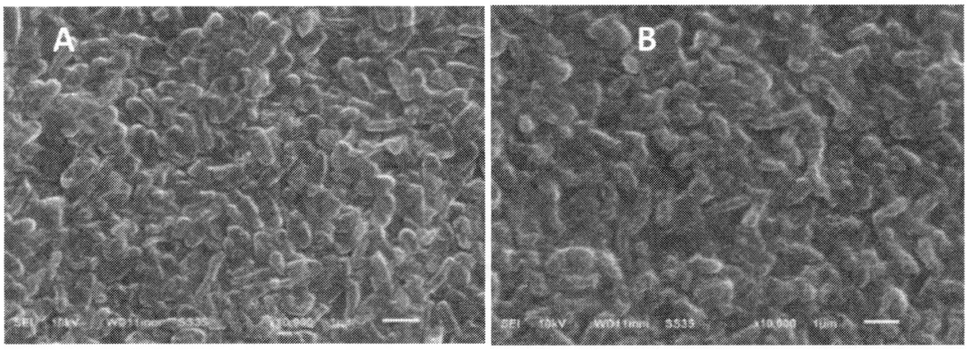

Fig 3. Cultured in liquid NB medium in the following treatments: A. control B. absence of metals Cu 50 mg/L

4 Conclusion

Alcaligenes faecalis and *Delftia tsuruhatensis* are indigenous Cu-resistant bacteria isolated from Sayung Demak. The ability of *Alcaligenes faecalis* to remove copper was 34% and *Delftia tsuruhatensis* was 45% at the highest concentration. The results showed that *Delftia tsuruhatensis* has the highest capacity for copper biosorption and condition bacterial cell added copper had an irregular cell shape, and there was a build-up of heavy metal molecules.

References

[1] C. N. Elizabeth, M. Y. Victoria, E. E. Nkechi, and B. O. Godwin, *African J. Biotechnol.*, **vol. 16**, no. 1, pp. 32–40, (2017), doi: 10.5897/ajb2016.15676.

[2] A. M. Elbedwehy, A. M. Abou-Elanwar, A. O. Ezzat, and A. M. Atta, *Polymers (Basel).*, **vol. 11**, no. 12, (2019), doi: 10.3390/polym11121938.

[3] S. Siddiquee, K. Rovina, and S. Al Azad, *J. Microb. Biochem. Technol.*, **vol. 07**, no. 06, pp. 384–393, (2015), doi: 10.4172/1948-5948.1000243.

[4] A. Abbas, S. Mohamed, and H. Zahir, *Int. Res. J. Eng. Technol.*, **vol. 03**, no. 08, pp. 1446–1450, (2016).

[5] O. B. Ojuederie and O. O. Babalola, *Int. J. Environ. Res. Public Health*, **vol. 14**, no. 12, (2017), doi: 10.3390/ijerph14121504.

[6] S. Srivastava, S. B. Agrawal, and M. K. Mondal, *Environ. Sci. Pollut. Res.*, **vol. 22**, no. 20, pp. 15386–15415, (2015), doi: 10.1007/s11356-015-5278-9.

[7] O. Abdi and M. Kazemi, *J. Mater. Environ. Sci.*, **vol. 6**, no. 5, pp. 1386–1399, (2015).

[8] E. A. Perpetuo and C. B. Souza, *Curr. Opin. Biotechnol.*, **vol. 11**, no. 3, pp. 262–270, (2011), doi: 10.1016/S0958-1669(00)00094-X.

[9] S. L. R. K. Kanamarlapudi, V. K. Chintalpudi, and S. Muddada, *Biosorption*, (2018), doi: 10.5772/intechopen.77315.

[10] M. Fomina and G. M. Gadd, *Bioresour. Technol.*, **vol. 160**, no. January 2014, pp. 3–14, (2014), doi: 10.1016/j.biortech.2013.12.102.

[11] H. Cerlyawati and S. Isworo, *Asian J. Biol.*, **vol. 13**, no. 115, pp. 34–51, (2021), doi: 10.9734/ajob/2021/v13i430194.

[12] L. Waluyo, *Pros. Semin. Nas. Pendidik. Biol.*, no. 2014, pp. 236–241, (2018).

[13] W. Irawati, R. Pinontoan, and T. Yuwono, *Biodiversitas*, **vol. 21**, no. 11, pp. 5077–5084, 2020, doi: 10.13057/biodiv/d211112.

[14] N. Z. S. Mazalan, A. Oyeleye, R. N. Z. R. A. Rahman, A. Z. Aris, A. B. Salleh, and Y. M. Normi, *Beni-Suef Univ. J. Basic Appl. Sci.*, **vol. 9**, no. 1, pp. 1–12, (2020), doi: 10.1186/s43088-020-00051-1.

[15] W. Irawati, *J. Biol. Papua*, **vol. 11**, no. 2, pp. 80–86, (2019), doi: 10.31957/jbp.878.

[16] G. Vasas *et al.*, *Toxins (Basel).*, **vol. 5**, no. 12, pp. 2434–2455, (2013), doi: 10.3390/toxins5122434.

[17] D. Ramya and A. J. Thatheyus, *Asian J. Biol. Sci.*, **vol. 12**, no. 4, pp. 869–876, 2019, doi: 10.3923/ajbs.(2019).869.876.

[18] A. Ahmad, R. Ghufran, and W. M. Faizal, *Clean - Soil, Air, Water*, **vol. 38**, no. 2, pp. 153–158, (2010), doi: 10.1002/clen.200900202.

[19] W. Mwandira *et al.*, *Sci. Rep.*, **vol. 10**, no. 1, pp. 1–9, (2020), doi: 10.1038/s41598-020-78187-4.

[20] S. Hussain *et al.*, *Front. Microbiol.*, **vol. 13**, no. May, (2022), doi: 10.3389/fmicb.2022.900740.

[21] É. B. Felestrino *et al.*, *PLoS One*, **vol. 15**, no. 11 November, pp. 1–25, (2020), doi: 10.1371/journal.pone.0241546.

[22] Z. Basharat, A. Yasmin, T. He, and Y. Tong, *Sci. Rep.*, **vol. 8**, no. 1, pp. 1–10, (2018), doi: 10.1038/s41598-018-21919-4.

[23] S. H. Al-Mijalli, *Appl. Environ. Soil Sci.*, **vol. 2022**, pp. 18–21, (2022), doi: 10.1155/2022/4316954.

[24] C. Cheng *et al.*, *Front. Cell. Infect. Microbiol.*, **vol. 11**, no. June, pp. 1–10, (2021), doi: 10.3389/fcimb.2021.663933.

[25] M. Vandenbossche, M. Jimenez, M. Casetta, and M. Traisnel, *Crit. Rev. Environ. Sci. Technol.*, **vol. 45**, no. 15, pp. 1644–1704, (2015), doi: 10.1080/10643389.2014.966425.

[26] V. Mishra, *Appl. Water Sci.*, **vol. 4**, no. 4, pp. 311–332, (2014), doi: 10.1007/s13201-013-0150-x.

[27] B. L. Fibriarti, N. P. Sari, and R. Fatzuarni, *Proceeding Biol. Educ. Conf.* , **vol. 15**, no. 1, pp. 880–882, (2018).

[28] F. S. Purnamawati, T. R. Soeprobowati, and M. Izzati, *Bioma*, (vol. 16), no. 2, pp. 102–113, (2015).

[29] M. D. P. G. Baltazar *et al.*, *J. Mater. Res. Technol.*, **vol. 8**, no. 1, pp. 475–483, (2019), doi: 10.1016/j.jmrt.2018.04.006.

[30] G. Pagnucco *et al.*, *Front. Microbiol.*, **vol. 14**, no. October, pp. 1–21, (2023), doi: 10.3389/fmicb.2023.1278886.

[31] S. Kumari *et al.*, *Separations*, **vol. 10**, no. 7, (2023), doi: 10.3390/separations10070393.

Coral reef health index on Tunda Island, Banten, Indonesia

M. *Rakha* Ubaidillah Faqih[2], *Idris*[1], *Mauludiyah*[2*], *Fakhrurrozi*[1], *Dian* Sari Maisaroh[2], *Marita* Ika Joesidawati[3]

[1]Indonesian Coral Reef Foundation (Yayasan TERANGI), Jalan Asyibaniyah No. 105 – 106 Pondok Jaya Cipayung Depok Jawa Barat 16438
[2]Department of Marine Science, Faculty of Science and Technology, State Islamic University of Sunan Ampel Surabaya, East Java, Indonesia
[3]Marine Science Study Program, Faculty of Fisheries and Marine Sciences, PGRI Ronggolawe University, Indonesia

Abstract: Tunda Island in Serang Regency, Banten Province, has rich marine ecosystems, including mangroves, seagrass beds, and coral reefs. However, increasing environmental pressures threaten coral reef health, necessitating an assessment of their current condition. This study aimed to evaluate the status of coral reefs in Tunda Island using a coral reef health index based on benthic and fish components. Data were collected at six sites using the Underwater Photo Transect (UPT) method for coral assessment and the Underwater Visual Census (UVC) method for reef fish. Coral cover, fleshy seaweed, and coral rubble were analyzed using the CPCe software, while fish biomass and abundance were calculated with standard fishery assessment methods. Results showed significant variations in coral cover, ranging from 2.74% to 50.12%. Fleshy seaweed cover was between 6.89% and 43.94%, whereas the coral rubble ranged from 1.44% to 37.22%. Reef fish biomass varied from 2.50 kg/ha to 85.31 kg/ha, with fish abundance between 80 and 840 individuals per hectare. The coral reef health index ranged from 1 to 4, indicating a low recovery potential due to highly fleshy seaweed and coral rubble. These findings highlight the urgent need for conservation measures to enhance coral reef resilience and recovery in Tunda Island.

1 Introduction

Tunda Island is a small island located within the administrative area of Serang Regency. Serang Regency is an administrative region in Banten Province that has the potential for non-biological resources and can be developed as a marine tourism destination to support regional and community development [1]. The intertidal zone of Tunda Island consists of three typical tropical marine ecosystems: mangrove, seagrass, and coral reef ecosystems, which are generally still in very good condition. A well-preserved aquatic ecosystem has the potential to influence the diversity of its associated biota [2].

* Corresponding author: mauludiyah@uinsa.ac.id

Coral reefs are fascinating and important marine ecosystems formed by colonies of marine microorganisms, primarily corals [3]. They create beautiful underwater labyrinths with a variety of vibrant colors and shapes. This ecosystem serves as a habitat for diverse marine species, including reef fish, mollusks, and other marine biota. Coral reefs play a crucial role in maintaining the sustainability of marine environments. In addition to providing a habitat for marine life, they act as natural wave breakers that protect coastlines from erosion. The existence of coral reefs is also important for the fishing industry because they support fisheries and provide resources for fishermen and coastal communities [4,5].

However, coral reefs worldwide face serious threats such as bleaching due to climate change, marine pollution, and destructive human activities [6]. The Directorate General of Marine Spatial Management of the Ministry of Marine Affairs and Fisheries noted that the area covered by coral reefs in Indonesian waters currently reaches 2.5 million hectares, with a total of 569 species. Unfortunately, approximately 33.82 percent of them are damaged or in poor condition [7].

Previous research on the coral reefs in the shallow waters of Tunda Island classified their health status based solely on coral cover. The study found that corals on the East Side and the North toward the East Side were in moderate to good condition, while those on the Central North Side and the North toward the West Side were categorized as poor to moderate. In contrast, the coral reefs at the reef slope were classified as damaged to poor, with the moderate category located on the North toward the East Side and the damaged to poor category on the Central South and South toward the West Side [8]. However, relying only on coral cover may not provide a comprehensive understanding of reef health.

The primary parameter traditionally used to assess coral conditions is the percentage of live coral cover, where a higher percentage indicates a healthier coral reef [9]. However, field observations have revealed cases where locations with low coral cover still had high reef fish abundance and vice versa. Therefore, to gain a more holistic understanding, a coral reef health index has been developed, incorporating both benthic and fish components. The benthic component is determined by live coral cover and the level of resilience or recovery potential, while the fish component is based on the total biomass of the target reef fish [10].

This study was conducted to assess the condition of coral reefs and to determine the coral reef health index on Tunda Island. By analyzing both benthic and fish components, this study aimed to provide a more comprehensive evaluation of reef health. These findings are expected to contribute to coral reef conservation efforts and inform management strategies for protecting the marine biodiversity in the region.

2 Research Methods

2.1 Research Sites

The research location is on the coast of Tunda Island, Serang Regency, Banten Province at coordinates LS 5°48'36.99"S - 5°48'46.44"S and BT 106°15'14.10"E - 106°17'50.56"E. Field data collection was conducted from March 5 to 7, 2024. Data processing and analysis were performed at the Indonesian Coral Reef Foundation (TERANGI) Office in Depok, West Java. Six observational points were established in this study (Fig. 1).

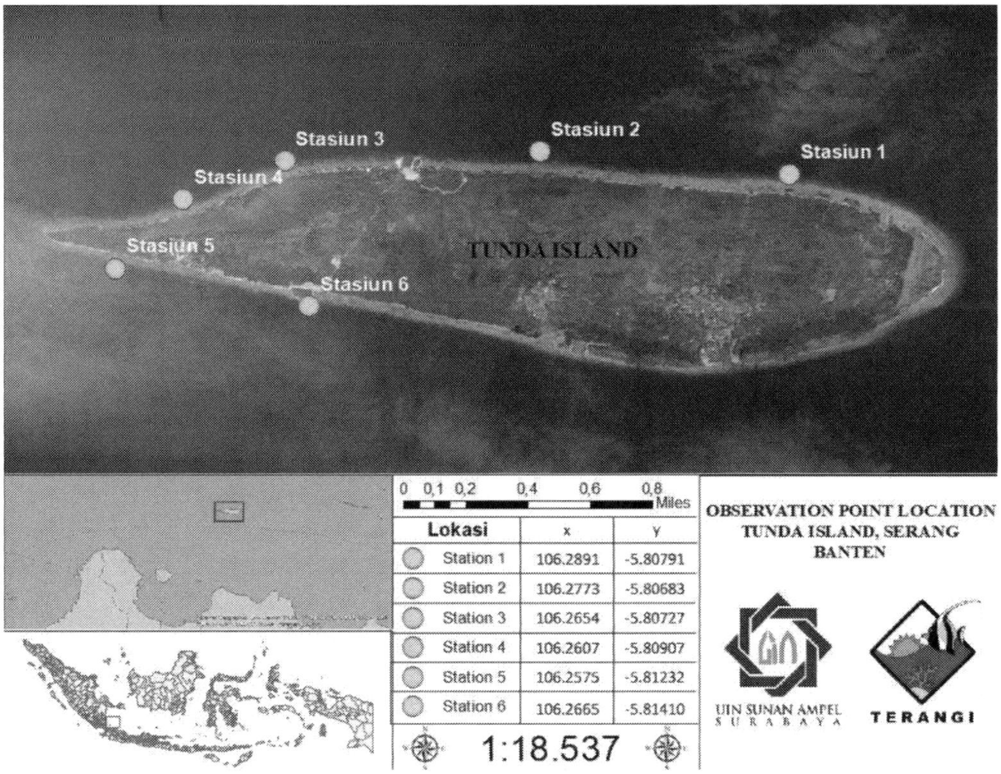

Fig. 1. *Research Location on Tunda Island, Banten*

2.2 Data Collection

2.2.1 Coral Reef Data Collection

Data collection was conducted using the Underwater Photo Transect (UPT) method (Fig. 2). This method involves SCUBA diving, capturing pictures with a high-resolution camera housed in a casing suitable for underwater operations, and using a rectangular metal frame painted in a highly visible color [11].

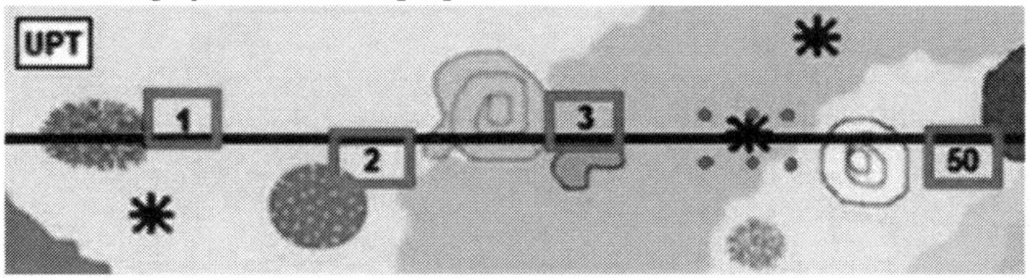

Fig. 2. *Illustration of data collection using the UPT method*

The UPT method used in this study included a slight modification: the transect length was set to 60 m, with photographs taken at 1-meter intervals using a 50 × 50 cm square transect as the photo frame. The identification method used was based on the percentage of cover assessed through snorkeling.

Coral reef data collection activities classify the data into three categories: hard coral, fleshy seaweed, and coral fragments (rubble), which serve as indicators of the coral reef health index [12]. Fleshy seaweed refers to macroalgal communities that can be seen without magnification, including species such as Sargassum, Gracilaria, and Padina. Data were collected from six stations across observation sites. The collected samples were then processed using MS Excel.

2.2.2 Reef Fish Data Collection

Fish data were collected using the Underwater Visual Census (UVC) method (Fig. 3). The UVC method was applied to count coral reef fish along a 20-meter transect, with three replications. The detection range was extended by 2.5 meters on both the right and left sides of the transect.

Data collected during the UVC survey were analyzed to estimate the population density [13]. In this study, only the target fish species were recorded, including seven families: Scaridae, Siganidae, Acanthuridae, Serranidae, Lutjanidae, Lethrinidae, and Haemulidae.

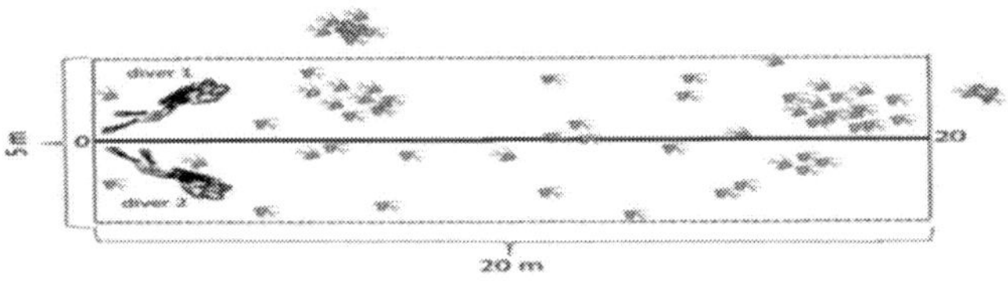

Fig. 3. *Illustration of reef fish data collection using the UVC method.*

2.3 Data Processing

2.3.1 CPCe Data Processing

The CPCe data-processing stage was used to determine the coverage of coral, fleshy seaweed, and rubble. Underwater photography data from each transect were analyzed using the CPCe software (Coral Point Count with Excel extension). The analysis of coral cover percentage followed the method described by English et al. (1997) [9] by calculating the percentage of coral coverage using the following formula. The categories of live coral covers are listed in Table 1.

$$\text{Coral coverage percentage} = \frac{Number\ of\ categories\ x}{Random\ number\ of\ points}$$

Table 1. Categories of live coral cover [14]

No	Category	Criteria
1	Poor	Live coral cover < 19%
2	Fair	19% ≤ Live coral cover ≤ 35%
3	Good	Live coral cover >35%

2.3.2 Reef Fish Data Processing

The processing stage for reef fish data using MS Excel was conducted to determine the total fish biomass and fish abundance as indicators of the coral reef health index. The methods used for calculating fish biomass and fish abundance were as follows:

1. Reef Fish Biomass

Reef fish biomass data were collected by recording and estimating the total length of each fish. The length-weight equation was used to estimate fish weight based on fish length by taking the average length (cm) of the fish and converting it to weight (g) using the length-weight index for each species, as derived from FISHBASE 2000 [15]. The equation used to calculate fish weight was as follows: The total biomass of the reef fish is shown in Table 2.

$$\mathbf{W} = \boldsymbol{a}\, L^b$$

Information :

W : Weight (g)

L : Total length per species (cm)

a dan b : Species specific index

Table 2. Total Biomass Category of Reef Fish [16]

No	Category	Criteria
1	Poor	Total biomass of reef fish < 970 kg/ha
2	Fair	970 kg/ha ≤ Total biomass of reef fish ≤ 1,940 kg/ha
3	Good	Total biomass of reef fish > 1,940 kg/ha

2. Abundance of Reef Fish

Fish abundance was calculated by comparing the number of individual fish to the total surveyed area. It is expressed as:

$$\text{Abundance of fish} = \frac{\textit{Number of Individuals}}{\textit{Transect Area}}$$

The transect area refers to the designated section of the reef surveyed during data collection. In this study, each transect was 20 m long, with a detection width of 2.5 meters on both sides, resulting in a total observation area per transect of 100 m^2.

2.3.3 Coral Reef Health Index Data Processing

The coral reef recovery potential was classified based on fleshy seaweed and coral rubble cover criteria, as shown in Table 3. The benthic component values were then determined using live coral cover and recovery potential categories [16] (Table 4). Finally, the overall coral reef health index was assessed by integrating benthic and reef fish components, following the criteria presented in Table 5. After calculating the index values, an index value map was generated using ArcMap software to visualize the results.

Table 3. Recovery Potential Criteria

No	Category	Criteria
1	Low	fleshy seaweed cover ≥3% or coral rubber cover >60% and live coral cover ≥5%
2	High	fleshy seaweed cover <3% or coral rubber cover ≤60% and live coral cover >5%

Table 4. Criteria for benthic component value

No	Live Coral Cover	Recovery Potential	Benthic Category	Benthic Component Value
1	Good	High	Healthy coral with high recovery potential in case of disturbance	6
2	Good	Low	The coral is healthy but if there is a disturbance it will be difficult to recover to its original condition	4
3	Fair	High	Corals are in fair condition and may be in the process of recovering from disturbance	5
4	Fair	Low	Corals are in fair condition but at risk of decline	2
5	Poor	High	Coral cover is low, but has potential to improve	3
6	Poor	Low	Coral cover is low, and it is difficult to improve its condition	1

Table 5. Criteria for coral reef health index values

No	Benthic Components		Fish Components	Coral Reef Health Index Value
	Live coral cover	Recovery Potential	Coral reef fish categories	
1	Good	High	Good	10
2	Fair	High	Good	9
3	Good	High	Fair	8
4	Good	Low	Good	8
5	Fair	High	Fair	7
6	Poor	High	Good	7
7	Good	High	Poor	6
8	Good	Low	Fair	6
9	Fair	Low	Good	6
10	Fair	High	Poor	5
11	Poor	High	Fair	5
12	Poor	Low	Good	5
13	Good	Low	Poor	4
14	Fair	Low	Fair	4
15	Poor	High	Poor	3
16	Poor	Low	Fair	3
17	Fair	Low	Poor	2
18	Good	Low	Poor	1

3 Result and Discussion

3.1 Live Coral Cover

The condition of coral reefs at six stations on Tunda Island varied in coverage, ranging from poor to fair and good. As shown in Fig.4, the lowest percentage of coral cover was at Station 6, with a value of 2.74%, which is categorized as low according to [13], as it falls below 19%. In contrast, the highest percentage of coral cover is at Station 4, with a value of 50.12%, which is categorized as high because it exceeds 35%.

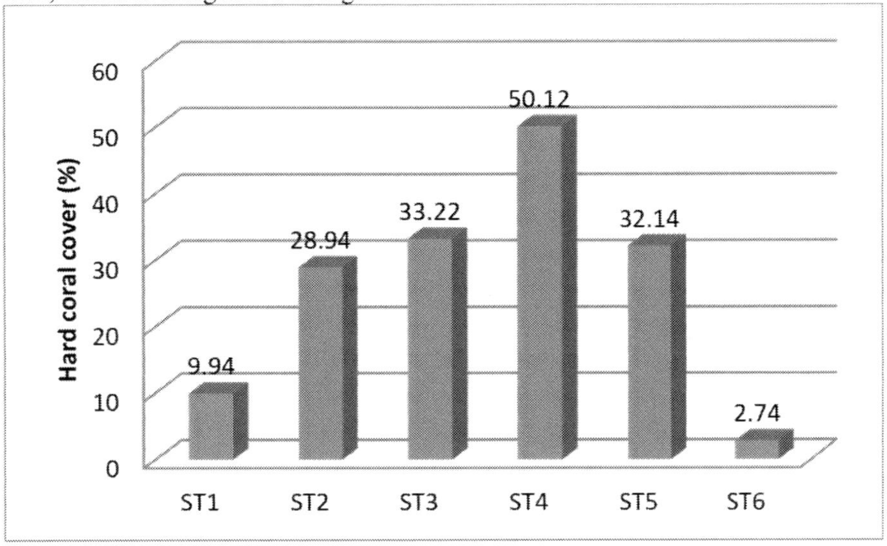

Fig. 4. *Percentage of hard coral cover in each station*

The variation in coral coverage at each station was influenced by multiple factors, such as high fleshy seaweed cover or extensive rubble accumulation. Other factors that affect coral coverage include the abundance of reef fish and human activities that harm the marine environment, such as waste pollution and sedimentation.

According to [17], fishing activities conducted by fishermen from outside Tunda Island, particularly at night in coral reef areas using non-environmentally friendly fishing gear, can disturb and damage coral reefs, making recovery difficult and leading to coral death. Another issue is the presence of marine debris, which is suspected to originate from the mainland and carried by the east and west monsoon winds [18].

3.2 Fleshy Seaweed

Fleshy seaweed cover is an important indicator of coral reef health, as excessive growth can outcompete corals and hinder reef recovery. As shown in Fig. 5, fleshy seaweed cover varied across stations, with the lowest percentage recorded at Station 1 (6.89%) and the highest at Station 6 (43.94%).

Fleshy seaweed competes with coral reefs, inhibiting their growth and preventing coral attachment owing to its secondary metabolites. The presence of algae in large areas causes physical disturbances to young corals, increases sediment deposition, reduces sunlight penetration, and lowers the success rate of coral larval settlement. Macroalgae can overgrow corals in nutrient-rich waters under abundant sunlight. An increase in macroalgal growth indicates a potential threat to coral development as macroalgae can grow on both living and dead corals [15].

According to [19], even a small percentage of macroalgal cover can affect coral reef resilience. A 0.5% increase in macroalgal cover can negatively affect coral reefs because algae grow faster than corals do. Therefore, even minimal algal growth could significantly hinder coral reef recovery.

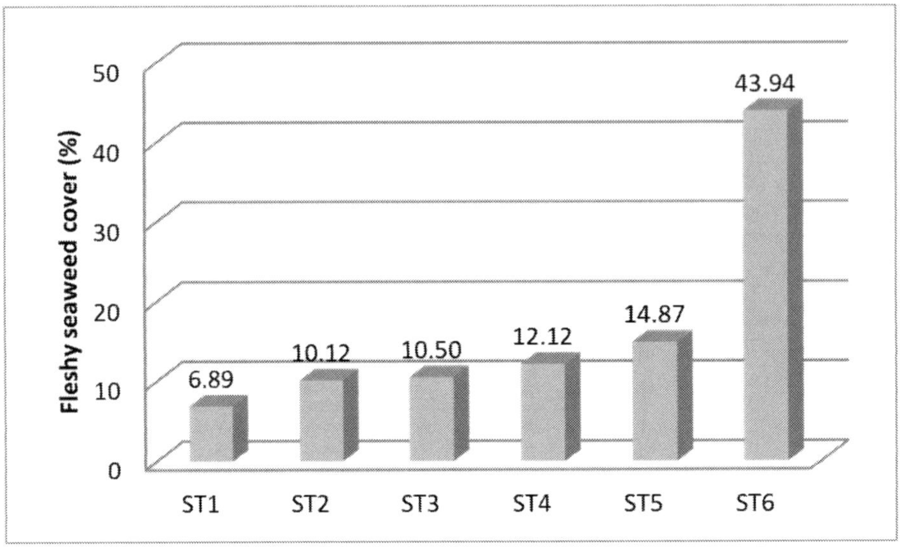

Fig. 5. *Percentage of fleshy seaweed cover in each station*

3.3 Rubble (Coral Fragments)

The lowest percentage of coral rubble cover was recorded at Station 3 (1.44%), whereas the highest was at Station 6 (37.22%) (Fig. 6). High coral rubble cover reduces the potential for coral reef ecosystem recovery, as coral rubble presents a challenging environment for coral larvae to grow.

Several studies have shown that coral rubble is not a suitable substrate for coral larval settlement and development because it remains dynamically unstable [9]. Similarly, [20] stated that substrates dominated by coral fragments are highly influenced by alternating currents and wave action, rendering them unsuitable for coral larval growth and development.

Additionally, coral fragments contribute to the physical damage to coral reef structures, which serve as critical habitats for various marine species. Habitat degradation can disrupt coral reef ecosystems, reducing the availability of shelter, food sources, and spawning sites for marine organisms.

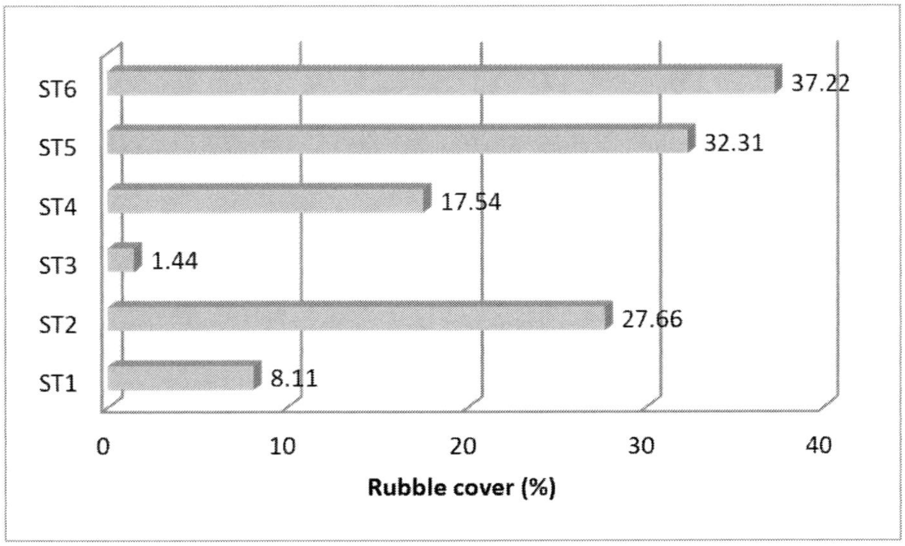

Fig. 6. *Percentage of rubble coverage in each station*

3.4 Reef Fish

Reef fish are economically important and serve as a key target species for fishermen. In this study, observations of the target reef fish identified only four families: Scaridae, Siganidae, Acanthuridae, and Serranidae. The recorded biomass values ranged from 2.50 to 85.31 kg/ha, whereas fish abundance varied between 80.00 and 840.00 individuals per hectare (Ind/ha). The relatively low fish biomass (below 970 kg/ha) indicates a limited fish population, which may affect both ecosystem balance and fishery sustainability.

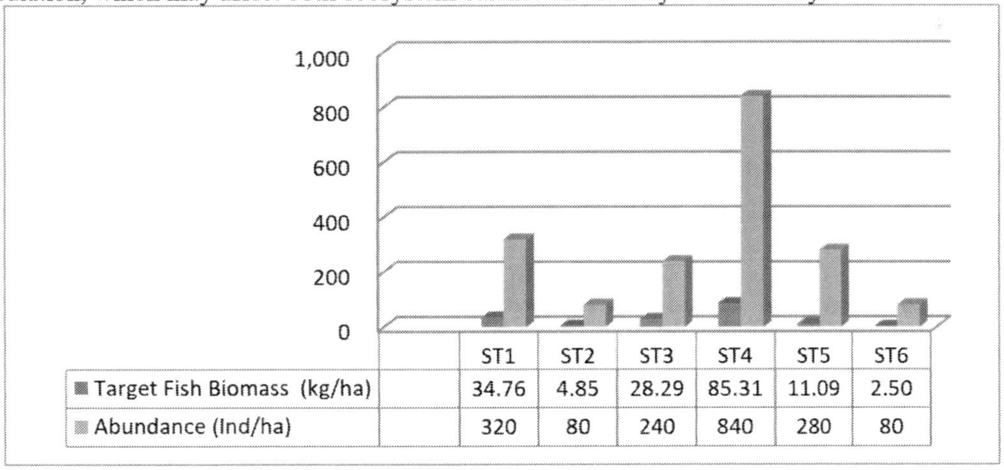

Fig. 7. *Biomass and abundance of reef fish in each station*

The lowest biomass value was recorded at Station 6 (2.50 kg/ha, whereas the highest biomass value was observed at station 4 (85.31 kg/ha). The lowest fish abundance was found at stations 2 and 6, with 80.00 individuals per hectare (Ind/ha), whereas the highest fish abundance was recorded at station 4 with 840.00 Ind/ha (Fig. 7).

The data indicated that biomass and abundance values varied across stations; some stations had high biomass but low abundance, and vice versa. At all six stations, the biomass values were lower than the abundance values. This discrepancy is likely due to the presence of small fish in greater numbers, resulting in a lower biomass value despite higher abundance.

According to [21], Tunda Island fishermen believe that this ecosystem crisis was a consequence of sand mining activities in Banten waters. Fishermen have observed these activities for several years but are uncertain when they first began. They believed that marine sand mining has led to declining fish catches and environmental degradation, contributing to the low population of reef fish. Additionally, fishermen from outside Tunda Island use destructive fishing practices, such as anesthetics and other harmful tools, which further affect reef fish populations.

Reef fish play a crucial role in coral reef ecosystems, particularly in controlling macroalgae that compete with corals. Herbivorous reef fish help protect corals by reducing excessive macroalgae growth, which could otherwise lead to coral mortality. Therefore, reef fish abundance is a key indicator of the coral reef health index. However, a high abundance of reef fish alone is not sufficient to maintain coral reefs if destructive human activities persist.

3.5 Coral Reef Health Index

The coral reef health assessment on Tunda Island revealed low recovery potential (Table 6), with index values ranging from 1 to 4 (Fig. 8). The combination of low live coral cover, high fleshy seaweed cover, and substantial rubble accumulation contributed to limited reef resilience. Additionally, the consistently low fish component values across all stations further constrained the reef's ability to recover.

Stations 1 and 6 recorded the lowest index value of 1, indicating severe degradation. Station 6 had the lowest coral cover (2.74%), extremely high fleshy seaweed cover (43.94%), and significant rubble accumulation (37.22%), creating unfavorable conditions for coral regeneration. Station 1, with slightly higher coral cover (9.94%), also showed limited recovery potential due to excessive fleshy seaweed growth (6.89%) and inadequate fish populations (value: 2).

Stations 2, 3, and 5 had an index value of 2, indicating moderate coral cover but low recovery potential. Station 2 had coral cover of 28.94%, but high fleshy seaweed cover (10.12%) and substantial rubble cover (27.66%) constrained recovery. Station 3 exhibited slightly higher coral cover (33.22%), but recovery remained limited due to the presence of fleshy seaweed (10.50%). Despite having a low rubble cover (1.44%), the insufficient herbivorous fish population (value: 2) allowed macroalgal to proliferate unchecked. Station 5 had coral cover of 32.14%, but high fleshy seaweed cover (14.87%) and rubble accumulation (32.31%) negatively impacted recovery potential.

Table 6. Benthic and Fish Component Values

No	Location	Benthic Components			Fish Components	
		Live Coral Cover	Recovery Potential	Value	Fish Category	Value
1	Station 1	Poor	Low	1	Poor	2
2	Station 2	Fair	Low	2	Poor	2
3	Station 3	Fair	Low	2	Poor	2
4	Station 4	Good	Low	4	Poor	2
5	Station 5	Fair	Low	2	Poor	2
6	Station 6	Poor	Low	1	Poor	2

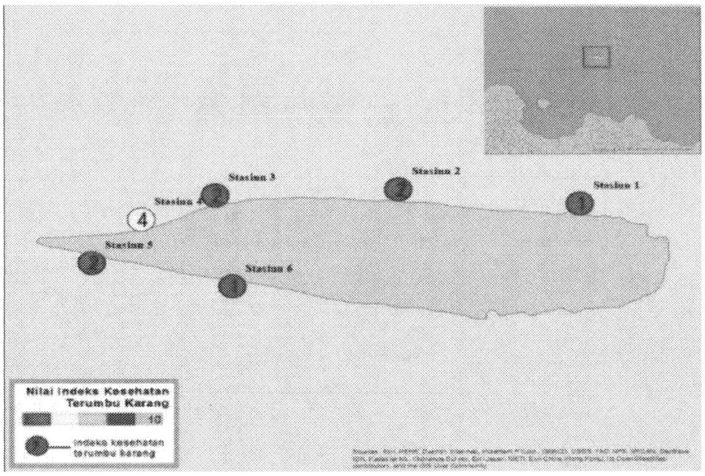

Fig. 8. *Coral Reef Health Index Map of Tunda Island*

Station 4 had the highest index value of 4, with the greatest coral cover (50.12%). However, recovery potential was still limited due to high fleshy seaweed cover (12.12%) and rubble accumulation (17.54%). Although the fish component value remained low (2), this station exhibited relatively better coral conditions compared to others.

Overall, the coral reef health index across all six stations remained low, with values not exceeding 4. The combination of high fleshy seaweed cover, substantial rubble accumulation, and low fish populations suggests that the coral reef ecosystem on Tunda Island faces significant challenges in recovery and resilience.

According to [22], reef fish play a crucial ecological role in supporting coral reef resilience by controlling macroalgal growth, which can be detrimental to coral communities. If macroalgal growth becomes too rapid and remains unchecked, it can dominate the reef ecosystem and compete with corals for space. This imbalance is worsened by low reef fish populations, which allow macroalgae to outcompete corals and disrupt the ecosystem.

4 Conclusion

This study found that live coral cover on Tunda Island ranged from 2.74% to 50.12%, fleshy seaweed cover varied between 6.89% and 43.94%, and coral rubble ranged from 1.44% to 37.22%. The total biomass of reef fish ranged between 2.50 and 85.31 kg/ha, while reef fish abundance varied from 80 to 840 individuals/ha. The coral reef health index across six observation stations ranged from 1 to 4, indicating that overall reef health remains relatively low.

Overall, Tunda Island's coral reefs are in poor condition, with a low potential for natural recovery. This limited resilience may hinder the regeneration of coral reefs if further degradation occurs.

To improve future research, long-term monitoring is recommended to track coral recovery trends over time. Additionally, analyzing water quality parameters and human-induced impacts could provide a more comprehensive understanding of the factors affecting coral reef health. Expanding the number of observation sites and incorporating advanced remote sensing technologies may also enhance the accuracy and scope of coral reef assessments.

References

[1] M. S. Legowo, A. Taofiqurohman, W. Pamungkas, and Subiyanto, *J. Perikanan dan Kelautan*, **10**, 2 (2019)

[2] E. S. Srimariana, M. Kawaroe, D. F. Lestari, and A. H. Nugraha, *J. Ilmu Pertanian Indonesia*, **25**, 1 (2020)

[3] A. Saputra *et al.*, *J. Kelautan dan Perikanan Terapan*, **4**, 2 (2022)

[4] G. Puspitasari, E. L. Widiastuti, H. Wijayanti, Tugiyono, *Jurnal Ilmiah Biologi Eksperimen dan Keanekaragaman Hayati*, **6**, 1 (2019)

[5] K. H. Rombe, D. Rosalina, G. Rahmawati, A. Surachmat, A. A. Sabilah, A. Rahman, R. Hermawan, *Jurnal Sumberdaya Akuatik Indopasifik*, **7**, 4 (2023)

[6] O. Hoegh-Guldberg, E. S. Poloczanska, W. Skirving, S. Dove, *Front. Mar. Sci.*, **4** (2017)

[7] A. C. Nugroho, *J. Terap. Tek. Inf.*, **1**, 2 (2018)

[8] Mujiyanto, M. G. Garcia, J. Haryadi, R. Rahayu, R. A. Budikusuma, *ILMU KELAUTAN: Ind. J. of Mar. Sci.*, **25**, 2 (2020)

[9] R. Aprilian, D. G. Bengen, and O. Johan, *ILMU KELAUTAN Indonesia J. Mar. Sci.*, **26**, 4 (2021)

[10] G. Giyanto *et al.*, *BIO Web Conf.*, **70** (2023)

[11] S. Munasik, M. Helmi, R.M Siringoringo, *J. Mar. Res.*, **9**, 3 (2020)

[12] O. L. Y. Rumkorem, R. Kurnia, and F. Yulianda, *J. Ilmu dan Teknol. Kelaut. Trop.*, **11**, 3 (2019)

[13] G. Giyanto *et al.*, *Status Terumbu Karang Indonesia,* (COREMAP-CTI Pusat Penelitian Oseanografi – LIPI, Jakarta, 2017)

[14] R. Froese and D. Pauly, *FishBase 2000, Concepts, Design and Data Sources,* (International Center for Living Aquatic Resources Management (ICLARM), Los Banos, Laguna, Philippines, 2000)

[15] S. Suparno *et al.*, *J. Kelaut. Trop.*, **24**, 1 (2021)

[16] G. Giyanto, P. Mumby, N. Dhewani, M. Abrar, and M. Y. Iswari, *Indeks Kesehatan Terumbu Karang* (Pus. Penelit. Oseanografi-Lembaga Ilmu Pengetahuan Indonesia, Jakarta, 2017)

[17] M. Mujiyanto *et al.*, *Rencana Pengelolaan dan Zonasi (RPZ) Kawasan Konservasi Pulau Tunda dan Pulau Pamujan Besar Kab. Serang Prov. Banten* (Dinas Kelautan dan Perikanan Provinsi Banten, Banten, 2018)

[18] N. E. Setyahandani, F. Yulianda, and G. Yulianto, *J. Ilmu dan Teknol. Kelaut. Trop.*, **13**, 1 (2021)

[19] O. R. Simarangkir, F. Yulianda, and M. Boer, *J. Ilmu Pertan. Indonesia*, **20**, 2 (2015)

[20] Giyanto, A. E. Manuputty, M. Abrar, R. M. Siringoringo, S. R. Suharti, K. Wibowo, *Panduan Monitoring Kesehatan Terumbu Karang* (CTI – LIPI, Jakarta, 2014)

[21] W. Safitri & J. Triadi. *J. Kebijakan Sos. Eko. Kelautan dan Perikanan*, **13**, 2 (2023)

[22] Idris, Fakhrurrozi, and C Adiwijaya, *IOP Conf. Ser.: Earth Environ. Sci.* **744 012030** (2021)

Coral growth rate in branching and massive coral in Sangiang Nature Tourism Park, Indonesia

Idris[1], *Fakhrurrozi*[1], *Muhammad* Erwin[2], *Mauludiyah*[2*], *Misbakhul* Munir[2]

[1]Indonesian Coral Reef Foundation (Yayasan TERANGI), Jalan Asyibaniyah No. 105-106 Pondok Jaya Cipayung Depok Jawa Barat 16438
[2]Department of Marine Science, Faculty of Sains and Technology State Islamic University of Sunan Ampel Surabaya, East Java, Indonesia

Abstract. Indonesia's coral reefs are deteriorating because of human activities, making rehabilitation crucial. Sangiang Island, a designated conservation area, has experienced significant coral degradation, necessitating restoration efforts. This study aimed to analyze the growth and survival rates (SR) of transplanted corals and determined the optimal seedling size for transplantation to aid coral reef rehabilitation. The study was conducted over seven months, from August 2019 to February 2020, on Sangiang Island. The transplanted coral species included *Acropora sp.*, *Platygira sp.*, and *Favites sp.*, which represent branching and massive coral forms. Coral growth was measured using underwater photography and analyzed using CPCe software. *Acropora sp.* showed the highest growth rate (2.47 cm/month), while *Platygira sp.* and *Favites sp.* had average growth rates of 0.36 cm/month and 0.44 cm/month, respectively. SR was 70.7% for *Acropora sp.*, 59.1% for *Platygira sp.*, and 88.2% for *Favites sp.* The optimal fragment size for *Acropora sp.* was 11–20 cm, whereas smaller fragments (<10 cm) showing higher mortality. Coral transplantation effectively supports reef rehabilitation with survival rates above 50%. However, environmental challenges, such as sedimentation, macroalgae competition, and predation, must be addressed to improve the success rates.

1 Introduction

The coral reef ecosystem, with its diverse types and habitat characteristics, not only covers almost all the islands in Indonesia, but also spreads across the open seas far from the mainland. The current condition of Indonesia's coral reefs is only 6.56% under very good conditions, 22.96% under good conditions, 34.3% under fair conditions, and 36.18% under poor condition [1]. One location where coral reefs are found is on Sangiang Island. Geographically, Sangiang Island is located between 5°56'00" - 5°58'00" S and 105°49'30" - 105°52'00" E. The conservation status of Sangiang Island under the Ministry of Forestry has been designated as the Sangiang Island Nature Tourism Park through the Decree of the

* Corresponding author: mauludiyah@uinsa.ac.id

Minister of Forestry No. 55/Kpts-II/1993 dated February 8, 1993. As a conservation area, the Sangiang Island Nature Tourism Park has potential marine biodiversity, including soft corals, stone corals, blue corals, and fire corals of Meliophora species. It also has potential reef fish, such as those from the Chaetodontidae, Pomacentridae, and Acanthuridae families. Human activities utilizing coral reefs are increasing, resulting in significant damage to coral reefs [2]. Efforts to preserve coral reefs
in Indonesia are necessary, in line with the International Union for Conservation of Nature and Natural Resources (IUCN) declaration that coral reefs are protected tropical ecosystems [3].

Coral reef ecosystems are highly vulnerable to environmental change, making them susceptible to damage. Worldwide, coral reefs are experiencing alarming declines, and calcification rates have decreased owing to natural disturbances and human activities [3, 4]. Human activities that have the potential to damage coral reefs include coastal development, marine pollution, sedimentation from land, overfishing, and destructive fishing. Damaged ecosystems are very slow to recover and require a long time to return naturally to their original state. This necessitates the development of more effective restoration systems.

One of the efforts to conserve coral reefs is to conduct rehabilitation that is adapted to the interests and environmental conditions [5]. Commonly used methods include the direct transplantation of coral fragments or colonies to degraded reefs [6]. A common rehabilitation technique is transplantation or fragmentation. Coral transplantation is a technique of planting new corals by fragmenting seeds obtained from the parent corals [7]. Coral transplantation activities are very suitable for locations affected by currents and waves, ensuring that the transplanted corals always receive oxygen and food in the form of plankton, and can help in combating the sedimentation that occurs on the corals [8,9].

Coral growth refers to an increase in the length, weight, volume, or surface area of a coral over time. Generally, coral growth rates are based on the rate of calcification of coral animals that secrete calcium carbonate (CaCO3) [10]. In this study, the transplanted corals were from the life forms of branching coral (*Acropora sp.*) and massive corals (*Favites sp.* and *Platygira sp.*). The aim of this study was to determine the growth rate, survival rate, and optimal size of coral fragments based on the life form of branching and massive, which were transplanted on pyramid-shaped modules on Sangiang Island, Banten Province.

2 Methods

2.1 Study site

The study was conducted from August 2019 to February 2020 (a duration of seven months) on Legon Bajo, Sangiang Island. The research location map is shown in Fig.1.

https://doi.org/10.1051/bioconf/202516801003

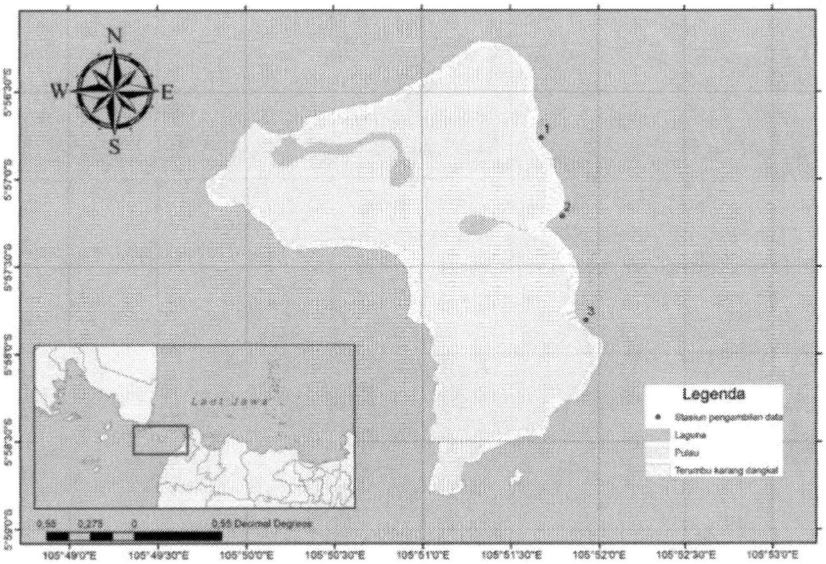

Fig 1. Study site for taking rehabilitation steps in the Sangiang Island area, Banten. The circle was the stations site for coral transplantation.

2.2 Data collection procedure

This study utilized 13 artificial reef modules in the shape of pyramids made from polyvinyl chloride (PVC) pipes with dimensions of 80 cm in length, 80 cm in width, and 70 cm in height (Fig.2). PVC is used because it is anti-rust, easy to shape, and more durable than other materials such as iron or concrete. Pyramid shapes provide ecological niches for other organisms. The modules were placed at a depth of 7–9 m. Coral fragment collection was performed near the rehabilitation site. Fragments were obtained from live coral fragments found on the substrate. A total of 144 fragments from *Acropora sp.*, 51 fragments from *Favites sp.*, and 46 fragments from *Platygira sp.* were used. The coral fragments were attached to the PVC pipes with an adhesive and then mounted on artificial reefs. The dimensions of the fragments, either their length or height, were measured. The measurement of coral fragments was conducted during monitoring every two months over a 7-month observation period.

Fig. 2. Pyramid-shaped for artificial reef modules as a media planting and fragments of *Acropora sp.*, *Favites sp.*, and *Platygira sp.*

There was a time interval between the transplantation of coral fragments and the measurement of coral fragments to allow corals to recover after the transplantation process. Data were collected by taking direct photographs of the coral fragments with a ruler as a scale reference (Fig.3). During monitoring, the data collection technique involved measuring the length or width of the fragments on the transplantation modules in the same sequence each time, with photographs of the samples taken. Data processing utilized the coral point count for Excel extension (CPCe) software [11] to obtain coral length units. To measure survival and determine transplantation success, dead corals were recorded and analyzed using the survival rate formula. In addition to measuring coral growth and survival rate, this study also measured water quality, such as salinity, temperature, pH, current, and visibility, as supporting factors.

Fig. 3. Photograph of coral fragment measurement using a ruler as a scale.

2.3 Data Analysis

2.3.1 Coral growth rate

The coral growth rate is the difference in coral growth over time divided by the study duration. Monthly coral growth rate was determined using the following formula:

$$\alpha = \frac{L_i+1-L_i}{t_i+1-t_1} \qquad (1)$$

Where,

α : Growth rate (cm/month)
L_{i+1} : Coral length at time i+1 (cm)
t_{i+1} : Time at i+1 (months)
l_i : Coral length at time i (cm)
t_i : Time at i (months)

2.3.2 Coral survival rate (SR)

Coral survival rate was calculated by subtracting the total number of living fragments from the total number of transplanted fragments at the end of the study. Survival rate is measured by calculated the number of survivors divided by the total transplanted corals and multiplied by 100% [12]:

$$SR = \frac{N_t}{N_0} \times 100\%$$

Where,

SR : Survival Rate
N_0 : Number of corals at first observation (colony)
N_t : Number of corals at final observation (colony)

2.3.3 *Statistical analysis*

Simple linear regression analysis was used for statistical analysis [13]. This analysis was able to analyze the correlation between the initial size of the seedlings and their absolute growth and growth rate.

3 Results and Discussion

3.1 Water quality

The presence of coral reefs is closely related to temperature, salinity, visibility, current, and pH. Water quality parameters tend to be homogeneous at the observation location over several years. In this study, the water quality was suitable for coral growth (Table 1). This is in accordance with the sea water quality standards for coral reefs (Decree of the Minister of the Environment, no. 51 of 2004). However, several parameters, such as the current speed, do not comply with these quality standards. For current speed can be grouped into three categories, namely very fast flowing waters (>1 m/sec), fast (0.5–1 m/sec), medium (0.25–0.5 m/sec), slow (0.1–0.2 m/sec) and very slow.
Based on these categories, the current speed at this location ranges from slow to medium and complies with coral growth.

Table 1. Water Quality on Sangiang Island from 2016-2020

Parameter	Legon Bajo			
	2016*	2018*	2019	2020
Temperature (°C)	29	29	30	31
Salinity (ppt)	30	32.5	30	30
pH	8	7.9	7.9	7.9
Visibility (m)	5	8	4	6
Current (m/s)	0.1	0.16	0.06	0.4

3.2 Coral growth and absolute growth

Coral growth can be determined by the coral's ability to form new polyps or increase in dimensions such as length, height, or width. Therefore, coral growth and development is generally based on colony formation. Coral growth is also referred to as a linear extension, meaning that corals exhibit relatively constant growth [14]. The growth dimensions measured in this study were coral length, width, and height. Each coral type exhibits different growth patterns based on its life form.

Based on the results of the monthly growth study of *Acropora sp.*, the growth ranged from 0.002 cm to 4.78 cm, with an average of 0.88 cm. The growth of *Acropora sp.,* fluctuated over time and depended on the initial fragment size at the time of planting, with the highest average growth recorded in February 2020, reaching 4.03 cm (Fig. 4). The growth of *Favites sp.* in each fragment sample was highly fluctuating, ranging from 0.01 cm to 0.16 cm per month, with an average of 0.53 cm per month. The highest growth of *Favites sp.* was observed in December 2019, reaching 0.53 cm (Fig. 5).

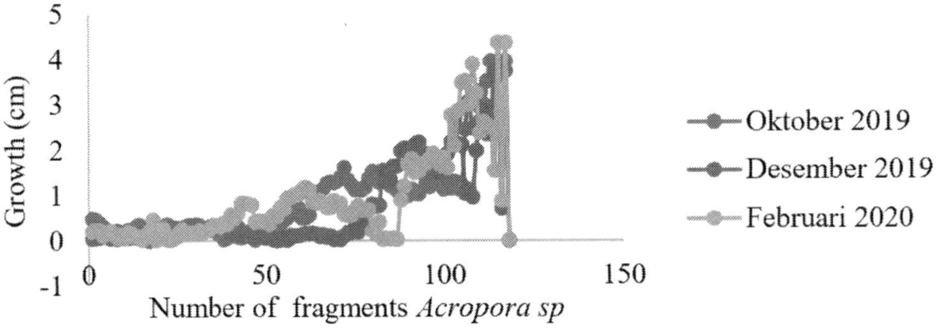

Fig. 4. Growth of *Acropora sp.* over three periods between October 2019 to February 2020 (n=144).

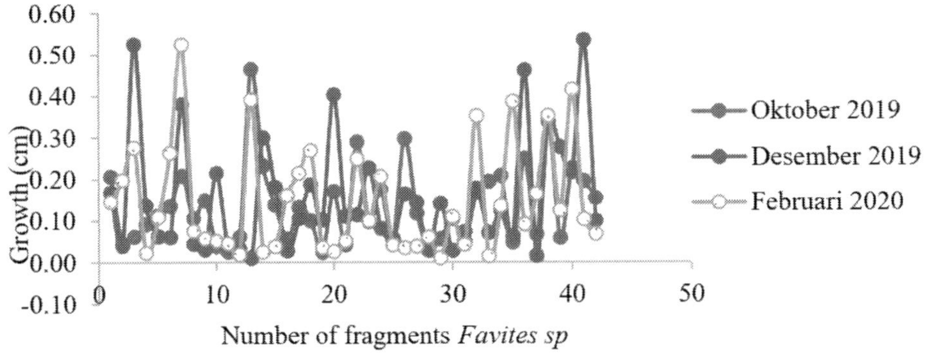

Fig. 5. Growth of *Favites sp.* over three periods between October 2019 to February 2020 (n=45).

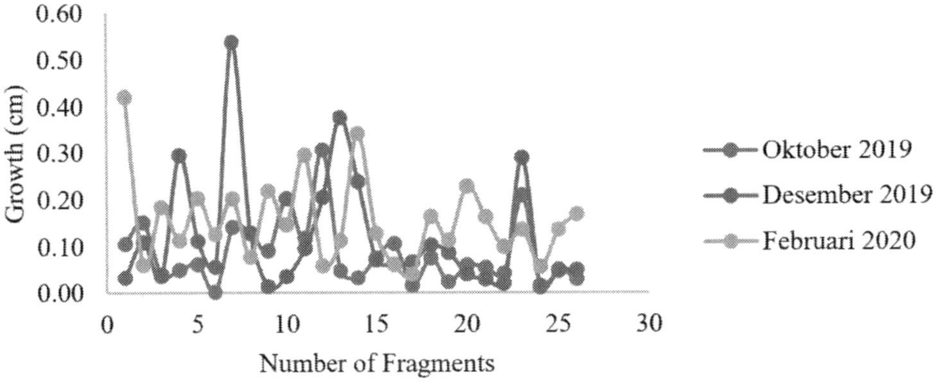

Fig. 6. Growth of *Platygira sp.* over three periods between October 2019 to February 2020

(n=26).

Furthermore, the growth of *Platygira sp.* coral in each fragment sample varied from 0.001 cm to 0.54 cm per month, with the highest average reaching 0.16 cm per month. This coral growth peaked in December 2019, with a maximum growth of 0.54 cm (Fig. 6).

In addition to studying coral growth every month, this study examined the absolute growth of each coral fragment used as a sample. The absolute growth of coral fragments was defined as the difference in the height or diameter of the coral between the end of the study and the beginning of planting (Fig.7).

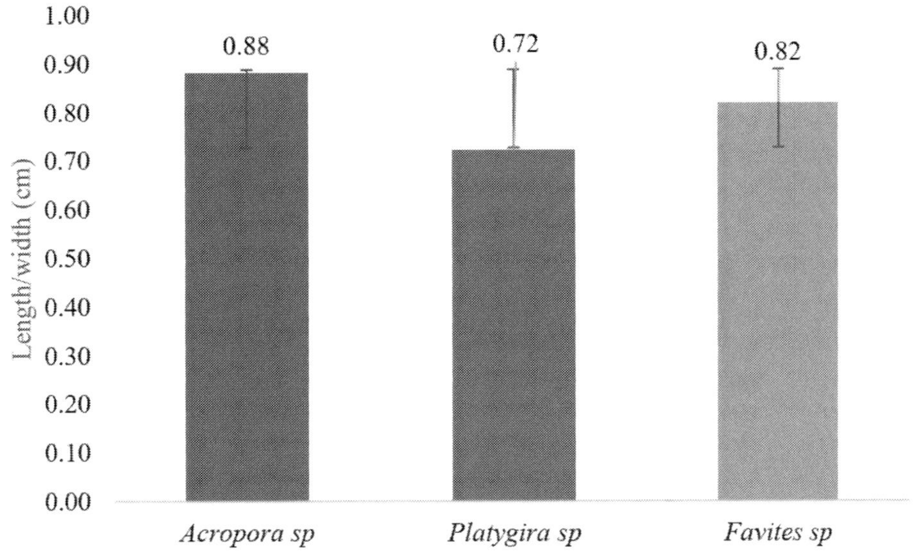

Fig. 7. The average of absolute growth in length or width of corals fragments over three time observations between October 2019 to February 2020.

During research and monitoring, the absolute growth length of *Acropora sp.* varied from 0.12-1.72 cm, with an average of 0.88±0.38 cm; *Platygira sp.* from 0.16-1.76 cm, with an average of 0.72±0.37 cm; and *Favites sp.* from 0.21-2.23 cm, with an average of 0.82±0.51 cm. *Acropora sp.* exhibited the highest absolute growth compared to *Favites sp.* and *Platygira sp.* This is because *Acropora sp.* have a branching life form, whereas *Platygira sp.* and *Favites sp.* are massive life forms. This finding is consistent with that of several studies conducted in other locations. For example, [15] reported that the average total growth of *Acropora sp.* corals is better than that of *Porites sp.* corals because of the branching life forms of *Acropora sp.* and the massive life forms of *Porites sp.* The growth of *Porites sp.* requires a long time because of its dense skeletons [16]. An *Acropora sp.*, which has a more fragile calcareous skeleton than others, has a faster growth rate than massive corals with very dense and hard calcareous structures, as stated by [17]. This is because corals of the genus Porites (massive corals) have the ability to adapt to temperature and salinity. Porite corals have tissues located far from the coral surface [18], which are believed to provide additional protection against several growth obstacles, one of which is solar radiation [19].

3.3 Absolute of coral growth rate

The rate of coral growth refers to the increase in the length, height, weight, or volume of corals over a unit of time. It is calculated as the difference in coral height between the initial planting and the end of the observation period divided by the duration of planting or observation. The rate of coral growth is significantly influenced by environmental factors and the reduction in stressors, resulting in variations in the growth of the same coral species at different locations [20]. The monthly growth rate of coral fragments on Sangiang Island is shown in Fig.8.

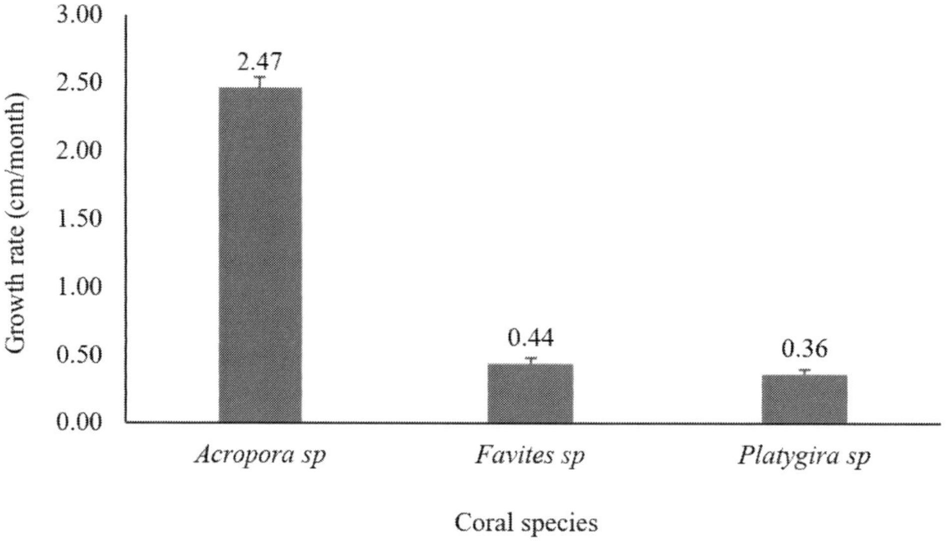

Fig. 8. The average of absolute growth rate of corals frgaments over three time observations between October 2019 to February 2020.

The results of the observations showed that the lowest growth rate of *Acropora sp.* coral fragments was 0.09 cm/month, and the highest reached 4.03 cm/month, with an average growth rate of 2.47 ± 0.08 cm/month. Meanwhile, the growth rate of *Favites sp.* coral fragments ,ranged from a low of 0.11 cm/month to a high of 1.11 cm/month, with an average growth rate of 0.44±0.04 cm/month. The average growth rate of *Platygira sp.* was 0.36 ± 0.04 cm/month, with the lowest growth rate being 0.08 cm/month and the highest reaching 0.88 cm/month. The growth rate of corals is greatly influenced by environmental factors and the reduction of stressors, leading to variations in the growth of similar corals at different locations [20]. Factors such as coral age, environmental conditions, temperature, sedimentation, light, and depth affect the growth rate of corals [9].

The results of this study showed that the growth rate of *Acropora sp.* was higher than that of the other two species. This is in line with the results of previous studies, which stated that *Acropora sp.* had a higher growth rate than other hermatypic corals [21, 22]. Other researchers such as [23] also stated that the growth of *Acropora sp.* 0.22±0.03 cm/month, [15] stated 0.51 cm/month, [24] stated 1.03 cm/week, [25] stated 1.07 cm/month, [26], states 7.8 cm/year and [27] states 2.27-3.37 cm/month. Meanwhile, the growth rate of *Favites sp.* and *Platygira sp.* corals is slower than *Acropora sp.* because the life forms of both species are massive corals. As stated by [28], the growth of massive coral lifeforms is generally very slow because massive life forms have a denser and more compact calcareous skeleton structure, resulting in relatively slower growth. This is also in line with the results of the study by [23] on Semut Island and [15] in North Lombok Island who measured the growth of the massive coral species Porites sp. get growth rates of 0.08±0.02 cm/month and 0.37 cm/month respectively.

Apart from the life form, several other factors influence the growth rate of coral fragments, such as environmental factors (temperature, salinity, pH, and turbidity), handling when cutting the fragments, and the initial size of the coral fragments. Several researchers have shown that high sedimentation and handling processes during transplantation can cause low coral growth rates and slow coral growth rates [29,12].

3.4 Survival rate (SR)

The coral survival rate was determined by the difference in the number of fragments from the start of planting to a certain time limit during the research. The survival rate of coral indicates the success of rehabilitation activity. Several factors can influence coral survival, including variations in temperature, light intensity, bleaching, and competition for territory. The distance between the parent coral and coral fragment also has an impact; the closer the location of the coral fragment to the parent coral, the greater the coral's ability to secrete mucus [30]. The survival rates are shown in Fig.9.

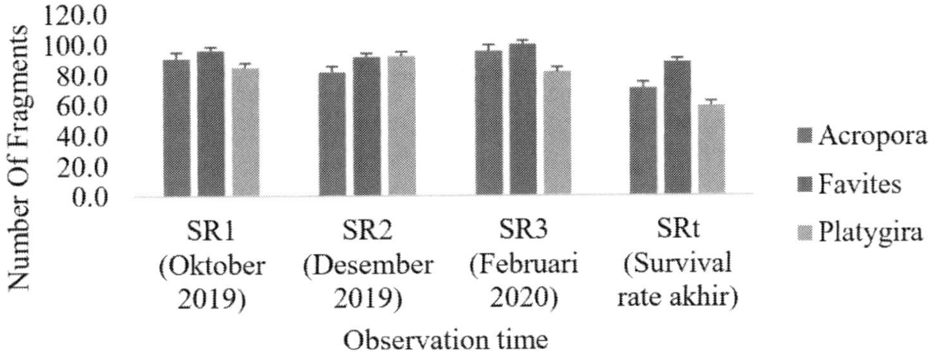

Fig. 9. Survival rate of *Acropora sp., Favites sp.*, and *Platygira sp.* (no. *Acropora sp.* = 239, no. *Favites sp.*= 51, no. *Platygira sp.* = 44)

The results of observations over 7 months showed that *Acropora sp.* and *Favites sp.* corals had the highest percentage of mortality that occurred in the second monitoring (December), with a mortallity percentage rate of 18.43% (40 colonies) for *Acropora sp.* and species respectively 8.16% (4 colonies) for the species *Favites sp.* Another thing was the *Platygira sp.* which experienced the highest mortality during the third monitoring activity (February) 18.18% (8 colonies). The final results obtained were that the survival rate for *Acropora sp.* during observations reached 90.8% (October 2019) with a survival rate at the end of the observation of 70.7%. The SR level of the *Favites sp.* type reached 100% (February 2020) with a final SR of 88.2%. The highest SR for the *Platygira sp.* was 92.1% (December 2019), with a final SR of 59.1% (lowest). The average survival rate of massive corals is higher than branching corals due to their stronger ability to withstand and adapt to environmental stresses such as sedimentation and climate change.The results of previous studies in other locations in Indonesia, such as on Tunda Island, showed that the SR level of Porites (Massive coral) was 52.27% and Montipora was 56.25% [31]; in Semut Island-Anambas, the survival rate was 64% [23], and in North Lombok, *Acropora sp.* had an SR of 96.67% and *Porites sp.* reached 86.67% [15]. Meanwhile, in other countries such as in the Red Sea, the survival rate of coral after 12 months was 67.2%, increased to 68.3% after 24 months [12], and in Maldives, the survival rate of 70.2% was recorded after one year of monitoring [32].

Differences in SR can occur due to several factors, such as transplantation technique, initial fragment size, and water quality parameters. The occurrence of coral mortality can be caused by several factors, such as sedimentation, competition, predation, and low tolerance of corals to the environment. This can be seen from the sedimentation and salinity on Sangiang Island, which is quite high due to runoff from the mainland and greatly influences coral growth [33-35]. In addition, high sedimentation causes bio-erosion and has the potential to disrupt the symbiotic processes between coral and zooxanthellae [36]. Another cause of coral death at the research location is thought to be the dominant growth of macroalgae in coral reef ecosystems, which have a greater biomass than coral, causing macroalgae to tend to be superior if competition for space occurs [37]. The ability to grow more quickly means that macroalgae can easily dominate coral reef ecosystems and ultimately threaten coral growth [38].

A rehabilitation activity using a transplant approach is declared successful if the survival rate of transplanted coral is in the range of 50–100% because it can function like a natural ecosystem in general [39,9]. So it can be said that coral transplantation using these three types of coral (*Acropora sp.*, *Favites sp.* and *Platygira sp.*) has been successfully used for the rehabilitation of coral reef ecosystems, because it has a survival rate above 50%.

3.5 Optimal size of coral fragments for transplantation

The relationship between the initial size of the fragments and the average growth rate and absolute growth of corals in the *Acropora sp.* for each month had a fairly strong correlation of 62%, so it can be assumed that the initial size of the seedlings had a quite significant influence on the absolute growth and average growth rate, because the percentage value was more than 50% (Fig.10). On the other hand, the relationship between the initial size of the fragments and the average growth rate and absolute growth of *Favites sp.* and *Platygira sp.* did not have a strong enough correlation, only 11.77% and 1.71%, so the initial fragment size data cannot be used to obtain the size of coral fragments are optimal for transplantation.

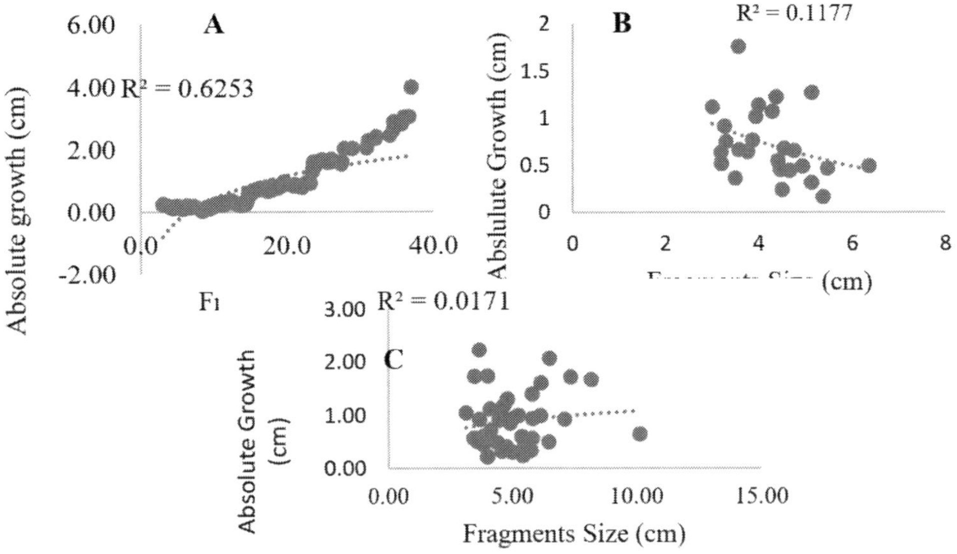

Fig. 10. Correlation of Fragments size and absolute growth of (A) *Acropora sp.* (n=142), (B) *Platygira sp.* (n=26) and (C) *Favites sp.* (n=42)

The size of the coral fragments determines the success of coral transplantation for rehabilitation. In this study, data on the size of *Acropora sp.* corals obtained will be used to determine the optimal size of coral seedlings for transplantation activities. The size of *Acropora sp.* coral seeds is divided into four classes according to size: class 1, size–0-10 cm, class 2, size–11-20 cm, class 3, size–21-30 cm, and class 4, size–31-40 cm. cm. A simple linear regression approach was used to determine the relationship between the initial seedling size and the absolute growth and growth speed of *Acropora sp.*

The results of this observation showed that size classes 2, 3, and 4 are the optimal size classes for transplantation activities because they have an R square above 70% (78.46%, 77.88%, and 72.75%). Meanwhile, size class 1 is a size that is not optimal to use as an initial size for coral seedlings for transplantation because the relationship is only 2.27% (Fig.11). Based on the results of this study, it can be said that the optimal initial coral seedling size for use is a maximum of 11-20 cm. This is because coral fragments that are quite large have more polyps; therefore, they have a more optimal growth rate. This assumption is also supported by other researchers, who state that smaller coral sizes tend to have a higher mortality rate than larger fragment sizes, and larger coral fragments have higher survival rates [10,40-44]. This is because large seeds have more energy and tissue; therefore, they adapt more quickly, have a greater reproductive capacity, and are more resistant to environmental stress.

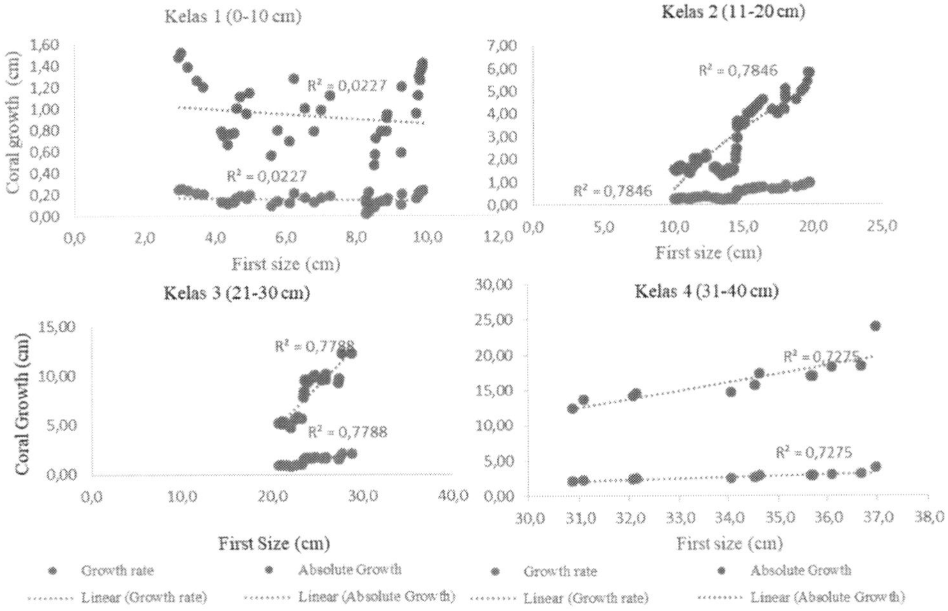

Fig. 11. Correlation of first size, growth rate and absolute growth of *Acropora sp.* (n=142)

4 Conclusions

The growth, absolute growth rate, survival rate, and optimal size of fragments of branching and massive corals transplanted at Legon Bajo, Sangiang Island, Banten, varied considerably. The average growth rate of branching corals (*Acropora sp.*) was faster than that of the two other types of massive corals (*Favites sp.* and *Platygira sp.*) because massive life forms have a denser and more compact calcareous skeleton structure, resulting in relatively slower growth. However, the average survival rate of massive corals is higher than that of branching corals because of their stronger ability to withstand and adapt to environmental stresses, such as sedimentation and climate change. The optimal coral size of the fragment for transplanting *Acropora sp.* (branching coral) was in the range of 11-20 cm. For coral reef conservation, especially in rehabilitation, the use of branching growth forms can be reduced by adding massive growth forms, so that variations in growth forms during rehabilitation can provide more ecological niches for other marine biota, such as coral reef fish.

This research was funded by PT. Asahimas Chemical collaborated with the KEHATI Foundation and TERANGI Foundation. The authors extend their gratitude to all parties for their assistance with the writing and compilation of this article. Special thanks are also given to the diving team for their cooperation in the data-collection process.

References

1. T. A. Hadi, M.Abrar Giyanto, B. Prayudha, O. Johan, A. Budiyanto, A. R. Dzumalek, La Ode Alifatri., S. Sulha, Suharson, Puslit Oseanografi – LIPI, vi.88 (2019)
2. N. D. Uar, S. H. Murti, S. Hadisusanto, *Majalah Geografi Indonesia*, **30**, 1 (2016)
3. Taufina, Faisal, Lova SM., *Jurnal Pengabdian Kepada Masyarakat*, **24,** 2 (2018)
4. D. M. Spalding and B. E.Brown, *Science*, **350,** 6262 (2015)
5. D. L., Lizcano-Sandoval,E. Londono-Cruz, and F. A. Zapata, *Marine Biology Research*, **14,** 8 (2018)
6. H. Madduppa, B. Subhan, D. Arafat , N. P. Zamani, *Risalah Kebijakan Pertanian dan Lingkungan*, **3,** 2 (2016)
7. W. D.Dela Cruz, B . Rinkevichc, E. D. Gomez, and H. T. Yap, *Ecological Engineering*, **84**, 408-415 (2015)
8. D. Soedharmaand D. Arafat, *Prosiding Seminar Transplantasi Karang*, 5-14 (2007)
9. D. D. Pelasula, M. P. Patria, S. Wouthuyzen, A. Syahailatua, *AACL Bioflux*, **16**, 1 (2023)
10. K. Mompala , A. B. Rondonuwu, U. N. W. J. Rembet, *Jurnal Ilmiah Platax*, **5,** 2 (2017)
11. S. R. M. Tabugo, D.L. Manzanares, A.D. Malawani, *Computational Ecology and Software*, **6**, 1 (2016)
12. M. A. J. Karauwan, J. A. Solang, *Int. J. of Multidisciplinary Research & Analysis*, **7**, 12 (2024)
13. M.A.M., Mahmoud, M. A.Dar, H. N. M. Hussein, M. E. A. El-Metwally, M. M.Maaty, M. Y. Omar, M. R. Seraj, T. A. A. Mohammed, *Egypt J Aquat Biol Fish*, **23,** 2 (2019)
14. L. K. Mansur, M. Kasim, R. D. Palupi, *Jurnal Kelautan*, **16**, 2 (2023)
15. Suharsono, *Bercocok Tanam Karang dengan Transplantasi*, LIPI Press, Jakarta (2008)
16. S. Hariyanto, R. Ibadur, R. H. Mahardika, *Jurnal Biologi Tropis*, **23,** 2 (2023)
17. Rudiansyah, D. Kurniawan, T. Apriadi, R. Kurniawan, *Journal of Marine Research*, **13**, 4 (2024)
18. A. Haris, R. Chair, T. Akbar, I. B. Andi, Muhammad F. S., Rahmadi T., Shinta W., Arniati, Ahmad F., *Spermonde,* **2**, 3 (2017)
19. A. Waris, *Thesis*, Fakultas Ilmu Kelautan dan Perikanan, Universitas Hasanuddin, Makassar (2016)
20. B. S. Barus, T. Prartono, D. Soedarma, *Jurnal Ilmu dan Teknologi Kelautan Tropis*, **10**, 3 (2018)

21. M. O. Luthfi, Rahmadita V. L., Setyohadi D., *Jurnal Ilmu Lingkungan*, **16,** 1 (2018)
22. Munasik, Nugroho A. A., Hartati R., Sabdono A., Sugiyanto, Sugianto D.N. *Jurnal Kelautan Tropis*. **23,** 3 (2020)
23. Y. Nozawa, Villanueva R. D., Munasik M., Roeroe K. A., Mezaki T., Kawai T., Guest J., Arakaki S., Suzuki G., Tanangonan J. J. B., Ang P. O., Edmunds P. J. *Coral Reefs*. **40**,1463-1471 (2021)
24. N. P. Putri, Muhammad R. M., Abdulrahman R., Corina D. R., Dian A., Munasik, *Biodiversitas*, **25,** 1 (2024)
25. P. A. P., Putra, Restu I. W., Ernawati N. M. *Aquatic Science*. **3,** 1 (2020)
26. F. K., Muzaki, Hanifa R., Akhwady R., Saptarini D., Buharianto. *Biodiversitas*, **20,** 6 (2019)
27. M. Nithyanandan, Le Vay L., Raja D. K., Kesavan R., Pereira D. *Cogent Environ Sci*, **4,** 1 (2018)
28. I. N. D. Adi, Damar A., Adrianto L., Soedharma D., Subhan B., Kusumastanto T., Rikardi N. *Intl J Sci: Basic Appl Res*, **29**, 3 (2016)
29. M. Utami, I. W. Arthana, N. M. Ernawati, *Current Trends in Aquatic Science*, **4**, 2 (2021)
30. C. S. Rogers. *Marine Ecology Progress*, **62**,185-202 (1990)
31. F. H. Nurman, Sadarun B., Palupi R. D. *Jurnal Sapa Laut*, **2**, (4 .(2017)
32. D. B. Suhendar, Ofri J., Idris, Safran Y., dan Hawis H. M. *Prosiding Seminar Nasional Kelautan dan Perikanan Ke -Vll*, Fakultas Kelautan dan Perikanan Universitas Nusa Cendana-Kupang (2020)
33. I. Pancrazi, Feairheller K., Ahmed H., Napoli C., Montefalcome M., *Diversity*, **15,** 9 (2023)
34. S. Yusri, M. Prastowo S. W., Idris, Hedra A. *Kondisi Ekosistem Terumbu Karang Pulau Sangiang*, Yayasan TERANGI, Jawa Barat (2017)
35. Idris, Aulia, R. P., Adiwijaya, C., Gilang, M., Santoso, P., Prabowo, B., Muhammad, F., Andriyani, W., Lestari, D.F., Setyaningsih,W.A. and Zamani, N.P. *IOP Conference Series: Earth and Environmental Science*, **429**, 1 (2020)
36. M. A. Prawira, Ofri J., Idris, dan Hawis H. M. *Prosiding Seminar Nasional Kelautan dan Perikanan Ke-Vll*, Fakultas Kelautan dan Perikanan Universitas Nusa Cendana-Kupang (2020)
37. O. M. Luthfi, I. N. Januarsa, *Jurnal Kelautan*, **11**, 1 (2018)
38. C. K. Wijaya, R. Komala, Giyanto, *Bioma*, **13**, 2 (2017)
39. A. Palallo, *Thesis*, Universitas Hasanuddin (2013)
40. J. N. Awaya, *Thesis*, Global Environmental Science (2023)
41. Z. Arifin, Luthfi M., *Seminar Nasional Perikanan dan Kelautan VI*, 556-561 (2016)
42. K. K. I. Babu and S. Sureshkumar, *International Journal of Applied and Pure Science and Agriculture,* **2,** 5 (2016)
43. D.A. Bunga, I. W.Arthana, N. M. Ernawati, *Current Trends in Aquatic Science*, **6,** 1 (2023)
44. K. Soong and T. Chen, *Restoration Ecology*, **11**, 62-71 (2023)

Biodiversity of coral reefs ecosystem at Tunda Island, Banten, Indonesia

Fakhrurrozi Fakhrurrozi [1], *Idris* Idris [1], *Hafidz* Permana Sereni[2], *Mauludiyah* [2*] , *Sutan* Mukh. Zakyrudin Ma'ruf [2], *Ruly* Isfatul Khasanah[2], *Wiga* Alif Violando[2], *Misbakhul* Munir[2]

[1]The Indonesian Coral Reef Foundation (Yayasan TERANGI), West Java, Indonesia
[2]Department of Marine Science, Faculty of Sains and Technology, State Islamic University of Sunan Ampel Surabaya, East Java, Indonesia

Abstract: Coral reef ecosystems play a crucial role in marine biodiversity, providing habitat, food, and ecological stability. However, increasing anthropogenic pressures and environmental changes threaten their sustainability, necessitating research to inform conservation efforts. This study examines the biodiversity of coral reefs around Tunda Island, Banten Province, Indonesia, with a focus on the diversity of corals and reef fishes to support conservation efforts. Fieldwork was conducted in November 2019 across six stations using scuba diving. Data collection involved the belt transect method to assess coral community structures and the Underwater Visual Census method to evaluate reef fish populations. The study recorded 25 coral genera, with Porites being the most dominant. Omnivorous fish constituted the highest biomass, while herbivorous fish, particularly Caesio cuning, indicated significant fisheries potential. The station located farthest from the community and adjacent to seagrass and mangrove ecosystems exhibited the best overall ecological condition. The findings highlight the ecological significance of Tunda Island's coral reef ecosystem and underscore the importance of adjacent habitats in maintaining biodiversity. Conservation efforts should focus on protecting areas with high biodiversity and promoting sustainable fisheries management.

1. Introduction

As an archipelagic country with the second longest coastline in the world, Indonesia spans more than 95,000 km². Over 60% or two-thirds of its territory is the ocean, extending from Sabang to Merauke. Its geographic location between the Indian and Pacific Oceans creates a rich and diverse marine biodiversity. This region, often referred to as the "Amazon of the Seas," harbours approximately 30% of the world's coral reefs and supports 75% of known coral species (Burke et al., 2012). [1] Fertile tropical marine areas with a climate suitable for the growth of various marine biota have enhanced the potential of Indonesia's marine and coastal resources [2].

* Corresponding author: mauludiyah@uinsa.ac.id

Coral reefs are shallow marine ecosystems composed of calcium carbonate (CaCO3) structures produced by marine organisms, such as corals and algae [3]. These highly productive ecosystems thrive in tropical marine waters and support a wide range of marine life due to their complex physical structure, which includes branching formations, crevices, and tunnels. Such intricate habitats offer shelter, feeding grounds, and breeding sites for numerous marine biota, including reef fish species, which rely on coral reefs for survival [4] [5]

Tunda Island, located in Banten, Indonesia, is recognized for its natural potential, featuring an extensive sea area, white sandy beaches, beautiful underwater biota, and various species of ornamental fish, coral reefs, and seagrass beds [6]. However, despite its ecological richness, the coral reef ecosystem around Tunda Island has experienced degradation. The high level of sand mining activity around the waters of Tunda Island puts pressure on coral reef ecosystems [7]. This ongoing pressure has led to a continuous decline in coral reef conditions around Tunda Island, thereby reducing the overall health and resilience of the ecosystem. The live coral cover in the waters surrounding Tunda Island is reported to range between 8% and 42%, classifying it as poor to moderate conditions [8]. Such low coral cover reduces habitat availability for reef-associated species, diminishes reef productivity, and alters the species composition. As specialized reef-dependent organisms decline, opportunistic species may dominate, leading to an imbalance in the ecosystem. Furthermore, reefs with poor to moderate coral cover are more susceptible to further degradation by environmental stressors, weakening their overall resilience.

The decline in coral cover not only affects the ecological balance but also poses a significant threat to the long-term sustainability of marine biodiversity on Tunda Island. This shift in species composition and habitat degradation underscores the urgent need for conservation efforts. Given these challenges, this research aims to identify the biodiversity conditions of coral reefs, reef-associated fish species, and reef fishes on Tunda Island. By understanding the current state of marine biodiversity, this study contributes to conservation efforts and informs sustainable management strategies for coral reef ecosystems on islands.

2. Methods

2.1. Study Site

The study was conducted from 29-30 November 2019. It is located in the waters of Tunda Island, Serang Regency, Banten Province. The observation stations were spread over six stations in the coral reef ecosystem areas. Scuba diving equipment was used to collect data at depth of 5-7 m. Figure 1 shows the detailed point locations of the survey area.

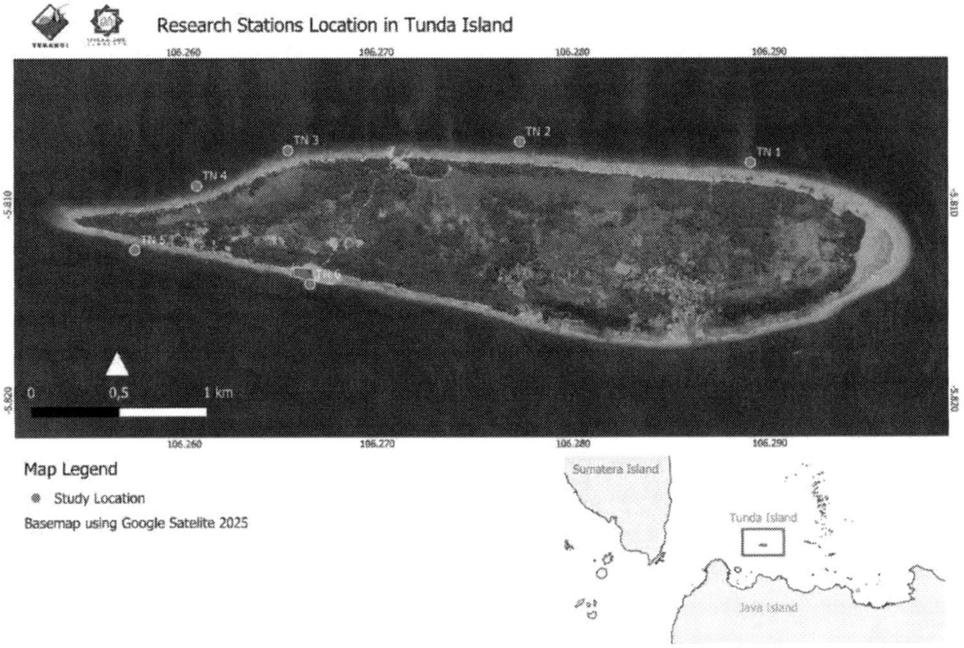

Fig 1. Map of research station location in Tunda Island

2.2. Data Collection

2.2.1. Coral Community Structure

Data were collected using the Belt Transect method with a transect length of 50 × 1 m [9] . This method allowed for the systematic recording of coral genera, colony sizes, and their abundances to assess the coral reef ecological index. Coral identification followed the taxonomic classifications established by [3] and [10]to ensure accuracy in genus identification. To ensure consistency in the classification, a ruler was used to measure the maximum diameter of each coral colony within the belt transect. Colony size was categorized into three classes: small (<5 cm), medium (5-25 cm), and large (>25 cm), following established studies on coral population structure and ecological resilience [11]. This classification provides a standardized approach for evaluating coral reef health and demographic trends across different reef ecosystems. Due to technical errors, only data from four of the six surveyed locations were suitable for further analysis of the coral community structure.

2.2.2. Reef fishes community structure

Data collection was conducted using the Underwater Visual Census (UVC) method, a widely used nondestructive survey technique in marine research for assessing reef fish diversity, distribution, and abundance. The transect was laid out along a 50-meter line parallel to the coastline, with observers recording fish species within a 5-meter wide belt (2.5 meters on either side of the transect line). Fish length estimation followed the methodology described in [12], which estimates fish length and categorizes them into predefined size classes. To enhance accuracy, divers were calibrated using reference objects before the field surveys, and size estimation was standardized into 5 cm interval classes. Observers also utilized known-size objects, such as rulers or fixed markers, to improve measurement reliability. Identification of reef fish species was based on [13] ensuring precise taxonomic classification, and data collection was conducted using the Underwater Visual Census (UVC) method. UVC is a survey technique that is used in marine research to gather data on the diversity, distribution, and abundance of underwater organisms. The transect was laid out along a 50 meter line parallel to the coastline. Observers surveyed the fish 2.5 meters to the left and 2.5 meters to the right of the transect line.

2.3. Data analysis

The Community structure of coral and reef fishes was analyzed to provide information about Richness, Diversity Index, Evenness, Dominance Index, and Abundance.

A. Diversity Index using Shannon-Wiener (H') [14]
   $H' = \sum Pi \ Ln \ Pi$
   Pi = proportion abundance of species
B. Evenness (E) [14]
   $E = H'/H \max$
   $H \max = H'/Ln \ S$ (S for species richness)
C. Dominance (D) [14]
   $D = - \sum Pi2$
D. Abundance (N) [14]
   $N = ni/A$
   ni = sum of each species
   A = survey area

Reef fish biomass was analyzed using an allometric Length-Weight (L-W) relationship that estimated biomass from the weight (W) data of a species.
$W=a \ L \ b$
W= weight a and b = constant
L = fork length

3. Result and discussion

3.1. Coral community structure

The total number of coral genera found on Tunda Island was 25, with Porites having the highest density of 57 colonies per 50 m^2 (Figure 2). Porites are known for their slow growth, but are more resistant to thermal-induced bleaching [15]. Additionally, Porites can grow in various environments, including both coral reef-dominated and sandy areas [16]. The dominance of Porites coral abundance is attributed to high sedimentation, as Tunda Island is geographically close to the northern Java mainland, which contains several river estuaries. The dominance of Porites coral has also been recorded in other locations with extreme wave conditions, such as Sabu Raijua and Rote in the East Nusa Tenggara [17] [18]. Therefore, the presence of Porites coral on Tunda Island is crucial for enhancing the resilience of coral reef ecosystems to the threats posed by climate change and extreme environmental shifts.

The diversity index (H) provides an overview of coral community diversity based on the number of genera and abundance of colonies within each species (Table 1). According to the data, Station 3 had the lowest diversity index (3.25), whereas Station 4 had the highest diversity index (4.46). As shown in Table 1, all stations fell into the high-diversity category ($H' \geq 3$). In terms of evenness, the highest index was found at Station 1 (1.82), and the lowest at Station 3 (1.23). The evenness (E) values for all stations were categorized as fairly high, whereas the dominance index values were classified as low. The highest dominance (C) index was at Station 3 (0.06) compared to other station. (Table 1)

Based on Figure 3, there are contrasting values between TN 1 and TN 5: TN 1 had the lowest coral richness but higher abundance, whereas TN 5 had the lowest abundance but greater richness. This was due to the dominance of the coral genera Montipora and Porites in TN 1, with values of 94 and 48 colonies per 50 m², respectively. Maintaining both the abundance and diversity of corals is crucial for preserving the ecological functions of coral reef ecosystems. Diverse coral communities enhance reef resilience, support a wide array of marine life, and contribute to the overall health of the ocean. For instance, a study from [19] emphasizes that corals provide complex habitats that support fish and macroinvertebrate diversity, thereby playing a vital role in maintaining ecosystem stability. The highest genera richness was observed in TN 6, with 22 genera, while the lowest was observed in TN 1, with 10 genera. Regarding coral colony abundance, TN 3 and TN 6 had the highest at 369 colonies per 50 m², whereas TN 5 had the lowest at 53 individuals per 50 m². Therefore, TN 3 and TN 6 play a more crucial role in supporting ecological processes because of their high diversity of coral species and colony abundance compared to the other sites.

Table 1. Community Structure of Coral Population

Station	H	E	C
TN 1	4.20	1.82	0.02
TN 3	4.40	1.91	0.02
TN 5	3.25	1.41	0.06
TN 6	4.46	1.94	0.02

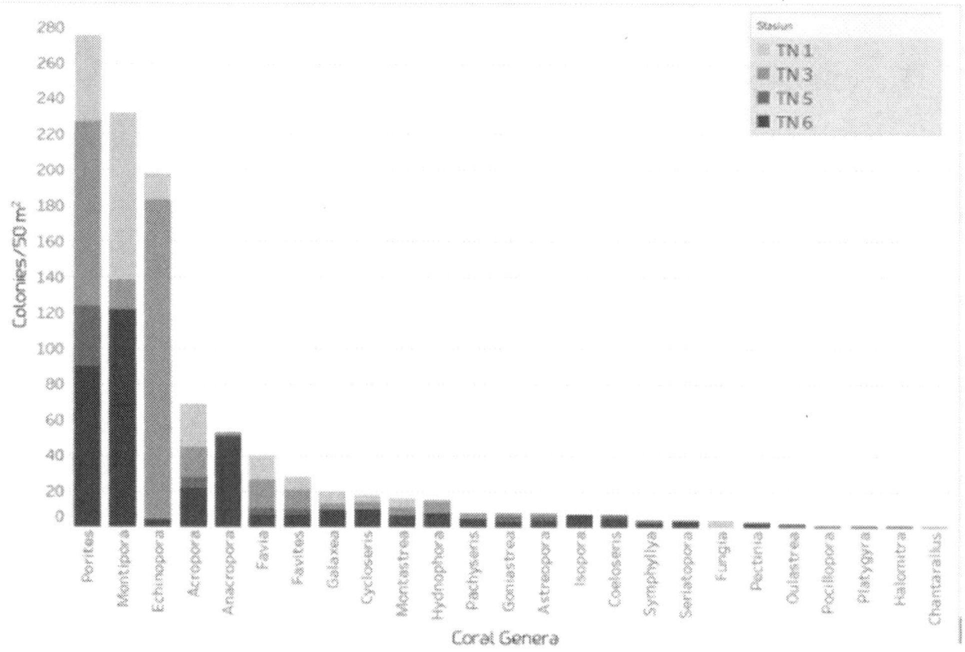

Fig 2. Abundance of coral genera recorded in each station

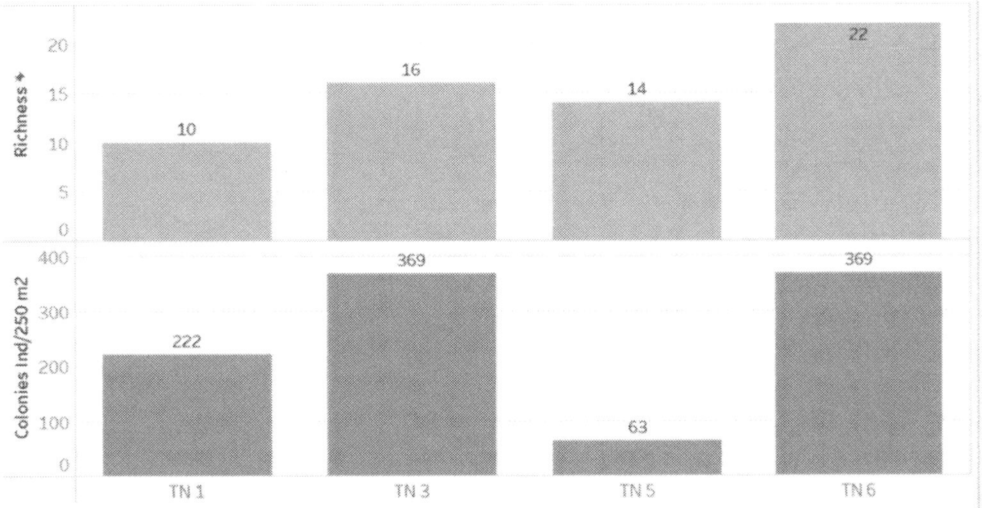

Fig 3. Richness genus and colonies abundance for each stations

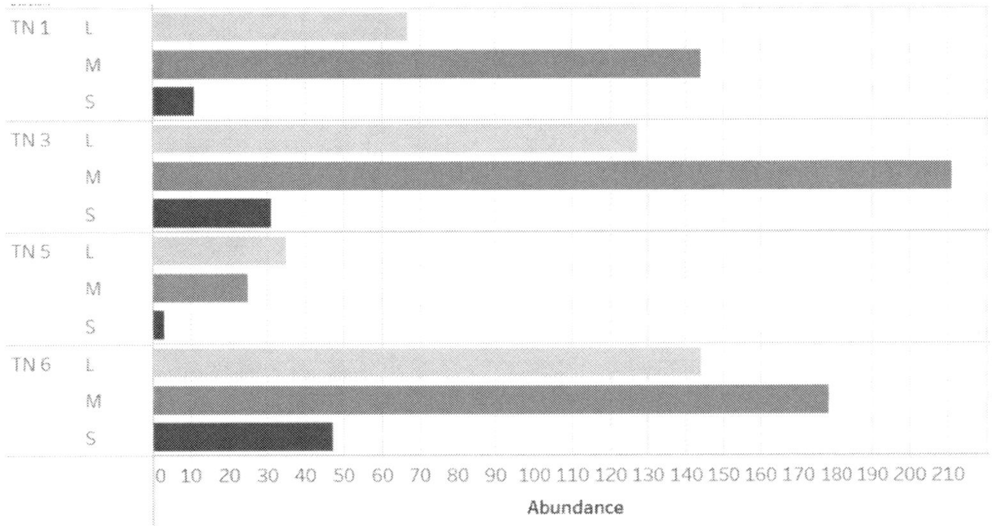

Fig 4. Size of coral population

The results of observations on Tunda Island show that medium-sized (M) coral colonies dominate all locations except TN 5 (Figure 4). These data indicate that coral colonies on Tunda Island are predominantly between 5-25 cm in size. Given the dominance of the Porites genus, it can be concluded that the age of coral reefs on Tunda Island ranges from 5 to 10 years. The presence of medium (M) and large (L) corals is crucial for enhancing ecosystem resilience to environmental pressure. Small-sized (S) corals, which are juveniles or recruits, play a crucial role in the attachment and sustainability of coral populations. Corals measuring up to 5 cm in length are considered juveniles, typically indicating that the corals are approximately three years old [20]. From the data in Figure 4, we can see that the abundance of coral recruits averaged 23 colonies at each observation station. TN 6 had the highest recruitment abundance, with 46 colonies, whereas TN 5 had the smallest abundance, with three colonies. The low number of coral recruits in TN 5 indicates that the coral population in this location is more vulnerable than that in other areas.

Based on the abundance, richness, and recruitment of the coral reef community, TN 6 exhibits the best values compared to the other locations. This superior performance is likely due to the presence of well-preserved mangrove and seagrass ecosystems in the surrounding areas. These ecosystems provide essential nursery habitats, reduce sedimentation, and offer protection against coastal erosion, thereby supporting coral health and diversity. Furthermore, the relative isolation of TN 6, which is far from the village, minimizes human interaction and direct community intervention, reducing anthropogenic stressors, such as pollution, overfishing, and physical damage. This combination of natural protection and limited human impact makes coral recruitment in TN 6 a vital and resilient part of the coral reef ecosystem on Tunda Island, highlighting the importance of preserving adjacent ecosystems and maintaining a low human impact for coral conservation.

3.2. Reef fishes community structure

Observations of coral fish conditions across six locations on Tunda Island have recorded a total of 86 species. It's quite low if compare to other location such as Lembeh Island 342 species and Raja Ampat 197 species [21] [22]. Among these locations, TN 4 had the highest species richness with 48 species, whereas TN 2 had the lowest speceis richness 14 species (Figure 5). Variations in the number of species and fish abundance among these locations can be attributed to several environmental factors. However, [11] highlighted that a decline in coral cover is the primary cause of loss of fish biodiversity. The fish diversity index ranged from 2.1 (low) to 3.6 (high), indicating that fish communities were under pressure to maintain stable conditions (Tabel 2). The dominance index ranged from low to moderate, reflecting varying degrees of dominance by certain species (Tabel 2).

High dominance values can affect the stability and diversity of the fish communities [14]. The average fish abundance across all observation stations was 674 individuals per 250 m^2, with TN 4 showing the highest abundance at 2.517 individuals per 250 m^2 and TN 3 the lowest at 163 individuals per 250 m^2 (Figure 5). A previous study in 2017 showed that the average abundance of fish on Tunda Island was 176 individuals per 250 m^2 [8]. This value indicates that there has been an increase in the average abundance of fish on Tunda Island over the past two years.

Figure 6 shows the top ten families of reef fishes with the highest richness based on family and abundance based on species. Among these, the Pomacentridae family exhibited the highest number of species, with a total of 65 species per 1,500 m^2. Frequently recorded Pomacentridae species during data collection included *Cirrhilabrus cyanopleura* 480 and *Caesio cuning* 130 individuals from the six observation areas. The Pomacentridae family was evenly distributed across all observation stations, with the highest abundance observed at TN 4, where 1,948 individuals per 250 m^2 were recorded. This widespread distribution and high abundance underscores the ecological significance and adaptability of the Pomacentridae family in the coral reef ecosystems of Tunda Island. Generally, Pomacentridae fish are known for their attractiveness, bright coloration, distinctive body shapes, and agile movements [23]. Furthermore, the coral fish populations on Tunda Island, particularly in TN 4, have significant economic potential through ornamental fish trade owing to these appealing characteristics.

Table 2. Community Structure of Reef Fishes

No.	Location	Diversity Index (H')	Evenness (E)	Dominance Index (D)
1.	TN 1	3,0	1,2	0,07
2.	TN 2	2,1	0,7	0,05
3.	TN 3	2,6	0,9	0,09

No.	Location	Diversity Index (H')	Evenness (E)	Dominance Index (D)
4.	TN 4	3,6	1,06	0,03
5.	TN 5	2,4	0,9	0,15
6.	TN 6	2,4	1,19	0,11

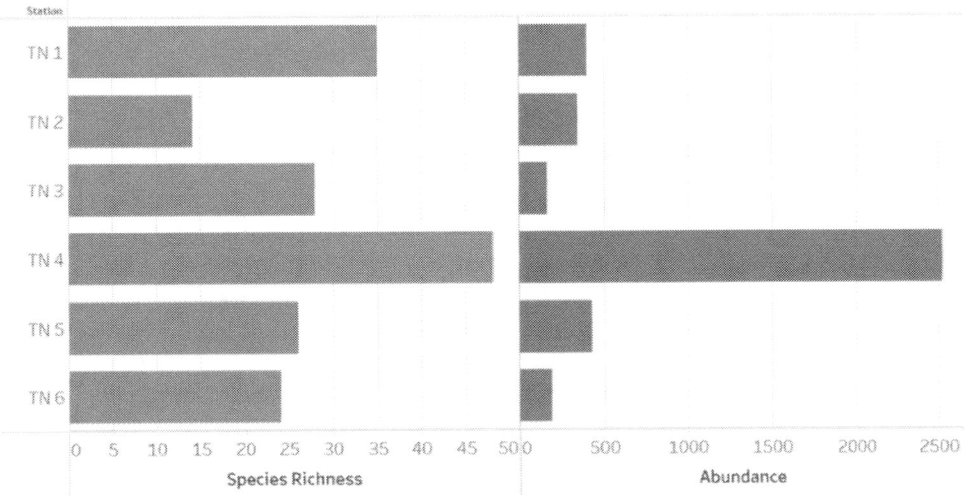

Fig 5. Species richness and abundance in each station (individuals per 250 m^2)

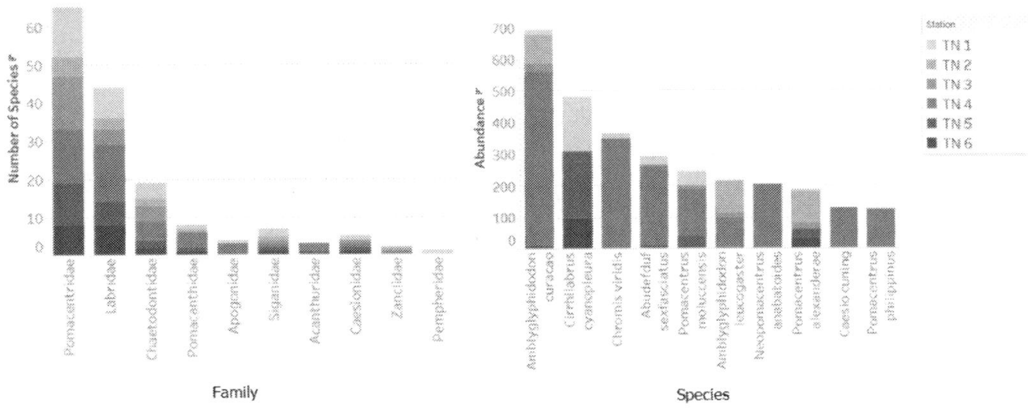

Fig 6. Top 10 Richness species based on family (left) and total abundance based on Species (right) (individuals per 1,500 m^2)

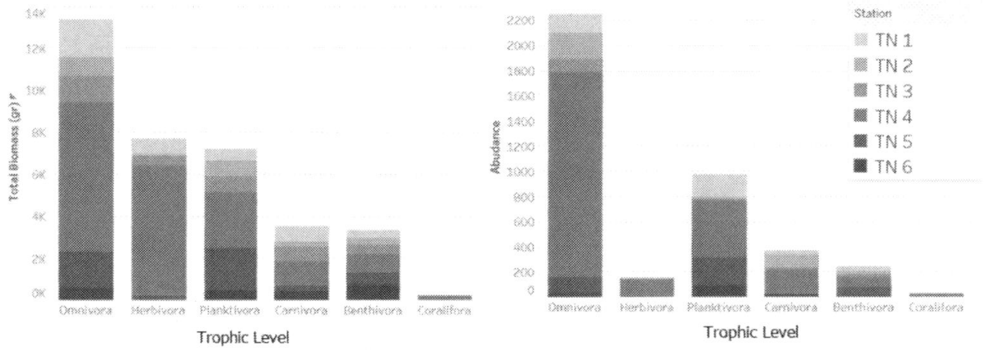

Fig 7. Total biomass gr (left) and abundance individuals (right) based on trophic level (per 1,500 m^2)

The results based on Fig 7 shows the biomass and abundance data based on trophic levels found that omnivores had the highest values compared to other trophic levels. This distribution indicates that omnivorous fish, which have a more varied diet, dominate the biomass on Tunda Island, whereas corallivorous fish, which feed specifically on corals, are less prevalent. Interestingly, the biomass data of herbivorous fish showed a significant inverse relationship with their abundance. Despite their small numbers (139 individuals per 1500 m^2), herbivorous fish exhibit a substantial biomass of 7,705 g per 1500 m^2. This indicates that herbivorous fish on Tunda Island are larger than those at other trophic levels. Notably, 80% of the herbivorous fish species were represented by *Caesio cuning*. These findings indicate that Tunda Island has significant fishery potential because of the large size of the recorded *Caesio cuning* fish.

4. Conclusion

The study results indicate that Tunda Island's coral reef ecosystem exhibits diverse biodiversity, with Porites being the dominant coral genus owing to its resilience against environmental stressors and high sedimentation levels. Variations in coral richness and abundance among observation stations suggest the localized impacts of environmental conditions and human activities. The presence of well-preserved mangrove and seagrass ecosystems near TN 6 supports higher coral recruitment and fish diversity, highlighting the importance of adjacent habitats for reef resilience. Reef fish diversity was high, with the Pomacentridae family being the most abundant, reinforcing its adaptability to various reef conditions. The dominance of omnivorous fish suggests shifts in trophic dynamics, potentially indicating changes in the ecosystem stability. Encouragingly, the increased abundance of reef-associated fish compared with previous studies suggests positive trends, but continued conservation efforts are essential. Strengthening habitat protection, minimizing human impact, and promoting sustainable fisheries and ecotourism will be critical for ensuring the long-term health and biodiversity of coral reefs on Tunda Island. Future research should focus on the long-term monitoring of coral recruitment and fish population dynamics to better understand the resilience of ecosystems under changing environmental conditions.

The author would like to thank the TERANGI Foundation and UIN Sunan Ampel Surabaya for facilitating this research activity as well as the YIIM Foundation for providing financial assistance in this study process. The authors also do not forget to thank their friends at Tunda Island who have helped a lot in the data collection process to complete this study.

References

1. Burke L. K Reytar, M Spalding, A Perry., World Resources Institute, 2012.
2. Mohammad Adrim, Syawaludin Alisyahbana Harahap, Kunto Wibowo, *ILMU KELAUTAN,* vol. 17, no. 3, pp. 154-163, 2012.
3. Suharsono, Jakarta: Pusat Penelitian Oseanografi -LIPI, 2008.
4. H. F. Bahrul Ulumi and M. Syafar, *Jurnal Antropologi: Isu-Isu Sosial Budaya,* vol. 23, no. 118, p. 1, 2021.
5. D. Dedi, P. Z. Neviaty, and A. Taslim, *Jurnal Kelautan Nasional,* vol. 11, no. 2, pp. 105-118, 2016.
6. MPS Widodo, Idris Idris, Nabila Aprilya, Fakhrurrozi, Safran Yusri, Paryono, WA Lestariningsih,and AA Damayanti. *BIO Web of Conferences MaCiFIC 2023*, 2023.
7. Pramewara B, dan IB Suryawan, *Jurnal Destinasi Pariwisata,* vol. 7, no. 1, 2019.
8. Karizma Fahlevy, Budi Prabowo, Prakas Santoso, Beginer Subhan, Hawis Madduppa, in *Prosiding Seminar Nasional Ekosistem Perairan Teluk Lada dan Pulau Tunda Tahun 2017*, 2017.
9. English SC, WIlkinson and V Baker, Australian Institute of Marine Science, 1997.
10. Veron JEN, Stafford Smith MG, Turak E, and DeVantier LM, " 2016. [Online]. Available: www.coraloftheworld.org. [Accessed 30 11 2024].
11. Estradivari, Muh. Syahrir, N. Susilo, S. Yusri dan S. Timotius, Yayasan TERANGI, 2007.
12. JD. Bell, G. J. S. Craik David Pollard, Barry C Russell, *Coral Reefs,* vol. 4, no. 1, pp. 41-44, 1985.
13. Allen, Gerald, R. Steene, Paul Humann dan Ned Deloach, New World Publication, Jacksonville, 2003.
14. Odum E. P.. and. G. W. Barrett, Philadelphia: Cengage Learning, 1971.
15. Maynard, *Mar biol,* vol. 155, no. 2, pp. 173-182, 2008.
16. S. Yusri, V. P. Siregar, and Suharsono, in *IOP Conf. Series: Earth and Environmental Science*, 2019.
17. F. Fakhrurrozi, S. Idris, Y. S. Adiwijaya, N. S. Wijoyo, M. A. Prawira, S. S. Agustami, Z. Yuniar, and C. A. G. Alisa, in *Conference Series: Earth and Environmental Science*, 2022.
18. Idris, Fakhrurrozi, Ofri Johan, Jotham S R Ninef, Edwin Jefri, Mahardika Rizqi Himawan, *Jurnal Biologi Tropis,* vol. 23, no. 2, pp. 434-441, 2023.
19. Bruno S. Carturan, Lael Parrott, and Jason Pither. *Frontiers in Ecology and Evolution,* vol. 10, no. 780406, 2022.
20. C. S. Rogers, H. Fitz III, M. Gilnack, J. Beets, and J. Hardin, *Coral Reefs,* vol. 3, pp. 69-76, 2003.
21. N P M Dwita, Idris and N S Widjoyo in *ICMMBT IOP Conf. Series: Earth and*

Environmental Science, 2021.

22. Muhammad Rizky Adipratama, Mauludiyah, Dian Sari Maisaroh, Wiga AlifViolando, Ofri Johan, Idris, Paryono, Sadikin Amir, *Jurnal Biologi Tropis,* vol. 23, no. 2, pp. 223-234, 2023.
23. P. B. Moyle and J. J. Cech, Jr, Fishes: Englewood Cliffs: NJ: Prentice Hall, 1998.

Oceanic index variability and its role in mangrove dieback events in East Lampung

Rizki Dimas Permana[1*], *Febri Dwi* Irawati[2], *Mohammad Ashari* Dwiputera[1], *Dea* Annisa[1], and *Angger* Firmanto[1]

[1] Marine Environmental Science Study Program, Institut Teknologi Sumatera (ITERA), South Lampung, Lampung 35365, Indonesia
[2] Data Science Study Program, Institut Teknologi Sumatera (ITERA), South Lampung, Lampung 35365, Indonesia

Abstract. A significant dieback event of Avicennia marina was observed in the coastal area of East Lampung, beginning in 2020, with its cause remaining unidentified. This study aimed to assess the rate of mangrove degradation and investigated the influence of the ocean index on mangrove dieback event. Sequential satellite imagery analysis (2019–2022) was performed to evaluate changes in mangrove cover, focusing on a 200-hectare area within the total 528.69-hectare mangrove forest, utilizing using Normalized Difference Vegetation Index (NDVI) and Mangrove Health Index (MVI). Field data were collected at 30 research stations, encompassing live mangroves, partially affected areas, and fully dieback sites, both before and during the dieback event (2020 and 2021). These efforts provide critical insights into the extent and drivers of mangrove decline in the region. Based on the results, the mangrove dieback caused a rapid and significant loss of healthy stands (results of the MVI) and very-high-density stands (results of the NDVI). With reduction in mangrove area was 117.64 ha during 2020-2021. A pronounced decline in 2021 coincided with a prolonged La Niña phase, leading to excessive freshwater influx and prolonged inundation, which likely decreased the salinity and sediment dynamics required for mangrove survival. Additionally, a negative ENSO phase (-1.2) and a concurrent positive IOD phase (+1.5) were associated with increased drought conditions, elevated salinity, and reduced freshwater availability, further stressing Avicennia marina populations. These findings underscore the strong link between oceanic variability and mangrove health, highlighting the need for continued monitoring and improved predictive models to assess the long-term impacts of climate change on coastal ecosystems.

1 Introduction

Mangroves are critical coastal ecosystems that provide numerous ecological, economic, and social benefits including shoreline protection, carbon sequestration, and habitat for marine biodiversity[1]. Despite their importance, mangrove forests face significant threats

* Corresponding author: rizki.permana@sll.itera.ac.id

worldwide owing to both natural and anthropogenic factors. In recent years, East Lampung in Indonesia has experienced severe mangrove dieback, particularly affecting the Avicennia species[2]. This phenomenon has raised concerns among researchers and environmentalists regarding its underlying causes and long-term implications for the region's coastal environment.

One major factor contributing to mangrove dieback is climate variability, which influences sea surface temperature, rainfall patterns, and sea level changes. Inter-annual ocean-climate anomalies, such as the El Niño-Southern Oscillation (ENSO) and Indian Ocean Dipole (IOD), are key indicators of climate variability. These indices affect regional and global climate patterns and can lead to extreme conditions, such as prolonged droughts or excessive rainfall, both of which can be detrimental to mangrove health. [3,4].

Previous studies have demonstrated the sensitivity of mangroves to changes in salinity, water level, and nutrient availability, all of which are influenced by climate variability[5–7]. For instance, El Niño events are generally associated with reduced precipitation, leading to increased salinity and water stress in mangrove ecosystems[3]. However, in Indonesian Seas, El Niño is linked to lower seawater temperatures, which may have additional implications for mangrove health[8].

Triple-dip La Niña events that occurred from 2020 to 2022 had profound impacts on global and regional climate systems[9]. These events induce significant alterations in local hydrological conditions, exacerbating environmental stress factors in mangrove ecosystems. [2,10,11]. Understanding the role of oceanic indices in driving mangrove dieback is crucial for developing effective conservation and management strategies. In East Lampung, an active mangrove rehabilitation program has been ongoing since 2005, led by local communities that have planted over 300 hectares of mangroves, predominantly consisting of Avicennia species. However, between 2020 and 2021, more than 100 ha of these rehabilitated mangroves experienced mass dieback, raising concerns about the resilience of such ecosystems under changing climatic conditions.[12]. Many mangroves planted in East Lampung consist of monospecific stands of Avicennia species, particularly Avicennia marina and Avicennia alba. These species are well known for their high salinity tolerance, attributed to their physiological adaptations such as salt excretion through specialized glands and ultrafiltration at the root level[13,14]. However, despite their resilience to saline environments, Avicennia species exhibit limited tolerance to extreme environmental changes including prolonged inundation, hyper salinity, and abrupt shifts in hydrological regimes[15,16]. Such stressors can disrupt physiological functions, leading to reduced growth, leaf senescence, and, in severe cases, mass dieback.

This study aimed to investigate the relationship between oceanic index variability and mangrove dieback in eastern Lampung, Indonesia. By integrating multi-temporal satellite remote sensing data, conducting systematic field surveys, and employing advanced statistical analyses, this study sought to elucidate the mechanisms through which climate variability influences mangrove ecosystem health. This study hypothesizes that variability in oceanic indices, such as the Indian Ocean Dipole (IOD) and El Niño–Southern Oscillation (ENSO), significantly drives mangrove dieback events. By rigorously testing this hypothesis, this research not only advances the scientific understanding of mangrove ecosystem dynamics under climatic stress but also provides actionable insights for mitigating future dieback events and enhancing sustainable mangrove management practices.

2 Method

2.1 Study area

This study was conducted in the mangrove ecosystems of East Lampung, Indonesia, at approximately 5°21' S latitude and 105°31' E longitude, which was chosen for its significant mangrove dieback events and history of rehabilitation efforts (Fig. 1). The region is characterized by extensive mangrove forests, which are crucial for coastal protection, biodiversity, and local livelihoods. This area has been the focus of various mangrove rehabilitation efforts owing to the significant dieback events observed in recent years. The mangrove ecosystems are influenced by climatic factors, such as sea surface temperatures, rainfall patterns, and sea level changes, making it an ideal location for studying the impacts of climate variability on mangrove health.

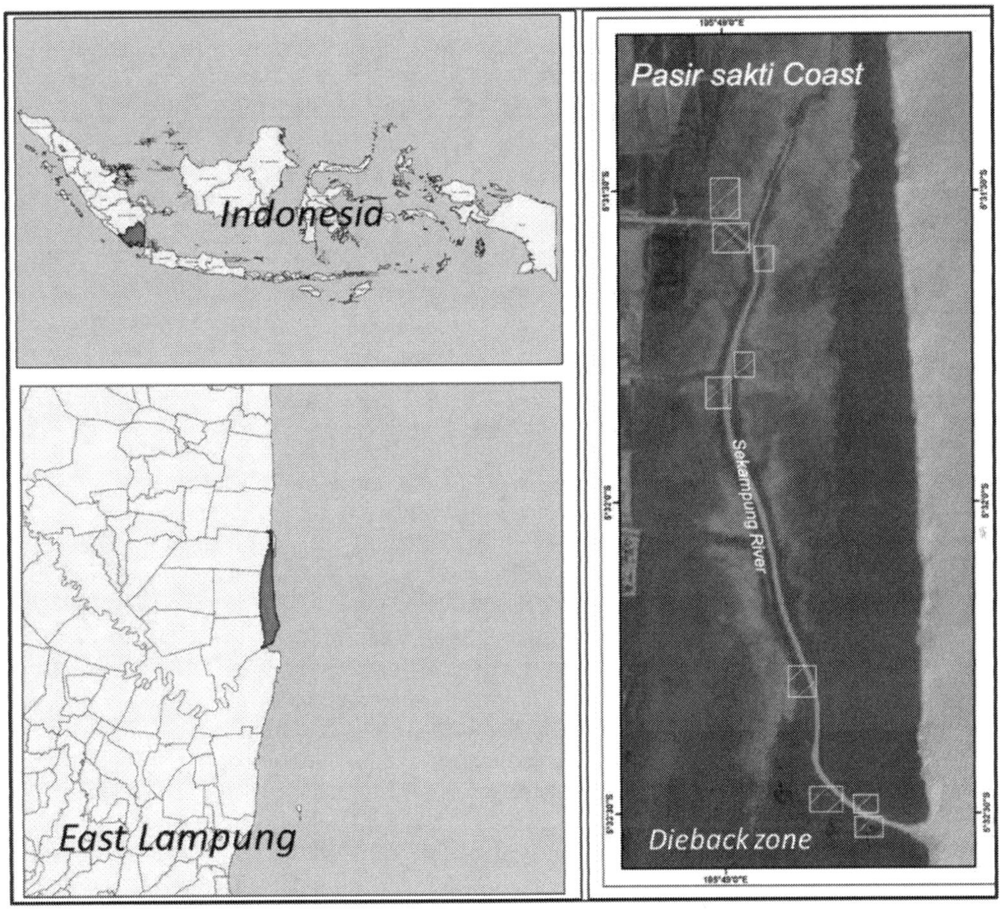

Fig. 1. The study area is in the Pasir Sakti Coast, East Lampung, Indonesia, specifically within the dieback zone. The red dots represent the observation stations for mangrove vegetation conditions and oceanographic data collection.

Table 1 summarizes the data collected, providing a comprehensive understanding of the factors contributing to mangrove dieback in East Lampung and aiding in the development of conservation strategies. The data include salinity, temperature, dissolved oxygen, and pH, which were collected through in-field measurements and samples by the research team from May 2024 to June 2024. Sea surface height data, covering the period from 2019 to 2024, were obtained from the Marine Copernicus Ocean Products. Ocean indices such as ENSO and IOD, spanning from 2019 to 2024, were sourced from the Niño 3.4 SST Index and the Dipole Mode Index (DMI). Additionally, NDVI and MVI data derived from Sentinel-2 satellite imagery for the same period were accessed via the Copernicus Data Space Ecosystem (Table 1).

Table 1. Data used in the research.

Data Type	Time Range	Data Source
Salinity, temperature, dissolved oxygen, pH	May 2024-June 2024	In-field measurements and samples collected by the research team
Sea Surface Height	2019-2024	Marine Copernicus Ocean Products. SEALEVEL_GLO_PHY_L4_MY_008_047 (Product ID)
Ocean index: ENSO, IOD	2019-2024	Niño 3.4 SST Index; Dipole Mode Index (DMI)
NDVI, MVI From (Sentinel-2)	2019-2024	https://dataspace.copernicus.eu/

2.2 Data Analysis

The analysis of the Mangrove Vegetation Index (MVI) and Normalized Difference Vegetation Index (NDVI) using Sentinel-2 data involves several key steps. Sentinel-2 provides high-resolution optical imagery with multiple spectral bands, including those required for vegetation analysis, such as Red, Near-Infrared (NIR), and Short-Wave Infrared (SWIR) bands[17,18]. The NDVI was calculated using the formula (NIR − Red) / (NIR + Red), where higher values indicate healthier and denser vegetation. MVI is derived from the formula (NIR-SWIR) / (NIR + SWIR), which is particularly useful for assessing mangrove health and distinguishing it from other vegetation types[17]. After extracting the relevant bands from Sentinel-2 imagery, these formulas were applied to calculate the NDVI and MVI. The resulting indices were visualized to highlight areas of high vegetation health and mangrove density. Temporal and spatial analyses of these indices can reveal changes in vegetation and mangrove cover over time, aiding land-use planning, conservation efforts, and environmental research[19]. By leveraging Sentinel-2 high-resolution data, researchers and policymakers can gain valuable insights into vegetation dynamics and ecosystem health[20].

Geospatial techniques using GIS were employed to map the spatial extent of the mangrove dieback. Satellite imagery from Landsat and Sentinel missions provided temporal data, enabling us to analyze changes in mangrove cover over time. The spatial distribution of mangrove dieback was correlated with environmental and climatic variables. This involved overlaying maps of the dieback areas with maps of soil salinity, water quality, and climate data to identify patterns and relationships.

This study employed correlation and regression analyses to examine the relationships between climatic variables (ENSO, IOD, and rainfall) and mangrove health indicators (MVI and NDVI). These analyses were instrumental in identifying the key climatic factors influencing mangrove dieback. Additionally, Principal Component Analysis (PCA) was used to reduce the dimensionality of the dataset and pinpoint the primary factors contributing to mangrove dieback. This statistical approach facilitates a deeper understanding of the variance within the data, and highlights the most significant variables affecting mangrove health[21,22].

3 Result and Discussion

3.1 Oceanic indices and Sea Surface height Variability

The temporal dynamics of ENSO and IOD indices from January 2019 to January 2024 are shown in the graph (Fig. 2). Pronounced oscillations were exhibited by both indices, indicating their influence on regional climate variability. A mangrove dieback event was marked at a specific point within the time series, suggesting a potential link between extreme climate conditions and mangrove ecosystem health. The positive phase of the IOD, as indicated by the red line, along with the concurrent [positive/negative] phase of ENSO, as shown by the blue line, is hypothesized to have contributed to the stressful conditions that triggered the mangrove dieback event. However, further analysis incorporating additional environmental factors and ground truth data is required to establish a definitive causal relationship[4,6].

The ENSO index (blue line) showed a negative phase (~ -1.5) in early 2021, indicating a La Niña event. The IOD index (red line) was also negative (~ -0.5 – -1) in early 2021, representing a negative IOD phase. The mangrove dieback event was marked from early to mid-2021. This period coincides with a strong La Niña and negative IOD, suggesting potential climatic stressors, such as cooler sea surface temperatures, altered precipitation patterns, and regional hydrological impacts[9]. La Niña and a negative IOD can cause increased precipitation and prolonged flooding, leading to salinity stress and affecting mangrove health. This can lower salinity, which is a critical factor for Avicennia marina, reducing its resilience. Excess water can limit the oxygen availability in mangrove soils, leading to root stress and mortality. A negative IOD typically weakens monsoon patterns, potentially altering tidal dynamics. The ENSO phases influence regional sea levels. La Niña can elevate sea levels and increase submersion stress on mangroves. Temperature anomalies during ENSO events may alter the metabolic and physiological responses of the mangrove species[16]. The correlation between ENSO, IOD, and mangrove dieback is likely to be driven by hydrological stress (low salinity and flooding).

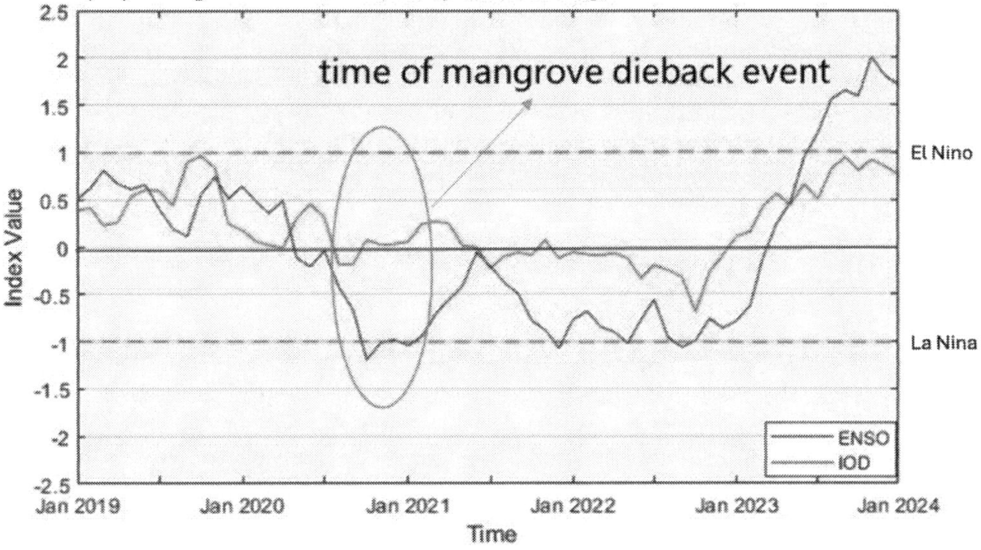

Fig. 2. The variability of the Indian Ocean Dipole (IOD) and El Niño-Southern Oscillation (ENSO) indices over time, from January 2019 to January 2024.

The annual variability of the Sea Surface Height (SSH) in East Lampung from 2019 to 2024 exhibited fluctuations that reflected dynamic oceanographic conditions (Fig. 3). The observed trends suggest potential influences of climatic events, as depicted in Figure 2. Elevated SSH in certain years might have been linked to climatic anomalies such as those caused by IOD and ENSO events. A high SSH could have led to coastal inundation, thereby affecting mangrove health and distribution[3,23].

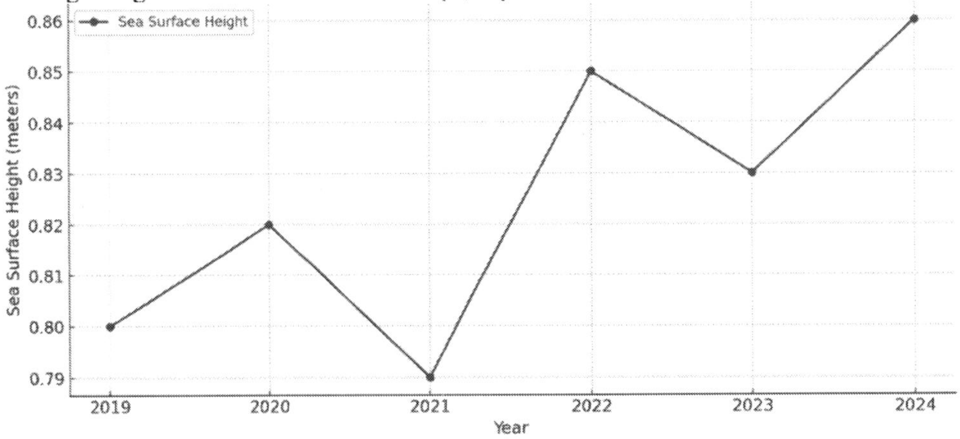

Fig. 3. Annual variability of Sea Surface Height in East Lampung (2019-2024)

The higher SSH in the study area, often linked to the positive phase of the IOD and El Niño, may be influenced by regional oceanographic and atmospheric mechanisms. While El Niño typically leads to a lower SSH in the western Pacific due to enhanced upwelling, regional effects can vary depending on monsoonal wind patterns, steric sea level changes, and oceanic circulation. In the case of East Lampung, positive IOD and El Niño events can contribute to an SSH rise through increased westerly winds, reducing coastal upwelling, and causing seawater accumulation along the coastline. Conversely, during negative IOD and La Niña periods, stronger easterly winds enhance upwelling, leading to SSH decline, increased exposure of mangrove roots, and higher susceptibility to desiccation and diseases[24,25]. The observed temporal variability in the SSH from 2019 to 2024 highlights these dynamic interactions. A more detailed analysis of SSH anomalies, particularly in 2021, when a strong La Niña and negative IOD coincided with mangrove dieback, is essential to establish a clear link between these climatic drivers and mangrove ecosystem responses.

3.2 Accumulative area of mangrove dieback dynamics

NDVI values fluctuated slightly over the years, with the lowest value in 2021 (0.66) and the highest in 2024 (0.72). This indicates a generally stable trend in mangrove vegetation health, with a slight dip in 2021, followed by recovery and improvement in the subsequent years (Table 2). Similarly, the MVI values showed a consistent increase, reflecting the overall positive trends in mangrove health and density. The slight decrease in 2021 aligns with the observed dip in NDVI, suggesting that this year might have experienced environmental stress affecting mangrove vegetation. Mangrove dieback in this region is likely to be influenced by a combination of environmental stressors and human activities. Research has shown that climate anomalies, such as the Indian Ocean Dipole (IOD) and El Niño-Southern Oscillation (ENSO), can exacerbate conditions leading to mangrove dieback, particularly through prolonged droughts, increased salinity, and sea level rise[11]. These climatic factors can cause physiological stress in mangrove trees, which makes them more susceptible to disease and mortality[8]. Moreover, human activities, including coastal development and deforestation, may compound these effects by disrupt natural hydrology and increase the vulnerability of mangroves to environmental stressors.

Figure 4 illustrates the progression of mangrove dieback in Pasir Sakti, East Lampung, from 2020 to 2023, highlighting the significant changes in mangrove coverage over this period. The dieback, indicated by the increasing area of barren land (marked in red), reflects the severe degradation of mangrove forests. This phenomenon is particularly evident when comparing the extent of dense mangrove cover (dark green) between 2020 and 2023, where the overall area of healthy, dense mangroves has fluctuated, suggesting periods of both recovery and further loss (fig. 4).

Table 2. Annual NDVI and MVI Values for Mangrove Vegetation in East Lampung the bold text indicates changes related to mangrove dieback.

Year	NDVI	MVI	Mangrove Area (ha)
2019	0.68	0.62	301.00
2020	0.70	0.64	308.82
2021	**0.66**	**0.60**	**191.18**
2022	0.71	0.65	213.24
2023	0.69	0.63	234.41
2024	0.72	0.66	317.65

The results indicated significant temporal variability in mangrove health and spatial extent in Pasir Sakti from 2019 to 2024. The NDVI and MVI values, which serve as indicators of vegetation health and density, show a general improvement from 2019 to 2024, apart from a marked decline in 2021. This decline corresponds to a substantial reduction in mangrove area, which dropped from 191.18 hectares to 308.82 hectares in 2020. The sharp decrease in 2021 is likely attributable to adverse climatic conditions, particularly the triple La Niña event, which resulted in prolonged periods of inundation, leading to stress and subsequent dieback of the mangrove ecosystem. This period of stress highlights the vulnerability of mangroves to extreme climatic events and their impact on both the health and extent of these critical coastal habitats[8].

Despite the significant losses in 2021, the data show a recovery in both vegetation health and mangrove area in the following years, with the area reaching 317.65 hectares by 2024. The steady increase in NDVI and MVI values during this period suggests that mangroves have not only recovered but potentially expanded beyond their pre-2021 extent. This recovery may be attributed to a combination of improved climatic conditions, natural resilience of the mangrove ecosystem, and effective conservation efforts. These results underscore the importance of understanding and mitigating the impacts of climate variability on mangrove ecosystems, as well as the need for continuous monitoring and adaptive management to ensure long-term health and sustainability. [11,26]

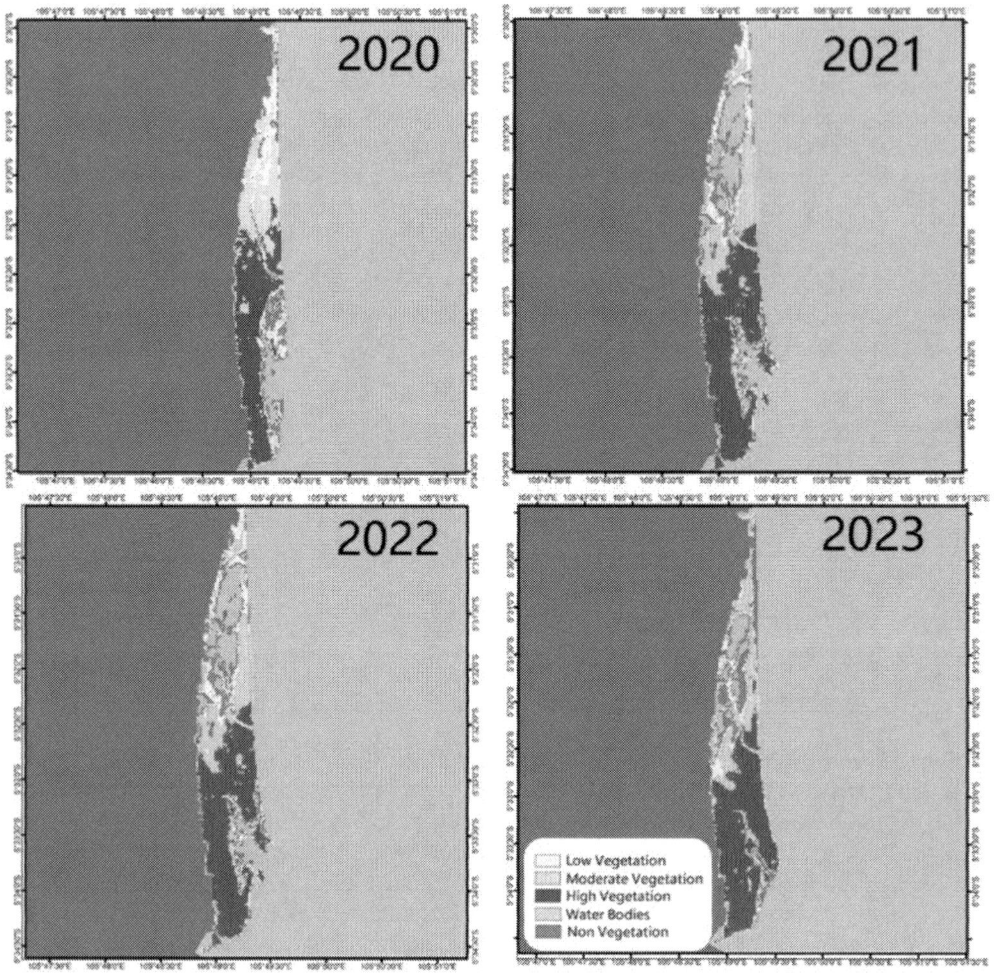

Fig. 4. Temporal Distribution of Mangrove Vegetation in East Lampung

3.3 Mangrove Dieback causes

The Indian Ocean Dipole (IOD) is a critical climatic driver in this region and is known for its significant impact on sea surface temperature (SST) anomalies and subsequent atmospheric circulation patterns. In East Lampung, fluctuations in IOD have been closely associated with variations in sea surface height (SSH) and salinity, both of which directly affect the health of mangrove ecosystems. During the positive IOD phases, warmer SSTs in the western Indian Ocean led to a higher SSH and increased moisture transport to the region. This can result in enhanced rainfall and freshwater input, creating favorable conditions for mangrove growth. However, the negative IOD phase, characterized by cooler SSTs in the western Indian Ocean and reduced rainfall, has been correlated with decreased freshwater availability and higher salinity levels, both of which impose osmotic stress on mangroves, thereby contributing to dieback[4,6].

The ENSO phenomenon, encompassing both El Niño and La Niña phases, exerts a profound influence on climate variability in Southeast Asia, including East Lampung. El Niño events are typically associated with prolonged dry conditions, reduced rainfall, and elevated temperatures, all of which exacerbate the water stress in mangrove ecosystems. The absence of sufficient freshwater input during El Niño events can lead to increased salinity in estuarine environments, further stressing mangroves and contributing to dieback[8]. Conversely, La Niña events are characterized by cooler SSTs and above-average rainfall, which can alleviate salinity stress through increased freshwater inflows. However, excessive water input during La Niña can also result in prolonged inundation periods, which may drown mangrove roots, lead to hypoxia, and ultimately cause dieback[24,27].

PCA analysis indicated that the first principal component explained approximately 87.18% of the variance in the data, while the second principal component explained approximately 7.41%. This suggests that most of the variability in mangrove density is influenced by a combination of factors (temperature, salinity, and pH) that most strongly align with the first principal component (Fig. 5). Therefore, these factors together significantly impact mangrove density, with temperature likely being the most influential factor, based on its variance contribution.

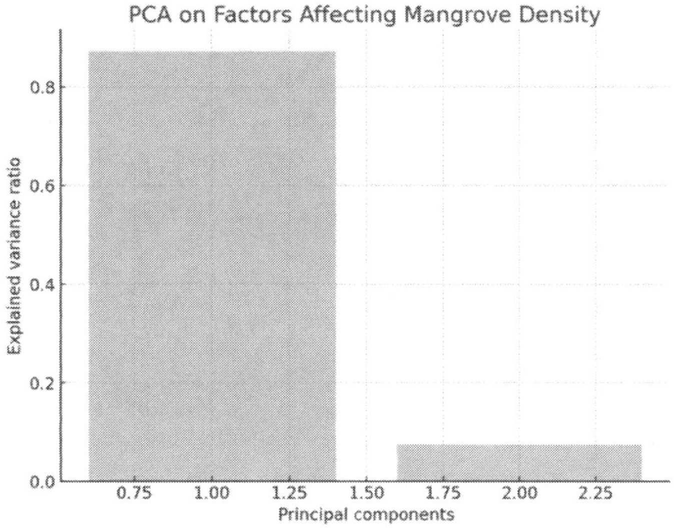

Fig. 5. Temporal Distribution of Mangrove Vegetation in East Lampung

From a physical perspective, mangrove dieback is the result of a complex interplay of environmental stressors, including sea level rise, salinity changes, temperature fluctuations, storm surges, and human-induced disturbances[28]. Each of these factors can weaken mangroves, leading to widespread dieback if stress is severe or prolonged. Understanding these physical factors is critical for developing effective management and conservation strategies to protect and restore mangrove ecosystems.

4 Conclusion

These findings indicate that mangrove dieback in East Lampung is driven by a complex interaction of environmental factors, with significant contributions from climate variability linked to the Indian Ocean Dipole (IOD) and El Niño-Southern Oscillation (ENSO), along with sea-level rise, salinity fluctuations, and temperature variability. The IOD and ENSO indices were particularly influential as they modulate critical environmental conditions that can induce osmotic stress, alter freshwater availability, and affect sea surface height, all of which are crucial to mangrove ecosystem stability. Principal Component Analysis (PCA) further elucidated that temperature, salinity, and pH are the primary determinants of mangrove density, with temperature emerging as the most significant variable. A comprehensive understanding of these physical drivers is imperative for formulating effective management and conservation strategies to safeguard and restore mangrove ecosystems.

References

1. H. Akram, S. Hussain, P. Mazumdar, K. O. Chua, T. E. Butt, and J. A. Harikrishna, Forests **14**, (2023)
2. Y. Apriliyani, R. Safei, H. Kaskoyo, C. Wulandari, I. Gumay, F. Program, S. Magister, I. Kehutanan, and F. Pertanian, Jurnal Hutan Tropis **8**, (2020)
3. S. Abhik, P. Hope, H. H. Hendon, L. B. Hutley, S. Johnson, W. Drosdowsky, J. R. Brown, and N. C. Duke, Sci Rep **11**, (2021)
4. W. Han, L. Zhang, G. A. Meehl, S. Kido, T. Tozuka, Y. Li, M. J. McPhaden, A. Hu, A. Cazenave, N. Rosenbloom, G. Strand, B. J. West, and W. Xing, Nat Commun **13**, (2022)
5. M. A. Marfai, Quaestiones Geographicae **33**, (2014)
6. C. E. Lovelock, I. C. Feller, R. Reef, S. Hickey, and M. C. Ball, Sci Rep **7**, (2017)
7. Y. Harada, R. M. Connolly, B. Fry, D. T. Maher, J. Z. Sippo, L. C. Jeffrey, A. J. Bourke, and S. Y. Lee, Biogeosciences **17**, (2020)
8. I. Valiela, L. Camilli, T. Stone, A. Giblin, J. Crusius, S. Fox, C. Barth-Jensen, R. O. Monteiro, J. Tucker, P. Martinetto, and C. Harris, Glob Planet Change **92–93**, (2012)
9. at Alhadid and A. Budi Nugroho, Sci Educ (Dordr) **3**, 707 (2024)
10. R. Safe'i, I. R. Maulana, F. Ardiansyah, I. S. Banuwa, S. P. Harianto, S. B. Yuwono, and Y. Apriliyani, International Journal of Sustainable Development and Planning **17**, (2022)
11. Duke and Schmitt, Tropical Forestry Handbook (2015)
12. B. Budiadi, A. Pertiwiningrum, L. D. Lestari, A. N. Jihad, B. A. Marpaung, and S. Prasetyo, Frontiers in Forests and Global Change **6**, (2023)
13. P. F. Scholander, H. T. Hammel, E. Hemmingsen, and W. Garey, Plant Physiol **37**, 722 (1962)
14. M. C. Ball, Trees **2**, 129 (1988)
15. D. M. Alongi, Estuar Coast Shelf Sci **76**, 1 (2008)
16. N. C. Duke, M. C. Ball, and J. C. Ellison, Global Ecology and Biogeography Letters **7**, 27 (1998)
17. M. Bayad, H. W. Chau, S. Trolove, K. Müller, L. Condron, J. Moir, and L. Yi, ISPRS Journal of Photogrammetry and Remote Sensing **169**, 292 (2020)
18. Z. Chen, M. Zhang, H. Zhang, and Y. Liu, IEEE Transactions on Geoscience and Remote Sensing **61**, (2023)
19. A. B. Baloloy, A. C. Blanco, S. Sharma, and K. Nadaoka, Frontiers in Remote Sensing **2**, (2021)
20. D. Radočaj, J. Obhođaš, M. Jurišić, and M. Gašparović, Land (Basel) **9**, (2020)
21. C. Giri, E. Ochieng, L. L. Tieszen, Z. Zhu, A. Singh, T. Loveland, J. Masek, and N. Duke, Global Ecology and Biogeography **20**, 154 (2011)
22. D. M. Alongi, Curr Clim Change Rep **1**, 30 (2015)

23. D. A. Sánchez-Núñez, J. A. Rodríguez-Rodríguez, and J. E. Mancera Pineda, Journal of Applied Ecology **60**, (2023)
24. S. M. Hickey, B. Radford, J. N. Callow, S. R. Phinn, C. M. Duarte, and C. E. Lovelock, Sci Rep **11**, (2021)
25. N. C. Duke, J. R. Mackenzie, A. D. Canning, L. B. Hutley, A. J. Bourke, J. M. Kovacs, R. Cormier, G. Staben, L. Lymburner, and E. Ai, PLOS Climate **1**, (2022)
26. E. F. Asbridge, R. Bartolo, C. M. Finlayson, R. M. Lucas, K. Rogers, and C. D. Woodroffe, Estuar Coast Shelf Sci **228**, (2019)
27. J. Touza, C. Lacambra, A. Kiss, R. M. Amboage, P. Sierra, M. Solan, J. A. Godbold, T. Spencer, and P. C. L. White, Environ Manage **68**, (2021)
28. R. Djamaluddin, D. Lamb, N. Duke, E. Moll, and J. Davie, Botany in School of Intergrative BiologyNatural and Rural Systems Management Department (2004)

Assesing ecological condition and diversity of mangrove and seagrass in Kepala Madan, South Buru, Maluku

Julia Intan Stevhani Br Siahaan[1*], *Fery* Kurniawan[1], *Reza Miftahul* Ulum[1], *Rizkia Damayanti* Putri[1], and *Aron Azis* Meliala[1]

[1]Department of Aquatic Resources Management, Faculty of Fisheries and Marine Sciences, IPB University, 16680, Indonesia

Abstract. Mangrove and seagrass ecosystems are critical components of the marine environment, providing essential functions for coastal ecosystems and contributing to climate-change mitigation. This study analyzed the biodiversity and ecological conditions of mangrove and seagrass ecosystems in the Kepala Madan District, South Buru, Maluku Province. Mangroves were assessed using line transects, while seagrasses were evaluated using the seagrass-watch methodology, with data collected directly from the field. Nine mangrove species were identified: *Rhizophora stylosa* (the dominant species), *Rhizophora mucronata*, *Sonneratia alba*, *Sonneratia ovata*, *Bruguiera gymnorrhiza*, *Bruguiera parviflora*, *Avicennia alba*, *Avicennia lanata*, and *Avicennia officinalis*. Seven species were found in the seagrass ecosystems: *Cymodocea rotundata* (the dominant species), *Enhalus acoroides*, *Halodule ovalis*, *Halodule uninervis*, *Halophila spinulosa*, *Halophila decipiens*, and *Syringodium isoetifolium*. Vegetation density varied across locations owing to multiple influencing factors. Despite this variation, water quality assessments have revealed favorable conditions for both mangrove and seagrass ecosystems, supporting their ecological health and sustainability.

1 Introduction

Buru Island is the second largest island in Maluku Province and is divided into two regencies: Buru Regency and South Buru Regency. This island has high and unique biodiversity, particularly in the South Buru Regency [1]. The coastal ecosystems of Buru Island consist of coral reefs, seagrass, and mangroves along its coastline. Coral reefs serve as habitats for various fish species, including small fish, which utilize reefs as shelters [2]. Seagrass acts as a habitat that connects marine and terrestrial ecosystems, providing refugia for small fish and other organisms. Mangroves function as spawning areas and nurseries for fish larvae [33]. These three ecosystems are interconnected, serve as natural protectors for the coast, and play

* Corresponding author: juliasiahaan@apps.ipb.ac.id

a crucial role in carbon storage. Through photosynthesis, mangroves and seagrass absorb carbon dioxide, aiding climate change mitigation and maintaining ecosystem balance [34].

Mangroves support coastal ecosystems because of their ecological, social, and economic roles [3]. Bengen [4] explained that mangrove forests have ecological functions, such as dampening waves and winds, protecting coastlines from abrasion, retaining mud, trapping sediments carried by water flow, serving as nursery areas and feeding grounds for various aquatic organisms, and providing spawning locations for diverse species. The biodiversity of mangroves on Buru Island has been reported to reach 25 species [5]. Mangrove forests are often utilized by coastal communities; for example, the communities in South Buru Regency frequently use mangrove forests as a source of firewood for expanding settlements, aquaculture areas, and fishing as well as for harvesting shellfish. Therefore, information regarding the ecology of mangroves is essential for understanding their management and utilization.

Seagrass beds are expanses of seagrass that form ecosystems along coastlines and play a crucial role in both ecological and economic sectors. Seagrass ecosystems are among the most important coastal ecosystems because they serve as feeding grounds and shelters for aquatic biota [6]. A healthy seagrass ecosystem can significantly affect carbon absorption, climate mitigation, nutrient cycling, and the ecosystem services it provides. A total of 60 seagrass species have been identified worldwide, of which 15 can be found in Indonesia, including *Enhalus acoroides, Cymodocea rotundata, Cymodocea serrulata, Halophila decipiens, Halophila ovalis, Halophila minor, Halophila spinulosa, Halodule pinifolia, Halodule uninervis, Syringodium isoetifolium, Thalassia hemprichii, Thalassodendron ciliatum, Halophila sulawesii, Halophila becarii*, and *Ruppia maritima* [7]. The benefits of seagrass ecosystems can be categorized into direct and indirect benefits. Direct benefits include the use of seagrass as a medicine and raw material for handicrafts [8]. The indirect benefits of seagrass include sediment trapping, stabilization of the waterbed, current mitigation, nutrient cycling control, oxygen production, and serving as nursery grounds, spawning grounds, and feeding grounds [9].

Because ecosystems are located in shallow waters, both mangroves and seagrasses are highly vulnerable to various disturbances. These disturbances can arise from natural events or human activity. Effective management of ecosystem areas is essential to ensure the sustainability of these two ecosystems and to maintain their benefits and continuity. Furthermore, there is currently a significant lack of updated information regarding the conditions of the ecosystems in southern Buru. Therefore, a study on the ecological condition profile of seagrass is necessary as a source of updated information and as a foundational reference for the management and implementation of conservation areas in Southern Buru waters.

2 Materials and methods

2.1 Study area and data collection

This research was conducted from September to October 2024 in the Kepala Madan District, South Buru Regency, Maluku Province. The study was carried out at five sites to determine the mangrove profile and at six sites to determine the seagrass profile (**Table 1**). The primary data observed in the field included the presence of mangroves at five sites located in North Pasir Putih, Tomahu, South Pasir Putih, Nanali, and Biloro. Primary seagrass data were obtained from six sites located in North Pasir Putih, Tomahu, South Pasir Putih, Nanali,

Biloro, and Fogi. The sites in this study were selected using purposive random sampling, where sampling points were chosen based on the representativeness of each location.

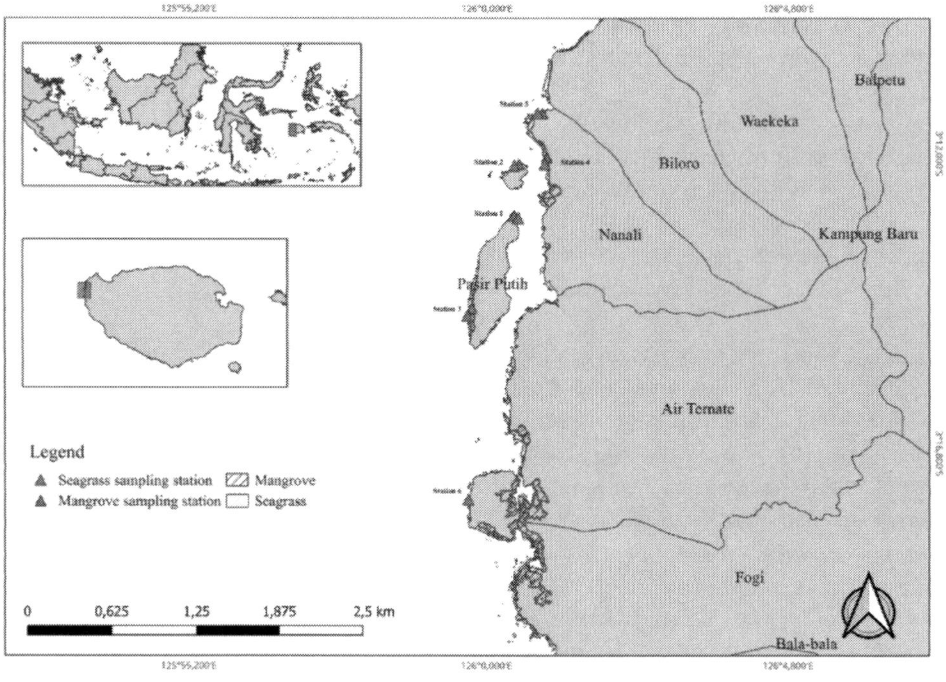

Fig. 1. Location of the research area

Table 1. Research sites

Site	Location	Mangrove	Seagrass
1	North Pasir Putih	✓	✓
2	Tomahu	✓	✓
3	South Pasir Putih	✓	✓
4	Nanali	✓	✓
5	Biloro	✓	✓
6	Fogi		✓

The method for data collection on the structure of the mangrove community utilized the line transect method, with the placement of plots measuring 10×10 m^2 extending inland from the shoreline, with a distance of 10 m between observation plots (**Fig.4**). The determination of the observation transects was based on the representativeness of mangrove zoning. The collected data included primary data such as the number of individuals at each tree level (trees, saplings, and seedlings), species identification, diameter measurement at each level, species frequency, and substrate characteristics. The diameter of the stands was classified into trees, saplings, and seedlings based on Dharmawan and Pramuji [37]. Additionally, water quality measurements, including pH, salinity, dissolved oxygen (DO), temperature, turbidity, and total dissolved solids, were taken at each observation plot.

The method for data collection on the structure of the mangrove community utilized the line transect method, with the placement of plots measuring 10×10 m^2 extending inland from the shoreline, with a distance of 10 m between observation plots. The determination of the observation transects was based on the representativeness of mangrove zoning. The collected data included primary data, such as the number of individuals at each tree level (trees, saplings, and seedlings), species identification, diameter at each level, species frequency, and substrate characteristics. Additionally, water quality measurements, including pH, salinity, dissolved oxygen (DO), temperature, turbidity, and total dissolved solids (TDS), were taken at each observation plot.

Seagrass data collection was conducted using the seagrass-watch method by establishing several line transects extending toward the seafloor (**Fig.4**). The number and length of line transects at each site were adjusted according to the area of seagrass that had been observed. The distance between the plots and line transects was tailored based on the size of the seagrass area. Each line transect consisted of 11 plots measuring 50 × 50 cm. The data collected from each plot included substrate type, seagrass species, number of stands per species, percentage cover, and presence of algae and epiphytes. Water quality data, including temperature, pH, salinity, dissolved oxygen (DO), turbidity, and total dissolved solids (TDS), were collected in each line transects.

2.2 Data analysis

The vegetation and ecological data of mangroves and seagrasses collected from the field were analyzed for relative frequency, relative density, relative dominance, and coverage. Subsequently, advanced data analysis was performed using the following formulae:

2.2.1 Species Diversity Index (H')

The species diversity index is a calculation of the composition of each individual species in a community. The species diversity index was calculated using the Shannon-Wiener approach [10] as follows:

$$H' = -\sum_{i=1}^{n} P_i \ln P_i$$

where H' is the species diversity index, Pi=Ni/N, Ni : the number of individuals of species *ii* and N = total number of individuals of all species. The Shannon-Wiener diversity index was classified into three categories: high ($H'>4$), moderate ($2<H'<4$), and low ($H'<2$).

2.2.2 Species Evenness Index (J)

The evenness index was used to measure the distribution of a species within a community. Evenness can be expressed by calculating the Evenness Index (J), which relates evenness to diversity. The evenness index is calculated based on the following formula [11]:

$$J' = \frac{H'}{\ln S}$$

where J is the evenness index, H is the species diversity index, and S is the total number of species. The evenness index was classified into three categories: high (0.6-1), moderate (0.4 $J' < 0.6$), and low (0-0.4) [12].

2.2.3 Species Dominance Index (D')

The species dominance index describes the extent to which a species dominates an ecosystem. This index is used to aid in the understanding of community structure and species interactions. The species dominance index was calculated using the following formula [10]:

$$D' = \left(\sum Pi\right)^2$$

where D' is the dominance index, Pi=Ni/*N* where Ni is the number of individuals of species *ii* and N is the total number of individuals of all species. The dominance index was classified into two categories: no species dominated (D'<0.5) and some species-dominated (D'>0.5).

3 Result and discussion

3.1 Ecological conditions of mangrove and seagrass in Kepala Madan Regency

Approximately 13 species of mangroves were found in Kepala Madan, South Buru, and Maluku, out of 240 species of mangroves spread across Indonesia [35]. Nine mangrove species were observed in the plots. These species included *Avicennia alba, Avicennia lanata, Avicennia officinalis, Bruguiera gymnorrhiza, Bruguiera parviflora, Rhizophora mucronata, Rhizophora stylosa, Sonneratia alba,* and *Sonneratia ovata* (Table 2). However, based on field surveys, other mangrove species are located outside the observation sites, specifically in Fogi. In the mangrove area of Fogi, there are various mangrove species that are not found at other sites, namely *Ceriops tagal, Xylocarpus granatum, Aegiceras corniculatum,* and *Nypa fruticans*. The distribution of mangrove species is shown in Table 2.

Table 2. Distribution of mangrove

Species	Site 1	Site 2	Site 3	Site 4	Site 5
Avicennia alba	✓				
Avicennia lanata	✓	✓		✓	✓
Avicennia officinalis	✓			✓	✓
Bruguiera gymnorrhiza			✓	✓	✓
Bruguiera Parviflora			✓	✓	✓
Rhizophora mucronata	✓	✓	✓	✓	
Rhizophora stylosa	✓	✓	✓	✓	✓
Sonneratia alba	✓	✓	✓	✓	✓
Sonneratia ovata					
Total	6	4	5	7	6

The most abundant species belong to the family Rhizophoraceae, namely *Rhizophora stylosa* and *Rhizophora mucronata*, which were found at all the observation sites. The species that were not very abundant were *Avicennia alba* and *Sonneratia ovata*. *Rhizophora* sp. can grow on hard substrates because of their ability to survive in such environments, although this species also thrives on muddy substrates. According to Noor *et al.* [13], *Rhizophora stylosa* can live in various habitats, from intertidal zones to rivers, such as muddy, sandy, and rocky substrates. Samson *et al.* [22] indicated that in another area of South Buru, specifically Kase Village in the Leksula District, the dominant mangrove species is the same as that of the family Rhizophoraceae. This is because of its rapid regeneration capability, with propagules maturing while still attached to the parent plant, and optimal growth in organic-rich muddy substrates. Its adaptation to tidal currents and salinity conditions makes it superior to that of other species. Meanwhile, in areas closer to the mainland, where the substrate tends to be sandy, *Avicennia sp.* was found. *Avicennia* sp. mangroves found at the observation sites that have ecological characteristics that allow them to grow in coastal areas, dry regions, and sandy substrates. This species also exhibits high tolerance to salinity.

Table 3. Distribution of seagrass

Spesies	Site 1	Site 2	Site 3	Site 4	Site 5	Site 6
Enhalus acaroides	-	+	+	+	+	+
Cymocodea rotundata	+	+	+	+	+	+
Halodule uninervis	+	-	+	-	+	+
Halophila ovalis	-	+	+	+	+	+
Syringodium isoetifolium	-	+	+	+	+	+
Halophila decipiens	-	-	+	+	+	+
Total	2	4	6	5	6	6

Six seagrass species were identified, namely *Enhalus acoroides, Halophila ovalis, Halophila decipiens, Halodule uninervis, Cymodocea rotundata*, and *Syringodium isoetifolium*. Approximately 10 seagrass species can be found in Maluku waters [36]. The distribution of the seagrass ecosystem at each site was of a multispecies type, where more than one seagrass species was present at each site (**Table 3**). This is consistent with Kawaroe *et al.* [14], who noted that one characteristic of the seagrass ecosystem in Indo-Pacific waters, particularly in Indonesia, is its multispecies nature. *Syringodium isoetifolium, Halophila ovalis,* and *Enhalus acoroides* are distributed across five sites, with *Enhalus acoroides* being one of the key species [15], and is widely distributed in tropical regions and the Pacific coast, including Indonesia [31]. Seagrasses from the genera Halophila and Halodule were present in all research locations as pioneer groups. Pioneer seagrasses have a high capacity to adapt to various aquatic environmental condition [16]. Species composition across all sites is influenced by the ability of seagrass species to adapt to environmental factors [17]. These environmental factors include temperature, substrate type, salinity, depth, and light [18].

3.2 Diversity of mangrove and seagrass in Kepala Madan Regency

3.2.1 Mangrove

The results of the calculation of each index for the mangrove ecosystem in the Kepala Madan Regency are shown in Fig.2. The diversity of mangrove species at each site ranged from 0.8141 ± 1.4562, with the highest diversity index found at Site I and the lowest at Site II (**Fig. 2**). A high diversity index indicated that the site had a large number of species and individuals. This suggests that the environmental conditions are mature and favorable [19]. However, based on the diversity index values, all the sites fell into the low category. The low diversity index may be because the environmental conditions at each observation site are dominated by muddy substrates, are almost always inundated by seawater, and have high canopy cover, which limits light penetration. As a result, only certain mangrove species thrive. Mangroves from the Rhizophoraceae family were present at every site and contributed significantly to the diversity index. According to Khairunnisa *et al.* [20], this is because Rhizophoraceae mangroves are well suited to muddy sandy substrates, allowing them to thrive at each site and sustain their growth. Substrate thickness is also a supporting factor for mangrove growth; in the presence of sediment and extensive land, mangroves have the potential to grow and expand in the area. Other mangrove species with high adaptability, besides *Rhizophora* sp., include *Bruguiera* sp. and *Avicennia* sp.

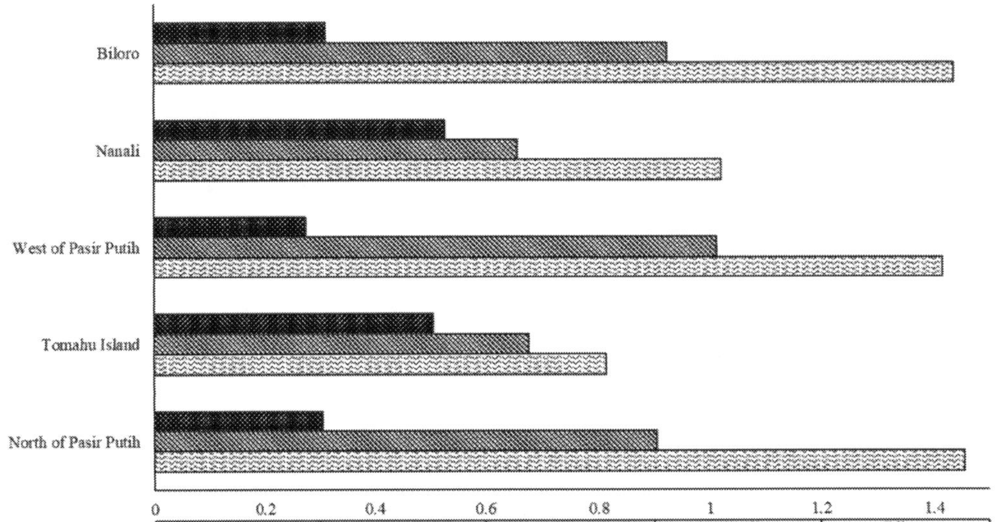

	North of Pasir Putih	Tomahu Island	West of Pasir Putih	Nanali	Biloro
D'	0.3050	0.5061	0.2742	0.5262	0.3108
J'	0.9048	0.6761	1.0127	0.6560	0.9226
H'	1.4562	0.8141	1.4157	1.0209	1.4358

Fig.2. Diversity, evenness, and dominance index in mangrove site

The evenness of the mangrove species at each site was measured based on the distribution of these species within the ecosystem. The evenness index of mangrove species at the observation sites ranged from 0.6560 ± 0.9226, with the highest value found at Site III and the lowest at Site IV. High evenness indicates that mangrove species in each ecosystem were distributed evenly, without any dominant species. In contrast, low evenness suggests the dominance of certain species or uneven distribution of species within the ecosystem [24]. All sites had a high evenness index category, except for site 4, which had a moderate category. The high evenness index at nearly all sites indicates that the mangrove community in the Kepala Madan Regency has a uniform distribution and a high level of adaptation to environmental conditions. Latumahina *et al.* [24] reported that important environmental conditions include humidity, air temperature, soil pH, salinity, and light intensity.

The dominance index indicates the level of dominance of a species within a community. Based on the calculation results, the dominant sites were sites II and IV, where the dominant species was *Rhizophora stylosa.* If dominance is centered on one species, it will increase the dominance index value; however, when dominance is spread to several species, the dominance index will be low. Dominance by one species indicated that the diversity of species at that location was low. The species with the highest dominance value at a particular site is *Rhizophora stylosa,* which can be explained by favorable environmental conditions such as inundation during normal tide conditions and deep and fine mud substrates [21]. This also shows that *Rhizophora stylosa* has good adaptability to the environment.

Table 4. The mangrove diversity

	Mangrove	
	Value	Class
H'	1.2285	Low
J'	0.8344	High
D'	0.3845	No domination

On average, mangrove ecosystems in the Kepala Madan Regency are unstable. This conclusion was drawn from the analysis of the diversity, evenness, and dominance indices **(Table 4)**. Low diversity values indicated that the variety of species in this ecosystem was limited. This could be because the ecosystem is more vulnerable to disturbances from natural factors and human activity [32]. Natural factors such as tidal phenomena, substrate type, and canopy cover contribute to this condition. Meanwhile, mangrove observation sites in Nanali, Biloro, and Pasir Putih are frequently accessed by communities to utilize natural resources. While the diversity index indicated instability and low species variation, the evenness index showed that species distribution was fairly even, while the dominance index indicated that no species dominated. In other words, although no single species dominates, mangrove ecosystems in the Kepala Madan sub-district remain low in diversity and are vulnerable to disturbance.

3.2.2 Seagrass

The results of the calculation of each index for the seagrass ecosystem are shown in Fig.3. The diversity of seagrass species at each site ranged from 0.2880 ± 0.7038, with the highest diversity index found at site V and the lowest at site I. All seagrass sites exhibited low diversity, with a diversity index value below one, indicating low diversity levels. The low diversity value at site I will affect the dominance and evenness indices, which is also observed in all research sites and aligns with the findings of Larasati *et al.* [23] that the high or low dominance index of seagrass in North Maluku will influence its diversity index. Variability in the seagrass diversity index can occur because of several influencing factors, including substrate homogeneity, the dominance of certain seagrass species, the number of species present, and the overall conditions of the seagrass ecosystem.

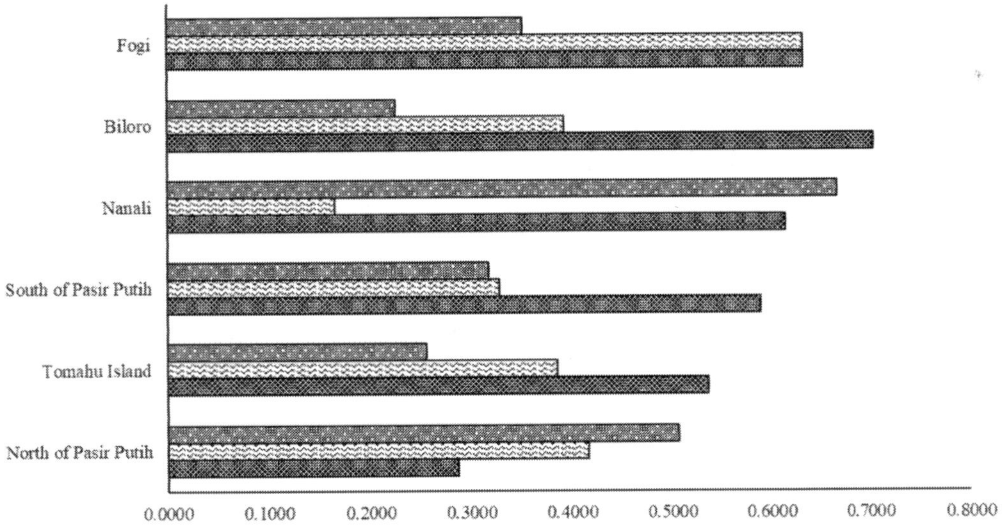

	North of Pasir Putih	Tomahu Island	South of Pasir Putih	Nanali	Biloro	Fogi
D'	0.5059	0.2563	0.3186	0.6656	0.2262	0.3524
J'	0.4155	0.3863	0.3290	0.1657	0.3928	0.6315
H'	0.2880	0.5356	0.5895	0.6139	0.7038	0.6315

Fig.3. Diversity, evenness, and dominance index in mangrove site

The six seagrass sites showed different evenness index values. The evenness index of the seagrass species in the Kepala Madan Regency ranged from 0.1657 ± 0.4155. The highest evenness index value is found at site 1, with a value of 0.4155, placing it in the moderate evenness category. In contrast, sites 3 to 6 fell into the low category. The lowest evenness index value is recorded at site 4, with a value of 0.1657. According to Sholiqin *et al.* [26], a low evenness index indicates that an ecosystem is under stress. A low evenness index is suspected to be caused by poor nutrient availability [27]. Among the six sites, Site 1 had the lowest diversity index, resulting in the highest evenness value. This is explained in a study by Sutia *et al.* [28], which states that high evenness is due to low diversity index values.

The seagrass dominance index at all sites ranged from 0.2262 ± 0.6656. The highest dominance index value was found at site 4, with a dominance index value of 0.6656, while the lowest dominance index value was found at site 5, with a dominance index value of 0.2262. From this analysis, it is known that the *Halophila ovalis* species is dominant at site 4 because it has a high dominance index value of 0.6341. At Site 1, a total dominance index value of 0.5059 was found, and there was no one species that dominated, but this point was filled by two seagrass species, namely *Halopila ovalis* and *Cymodocea rotundata.* For other species at each site, there was no high dominance index value or D' value of more than 0.5, so it can be concluded that there were no other seagrass species that dominated. According to Umawaitina *et al.* [29], the dominance index value indicates whether or not there is a dominant seagrass species at a location with the dominance value indicating that if more than 0.5, then there is a dominating species, and if less than 0.5, then it can be concluded that there are no species dominating the location. This value indicates that the *Halophila ovalis* species in the waters have a significant presence and ecological role, so that they have a high dominance [30].

Table 5. The seagrass diversity

	Seagrass	
	Value	**Class**
H'	0.5604	Low
J'	0.3868	Moderate
D'	0.3875	No domination

Based on the analysis of the seagrass ecosystem conditions (**Table 5**), it can be concluded that this area shows a relatively low level of diversity. A diversity index (H') value of 0.5604 indicates that species variation at this site is limited, potentially reducing the overall stability of the ecosystem. In addition, a dominance index (J') value of 0.3868 indicated moderate dominance by some species, although no species was highly dominant. This could reduce the functional diversity and resilience of ecosystems to environmental change. The species dominance index (D') value of 0.3875 confirmed that there was no clear dominance among species. Overall, the final condition of the seagrass ecosystem at this site indicates high vulnerability with the risk of further habitat degradation.

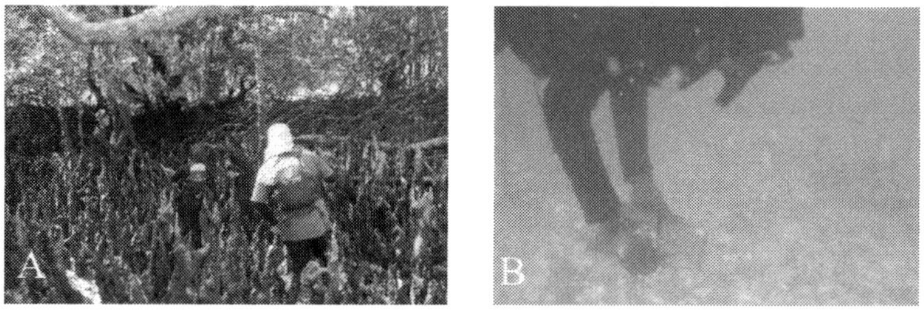

Fig.4. Mangrove and seagrass ecosystem condition in Kepala Madan sub-district, South Buru: **A.** mangrove in Nanali Village **B.** seagrass sampling **C.** water quality assesment in Biloro Village **D.** sea urchin from seagrass ecosystem **E.** Fish catching using net **F.** sea cucumber from Fogi Village's seagrass ecosystem **G.** breath root from *Sonneratia* sp. community **H.** seagrass sampling in Fogi **I.** measuring the GBH from *Rhizophora* tree **J.** seagrass ecosystem

4 Conclusion

The mangrove and seagrass ecosystems in South Buru host a variety of species, and our study identified 13 mangrove and seven seagrass species. However, biodiversity indices indicate low overall biodiversity. This low diversity suggests that these ecosystems are experiencing instability, likely due to environmental pressures such as pollution, habitat degradation, and climate change. The compromised health of these ecosystems not only affects species richness but also leads to imbalances in species evenness and dominance. Consequently, this situation highlights the urgent need for targeted conservation strategies to restore and protect the ecological integrity of mangroves and seagrasses in South Buru, ensuring their resilience and vital ecosystem services.

The writing and implementation of this research were supported by the CFI Indonesia GEF 6 Grant through the World Wildlife Fund (WWF), the Global Environmental Facility (GEF) Agency, and the Ministry of Marine Affairs and Fisheries for the Expedition of PIT. We extend our sincere gratitude to all the supporting organizations and individuals for their invaluable contributions to this project.

References

1. Soulisa, S., Tawainella, M. N., Subair, S. 2023. *HORIZON: Indonesian Journal of Multidisciplinary*, **1**(1), 01-18.
2. Sambrook, K., Hoey, A. S., Andréfouët, S., Cumming, G. S., Duce, S., & Bonin, M. C. 2019. *Fish and Fisheries*, **20**(5), 903-920.
3. Carlson, R,R., Evans, L,J., Foo, S, A., Grady, B, W., Li, J., Seeley, M., Xu, Y., Asner, G, P. 2021. *Global Ecology and Conservation,* **28**,1-21.
4. Bengen, D., Yonvitner, Y., Rahman, R 2023. Bogor: IPB Press.
5. Gazali, M. 2019. *Jurnal Laot Ilmu Kelautan,* **1**(1),1-12.
6. Akbar, N., Buamona, A., Tahir, I., Baksir, A., Effendi, R., Ismail, F. 2020. *Jurnal Sumberdaya Akuatik Indopasifik*, **4**(1),33-42.
7. Jalaludin, M., Octaviyani, I, N., Putri, A, N, P., Octaviyani, W., Aldiansyah, I. 2020. *Jurnal Geografi Gea,* **20**(1),44-53.
8. Manurung, N, D., Kondoy, K., Rondonuwu, A, B., Wantasen, A., Mantiri, R,O., Manengkey, H. 2022. *Jurnal Ilmiah PLATAX,* **10**(1),98-107.
9. Simpson, J., Bruce, E., Davies, K, P., Barber, P. 2022. *Remote Sensing*, **14**(3572),1-18.
10. Odum, E, P. 1993. Penerjemah: Tjahyono Samingan.
11. Pielou, C. 1975. Wiley-Interscience : New York.
12. Krebb, C, J. 1989. Harpercollins College : New York.
13. Noor, Y, R., Khazali, M., Suryadiputra, I, N, N. 2006. Bogor: Ditjen PKA & Wetlands International - Indonesia Programme, Bogor.
14. Kawaroe, M., Nugraha, A.H., Juraij, J., & Tasabaramo, I.A., 2016. *Biodiversitas,* **17**(2),585-591.
15. Nugraha, A, H., Idris, F., Apriadi, T., Dhevanda, C., Azis, M, Y., Hafsar, K. 2025. *Marine Pollution Bulletin,* **2**
16. Kilminster K, McMahon K., Waycott, M., Kendrick, G.A., Scanes, P., McKenzie, L., O'Brien, K.R., Lyons, M., Ferguson, A., Maxwell P, *et al.* 2015. Science of the Total Environment, **5**,97–109.
17. Vogel, M, A., Mason, O, U., Miller, T, E. 2021. *Climate Change Ecology*, **2**,1-9.
18. Kusumaningtyas, A, R., Suryono, Ambariyanto. 2023. *Journal of Marine Research,* **12**(2),230-239.
19. Supriadi, Romadhon, A., & Farid, A. 2015. *Jurnal Kelautan*, **8**(1), 44-51.
20. Khairunnisa, C., Thamrin, E., Prayogo, H. 2020. *Jurnal Hutan Lestar,* **8**(2),325-336
21. Amalo, F, L., Putra, M, D., Handayani, L, D, W., Sumpeno, Luturmas, R., Suryadi. 2024. *Media Konservasi,* **30**(1),84-95.
22. Samson, E., Sigmarlatu, V., Wakano, D. 2020. B*ioWallacea: Jurnal Penelitian Biologi (Journal of Biological Research),***7** (1),1055-1063,
23. Larasati, R,F., Jaya, M,M., Putra, A., Djari, A,A., Sako, K., Khairunnisa, A., Jatayu, D., Aini, S., Suriadin, H. 2022. *Journal of Indonesian Tropical Fisheries,* **5**(2),162-178.
24. Pototan, B., Capin, N., Delima, A. G., & Novero, A. 2021. *Biodiversitas Journal of Biological Diversity*, **22**(1),144-153.

25. Latumahina, F., Susilawati, Antonius, Tewoloipaky, W. 2024. *Agricultural Journal,* **7**(2),295-306.
26. Sholiqin, M., Pramadaningtyas, P, S., Solikah, I., Febriyanti, S., Pambudi, M, D., Mahartika, S, B., Umam, A, F., Liza, N., Setyawan. *INTL J Bonorowo Wetlands,* **11**(2),84-94.
27. Zachawerus, F. H., Kambey, I. A. D., & OSE, I. R. 2015. *Ilmiah Platax*, **3**(1), 16-22.
28. Sutia, A. A. P., Hamdani, H., Herawati, H., & Ismail, M. R. 2021. *Asian Journal of Fisheries and Aquatic Research*, **14**(5), 27-39.
29. Umawaitina, A., Wagey, B,T., Sinjal, C,A., Rumengan, I,F., Tilaar, S,O., Kumampung, D,R. 2023. *Jurnal Pesisir dan Laut Tropis*, **11**(2),160-169.
30. Fahruddin, M., Abdurachman, M,H., Suriyadin, A., Murtawan, H., Setyono, B,D,H., Saputra, A., Ilyas, A,P. 2022. *Jurnal Ilmu Kelautan dan Perikanan Papua*, **5**(2), 50-54.
31. Retawimbi, A,Y. 2018. *Kepariwisataan: Jurnal Ilmiah*, **12**(3),59-70.
32. Setyadi, G., Pribadi, R., Wijayanti, D,P., Sugianto, D,N. 2021. *Biodiversitas,* **22**(8),3562-3570.
33. Wahyuni, S., Sulardiono, B., Hendrarto, B. 2015. *Journal of Maquares*, **4**(4),66-70
34. Thorhaug, A., Gallagher, J, B., Kiswara, W., Prathep, A., Huang, X., Yap, T., Dorward, S., Berlyn, G. 2020. *Marine Pollution Bulletin.* **160**.
35. Rahman, Lokollo, F.F., Manuputty, G.D. *et al.* 2024. *Biodivers Conserv,* **33**, 875–903.
36. Rumahlatu, D., Tuapattinaya, P, M, J. 2019. *International Journal of Applied Biology,* **3**(1), 70-80.
37. Dharmawan IWE, Suyarso, Ulumuddin YI, Prayudha B, Pramudji. 2020. Bogor: PT. Media Sains Nasional.

Distribution and characteristics of internal waves observed during The Expedition of Jalacitra 3-2023 "FLORES"

I W Sumardana E Putra[1,5*], *Agus S* Atmadipoera[2], *Henry M* Manik[2], *Gentio* Harsono[3,6], *Adi* Purwandana[4] and *Budi* Purwanto[6]

[1]Department of Marine Science, Post-Graduate School, Faculty of Fisheries and Marine Science, IPB University, 16680, Bogor, Indonesia
[2]Marine Science and Technology Department, Faculty of Fisheries and Marine Sciences, IPB University, 16680, Bogor, Indonesia
[3]Faculty of Defence Technology, Republic Indonesia Defence University, Bogor, 16810. Indonesia
[4]Research Center for Oceanography, National Research and Innovation Agency, 10340, Central Jakarta, DKI Jakarta, Indonesia
[5]Department of Hydro-Oceanography, Indonesian Naval Technology College (STTAL), Indonesian Navy, 14240, Jakarta, Indonesia
[6]Naval Hydro-Oceanography Center (PUSHIDROSAL), Indonesian Navy, 14430, Jakarta, Indonesia

Abstract. Navigation safety and the advent of Internal Solitary Waves (ISWs) are inextricably linked. In the Flores Sea, the processes of generation, propagation, and shoaling of ISWs have both positive and negative effects. The objective of the Expedition Jala Citra (EJC) 3-2023 "Flores" is to identify the ISWs through hydro-oceanography survey acoustic apparatus on the Navy Ship, KRI Spica 934. The objective of this study was to qualitatively map the distribution and characteristics of ISWs around the Flores Sea, the Alor Strait, the Ombai Strait, and the Leti Strait. The results of the observation data processing indicate that solitary waves dominate internal waves, with amplitudes ranging from 20 m to 100 m. The distribution pattern in the study area was consistent with the appearance obtained from the satellite image recording. The results of the processing of the water column data on this expedition demonstrated the role of the background current in altering the high (low) amplitude of the ISWs. The occurrence of ISWs in the Leti Strait and the southward propagation of ISWs during the neap tide in the Ombai Strait provide new information for future ISW research. It is anticipated that this information will be used to update nautical chart information, thereby supporting underwater navigation safety.

1 Introduction

Indonesia has seven locations that display the appearance of Internal Solitary Waves (ISWs) using satellite imagery datasets (see Fig. 1); Synthetic Aperture Radar (SAR) and optical

*Corresponding authors: sumardanai@apps.ipb.ac.id

sensors play vital roles in the research and comprehension of ISWs dynamics in the ocean [1–3]. The presence of ISWs in Indonesian Seas (ISs) is attributed to three main factors: the existence of barotropic tidal wave current energy; the presence of topographical complexity such as sills, seamounts, narrow straits; and variations in the density of stratified water masses within the water column [4–8]. ISWs are distinguished by a longer period of minutes, higher amplitudes of tens to hundreds of meters, and a longer wavelength of kilometers, in contrast to surface waves, which are characterized by a short period of seconds, an amplitude of centimeters to a few meters (m), and a short wavelength of a few meters [9–12]. The sinking of the Indonesian Navy Submarine KRI Nanggala 402 in northern Bali on April 21, 2021, is attributed to the appearance of ISWs [6,13,14]. In addition to using a satellite imagery dataset to detect, measure, and characterize ISWs, ISW studies of marine biodiversity and hydrography data quality (bathymetry) are important. There is an imperative need for a comprehension of the characteristics of ISWs in order to support maritime safety navigation and naval operations [14,15], this aligns with the "Hydrographic Information - Enhancing Safety, Efficiency, and Sustainability in Marine Activities" theme of World Hydrography Day 2024 [16]. According to [17–20], understanding the behavior of ISWs in ISs requires the application and utilization of hydroacoustic technology, such as water column imagery (Water Column Imaging-WCI) using hydrographic acoustic equipment (Singlebeam Echosounders-SBES or Multibeams Echosounders-MBES) or seismic oceanography using multichannel seismic datasets [21]. This technique has been applied in different fields, including research, hydrography, mapping of marine ecosystems, military operations, fishing, oceanography, monitoring of climate change, and geology of the seabed [21,22].

The Hydrographic and Oceanographic Center of the Indonesian Navy (PUSHIDROSAL) conducted the 3-2023 Jala Citra Expedition (EJC) in Flores Waters on April 25, 2023, and has been applying and utilizing WCI MBES and SBES to identify not only for estimation of plankton [23] but also the appearance of ISWs on the water column during the expedition. This study aimed to qualitatively map the distribution and features of ISWs utilizing the EA 600 SBES and EM 302 MBES on the Indonesian Naval Vessel KRI Spica 934, which is operated by PUSHIDROSAL. Measurements were performed along the path of the Indonesian Archipelago Sea Lanes (IASLs) in the waters of the Flores Sea.

2 Materials and Methods

2.1 Water Column Imaging (WCI) datasets

The Jala Citra Expedition 3-2023 "Flores" gathered data in the ISs, especially in Flores Sea, from March 2023 to May 2023. Data collection was conducted using the Indonesian Naval Vessel KRI Spica 934, which operates under PUSHIDROSAL. The expedition coincided with the first Monsoon Break (MB-1), which changed from the wet to the dry season in the IS. The measurements were carried out around the IASLs path of the Flores Sea and Indonesian Throughflow (ITF) line, as proposed by [24,25]. The Kongsberg Maritime Type EA 600 SBES operating at a frequency of 50 kHz was used continually during the voyage while the vessel was in motion. SBES data files with the file extension *".raw"* were analyzed using Kongsberg Maritime EA640 Echosounder v.16.3.2. The detection was performed using a 30 kHz multibeam echosounder-MBES (with the file extension *".all"* and *".wcd"*) were analyzed using Echo Sounder Package3- ESP3 [26]. Fig. 2 displays the representation of the intricate underwater landscape stretching from the Flores Sea, which consists of steep slopes, cliffs, sills, and underwater seamounts. These data are the primary outcome of the MBES survey, covering a depth range of 200–3500 m.

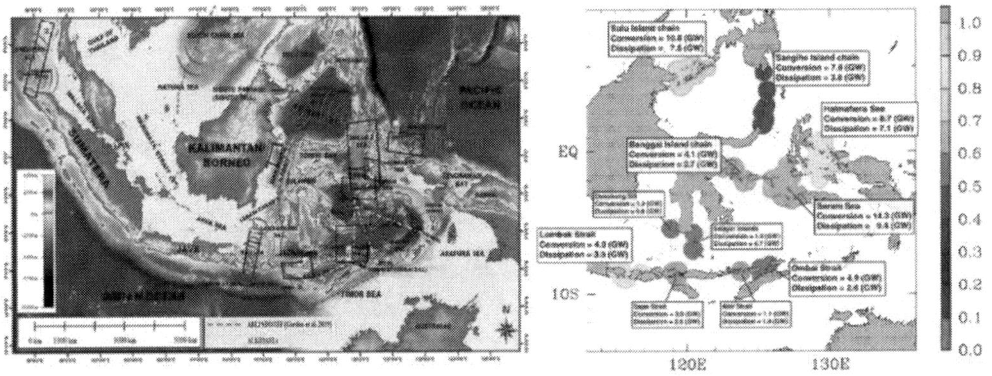

Fig. 1. Left panel is Schematic distribution of generating and propagating internal tidal waves region, such as in, a) North Aceh Waters (Andaman Sea) [2,3]; b) Bali Waters (Lombok Strait) [1,6,14,27]; c) Sulawesi Waters – Makasar Strait (Mangkaliat Cape) [28,29]; d) Sawu Sea - Sape Strait [9,11]; e) Banda Water – Ombai Strait [30]; f) Maluku Waters (Seram Sea – Buru Sea) [18,28]; and g) Halmahera Sea – North Papua Water (Waigeo Sea) [28]. The right panel displays the values of the conversion rate from barotropic to baroclinic energy and the dissipation of baroclinic energy in the Indonesian areas, as modelled by [5]. The Indonesian territorial waters depicted in the left panel (point a-g) correspond to the barotropic to baroclinic energy distribution illustrated in the right panel. The generation and propagation sites of internal waves in Indonesian regions align with areas exhibiting substantial energy conversion levels. This indicates that internal waves significantly contribute to the energy transformation process in this area, where energy from barotropic tides is changed into baroclinic energy, which subsequently undergoes dissipation during its propagation.

2.2 Satellite Image Dataset

We used the website https://zoom.earth/ to acquire data on ISWs's generation and propagation of ISWs derived from satellite imagery gathered in 2023. Subsequently, we employed the Global Mapper V.23 software to convert these photos into a digital format. Fig. 2 illustrates the formation of ISWs (red circular pattern) originating from the Ombai Strait and expanding into the Flores Sea in a northwest direction. We included data from [31], which indicated the southward spread of ISWs between Kambing and Wetar Islands. This study utilized satellite image data from the Sentinel-3 Ocean and Land Color Instrument (OLCI). From the above data, it was inferred that the occurrence of ISWs moving southward is likely due to the existence of a seamount/sill located on the northeast side of the Ombai Strait. This seamount/sill was identified during EJC 3-2023 in the Flores Sea (see **Fig. 2 point A-E**).

2.3 Data analysis

In this study, the focus is only at a depth of 0-300 m depth, where the appearance of internal waves can be detected in the Indonesian region. The initial intensity value of the decibels was set between -55 and -70 dB [7]. However, to provide more clear visual description of the ISWs in the water column, the decibel intensity value was adjusted to a range of -80 to -90 dB (see the red box in **Fig. 3**). The data analysis approach employed in this study to validate the properties and distribution of ISWs was conducted qualitatively. In general, ISWs appeared during the spring tide period, with the typical highest amplitudes observed during the peak period of the spring tide [1,7,9,14]. We replaced the missing water mass dataset from the CTD, which was used to characterize the type and stratification of water masses during the EJC, with model data from the Hybrid Coordinate Ocean Model (HYCOM), which we obtained from the website https://hycom.org/dataserver/gofs-3 pt1/analysis. The model data includes temperature, salinity, and current speed components "u" and "v" with a spatial resolution of 0.08^0. Ocean Data View (ODV) V.5.6 software, created by [32], was employed for data processing (**Fig. 4**). We used tidal prediction data from Oregon State University's Tidal Inversion Software (OTIS) developed by [33]. This software utilizes tidal harmonic constants, such as M2, S2, N2, K1, O1, P1, Q1, M4, Ms4, Mn4, Mm, and Mf. We used these data to determine the time period of the event (spring/neap tide conditions) when ISWs appeared in the Flores Sea (April-May 2023). This enables us to predict and establish a connection between the occurrence of ISWs and the circumstances of tidal events, which, as prior research indicates, often occur around the spring tide [34].

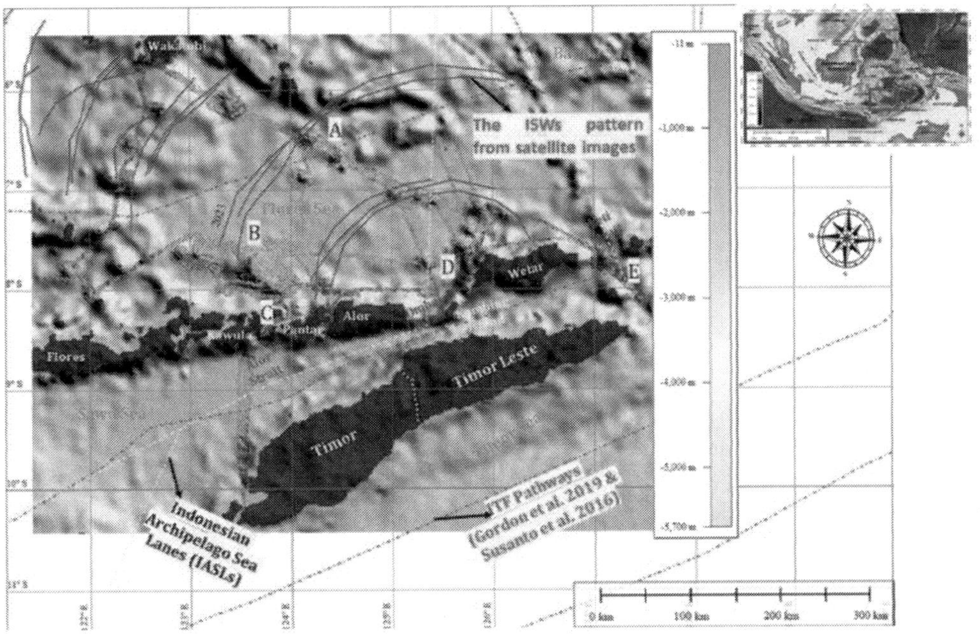

Fig. 2. Research area as well as The Jala Citra Expedition 3-2023 "Flores" vessel tracks. The red circular pattern (the leading waves of ISWs) is formed by digitizing the appearance of the ISWs from satellite images on April, 1st until May, 31st 2023, while the bathymetry data is acquired by MBES.

3 Results and discussion

The presence of steep bottom topography gradients in stratified seas leads to interactions between barotropic tidal currents and results in the generation of localized mixing in the water. This mixing eventually developed into ISW packets. The height of ISWs is defined by a tiny horizontal scale but exhibits significant changes in the water column, mostly due to isopycnal displacement. These oscillations are particularly pronounced near generating areas [35]. Model of [5] demonstrated that the Ombai Strait exhibits a conversion rate of ~4.9 Gigawatts (GW) from barotropic to baroclinic energy, as well as a dissipation rate of ~2.6 GW for baroclinic energy. The numbers for the conversion rate and energy dissipation in the Strait of Lombok are around 4.0 GW and 3.3 GW, respectively, which are comparable to the values mentioned (see **Fig. 1**, right panel). Hence, studies conducted by [30] in the Ombai Strait and [4] in the Lombok Strait have demonstrated that the two straits have the capability to generate and produce soliton packets, also known as ISWs. Nevertheless, when the energy dissipation value is very high or comparable to the energy conversion rate, it might hinder the creation of soliton packages in the water region. This study categorizes the ISWs in the Flores Sea during the EJC 3-2023's activity based on the direction of their propagation and emergence using a combination dataset, that is, SAR imagery and hydrographic acoustic equipment, in five locations (see **Fig. 2**). The amplitude values of the ISW characteristics depicted in **Fig. 5A (#1)** were approximately 56 and 31 m at a depth of 200 m, as determined from the SBES data. [36] derived depth using CTD casts from the INDOMIX voyage, [37] used CMEMS data, and [18] used seismic oceanography (SO), stating that a depth of 200 m is the thermocline layer. The water masses emanating from the Pacific Ocean in the thermocline layer (known as ITF) are characterized by high salt levels ranging from ~34.7 to 35.5 practical salinity units (psu) and isopycnal densities of ~24.5-25.2 kilograms per cubic meter (kg/m^3). According to [38,39], The interplay of ISW, thermocline layers, and ITF water masses in Indonesian waters reveals a complex and interrelated interaction that profoundly impacts the oceanographic attributes of the region. ISWs primarily develop and propagate within the thermocline layer, where pronounced density gradients serve as natural waveguides, facilitating extensive transit of these waves while preserving their energy. The intensity of stratification in the thermocline layer directly affects the amplitude and phase speed of ISWs, underscoring the significance of the vertical water column structure in wave dynamics. ITF water masses traversing Indonesian waters significantly influence the thermocline structure along their transport routes, resulting in seasonal and interannual variability that alters ISW characteristics through modifications in water mass stratification. This interaction is particularly critical because ISWs facilitate the vertical mixing of ITF water masses, thereby modifying the water characteristics throughout the transit pathways. ISWs facilitate intensified vertical mixing processes that influence the distributions of temperature, salinity, and nutrients, significantly altering the characteristics of ITF water masses as they traverse Indonesian waters. Vertical mixing caused by ISWs might alter the thermocline structure, thereby affecting the ITF transport patterns and intensities. The triadic interaction of ISWs, thermocline layers, and ITF water masses is a crucial element of oceanographic dynamics in Indonesian seas, with one component affecting and reacting to alterations in the other, thus shaping the overall properties of the water column. Understanding these linkages is crucial to understand the wider implications of ocean mixing processes and their effects on regional oceanographic conditions. A change in wave properties characterizes the connection between the background current (ITF) and the amplitude of the ISWs is characterized by a change in wave properties [40,41]. According to [24,25], this area is the ITF pathway (magenta line in **Fig. 5 A#2**). The ITF acts plays an important role as a background current, which can increase or decrease the amplitude value after the generation and propagation processes. The interaction between ISWs and the ITF as a background current can be understood through the fundamental wave-current interaction theory, which demonstrates significant implications in Indonesian waters. When internal waves encounter an ITF, their characteristics undergo modifications based on the principle of wave action conservation (E/ω = constant), where E represents the wave energy and ω denotes the intrinsic frequency. In regions of strong ITF flow, particularly through Indonesian straits where currents can reach 0.5-1.0 m/s, waves propagating against the current experience wavelength reduction and amplitude enhancement, while those traveling with the current exhibit increased wavelength and decreased amplitude. This interaction becomes particularly pronounced during the southeast monsoon (JJA) when ITF transport reaches its maximum, potentially enhancing internal wave amplitudes by 15-30% depending on the current strength and direction. The observed IW characteristics in Indonesian seas typically show amplitudes ranging from 20-100m, periods of 1-3 hours, and wavelengths of 2-10 kilometers (km), all of which can be significantly modulated by seasonal ITF variability.

https://doi.org/10.1051/bioconf/202516801007

During the northwest monsoon (DJF), when ITF transport weakens, these wave-current interaction effects become less prominent.

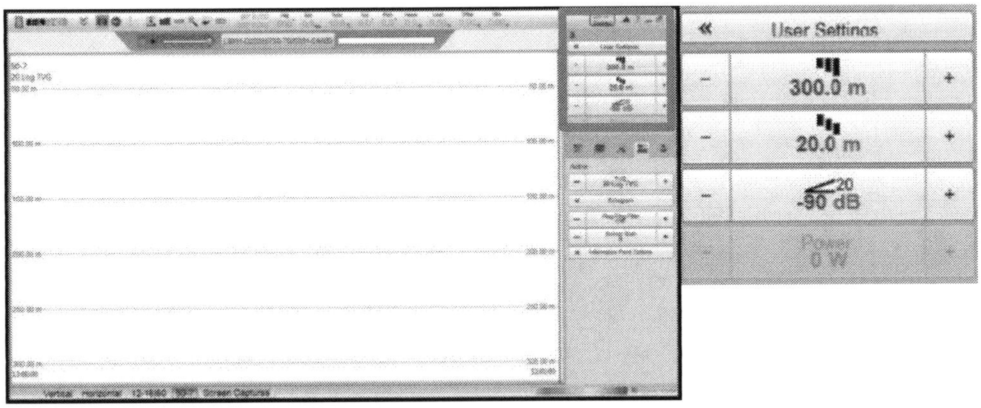

Fig. 3. Configuring settings in the Kongsberg Maritime EA 640 Echosounder v.16.3.2 software to detect ISWs.

The coordinates of the ISW appearance on April 21, 2023, are $6,42^0$ S and $124,12^0$ E. The ITF pathways (magenta line in **Fig. 5A (#2)**) proposed by [24,25] and the satellite imaging (ISWs pattern) findings (red line in **Fig. 5A (#2)**) align with this position. **Fig. 5A (#2)** confirms the presence of seabed topography in the form of seamounts or sills through the results of image data digitization and MBES data acquisition. In particular, the amplitude and length of the ISWs increased as they became more spread out and flat (as indicated by the red arrow in **Fig. 5A (#1)**) because of the weakening of the background current. Furthermore, the velocity at which the wave travels may decrease, thereby affecting the form and magnitude of the wave in the event that the underlying current decreases. This is due to the fact that ISWs rely on the speed of the background current for their propagation. Reducing the background current enhances energy dissipation [29], leading to a more rapid loss of energy in the wave. This might result in a faster decrease in amplitude and perhaps lead to wave breaking. However, external factors, such as the shape of the ocean floor, wind, and surface currents can also influence ISWs. The attenuation of the underlying current might enhance the impact of these external influences on ISWs, thereby altering their amplitude and stability. EJC data collection took place during the MB-1 period, which spanned from March to April. During this period, the ITF flow tends to diminish, which is in contrast to the eastern monsoon season when the ITF flow is rather significant [8,42,43]. The results of EJC 3 at this site show that the shape of the seabed and background currents can change the characteristics (amplitude and wavelength) of ISWs in the Flores Sea.

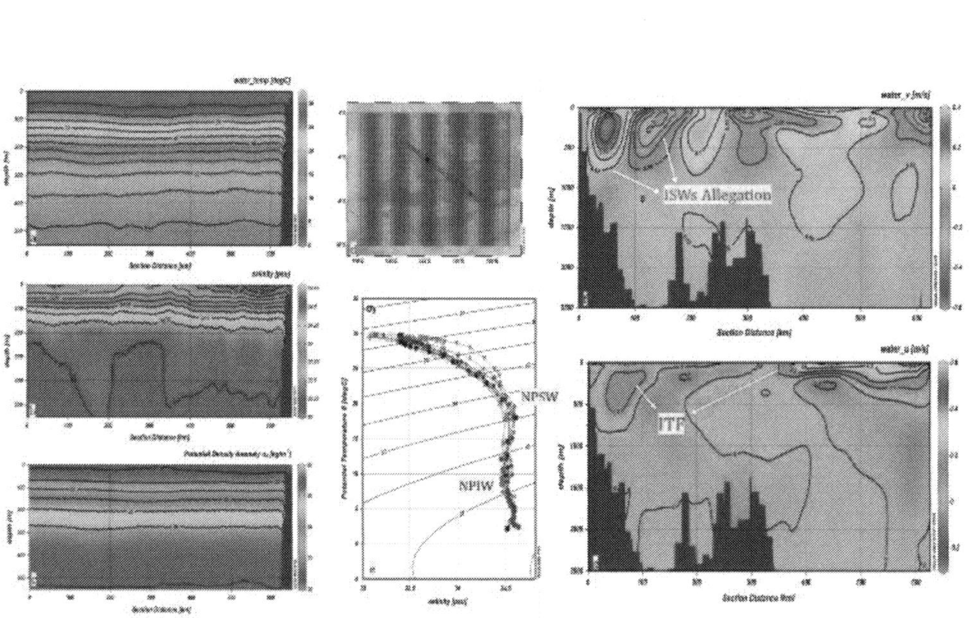

Fig. 4. The characteristics of water masses in the Flores Sea, as determined by HYCOM data from April 2023, include: a) temperature; b) salinity; c) density; d) transects drawn based on ISWs propagation locations as shown in Fig. **2**; e) T-S diagram illustrating the types of water masses in the Flores Sea, specifically the North Pacific Subtropical Water (NPSW) and North Pacific Intermediate Water (NPIW) [36]; f) the meridional current component (v) responsible for the ISWs allegation's propagation process in the Flores Sea, which primarily moves towards the north-northwest; and g) the zonal current component (u) indicating the eastward movement of ITF (Indonesian Through Flow) water masses into Flores waters through the bifurcation process of the Makassar Strait.

Point B (**Fig. 2**) also has the same characteristics. The amplitude and wavelength of the ISWs tended to be flat in the thermocline layer (approximately 100 m), with an amplitude of approximately 31 m and a wavelength of approximately 26 m at depths of approximately 100 m and 425 m, respectively (**Fig. 6B (#1)**). This influence is thought to be due to the role of the ITF, which originated from the weakening of the Makasar ITF bifurcation during the MB-1 period. Large energy dissipation values or equivalent energy conversion rates can also prevent the formation of soliton packages in a water area. This is supported by research results of [5], which show that in the Alor Strait, the energy dissipation value is approximately 1.8 Gigawatts (GW) and the energy conversion is approximately 1.1 GW (right panel in **Fig. 1**). On April 6, 2023, ISW was observed at point B, which was located at $7{,}99^{0}$ S and $123{,}11^{0}$ E. It is on the northwest side of the Alor Strait, approximately 50 km away. We suspect that refraction processes with underwater seamounts [9] formed ISWs at a depth of approximately 425 m (**Fig. 6B (#2)**).

At Point C, ISW was discovered in the Alor Strait between coordinates 8.29^{0} S and 123.88^{0} E on May 5th, 2023 (see **Fig. 7C (#1)**), at a depth of approximately 180 m and with an amplitude of around 100 m. Additionally, we saw wave depression with a magnitude of ~40 m at a depth of approximately 350 m. Furthermore, potential boluses of ~20 m and sand waves reaching a maximum height of over 20 m at a depth of approximately 500 m were identified. **Fig. 7C** displays the findings of the MBES base surface and the position of the observed sand wave occurrence, indicated as **(#2)**, **(#3)**, and **(#4)**.

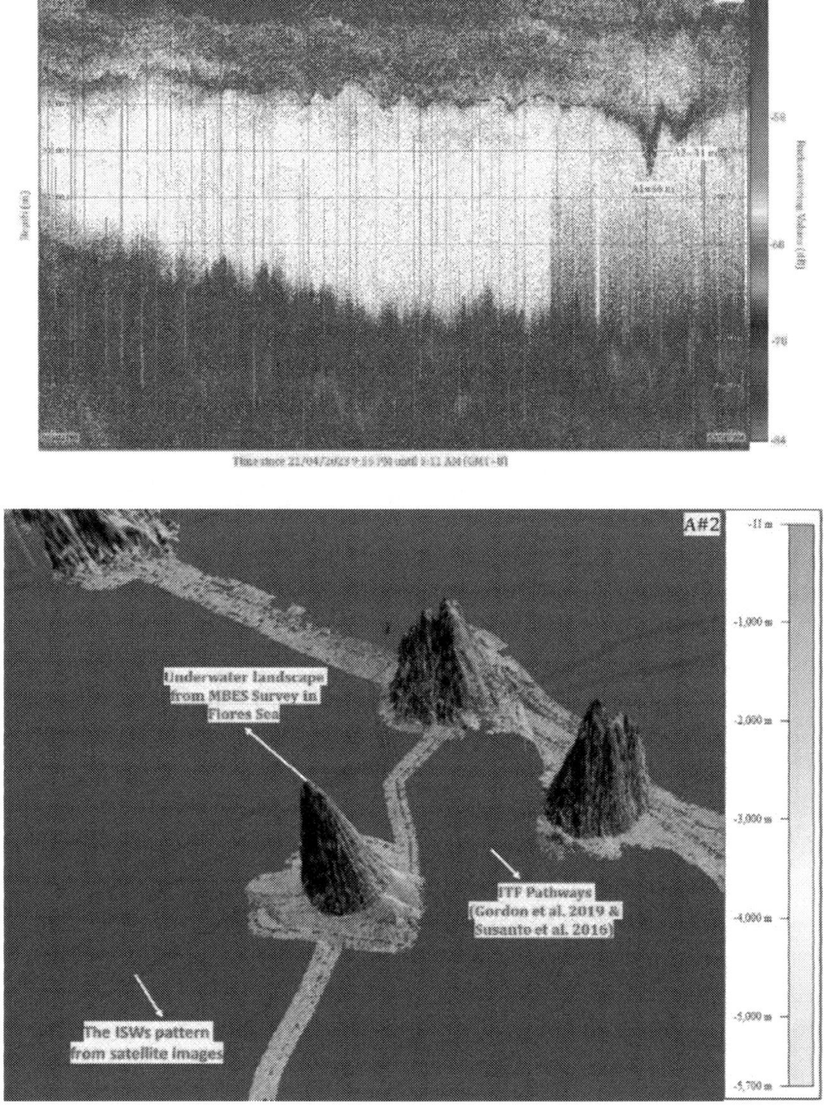

Fig. 5. The ISW (red dashed line in **A#1**) shows up in SBES data from EJC 3-2023 "Flores" on April 21, 2023, at 6.42^0 S and 124.12^0 E, which is where the alleged producing area starts from the Ombai Strait. The topography obtained from MBES data shows an underwater landscape in the form of sills or seamount with a depth of ~1600 m–2500 m. ISW packet amplitudes are ~ 56 m and ~31 m in depth ~200 m, respectively. The red line in **A#2** is the leading waves of ISWs (captured by satellite imagery from https://zoom.earth/) on 2023, as shown in **Fig. 2**.

https://doi.org/10.1051/bioconf/202516801007

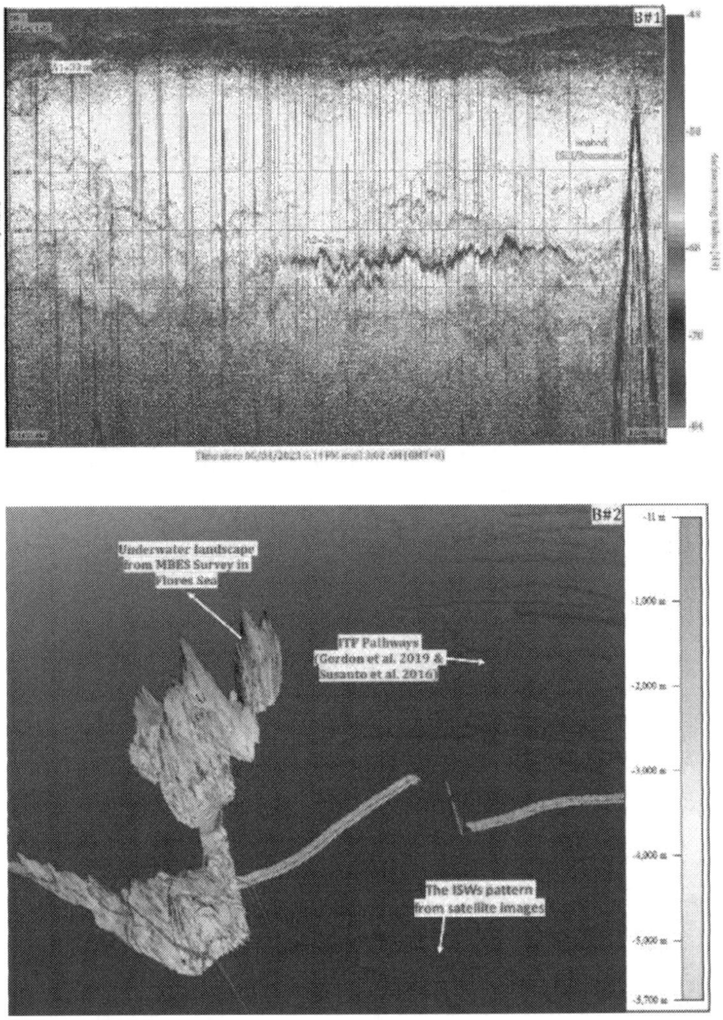

Fig. 6. On April 6, 2023, the ISWs appearance at Point B (see **Fig. 2**) will be at positions 7.99^0 S and 123.11^0 E. The ISW packet amplitudes are ~ 31 m and ~ 26 m in depth ~ 100 m and ~ 425 m, respectively (see panel **B#1**). Satellite imagery appearances in the right panel (**B#2**) indicate the propagation of ISWs in this area, thought to undergo a refraction process due to interaction with seabed topography.

According to [44,45] asserted that rugged topography, such as sills, straits, and internal wave action, might induce a process of vertical mixing. The water mass movement across the Alor Strait is around 1.07 ± 0.03 Sverdrup (Sv), with 1 Sv equivalent to 10^6 cubic meters per second (m^3/s). [46] found that seabed types in the Alor Strait mostly consist of sediments such as muddy sand, mud, and gravelly sand. In addition, studies conducted by [47–49] indicated that the presence of ISWs in a water body can lead to the development of sand waves on the seabed because of turbulence at the bottom, which causes forces acting on the seabed. Shear stress affects the movement of sediments at the bottom. The EJC findings provide evidence of the presence of ISWs in the Alor Strait and their ability to cause sediment movement at the bottom. We believe that a sill on the northern side of the Alor Strait (**Fig. 7 C#4** in the right panel) is responsible for the formation of the ISW. In addition, the appearance of sand wave patterns at the bottom of the water suggests that ISW contribute to seabed changes

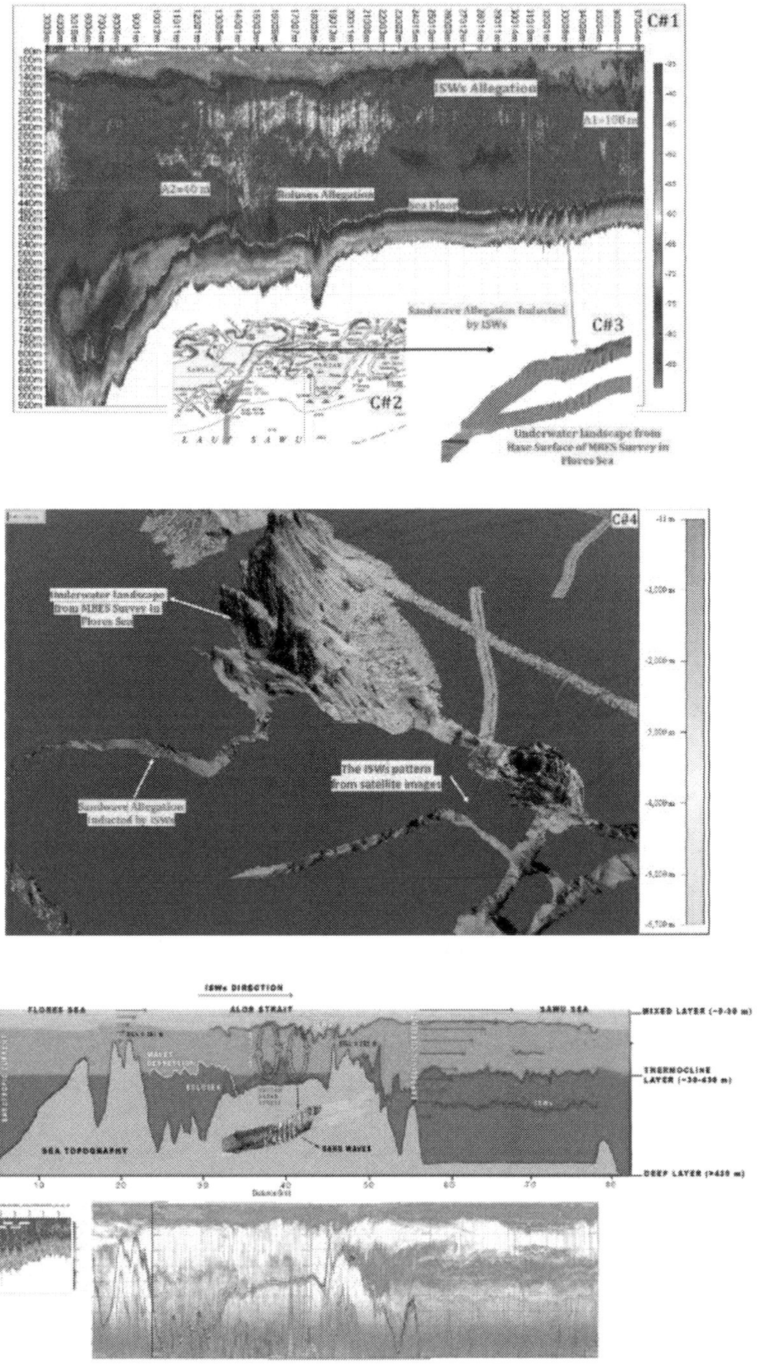

Fig. 7. The left panel (**C#1**) is a WCI-MBES catch in the Alor Strait, showing the appearance of ISW with amplitudes of ~100 m and ~40 m, respectively, at depths of ~180 m and ~350 m on May 5th, 2023,

at positions 8,31⁰ S and 123,86⁰ E. The Alor Strait and the results of the MBES sweep, processed using ESP3 software, are located in **C#2**. The results from the MBES base surface data in **C#3** indicate the presence of sand waves. In lower panel, a). The schematic of the mechanism of ISWs induced sand waves in the Alor Strait showcases the interaction between oceanographic processes and seabed features. The figure captures the dynamic environment influenced by barotropic and baroclinic currents, which drive the formation and propagation of ISWs (red lines). These waves significantly impact the sediment characteristics and distribution, leading to the development of sand waves. The thermocline layer plays a critical role in the propagation of ISWs, which in turn induce orbital wave patterns observable in the water column (black dash line orbital pattern). These patterns result in bottom shear stress (red arrow), reshaping the seabed into distinct sand waves (in b) and boluses (violet line). The MBES data provides a detailed view of the sand waves' formation, while the SBES data (c) reveals the presence and movement of ISWs within the water column. By understanding these processes, researchers can better predict sediment transport and seabed morphology in the Alor Strait, contributing to a comprehensive understanding of the region's oceanography and its implications for marine navigation and environmental management; b) Sand waves formation from base surface of MBES data; and c). SBES data showing IWSs in water coloum encountering continental shelf.

At Point D, Neither SBES nor MBES data provided us with data on the appearance of ISWs in the water column. However, using satellite imagery from https://zoom.earth/, we captured the leading waves of the ISWs in 2023, indicating their northward propagation. Meanwhile, [31], utilizing satellite image data from Sentinel-3 OLCI, confirmed the presence of ISWs propagating southward between Kambing Island and Wetar Island (**Fig. 8D (#1)**). We believe that a refraction process [9] is responsible for this, as it collides with the underwater mountain feature on the northern side of the Ombai Strait. This feature boasts seamount heights of 2100 m and 1000 m (**Fig. 8D (#2)**), derived from MBES base surface data collected during EJC activities in the Flores Sea. Internal waves experience refraction when encountering heterogeneous bathymetry, adhering to Snell's law, whereby the wave's phase velocity changes with changes in water depth, resulting in the bending of wave crests towards shallower areas of the seamounts (2100m and 1000m features). When internal waves encounter seamounts in the Ombai Strait, the segment of the wave in deeper water retains a greater phase speed, while the segment over the seamount decelerates, leading to wavefront curvature and possible wave concentration or amplification. Refracted internal waves generally display altered propagation direction, modified wavelength, and changed energy density, potentially becoming more energetic and nonlinear as they converge on seamount features. These results validate the findings of [30], who asserted the presence of contour formations measuring 3000 m and 1000 m on the north side and substantiated the propagation of ISWs towards the south, as reported by [50,51].

At Point E, on April 20, 2023, uncommon ISWs were observed in the Leti Strait between coordinates 7.67⁰ S and 127.04⁰ E. These waves occurred at a depth of 150 m and had an amplitude of 32 m (**Fig. 9 E(#1)**). Interestingly, this event coincided with the Total Solar Eclipse in Indonesia, as reported by [52,53]. The numerik model article of [54] found that the model findings show a circular pattern moving towards the Banda Sea, starting from the Ombai Strait and Leti Strait. This site was situated along the IASLs-3 route and on the eastern side of the ITF exit route. Regarding the eastern section of the ITF route proposed by [24,25,55], it is plausible to hypothesize that the movement of ITF water masses when they encounter a sill approximately 350 m high (**Fig. 9 E(#2)**) is facilitated by the influence and intensification of tides during solar eclipse conditions.

https://doi.org/10.1051/bioconf/202516801007

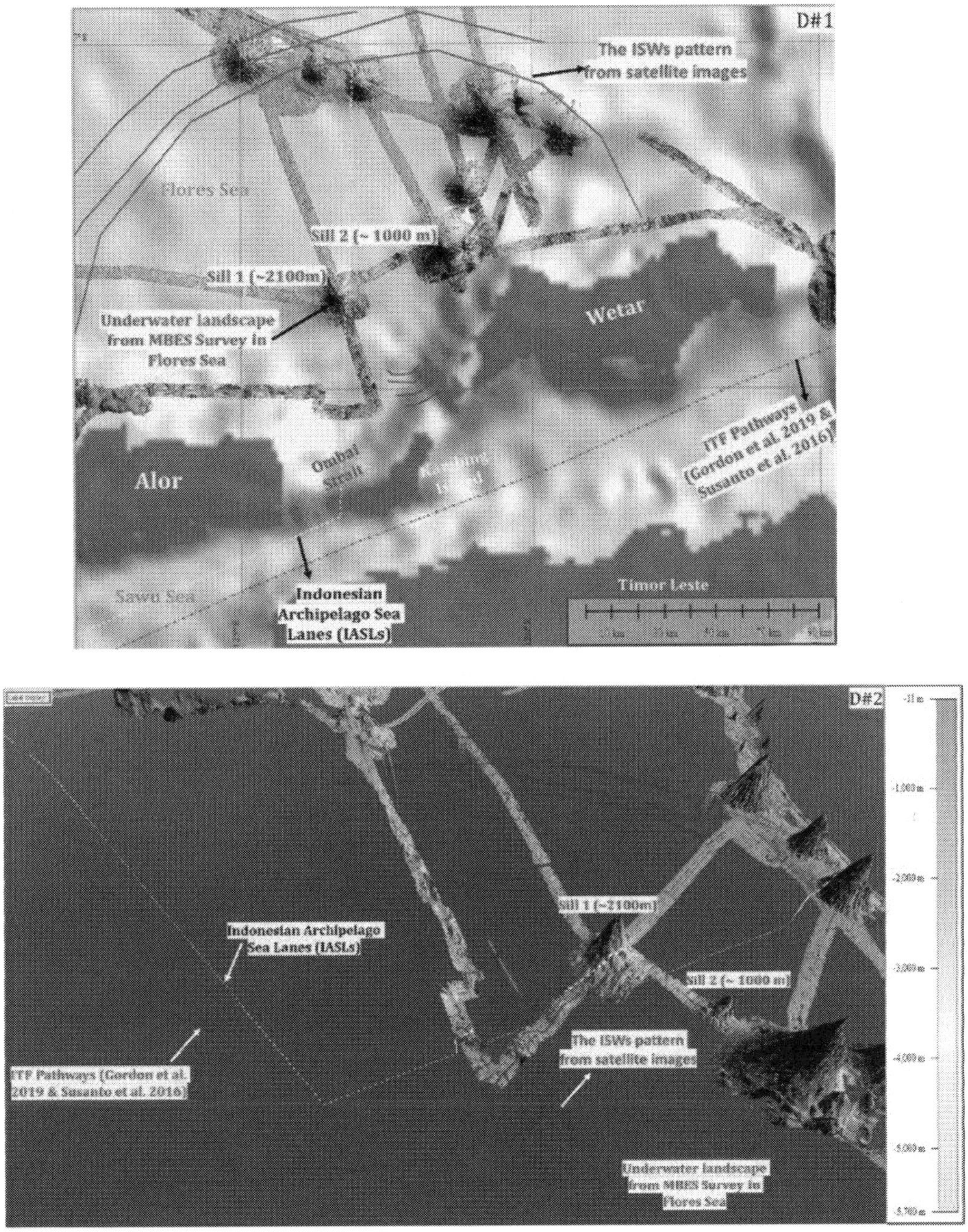

Fig. 8. According to captured satellite imagery, Panel **D#1** shows the propagation of ISWs heading north and south from the Ombai Strait. Panel **D#2**, a 3D view of the MBES base surface data results during EJC 3-2023, suggests that the presence of Sill 1 and Sill 2, each with a height of approximately 2100 m and 1000 m, respectively, is responsible for the propagation of ISWs heading south (between Wetar Island and Kambing Island).

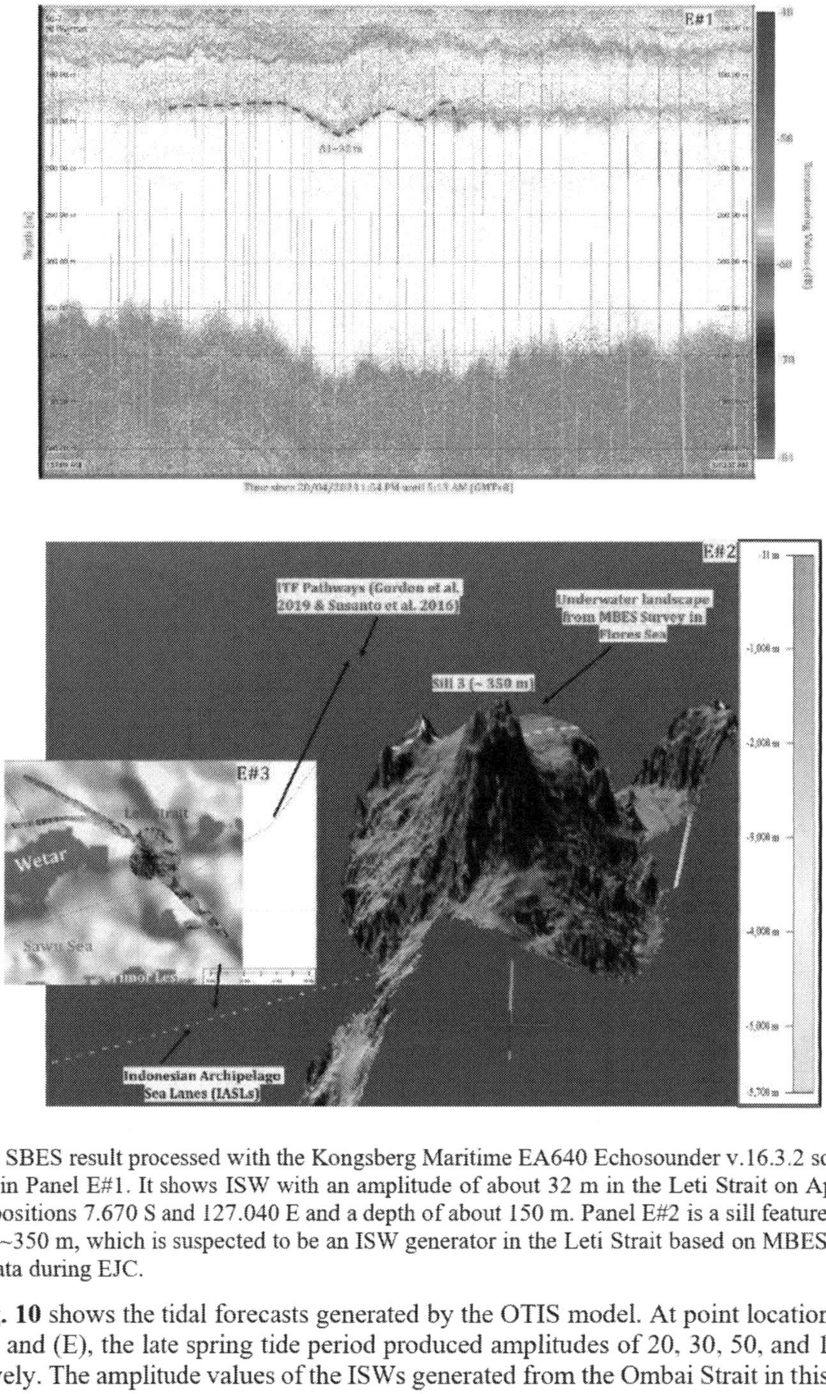

Fig 9. An SBES result processed with the Kongsberg Maritime EA640 Echosounder v.16.3.2 software is shown in Panel E#1. It shows ISW with an amplitude of about 32 m in the Leti Strait on April 20, 2023, at positions 7.670 S and 127.040 E and a depth of about 150 m. Panel E#2 is a sill feature with a height of ~350 m, which is suspected to be an ISW generator in the Leti Strait based on MBES's base surface data during EJC.

Fig. 10 shows the tidal forecasts generated by the OTIS model. At point locations (A), (B), (C), and (E), the late spring tide period produced amplitudes of 20, 30, 50, and 100 m, respectively. The amplitude values of the ISWs generated from the Ombai Strait in this study were relatively similar to the amplitude values of the ISWs generated from the Lombok Strait. This similarity is due to the greater energy conversion value compared to energy dissipation, resulting in the formation of soliton packages that spread over hundreds of kilometers from the generating area. This finding aligns with previous research conducted by [7,34]. Please refer to **Fig. 1** for further details. In contrast, neither the SBES nor MBES water column measurements captured the occurrence of ISWs at point position (D). An intriguing area of research is the occurrence of ISWs during the neap tide, which occurs when the tide approaches high to low tide. A hypothesis can be proposed to explain this phenomenon: the ITF begins to intensify, with the event at point (D) occurring at the end of MB-1, specifically on May 13, 2023, while the ITF intensifies during the eastern season (June-August) [24].

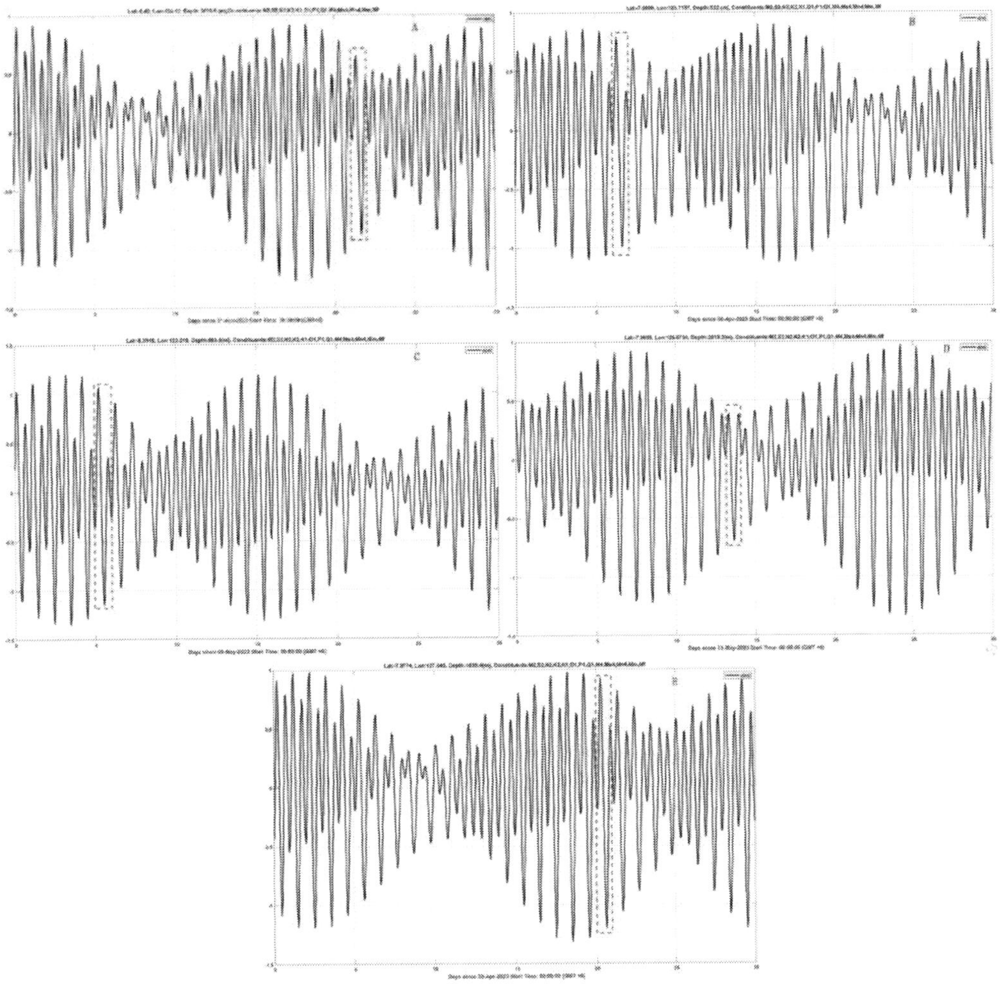

Fig. 10. Tidal height prediction in five research locations which shown the emergence of ISWs in Flores Sea and surrounding waters, derived from OTIS model in : (A) April ,21st 2023 (B) April, 06th 2023 (C) May, 5th 2023 (D) May, 13rd 2023 (E) April, 20th 2023. Red dash rectangle indicate the timing of ISW appearances in five SBES and MBES images analyzed in this study.

We suspect that Kelvin wave intrusions, which frequently occur in May and November, penetrate Indonesian territory through the Lombok, Sumba, and Ombai straits [56,57]. Additionally, the ISWs reflection process [9] occurs in a water column with underwater features (sills/seamounts). Kelvin waves across the Indonesian straits (Lombok, Sumba, and Ombai) induce vertical displacements of the pycnocline, subsequently generating internal waves as the stratified water column interacts with diverse topographical features, such as sills or seamounts. Internal waves may transition into ISWs as a result of the interplay between nonlinear steepening and dispersion effects, particularly in regions where the pycnocline meets abrupt bathymetric structures. When these ISWs contact underwater topographic features with critical slopes (where the slope angle corresponds to the characteristic angle of internal wave propagation), they may reflect them, potentially resulting in wave amplification or energy concentration in particular areas of the water column.

4 Conclusion

The examination of EJC 3-2023 "Flores" data demonstrated compelling evidence of internal wave dynamics in the Flores Sea, especially in the vicinity of the northern Ombai Strait seamounts (2100m and 1000m elevations), where notable refraction patterns were identified via satellite imagery and in-situ measurements. The SBES and MBES observations identified five distinct locations with solitary waves exhibiting significant amplitudes between 20 and 100 m, where these waves displayed crest bending towards shallower areas and showed interference patterns with incoming waves upon interacting with critical slope features. The refracted waves revealed concentrated energy around the seamount features, leading to wave amplification and potential energy concentration in specific water column regions, which may have contributed to the creation of sand waves similar to those observed in the Alor Strait. Future research must concentrate on comprehending ISW dynamics in neighboring areas, specifically their manifestation in the Leti Strait, southward movement during neap tides in the Ombai Strait, seasonal fluctuations, and interaction with the ITF, which will collectively enhance navigational safety by improving the mapping of these oceanographic characteristics.

Acknowledge to the analysis incorporates data from many data centers, namely the World Ocean Database (WOD), the General Bathymetric Chart of the Oceans (GEBCO), and the Alfred Wegener Institute (AWI); to Naval Hydro-Oceanography Center (PUSHIDROSAL) and Indonesian Naval Technology College (STTAL) in the context of datas use and software processing; to all scientists that cited journals. This paper is part of my doctoral thesis. The suggestions and comments from blind reviewers for the refinement of this manuscript are greatly appreciated. All authors contributed equally as the principal authors of this manuscript. The authors have declared that no competing interests exist.

5 References

1. Chonnaniyah, I. W. G. A. Karang, and T. Osawa, IOP Conf. Ser. Earth Environ. Sci. **944**, 1 (2021)
2. L. Sun, J. Zhang, and J. Meng, J. Oceanol. Limnol. **39**, 2195 (2021)
3. I. A. Prasetya, A. S. Atmadipoera, S. Budhiman, and U. C. Nugroho, IOP Conf. Ser. Earth Environ. Sci. **944**, 1 (2021)
4. R. D. Susanto, L. Mitnik, and Q. Zheng, Oceanography **18**, 81 (2005)
5. T. Nagai and T. Hibiya, J. Geophys. Res. Ocean. **120**, 3373 (2015)
6. Y. Gong, J. Xie, J. Xu, Z. Chen, Y. He, and S. Cai, Acta Oceanol. Sin. **41**, 109 (2022)
7. E. Situmorang, H. M. Manik, and A. S. Atmadipoera, Positron **12**, 132 (2022)
8. J. Sprintall, A. L. Gordon, S. E. Wijffels, M. Feng, S. Hu, A. Koch-Larrouy, H. Phillips, D. Nugroho, A. Napitu, K. Pujiana, R. Dwi Susanto, B. Sloyan, D. Yuan, N. F. Riama, S. Siswanto, A. Kuswardani, Z. Arifin, A. J. Wahyudi, H. Zhou, T. Nagai, J. K. Ansong, R. Bourdalle-Badié, J. Chanut, F. Lyard, B. K. Arbic, A. Ramdhani, and A. Setiawan, Front. Mar. Sci. **6**, (2019)

9. I. W. G. A. Karang, Chonnaniyah, and T. Osawa, Int. J. Remote Sens. **41**, 5726 (2020)
10. F. Syamsudin, A. Kaneko, and D. B. Haidvogel, Geophys. Res. Lett. **31**, 1 (2004)
11. K. Drushka, J. Sprintall, S. T. Gille, and I. Brodjonegoro, J. Phys. Oceanogr. **40**, 1965 (2010)
12. R. C. Lien, F. Henyey, B. Ma, and Y. J. Yang, J. Phys. Oceanogr. **44**, 1095 (2014)
13. Y. Stepanyants, **39**, 469 (2021)
14. T. Wang, X. Huang, W. Zhao, S. Zheng, Y. Yang, and J. Tian, J. Mar. Sci. Eng. **10**, (2022)
15. I. W. S. E. Putra, A. S. Atmadipoera, H. M. Manik, G. Harsono, A. Purwandana, M. R. Keulana, D. Handoko, J. Setiyadi, and W. S. Pranowo, Chrart Datum **8**, 95 (2022)
16. S. IHO, (2024)
17. K. Colbo, T. Ross, C. Brown, and T. Weber, Estuar. Coast. Shelf Sci. **145**, 41 (2014)
18. R. Firdaus, H. M. Manik, A. S. Atmadipoera, R. Zuraida, and C. Purwanto, J. Ilmu Dan Teknol. Kelaut. Trop. **13**, 151 (2021)
19. T. Hamilton and J. Beaudoin, Can. Hydrogr. Conf. **784**, 1 (2010)
20. H. M. Manik and S. Hadi, Int. J. Remote Sens. Earth Sci. **7**, 96 (2011)
21. H. Song, J. Chen, L. M. Pinheiro, B. Ruddick, W. Fan, Y. Gong, and K. Zhang, Deep. Res. Part I Oceanogr. Res. Pap. **177**, 1 (2021)
22. M. Z. Lubis, G. Surya, K. Anggraini, and H. Kausarian, Oseana **42**, 34 (2019)
23. I. H. Hariyanto, A. W. Putranto, N. Nurhidayat, B. Purwanto, D. P. Sobarudin, P. D. Saputro, D. R. Khair, and M. A. Wibowo, IOP Conf. Ser. Earth Environ. Sci. **1276**, 1 (2023)
24. A. L. Gordon, A. Napitu, B. A. Huber, L. K. Gruenburg, K. Pujiana, T. Agustiadi, A. Kuswardani, N. Mbay, and A. Setiawan, J. Geophys. Res. Ocean. **124**, 3724 (2019)
25. R. D. Susanto, Z. Wei, T. R. Adi, Q. Zheng, G. Fang, B. Fan, A. Supangat, T. Agustiadi, S. Li, M. Trenggono, and A. Setiawan., Oceanogr. Soc. 264 (2016)
26. Y. Ladroit, P. C. Escobar-Flores, A. C. G. Schimel, and R. L. O'Driscoll, SoftwareX **12**, 100581 (2020)
27. Chonnaniyah, T. Osawa, A. R. As-syakur, I. W. G. A. Karang, and J. C. B. da Silva, Int. J. Remote Sens. **00**, 1 (2023)
28. J. R. Apel, SAR Mar. User's Man. 189 (2004)
29. D. Nugroho, A. Koch-Larrouy, P. Gaspar, F. Lyard, G. Reffray, and B. Tranchant, Mar. Pollut. Bull. **131**, 7 (2018)
30. L. Mitnik and V. Dubina, Int. Geosci. Remote Sens. Symp. **3**, 2 (2009)
31. T. Aji, N. S. Adi, and A. I. Santoso, Analysis Of Variableity Of Chlorophyl-A And Sea Surface Temperature During The Occurrence Of An Internal Wave In The Ombai Strait (Analisis Variabelitas Klorofil-A Dan Suhu Permukaan Laut Saat Kejadian Internal Wave Di Selat Ombai), STTAL, 2024
32. R. Schlitzer, (2022)
33. G. D. Egbert and S. Y. Erofeeva, J. Atmos. Ocean. Technol. **19**, 183 (2002)
34. N. Jayanti, A. Wicaksono, and A. Purwandana, BIO Web Conf. **89**, 1 (2024)
35. Y. Gong, J. Xie, J. Xu, Z. Chen, Y. He, and S. Cai, Prog. Oceanogr. **202**, 102759 (2022)
36. A. S. Atmadipoera, A. Koch-Larrouy, G. Madec, J. Grelet, F. Baurand, I. Jaya, and I. Dadou, Deep. Res. Part I Oceanogr. Res. Pap. **182**, 1 (2022)
37. S. Zubaedah, H. Setiyono, C. D. Puspita, N. F. Gusmawati, and W. S. Pranowo, IOP Conf. Ser. Earth Environ. Sci. **750**, 1 (2021)

38. J. P. Matthews, H. Aiki, S. Masuda, T. Awaji, and Y. Ishikawa, J. Geophys. Res. Ocean. **116**, 1 (2011)
39. R. D. Susanto and R. D. Ray, Remote Sens. **14**, (2022)
40. Y. Bai, H. Song, Y. Guan, and S. Yang, Cont. Shelf Res. **143**, 9 (2017)
41. J. R. Apel, An Atlas Ocean. Intern. Solitary Waves 189 (2002)
42. T. Lee, S. Fournier, A. L. Gordon, and J. Sprintall, Nat. Commun. **10**, 1 (2019)
43. Apriansyah and A. S. Atmadipoera, IOP Conf. Ser. Earth Environ. Sci. **429**, 1 (2020)
44. A. Purwandana, M. Purba, A. S. Atmadipoera, P. Oseanografi-lipi, F. Perikanan, I. P. Bogor, and K. I. P. B. Darmaga, ILMU Kelaut. Indones. J. Mar. Sci. **19**, 43 (2014)
45. A. Koch-Larrouy, A. Atmadipoera, P. van Beek, G. Madec, J. Aucan, F. Lyard, J. Grelet, and M. Souhaut, Deep. Res. Part I Oceanogr. Res. Pap. **106**, 136 (2015)
46. P. Widianwari and Widianingsih, Ilmu Kelaut. **16**, 219 (2012)
47. G. La Forgia, T. Tokyay, C. Adduce, and G. Constantinescu, Adv. Water Resour. **135**, 103475 (2020)
48. G. La Forgia, R. Droghei, M. Pierdomenico, P. Falco, E. Martorelli, A. Bergamasco, A. Bergamasco, and F. Falcini, Sci. Rep. **13**, 1 (2023)
49. Y. Zhang, Z. Zang, Q. Yi, D. Liang, Z. Liu, and G. Li, in *Proc. 10th Int. Conf. Asian Pacific Coasts (APAC 2019) Hanoi, Vietnam, Sept. 25-28, 2019*, edited by N. T. Viet, D. Xiping, and T. T. Tung (Springer, Hanoi, 2019), pp. 457–462
50. C. R. Jackson, **112**, 1 (2007)
51. C. R. Jackson, J. C. B. da Silva, and G. Jeans, Oceanography **25**, 108 (2012)
52. O. Bosscha, Obs. Bosscha (Bosscha Sterrenwacht) (2023)
53. M. Khoeron, Minist. Relig. Repub. Indones. (2023)
54. R. Robertson, Commons, Wikimedia (2021)
55. H. M. van Aken, I. S. Brodjonegoro, and I. Jaya, Deep. Res. Part I Oceanogr. Res. Pap. **56**, 1203 (2009)
56. F. Syamsudin, H. M. van Aken, and A. Kaneko, Dyn. Atmos. Ocean. **50**, 129 (2010)
57. K. Pujiana, A. L. Gordon, and J. Sprintall, J. Geophys. Res. Ocean. **118**, 2023 (2013)

Diversity and community structure of seagrass ecosystem at Menjangan Island, West Bali National Park

Alfaiz Arselan[1], *Radixa Radhya Ramadhan* Haryanto[2], *Kesha Mufthiy Java* Tanaya[1], *Revita Ramadhani* Fitri[1], *Laura Natalya Elia* Sianturi[1], *Raihana Yasmin* Azzahra[1], *Caesar Aditya Wirayudha* Putra[2], *Dwiky Harris* Abimanyu[1], *Adriani* Sunuddin[3*]

[1]Undergraduate Course in Marine Science and Technology, IPB University
[2]Magister Course in Marine Science, IPB University
[3]Department of Marine Science and Technology, Faculty of Fisheries and Marine Science, IPB University

Abstract. Menjangan Island, an uninhabited island located within The West Bali National Park (TNBB), is a prominent marine tourism destination site in northwest Bali which features rich coastal biogenic ecosystems. This study assessed the status of seagrass beds in the northwest of Menjangan Island, as part of a student-led research expedition by HIMITEKA-IPB University. Five sampling sites were observed during August 7-11, 2023, and modified Seagrass Watch method was applied to record seagrass species, abundance, diversity, dominance, and associated biota. Four seagrass species were identified, with *Cymodocea rotundata* as the most abundant, followed by *Thalassia hemprichii* and *Halophila ovalis*, while *Halodule pinifolia* was absent in Site 2. Seagrass percent cover was considered poor with a range between 17.89% and 27.09%. Seagrass density was prominent for Cymodoceaceae (NCr= 173.58-319.39 ind/m2, NHp= 0-164 ind/m2), while two Hydrocharitacean species had lower density (NTh= 42-112 ind/m2, NHo= 37-131 ind/m2). Diverse seagrass associated biota were observed and recorded during rapid assessment, for example macroalgae, vertebrates (finfish, eel), and invertebrates (clams, sea cucumbers, worms), indicating resilient seagrass communities exist in the offshore environment of the West Bali National Park.

1 Introduction

Seagrass is a flowering plant that grows in shallow waters and is vital to coastal ecosystems. Seagrass ecosystems are essential for human well-being by providing various ecosystem services, from seafood provisioning, regulating local climate, supporting biogeochemical cycling, organic particle entrapment, wave soil sediment stabilisation, until preventing siltation in coral reefs and coastal erosion, yet global seagrass meadows are deteriorating [1]. Environmental pressures affecting seagrass are direct and indirect human activities which

* Corresponding author: adriani@apps.ipb.ac.id

modifies water quality, and climate change. Global rate of seagrass loss reached 27 km^2 per year, with an accelerated decline period in 1879-2006 due to various anthropogenic and natural factors [2]. In Indonesia, approximately 40-60% of seagrass meadows have been damaged and rehabilitation experiments are promoted to support its resilience [3].

Menjangan Island, located within the West Bali National Park, is a region known for its rich marine biodiversity, including extensive seagrass meadows. As part of an established national park since 1984, Menjangan Island functions as an important conservation area, serving both biodiversity preservation and ecological tourism [4]. However, like many other coastal ecosystems, the seagrass meadows at Menjangan Island face threats from humans, such as unsustainable fishing practices, irresponsible tourist activities, sewage or waste run-offs and different types of environmental pressures (heat wave, rising sea level, sub-aerial exposure) which may lead to seagrass die-off.

Studying seagrass community structure at Menjangan Island is important for quickly assessing its ecological status, particularly when the seagrass meadow is located within the marine protected area. Community structure evaluated are seagrass species diversity, coverage, density, and distribution, thus such indicators can be evaluated in comparison to previous assessments [5, 6]. HIMITEKA (Himpunan Mahasiswa Ilmu dan Teknologi Kelautan) is an official undergraduate student association in the Department of Marine Science and Technology at IPB University and organises annual expedition events to enhance academic and professional skills. One of many HIMITEKA functions is to promote marine science and technology through community outreach, environmental conservation projects, and scientific campaigns. Thus, HIMITEKA Expedition VIII aims to facilitate students networking with alumni and marine professionals, while at the same time serve as a pivotal leadership program in building teamwork and soft skills among student members. This paper aimed to study seagrass community structure and biodiversity at Menjangan Island, West Bali National Park.

2 Methods

2.1 Study Area

Seagrass assessment and associated biota observation were conducted on August 8-10, 2023, between 9:00 AM and 4:00 PM Central Indonesian Time (WITA). There are five observation sites in the northwest of Menjangan Island (**Fig. 1**). The study area features seagrass ecosystems which are located in the offshore environment of the West Bali National Park.

 https://doi.org/10.1051/bioconf/202516801008

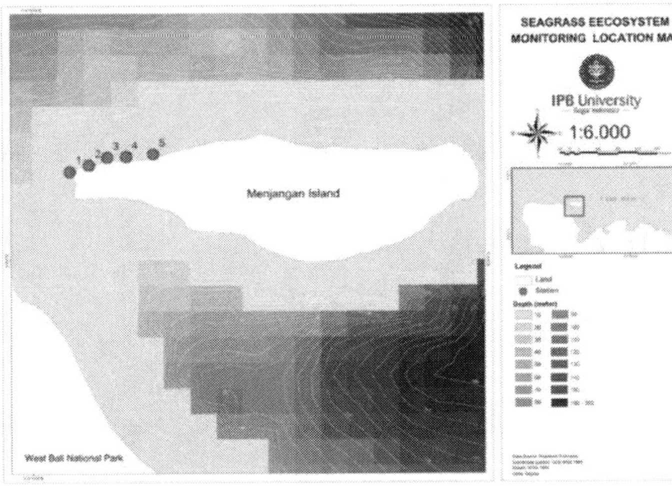

Fig. 1. Observation sites for seagrass diversity assessment in Menjangan Island

2.2 Seagrass Observation

In this study, several tools and materials were utilised to support field data collection. Roll meters were used to measure distances in the observation sites, while 50 cm × 50 cm quadrant transect and transect markers served as guides for sampling at each point. Data sheets, stationery, and slates were used to record direct field observations. Cameras were used to document environmental conditions and types of seagrass. Seagrass Watch identification keys were used to identify seagrass species and cover. Booties or mountain sandals were used for foot protection, while masks and snorkels assisted the observers in the water. GPS was used to accurately record the coordinates of the research location. These tools and materials worked together to ensure that the data were accurate and complete.

The quadrat transect method was used for seagrass data collection [6]. Data collection was conducted along three line transects, each 50 m long, with a 25-meter distance between the line transects. Quadrat transects were placed on the right side of each line transect at 5-meter intervals, resulting in 11 quadrats per line transect. The starting point of each line transect was 5-10 meters from the first seagrass encountered and positioned perpendicular to the shoreline. An illustration of the seagrass transect line method is shown in **Fig. 2**.

The data consisted of both biotic and abiotic parameters. The abiotic parameters considered were the pH, temperature, salinity, substrate type, depth, and water clarity. Parameters such as pH, temperature, and salinity were measured using measuring tools. Other parameters such as substrate type, depth, and water clarity were obtained through visual observation. The data collected from the seagrass community were the percentage of seagrass cover, number of seagrass stands, frequency of seagrass species, and associated biota. The percentage of seagrass cover was obtained using estimates based on the Seagrass Watch identification key. The number of seagrass species was recorded to calculate the density of seagrass species. The frequency of seagrass species was determined based on how often they were found during the survey.

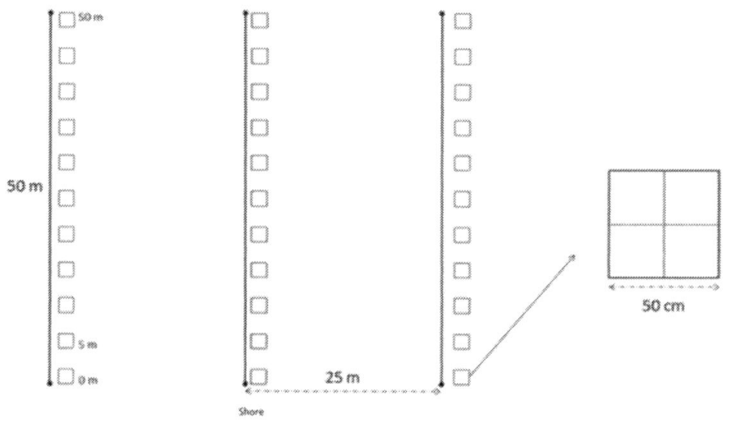

Fig. 2. Illustration of roll meter (left) and quadrant transect (right) deployment

2.3 Data Analysis

The analysis carried out in the processing of seagrass data included the value of species coverage, relative cover, species density, relative density, species frequency, relative frequency, diversity index, and dominance index. The analysis was performed using the following calculations:

Species Density

The density of the seagrass species is the total number of individuals per unit area. The formula used for the species density is as follows [7]:

$$D_i = \frac{N_i}{A}$$

Notes: Di: species density $(\frac{ind}{m^2})$; N_i: number of stands of each seagrass species; A: transect area (m^2)

Relative Density

Relative density is the comparison between the number of individuals of a species and the total number of individuals of all species, according to [7]

$$RD = \frac{n_i}{\sum n_i \times 100}$$

Notes: KR: relative density; N_i: number of stands of each seagrass species; $\sum N_i$: total number of stands.

Species Frequency

The frequency of seagrass species (Fi) describes the probability of a species being found at the observed sample points. The seagrass species frequency was calculated as follows [7]:

$$F_i = \frac{P_i}{\sum P_i}$$

Notes: F_i= Species frequency; P_i= number of sample plots per species; $\sum P_i$= Total number of plots observed

Relative Frequency

The relative frequency (RF) is the comparison between the frequency of the i-th type (Fi) and the total frequency for all types, using the following formula [7]:

$$RF = \frac{F_i}{\sum F_i \times 100}$$

Notes: FR= Relative frequency; F_i= Species frequency; $\sum F_i$= Number of observed frequencies

Species Coverage

Species cover (Ci) is the area covered by seagrass species. Seagrass species cover was calculated using the following formula [7]:

$$C_i = \frac{a_i}{A}$$

Notes: C_i: Seagrass species cover for each species; A_i: Total area covered by each seagrass species; A: Transect area

Relative Coverage

Relative Coverage is the comparison between the individual coverage of species i and the total coverage of all species. The relative cover of seagrass can be calculated using the formula [7]:

$$RC = \frac{C_i}{\sum C_i \times 100}$$

Notes: C_i= Seagrass species cover for each species; $\sum C_i$= Total number of species covered.

The criteria for evaluating seagrass ecosystem quality was referring to Indonesian Minister of Environment Decree No. 200 of 2004, which elaborated in Table below.

Coverage (%)	Condition	Quality	Index
≥ 60	Good/Rich/Healthy	Good	3
30 - 59,9	Damaged/Less Healthy	Moderate	2
≤ 29.9	Damaged/Poor	Poor	1

Importance Value Index

The Importance Value Index (IVI) of seagrass is an ecological parameter used to determine the dominance of seagrass species in an ecosystem. The IVI shows how large the role of a seagrass species is compared to that of other species in a community. This calculation can be performed using the following formula [7]:

$$IVI = RFi + RDi + RCi$$

Description:

RFi: relative frequency of the i-th species
RDi: relative density of the i-th species
RCi: relative coverage of the i-th species

Shannon-Wiener Diversity Index
The Species Diversity Index was used to determine whether seagrass has species variation. The diversity index was calculated using the formula below [7]:

$$H' = -\sum P_i \times ln \times P_i$$

Notes: H': Shannon-Wiener diversity index; P_i: $\frac{n_i}{N}$ (Composition of the i-th species); N_i: number of strands per species; N: Total number of strands

The obtained values were then compared with existing criteria for grouping [8]. An H' value > 3 indicated high species diversity. A value of $1 \leq H' \leq 3$ indicated moderate species diversity. An H' value ≤ 1 indicated low species diversity.

Evenness Index
The evenness Index is an ecological parameter used to measure the distribution of individuals of different species within a community. This index indicates whether a community is dominated by only a few species, or whether individuals are evenly distributed among the species present. The calculation was performed using the following formula [7]:

$$e = \frac{H'}{ln\, ln\, S}$$

Notes: e: Evenness index; S: number of species; *H'*: Shannon-Wiener diversity index

Dominance Index
The dominance index was used to determine the extent to which a group of biota dominated other groups. A sufficiently large dominance can lead to an unstable or stressed community. This dominance is obtained using the following formula [9]:

$$C = \sum \frac{n_i}{N}$$

Notes: C: Dominance index; N_i: number of strands per species; N: Total number of strands

The obtained values were then compared with the existing seagrass dominance index scale [9]. A value of $0 < C \leq 0.5$ indicates low dominance. A value of $0.5 < C \leq 0.75$ indicates moderate species dominance. A value of $0.75 < C \leq 1$ indicates high species dominance.

3 Results

3.1 Abiotic Parameters

Abiotic parameters collected from five stations showed temperature values ranging from 26.1-29.4°C, salinity ranging from to 29-33 ppt, and pH ranging from to 8.2-8.4. The values obtained from these measurements are included in the range of values in accordance with the Decree of the Minister of Environment No. 51 of 2004 Concerning Seawater Quality Standards for Marine Biota. According to [8], the optimum conditions for seagrass growth are a salinity of approximately 35 ppt and temperature below 38 °C. The types of substrates found in the Menjangan Island seagrass ecosystem mostly consisted of sand and coral fragments. The brightness at the observed location appears very bright because it is in an intertidal area with a depth of less than 1 m, which can still be penetrated by sunlight to the bottom of the water at the highest tide. The observed abiotic parameters produce values that describe the environmental conditions that tend to be ideal for seagrass survival.

3.2 Seagrass Community Structure

The species identified on Menjangan Island represent 33% of the 12 seagrass species recorded in Indonesia. Four species belonging to two families were identified: Hydrocharitaceae and Cymodoceaceae. The Cymodoceaceae family includes *Cymodocea rotundata* and *Halodule pinifolia*, whereas the Hydrocharitaceae family is represented by *Thalassia hemprichii* and *Halophila ovalis*. These findings indicate a species composition consistent with previous research. A study by [10] recorded the same species, namely *Thalassia hemprichii*, *Halodule pinifolia*, *Halophila ovalis*, and *Cymodocea rotundata*. This suggests that there were minimal changes at the study site during the observation period.

Seagrass species frequency provides insights into the proportion (%) of seagrass species found during field data collection [11]. The species with the highest frequency at each station was *C. rotundata* (**Fig. 3**). The high frequency of *C. rotundata* indicated that this species is adaptive, with long, relatively narrow leaves. This species thrives in sunlight-exposed waters and can grow across almost all habitat categories [12]. The highest relative frequency of seagrass at each station was observed for *C. rotundata*. The genus *Cymodocea* can survive in moderately disturbed environmental conditions. The adaptability of *C. rotundata* likely contributes to its higher relative frequency. This species has a robust root system that allows it to absorb nutrients and grow on various substrates [10].

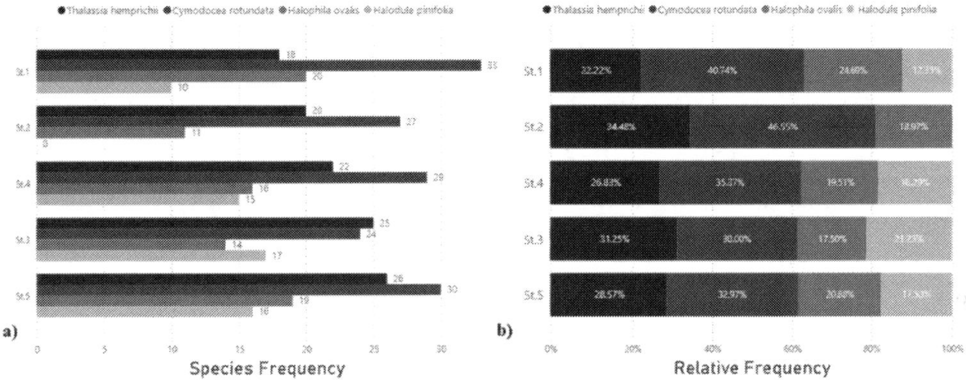

Fig. 3. Seagrass a) species frequency and b) relative frequency at Menjangan Island

Observations showed that the highest density was recorded for *Cymodocea rotundata*, with a density of 319.39 at ST-5, holding the highest density at all stations (**Fig. 4**). Seagrass species density can be used as an indicator of seagrass ecosystem conditions. According to [13], *the Cymodocea rotundata* density at all stations was classified as very dense (>175 ind/m^2). *Thalassia hemprichii* density was sparse (stations 1, 2, and 4) and moderately dense (stations 3 and 5). *The density of Halophila ovalis* was moderately dense at most stations, except for Station 3, which was sparse. *Halodule pinifolia* was absent at Station 2, but was moderately dense at Stations 1, 3, and 4 and dense at Station 5.

Based on the relative density calculations, the seagrass with the highest relative density was *Cymodocea rotundata*, with a density percentage exceeding 39% at all stations. The other three species, *Thalassia hemprichii, Halophila ovalis,* and *Halodule pinifolia*, showed varying and generally lower densities than *Cymodocea rotundata* at all stations. Seagrass density per unit area varies significantly depending on the species due to morphological differences. *Cymodocea rotundata* had the highest density across all stations, with values of 315.27 ind/m^2 at St-1, 271.76 ind/m^2 at St-2, 173.58 ind/m^2 at St-3, 271.27 ind/m^2 at St-4, and 319.39 ind/m^2 at St-5. *Cymodocea rotundata* thrives on sandy substrates, as found in the study area [14]. This environment supports the growth of *Cymodocea rotundata*, allowing it to proliferate. Three out of four seagrass species on Menjangan Island are small: *Cymodocea rotundata, Halophila ovalis*, and *Halodule pinifolia*, whereas *Thalassia hemprichii* is a larger species. Tidal patterns likely influence this composition in the intertidal zone, where water levels recede to the seabed during low tides.

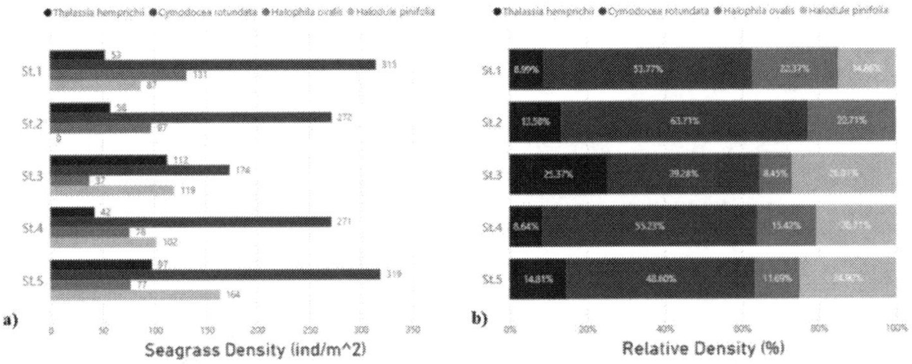

Fig. 4. Seagrass (a) density and (b) relative density at Menjangan Island

Based on **Fig. 5**, the average total seagrass cover at stations 1 and 2 was 20.48%, at Station 3 was 17.89%, at Station 4 was 22.1%, and at Station 5 was 27.09%, respectively. According to Indonesian Minister of Environment Decree No. 200 of 2004, the seagrass cover percentage at all stations falls within the poor category of ≤ 29.9%.

The relative cover percentages of seagrass at all stations were similar. *Cymodocea rotundata* had the highest relative cover percentages at all stations: 61.72% (St.1), 68.27% (St.2), 45.24% (St.3), 67.89% (St.4), and 51.93% (St.5). The other three species, *Thalassia hemprichii, Halophila ovalis*, and *Halodule pinifolia*, had lower cover percentages at all stations with values below 30%. The seagrass cover percentage represents the area occupied by the seagrass communities. Species density had a significant impact on the percentage of seagrass cover. Other factors, such as substrate, body size, and seagrass structure, can affect seagrass diversity and density [15].

The high cover percentage of *Cymodocea rotundata* across all stations was closely linked to its very high density, with an average of 270.25 ind/m² across all stations. Despite its small seagrass with a flat leaf morphology, this high density results in a high cover percentage for *Cymodocea rotundata*. Other small seagrasses, such as *Halophila ovalis* and *Halodule pinifolia*, have low cover percentages. In addition, several factors that influence the percentage of cover of a species include the density value and the width of the leaves [16]. This is because seagrass cover observations are based on leaf blades, whereas the number of stands measures the species density. Seagrass leaf width significantly influences substrate cover [17]. Leaf width affects the percentage of cover of a species [18]. *Thalassia hemprichii* also had low cover percentages, which is correlated with its low density.

 https://doi.org/10.1051/bioconf/202516801008

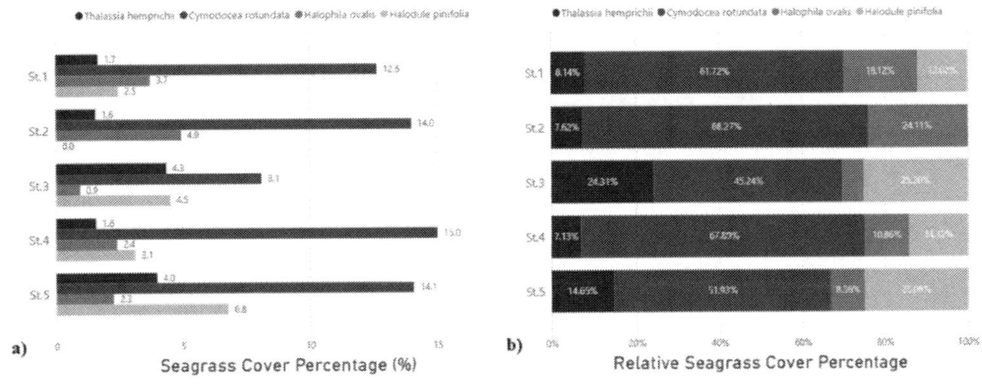

Fig. 5. Seagrass (a) cover percentage and (b) relative cover percentage at Menjangan Island

The ecological indices and Importance Value Index (IVI) in **Fig. 6** reveal distinct characteristics of the seagrass community across five stations on Menjangan Island. The diversity index values, ranging from 0.90 to 1.28, suggest moderate species variation, with station 3 showing the highest diversity (1.28) and evenness (0.92), indicating a balanced distribution of individuals among species. The dominance index at each station varied, with Station 2 showing the highest dominance (0.48), suggesting the strong presence of a particular species. At the same time, station 3 had the lowest dominance (0.30), indicating a more evenly structured community without significant dominance.

The IVI graph further highlights *Cymodocea rotundata* as the dominant species across all stations, with the highest IVI values (166 at St.1, 160 at St.2, 120 at St.3, 158 at St.4, and 154 at St.5), reflecting its strong adaptability to local environmental conditions. *Thalassia hemprichii* was notable at some stations, mainly St.3 and St.5, while *Halophila ovalis* and *Halodule pinifolia* contributed to diversity at lower and more variable IVI values. This dominance of *C. rotundata* aligns with studies that emphasize its adaptability to sandy substrates and high-light conditions, which likely gives it a competitive advantage [19].

Overall, Station 3 displayed a balanced and diverse community, whereas Station 2 was marked by a higher dominance of specific species. The variations in IVI and ecological indices across stations suggest that factors such as substrate type, light intensity, and salinity influence community structure, supporting the finding that habitat conditions play a crucial role in shaping seagrass communities [9]. The dominance of C. rotundata across the stations not only reflects its adaptability, but also indicates that the environmental conditions in Menjangan Island's seagrass ecosystem favor its growth and proliferation, consistent with previous studies on seagrass distribution and community dynamics [10, 19, 3].

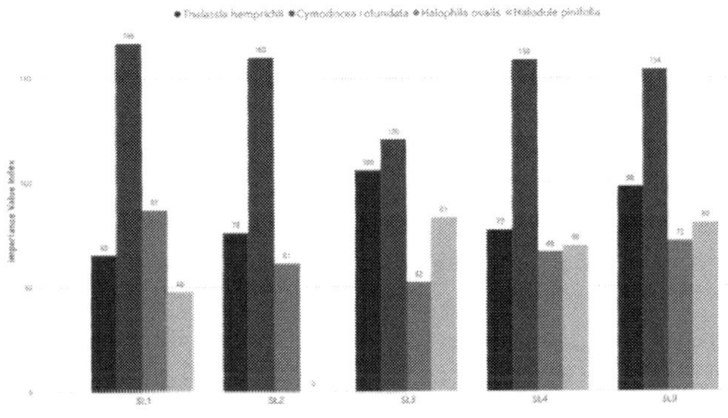

Fig. 6. Seagrass species Importance Value Index at Menjangan Island

Based on the ecological index graph in **Fig. 7**, illustrating diversity, evenness, and dominance indices across five stations (St. 1 to St. 5), it can be concluded that each station displays unique characteristics of the seagrass community. The diversity index for all stations ranges from 0.90 to 1.28, with station 3 achieving the highest value (1.28), indicating a relatively high level of species diversity. Overall, the diversity values fell within the moderate category, suggesting a reasonably good, though not exceptionally high, variation in species. This could be attributed to the dominance of certain species. The evenness index, which assesses the distribution of individuals among species, ranges from 0.81 0.92. Station 3 exhibited the highest evenness (0.92), signifying a more balanced distribution of individuals than at the other stations. This elevated evenness value indicates that no single species significantly dominates, leading to a more equitable distribution among various species. The dominance index across the stations ranges from 0.30 to 0.48, with Station 2 showing the highest dominance (0.48). This suggests that a specific species substantially dominated the seagrass community at this station. Conversely, Station 3 had the lowest dominance value (0.30), which is consistent with its higher diversity and evenness, indicating a seagrass community that is both diverse and well distributed without considerable dominance by any single species.

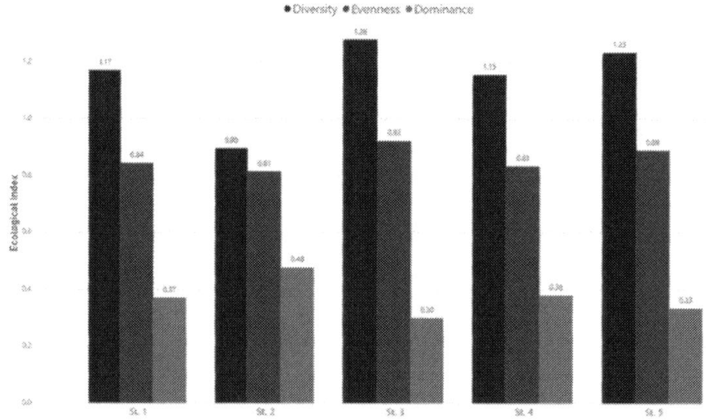

Fig. 7. Seagrass Ecological Index (Diversity, Dominance, and Evenness) at Menjangan Island

3.3 Associated Biota

The marine life observed alongside the seagrass at the study site comprised macroalgae, vertebrates, and invertebrates. The identified macroalgae species included *Ulva* sp., *Caulerpa* sp., *Halimeda macroloba*, and *Padina* sp. Among vertebrates, fish from the families Gobiidae, Muraenidae, and Pomacentridae were noted. The invertebrate community exhibited greater diversity, including *Synapta* sp., *Holothuria* sp., *Tridacna* sp., *Stichodactyla* sp., Ophiuroidea, and Polycladida (flatworms). The sandy substrate in the area facilitates both seagrass growth and survival of these associated organisms. Low turbidity permits light-dependent species to flourish, while other environmental factors, such as salinity and temperature, further support marine life in this region [20].

4 Conclusion

The seagrass ecosystem on Menjangan Island demonstrates optimal environmental conditions with suitable abiotic parameters, such as temperature, salinity, and pH, which support the growth of seagrass and a variety of associated marine biota. Four seagrass species were identified, with *Cymodocea rotundata* being the dominant species, with the highest frequency, density, and relative cover across all stations. Although the overall seagrass cover percentage was relatively low ($\leq$ 29.9%) and classified as poor, the ecosystem remained stable with a species composition similar to that in previous studies. The diversity of the associated biota, including macroalgae, fish, and invertebrates, indicates an ecosystem capable of supporting diverse marine life, making it an important area for seagrass conservation and preservation on Menjangan Island.

This research was conducted as a part of and supported by the 8th HIMITEKA Expedition. The authors would like to express their gratitude and appreciation to all parties involved in this research.

References

1. L. J. McKenzie, L. M. Nordlund, B. L. Jones, L. C. Cullen-Unsworth, C. Roelfsema, R. K. F. Unsworth. Environ. Res. Lett. **15**, 074041, (2020)
2. J. E. W. S. Sumbayak, A. Ambariyanto, Widianingsih. Jurnal Ilmiah Perikanan dan Kelautan, **15**, 84-94, (2023)
3. M. Waycott, C. M. Duarte, T. J. Carruthers, R. J. Orth, W. C. Dennison, S. Olyarnik, A. Calladine, J. W. Fourqurean, K. L. Heck Jr. Proceedings of the national academy of sciences, **106**, 12377-12381 (2009)
4. S. M. Wahyuningsih, Journal of Natural Resources and Environmental Management, **9**(2), 264 – 275, (2019)
5. M. D. K. Putra, S. Haryanti. Jurnal Kelautan, **8**, 23–30, (2015)
6. L. McKenzie, Seagrass Educators Handbook. www.seagrasswatch.org.
7. S. English, C. Wilkinson, & V. Baker. Australian Institute of Marine Science.(1997)
8. J. Wilhm, T. Dorris, Ecology, **49**, 880–881, (1968)
9. I. Setyobudiandi, Sulistiono, F. Yulianda, C. Kusmana, S. Hariyadi, A. Damar, A. Sembiring, Bachtiar, (Makaira FPIK IPB, Bogor, 2009)
10. H. K. Purnomo, Y. Yusniawati, A. Putrika, W. Handayani, Yasman, Pros sem nas masy biodiv indon, **3**, 236-240, (2017)
11. R. F. Larasati, M. M. Jaya, A. Putra, A. A. Djari, K. Sako, A. Khairunnisa, D. Jatayu, S. Aini, H. Suriadin, Journal of Indonesian Tropical Fisheries, **5**, 162-178, (2022)
12. A. Ramili, T. Hidayat, W. Suwarno, Biodiversitas, **19**, 2037-2043, (2018)
13. A. R. Kusumaningtyas, Suryono, Ambariyanto, Journal of Marine Research, **12**(2), 230-239, (2023)
14. P. T. Lefaan, M. Peday, B. Duwit, D. Orisu, S. N. Yoku, Y. Baab, A. W. Manumpil, Z. Mardiyadi, E. Manangkalangi, L. Sembel, Jurnal Perikanan Pantura, **6**, 315-332, (2023)
15. F. F. Lokollo, S. N. Sangaji, S. R. Loupatty, T. F. Pattiasina, AACL Bioflux, **15**(4), 2227-2237, (2022)
16. Erniati, Y. Andika, Imanullah, Imamshadiqin, Erlangga, Rahmad, T. A. Tauladan, F. R. Siregar, A. Fitri, G. H. Ritonga, Biodiversitas, **24**, 6621-6628, (2023)
17. Satriani, Z. Imran, F. Kurniawan, Perdinan, A. A. Digdo, Jurnal Penelitian Pendidikan IPA, **10**, 201–209, (2024)
18. F. F. Lakollo, S. N. Sangaji, S. R. Lopatty, T. F. Pattisiana TF, AACL Bioflux, **15**, 2227 – 2237, (2022)

19. N. D. M. Sjafrie, U. E. Hermawan, B. Prayudha, I. H. Supriyadi, M. Y. Iswari, Rahmat, K. Anggraeni, S. Rahmawati, Suryasi, (Pusat Penelitian Oseanografi LIPI, Jakarta, 2018)
20. B. Hamuna, R. H. Tanjung, Suwito, K. Hendra, Maury, Alianto, Jurnal Ilmu Lingkungan, **16**(1), 35 – 43, (2018)

Phylogenetic structure of tuna (Thunnus spp.) based on mitochondrial COI DNA in Indonesian and Australian waters

Jean Claudia De Soysa[1], *Satrio Hani Samudra[1], Siti Zanuba Aisyah[1], Alda Awayan Banjarsari[1], Radixa Radhya Ramadhan Haryanto[1], Dondy Arafat[1], Christopher Halashon Tambunan[1], Purwa Sulistya Ningrum[1],Inna Puspa Ayu[2] , Meutia Samira Ismet[1] ,Beginer Subhan[1*]*

[1]Department of Marine Science and Technology, Faculty of Fisheries and Marine Sciences, IPB University, Bogor, Indonesia
[2]Department of Aquatic Resources Management , Faculty of Fisheries and Marine Sciences, IPB University, Bogor, Indonesia

Abstract. Tuna (*Thunnus* spp.) is a seafood commodity that significantly contributes to a country's foreign exchange. Given the increase in fishing activities, genetic studies have become vital to ensure sustainable management. This study analyzed 65 mitochondrial cytochrome oxidase subunit I (COI) DNA sequences were obtained from GenBank and subjected to genetic and phylogenetic analyses using the Bootstrap Neighbor-Joining Model Kimura 2 Parameter Method. Phylogenetic analysis revealed that the 65 Thunnus spp. sequences clustered into five clades. Tuna had a bootstrap phylogenetic value of more than 50%, with the largest genetic distance recorded at 0.04 and the smallest at 0.00. Regarding tuna species and distribution zones, these findings imply that there are no significant genetic differences between Australia and Indonesia and indicating high genetic connectivity between these regions.

1 Introduction

Tuna is recognized as a significant fishery commodity in Indonesia, where the fishery sector is considered pivotal for economic growth in this developing nation (citation?). Indonesian tuna has a substantial market share and is renowned for its competitive edge against tuna products from other countries. According to [1], the export competitiveness of Indonesian tuna, skipjack, and mackerel in the United States market exhibited a Revealed Comparative Advantage (RCA) index value exceeding 1 from 2010 to 2019, signifying a comparative advantage in the target market. Therefore, maintaining the quality of tuna products is crucial for meeting consumer satisfaction at export destinations. This is substantiated by data indicating a 30.53% increase in tuna, skipjack, and mackerel exports during 2021-2022, contributing USD 280.74 million in export value [2]. Furthermore, Indonesia ranks among

* Corresponding author: beginersubhan@apps.ipb.ac.id

the largest producers among the 32 member countries of the Indian Ocean Tuna Commission (IOTC), with an average production ranging from 2019 to 2021 of 356,862 tons per year [3].

The increase in export value correlates with heightened exploitation of wild tuna [4]. This can lead to a decline in the genetic quality of tuna populations within a community. Genetic diversity analysis of various tuna species has been extensively researched [5] conducted a study on Thunnus spp using Gene Bank data from the Philippines and Spain, analyzing cytochrome oxidase subunit I (COI) sequences from 37 samples. The findings revealed no significant genetic differences between these geographically distant countries, a phenomenon likely attributed to the tuna's ability for long-distance migration. Ardiana et al. [6] asserted that species undergoing extensive migrations have significant opportunities for interbreeding with other populations, thereby promoting genetic flow and enhancing diversity within those populations. For instance, it is well known that bluefin tuna spawn off the southern coast of East Java and migrate seasonally to the southern coast of Australia. This aligns with Nikijuluw's research [7], which documented that, during the summer months (December-April), bluefin tuna tended to aggregate near the surface off Australia's southern coast, moving to deeper waters during winter. Given these factors, it is imperative to conduct genetic studies on tuna populations in both Indonesia and Australia. Genetic approaches are crucial for elucidating the kinship relationships among tuna species originating from these two countries.

2 Data and Material

This research commenced with the collection of tuna sequence information from the NCBI (National Center for Biotechnology Information (NCBI) website, which can be downloaded online from the link (https://www.ncbi.nlm.nih.gov/). The data consisted of mitochondrial DNA cytochrome oxidase subunit I sequences with a bootstrap length of above 500. The Thunnus spp. sequences were derived from five species, Thunnus albacares, Thunnus alalunga, Euthynnus affinis, Katsuwonus pelamis, and Thunnus obesus, totaling 65 sequences from Indonesia and Australia. All data were sourced from the GenBank, and the accession numbers are presented in Table 1.

A total of 65 DNA sequences were analyzed using the MEGA 11.0 software [8] to obtain Maximum Likelihood (ML) probability values, genetic distances, and phylogenetic trees, employing the bootstrap neighbor-joining approach with the Kimura 2-Parameter model [9].

Table 1. Acession numbers of COI DNA sequences for 65 *Thunnus* spp. species

No	Accesion number	Species	Keterangan
0	GU674430	*Thunnus albacares*	Indonesia
2	GU673849	*Thunnus albacares*	Indonesia
3	GU673630	*Thunnus albacares*	Indonesia
4	MN257555	*Thunnus albacares*	Indonesia
5	JN312307	*Thunnus albacares*	Indonesia
6	KF528380	*Thunnus albacares*	Indonesia
7	KF528379	*Thunnus albacares*	Indonesia
8	KF528378	*Thunnus albacares*	Indonesia
9	KF528377	*Thunnus albacares*	Indonesia
10	KF528376	*Thunnus albacares*	Indonesia
11	KF528375	*Thunnus albacares*	Indonesia
12	KF528374	*Thunnus albacares*	Indonesia
13	MH085917	*Thunnus albacares*	Indonesia
14	KF544951	*Thunnus alalunga*	Indonesia
15	GU673853	*Euthynnus affinis*	Indonesia
16	GU673850	*Euthynnus affinis*	Indonesia
17	JN313037	*Euthynnus affinis*	Indonesia
18	MH085918	*Euthynnus affinis*	Indonesia
19	MN257557	*Katsuwonus pelamis*	Indonesia
20	HQ564314	*Katsuwonus pelamis*	Indonesia

No	Accesion number	Species	Keterangan
21	HQ564313	*Katsuwonus pelamis*	Indonesia
22	KF528383	*Katsuwonus pelamis*	Indonesia
23	KF528382	*Katsuwonus pelamis*	Indonesia
24	MH085920	*Katsuwonus pelamis*	Indonesia
25	KF528371	*Thunnus obesus*	Indonesia
26	KF528370	*Thunnus obesus*	Indonesia
27	KF463146	*Thunnus obesus*	Indonesia
28	OM743318	*Euthynnus affinis*	Australia
29	OM743311	*Euthynnus affinis*	Australia
30	OM743310	*Euthynnus affinis*	Australia
31	DQ107699	*Euthynnus affinis*	Australia
32	DQ107698	*Euthynnus affinis*	Australia
33	DQ107686	*Euthynnus affinis*	Australia
34	DQ107685	*Euthynnus affinis*	Australia
35	DQ107684	*Euthynnus affinis*	Australia
36	EU014261	*Katsuwonus pelamis*	Australia
37	EU014260	*Katsuwonus pelamis*	Australia
38	EU014259	*Katsuwonus pelamis*	Australia
39	EU014258	*Katsuwonus pelamis*	Australia
40	DQ107683	*Katsuwonus pelamis*	Australia
41	DQ107682	*Katsuwonus pelamis*	Australia
42	DQ107681	*Katsuwonus pelamis*	Australia
43	DQ107669	*Katsuwonus pelamis*	Australia
44	DQ107668	*Katsuwonus pelamis*	Australia
45	DQ107644	*Thunnus obesus*	Australia
46	DQ107643	*Thunnus obesus*	Australia
47	DQ107642	*Thunnus obesus*	Australia
48	DQ107630	*Thunnus obesus*	Australia
49	DQ107629	*Thunnus obesus*	Australia
50	DQ107659	*Thunnus alalunga*	Australia
51	DQ107658	*Thunnus alalunga*	Australia
52	DQ107647	*Thunnus alalunga*	Australia
53	DQ107646	*Thunnus alalunga*	Australia
54	DQ107645	*Thunnus alalunga*	Australia
55	OM743312	*Thunnus albacares*	Australia
56	KX781891	*Thunnus albacares*	Australia
57	KX781879	*Thunnus albacares*	Australia
58	DQ107652	*Thunnus albacares*	Australia
59	DQ107651	*Thunnus albacares*	Australia
60	DQ107650	*Thunnus albacares*	Australia
61	DQ107649	*Thunnus albacares*	Australia
62	DQ107648	*Thunnus albacares*	Australia
63	EF609629	*Thunnus albacares*	Australia
64	EF609628	*Thunnus albacares*	Australia
65	EF609627	*Thunnus albacares*	Australia

3 Result and Discussion

Based on the results, the highest substitution probability was found for base T, followed by A, C, and G (Table 2). The composition of nucleotide bases is one of the genetic characteristics, with the arrangement of nucleotide bases being a key component that differentiates species within a DNA sequence. This indicates that a higher nucleotide value in tuna correlates with a higher likelihood of substitution. The substitution probability observed for (PT-A; T-C; T-C) was higher than for (PA-T; A-C; A-G), and so forth. Wijana et al. [5] explained that this occurs because transition substitutions, which occur between purines (or between pyrimidines), are more likely than transversion substitutions, which occur between purines and pyrimidines. This is due to the similarity of the molecular structure of tuna fish that is formed [10].

Table 2. Nucleotide Substitution with Maximum Likelihood (ML) Analysis

Maximum Likelihood Estimate of Substitution Matrix				
From\To	A	T/U	C	G
A	-	4.0113*	3.6065*	13.8498**
T/U	2.8957*	-	20.6352**	2.3948*
C	2.8957*	22.9516**	-	2.3948*
G	16.7468**	4.0113*	3.6065*	-

Phylogenetic analysis using the neighbor-joining method with the Kimura 2-Parameter model identified five clades: the first clade for yellowfin tuna (*Thunnus albacares*), the second clade for bigeye tuna (*Thunnus obesus*), the third clade for albacore tuna (*Thunnus alalunga*), the fourth clade for kawakawa (*Euthynnus affinis*), and the fifth clade for skipjack tuna (*Katsuwonus pelamis*). The constructed phylogenetic tree exhibited bootstrap values ranging from 47 to 100% for each branch within the population groups (Fig. 1).

The phylogenetic relationships demonstrated close genetic ties among the yellowfin tuna, bigeye tuna, and albacore tuna groups, whereas the kawakawa and skipjack tuna groups were more distantly related. This indicates genetic divergence between these three tuna species and skipjack and kawakawa Based on their regional distribution in Australia and Indonesia, there are no clade differences for each species [11]. This indicates a relatively similar between species, with the distance between Indonesia and Australia not too far for tuna to migrate.

Clade one represented the yellowfin tuna population, demonstrating a mixture of individuals from Indonesia and Australia. This suggests that these populations are genetically close and share similarities, likely because of the proximity of Indonesian and Australian waters. Similar findings have been reported in studies by Grewe and Hampton in the Pacific Ocean, Martinez and Zardoya, Martinez et al. in the Atlantic Ocean, Chiang et al. in the South China Sea, Philippines, and western Pacific Ocean, and Chiang et al. in the Indian Ocean [12] [13] [14] [15].

Clade two consisted of bigeye tuna, showing clustering based on shared individuals. This indicates a common ancestry among these individuals, which leads to genetic similarities. These findings are supported by Kunal and Kumar (2013) [16], who found that all haplotypes formed within the phylogenetic tree and showed no genetic differences in Indian waters.

Clade three represents albacore tuna and shows clustering similarities. This suggests geographical similarities among the individuals in the study. Similar patterns have been observed by Nakadate et al. [17], indicating population mixing from different locations, such as the Atlantic and Mediterranean seas. It is important to note that albacore tuna are not commonly found in Indonesian waters but may migrate into these waters and to Australian waters [18] showing genetic similarities among individuals found in other regions.

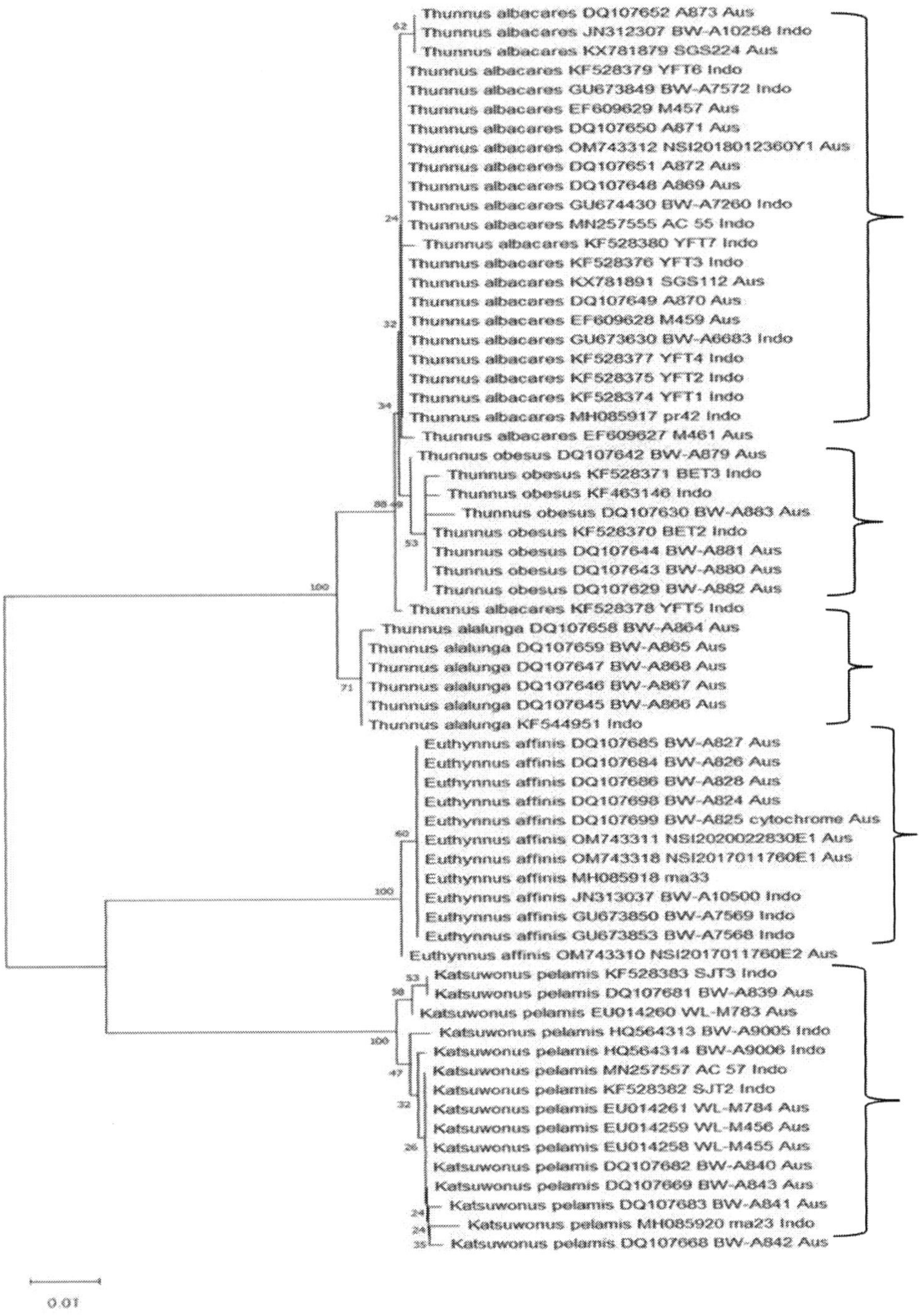

Fig. 1. Phylogenetic Tree Reconstruction using the Neighbor-Joining Method for Five Tuna Species

Clades four and five include bonito and skipjack tuna groups. Phylogenetic tree reconstruction places bonito and skipjack tuna outside the other species groups (*Thunnus* spp.) and is distantly related to them. This is consistent with findings by Collette and Chao [19], who noted that bonito species are distantly related to tuna species but are more closely related to other tuna species such as Auxis and Euthynnus. These findings align with those of Chow and Kishino [20], who identified three distinct clades with shared individuals in each clade. Chow et al. [21] further explained that bigeye tuna are genetically closer to yellowfin tuna than albacore tuna. Similar results were reported by Vinas and Tudela [22], indicating ten different clades among tuna species, including Atlantic bluefin tuna (T. *atlanticus*), yellowfin tuna (T. *albacares*), longtail tuna (T. *tonggol*), bigeye tuna (T. *obesus*), southern bluefin tuna (T. *moccoyii*), Pacific bluefin tuna (T. *orientalis*), bluefin tuna (T. *thynnus*), and albacore tuna (T. *albacore*).

The *Euthynnus* sp. and *Katsuwonus* sp. groups were placed in different clades from the *Thunnus* sp. group, consistent with Koladam's [17] findings that COI gene sequence variations in 13 tuna species revealed 248 nucleotide differences primarily between *Thunnus* sp. and other groups, such as *Euthynnus* sp. and *Katsuwonus* sp. Collette and Chao's [19] study, as cited in Matsumonto [24], also indicated that fish from the *Katsuwonus* sp. group are distant relatives of the *Thunnus* sp. group, but are closely related compared to groups like *Euthynnus* sp. and *Auxis* sp. Additionally, the phylogenetic tree demonstrated that fish from the same genus remained within the same clade, even when sampled from different waters, namely, Indonesia and Australia. This may be attributed to tuna being a pelagic species capable of extensive migration and widespread distribution [25][26].

4 Conclusion

Phylogenetic analysis of tuna (Thunnus spp.) indicated the presence of five distinct clades; however, no significant genetic differences were observed based on the species or geographic distribution of tuna. Overall, it can be concluded that tuna species share common ancestry and do not exhibit clear boundaries in their phylogeographic distribution.

References

1. Hartanto TR, Suharno, Burhanuddin. 2021. Journal JPHPI, 24(2): 227-235.
2. Kementerian Kelautan dan Perikanan. 2022.. Pusat Data, Statistik, dan Informasi Sekretariat Jenderal. KKP. Jakarta
3. Kementerian Kelautan dan Perikanan. 2015. Direktorat Sumber daya Ikan Direktorat Perikanan Tangkap. KKP. Jakarta.
4. Dişa, D., Akoglu, E. and Salihoglu, B., 2024. *Frontiers in Marine Science, 11*, p.1389941.
5. Wijana IMS, Mahardika, IGN. 2010. Journal Bumi Lestari, 10(2): 270-274
6. Ardiana SA, Astarini IA, Putra ING, Pertiwi PD, Sembiring A, Yusmalinda A, Malik DA. 2021. Journal Musamus Fisheries and Marine Journal, 3(2): 107-115.
7. Nikijuluw VPH. 2008. Journal Kebijakan Perikanan Indonesia, 1(1): 121-129.
8. Tamura K, Stecher G, Kumar S. 2021. Mega 11: Molecular Evolutionary Genetics Analysis Version 11. Mol. Biol. Evol, 28(7): 3022-3027
9. Yonatika NO, Widiasih N, Hamidah M, Nurhakim MD, Budiarto H, Bintang DM, Sani LM, Lestari DF, Setyaningsih WA, Subhan B, Madduppa H. 2021. IOP Conference Series: Earth and Environmental Science. 944(1):012028. IOP Publishing.
10. Dash M, Meher P, Kumar A, Satapathy SS, Namsa ND. 2024. PeerJ. 22(12):e16962.
11. Feutry P, Foster S, Grewe PM, Aulich J, Lansdell M, Clear N, Cooper S, Williams A, Johnson G, Dilrukshi T, Wudianto. 2025. ICES Journal of Marine Science. 82(2):fsae162.
12. Grewe P, Hampton J. 1998. Commonwealth Scientific and Industrial Research Organitation, Australia.

13. Martinez P, Zardoya R. 2005. Collective Volume of Scientific Papers, 57(1): 195-205.
14. Martinez P, Gonzales GE, Castilho R, Zardoya R. 2006. Molecular Phylogenetics and Evolution, 39(3): 404-416.
15. Chiang HC, Hsu CC, Lin HD, Ma GC, Chiang TY, Yang HY. 2006.79(2): 219-225. Chiang HC, Hsu CC, Wu GCC, Chang SK, Yang HY. 2008. Fisheries Research. 90(1): 305-312.
16. Kunal SP, Kumar G. 2013. International Journal of Bioinformatics Research and Applications 9(3): 301-309.
17. Nakadate, M, Vinas, J, Corriero, A., Clarke, S, Suzuki, N, Chow, S., 2005. Journal of Fish Biology, 66(4): 1545– 1557.
18. Setiawati MD, Rachman HA, Setiawan RY, Syahailatua A, Wouthuyzen S. 2024. Deep Sea Research Part II: Topical Studies in Oceanography. 1(216):105400.
19. Collette BB, Nauen CE. 1983. FAO species catalogue. Vol. 2. FAO Fisheries Synopsis 125(2):137.
20. Chow S, Kishino H. 1995. Journal of Molecular Evolution 41(2): 741-748.
21. Chow S, Nohara T, Tanabe T, Itoh S, Tsuji Y, Nishikawa S, Uyeyanagi K, Uchikawa. 2003. Bulletin of Fisheries Research Agency 8(3): 1-14
22. Vinas J, Tudela S. 2009. PLoS ONE, 4(10): 1- 10.
23. Kolondam BJ. 2020. 8(2): 70-75.
24. Matsumoto WM, Skiliman RA and Dizon AE. 1984.Synopsis of biological data on skipjack tuna (*Katsuwonus pelamis). FAO Fisheries Synopsis*. 136: 1-92.
25. Filous A, Friedlander AM, Toribiong M, Lennox RJ, Mereb G, Golbuu Y. 2022. ICES Journal of Marine Science.79(2):445-56.
26. Weist P, Kusche H, Tørresen OK, Hermida M, Lopes EP, Jentoft S, Hanel R. 2024. Global Ecology and Conservation. 1(54):03117.

Characteristics of beach and foreshore sediments at the coastal of Carita Beach, Pandeglang, Banten

Anak Agung Eka Andiani[1], *I Gusti Ayu Istri Pradnyandari Dewi*[1], *Ni Putu Pradnya Megasari*[1], *Yola Wulandari*[1], *Nur Almira Mahsa*[1], *Wulantika Irza Savira Putri*[1], *Santi Susanti*[1], *Tri Prartono*[1], *Rastina*[1*], *Mochamad Tri Hartanto*[1], *Erwin Maulana*[1], *Sylvie Siskawati*[1], *Teguh Nugraha*[1], *Dwi Ikhsan Yulianto*[1], *Yuli Naulita*[1], *Nyoman Metta Nyanakumara Natih*[1], *Santoso Budi Widiarto*[2], and *Kattia Setyani Widiastuti*[2]

[1]Department of Marine Science and Technology, Faculty Fisheries and Marine Science, IPB University, Bogor 16680, Indonesia
[2]Loka Pengelolaan Sumberdaya Pesisir dan Laut (LPSPL) Serang, Ministry of Marine Affairs and Fisheries, Pandeglang, Banten 42264, Indonesia

Abstract. Coastal morphology and dynamics influence the sediment transport process from land to coast and sea. A study of the distribution of sediment types was carried out on the beach and foreshore of Carita Beach, Pandeglang Banten, to provide an overview of the sediment transport and deposition processes based on sediment grain size distribution. Sediment samples were collected: nine from the foreshore and six from the beach. The sediment grain size was analysed using the fractionation method with sieving and pipetting, and the organic material content was analysed using the Loss on Ignition (LOI) method. Aggregations and statistics of the sediments were analysed using the Gradistat program. water content (%) of the sediments collected ranges from 10 - 65 %, the highest at the foreshore station (SL4) and the lowest at the beach station (SP1). The grain size (%) of the sediments collected had 11 categories, from gravel to fine clay. The total organic matter (%) in foreshore sediments is low, while that on the beach is very low. Statistically, the sediment distribution shows good sorting, high variation in skewness, and leptokurtic dominance

1 Introduction

Understanding the characteristics and dynamics of coastal areas is essential for effective management and sustainable development. Coastal characteristics and dynamics can be observed in various aspects, including the characteristics of sediments deposited in the area. Differences in coastal morphology such as the presence of river estuaries and morphology in the form of bays or capes, affect the sources and processes of sediment deposition. Coastal

* Corresponding author: rastina049@apps.ipb.ac.id

areas are generally dominated by the characteristics of rock sediments or coarse or medium sand [1], whereas open sea areas are generally dominated by fine or very fine sand sediments, as well as silt or clay.

Grain size is the most fundamental aspect of sediment particles and affects sedimentation, transportation, and deposition. Grain size distribution is influenced by other factors, such as distance from the coastline, distance from the source (such as a river), source of sediment material, topography, and sediment transport mechanisms [2]. Granulometric analysis is an analysis of sediment grain size, which is generally carried out to determine the level of resistance to the exogenic processes of sediment grains, such as weathering, erosion, and abrasion, from the origin of sediment transportation and deposition processes [3]. Furthermore, information on sediment grain size is related to the characteristics of materials or chemical compounds associated with particles, such as organic matter [4] and metal minerals [5, 6].

Carita Beach is a shallow coastal area located in Pandeglang on the west coast of the Banten Province. This beach has a sloping slope, white sand, and is supported by a coral reef area, which is often visited by local and foreign tourists. Sandy beaches are more comfortable for tourism than rocky or muddy beaches, particularly for swimming. Rocky beaches have the potential to cause accidents, such as collisions, whereas muddy beaches usually have murky water that interferes with visibility. Therefore, it is interesting to further identify the characteristics of sandy sediments on the coast (beach and foreshore) of Carita Beach, Banten, using granulometry and Gradistat statistical analysis, as well as their association with organic matter.

2 Materials and method

2.1 Collection of sediment and sampling location

Surface sediment samples from the foreshore of Carita Beach were collected once at nine observation areas using a Van Veen grab (~10-15 cm), and samples from the beach were collected at six points using a hand scope (~5-10 cm) (**Fig. 1**). The subsampled sediment was transferred into a polyethylene container, placed in a cool box during transportation to the laboratory, and stored at 4°C until analysis [7]. The sample analysis in the laboratory included basic characteristics of the sediment: grain size, water content, and organic content.

2.2 Sediment fractionation and organic compounds

Sediment grain size and organic matter content were used to determine possible correlations. The grain size analysis of the sediment was fractionated into 10 fractions using sieving (very coarse, coarse, medium, fine, and very fine sand) and pipetting (coarse silt, medium silt, fine silt, coarse clay, and fine clay) methods based on the Wentworth scale [8]. The organic matter in the sediment was analyzed using the Loss on Ignition (LOI) method [8, 9]. The sediment samples were heated at 60°C (1 h) and raised to 550°C (after they were left for 5 – 6 h). The samples were allowed to decrease in temperature in a desiccator between two stages of increasing temperature (initial and final weights). The Water content was calculated based on the difference between the initial and final weights of the sediment multiplied by 100% [10]. Organic matter content was calculated using the percentage of sediment organic matter lost during the combustion process. The carbon content in the sediment was calculated using a constant value (0.580) multiplied by the total organic matter content.

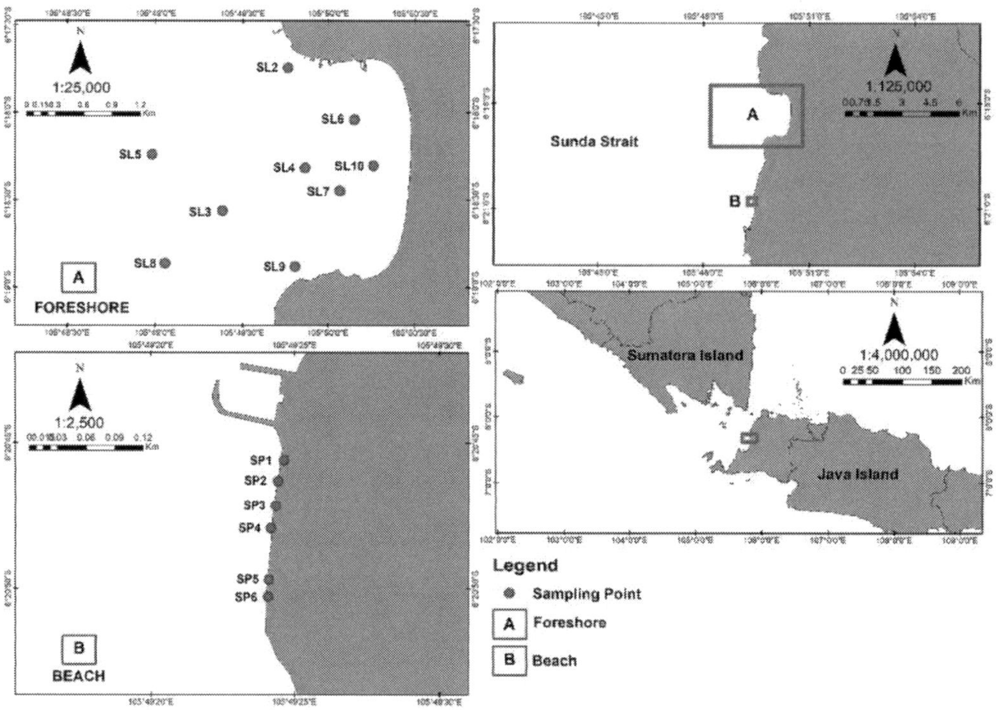

Fig. 1. Sampling stations for collecting sediment samples on the coast of Carita Beach.

2.3 Statistic of sediments

Analysis of the distribution characteristics of foreshore and beach sediments in the coastal region of Carita Beach using Gradistat focuses on discussing the statistical grain size distribution [11, 12]. The parameters used in the Gradistrat method include the mean, which is the average size of sediment particles. The median particle size divides the distribution into two equal halves. Sorting measures the homogeneity of the particle sizes in the sample, skewness describes the asymmetry of the particle size distribution, and kurtosis describes the peak and width of the particle size distribution. Then, we depicted the sediment distribution and created graphs and histograms to visualize the grain size distribution. Finally, we interpreted the results using statistical results to understand the sedimentation process, sediment sources, and environmental conditions.

3 Results and discussion

3.1 Physical characterization of sediments

The pore size distribution characteristics of sediments affect their water content [13]. The water content in the sediment affects particle aggregation and sediment transport processes. A high water content causes a decrease in particle aggregation stability (both physical and chemical) and increases the sediment transport process to the water column (sediment-water interface). The results show that foreshore sediments have a higher water content percentage than beach sediments (**Fig. 2**). The highest water content at the foreshore station is SL4 (66.07%) and the lowest at SL5 (30.66%). Meanwhile, The highest water content at the beach station was SP3 (28.55%), and the lowest was SP1 (9.58%). Foreshore areas generally have finer sediment characteristics than coastal areas do. Fine sediment particles have a better ability to hold water than coarse sediment particles [14].

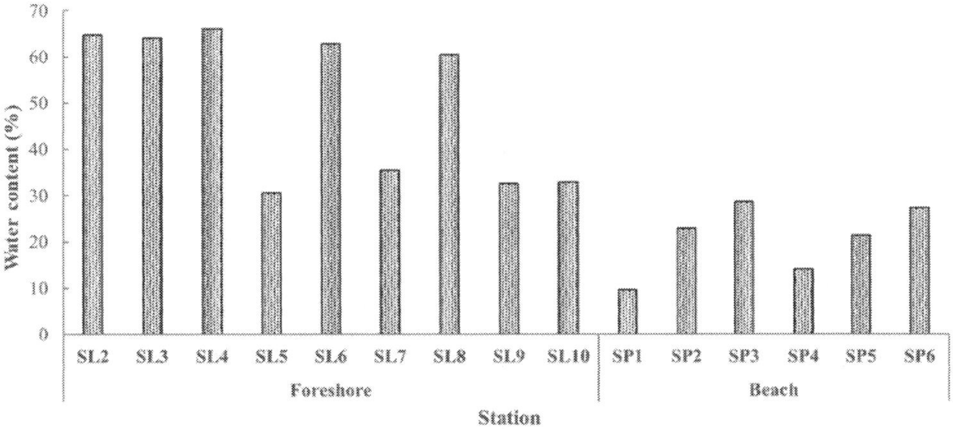

Fig. 2. Water content (%) of the sediments collected from foreshore (left) and beach (right) sampling stations of Carita Beach.

Based on the results of sieving and pipetting at the foreshore station, the sediment particle size was dominated by very fine sand and fine sand, except for SL9, which was dominated by fine clay, followed by SL7 and SL5, which were dominated by fine silt (**Table 1**). Unlike at the beach station, medium sand and fine sand were the dominant sediment contents (**Table 2**). Most of the large particles were deposited at a location not far from the sediment source [15]. In the marine environment, sediment grain size is often dominated by fine clay and very fine sand owing to various factors, such as hydrodynamic conditions and deposition processes [16].

Sediments in marine waters are formed by weathering of rocks in the sea. In addition, sediments also originate from land carried by rivers to the sea. Sediments are formed by the process of mineral accumulation due to deposition. The deposition process can continue when sediment loads enter the water column, and these conditions occur continuously [17]. The variation in deposition between coarse and fine sediments shows the difference in energy that occurs during the deposition of the sedimentary material [18]. Coarse sediments are carried by greater energy, whereas fine-sized materials are carried by weaker energies [19]. The sediment source in a body of water comes from a variety of particles that can affect the sediment composition in water [20]. The fraction in the coastal area generally has a larger and coarser size owing to the strength of currents and waves, which are still large in coastal areas [21; 22]. Tides are one of the factors that affect current velocity, and tides that vary and fluctuate affect the processes of sediment transport and deposition [23].

Table 1. Grain size (%) of foreshore sediments collected from the coast of Carita Beach.

Grain Size	SL2	SL3	SL4	SL5	SL6	SL7	SL8	SL9	SL10
Gravel	0	0	0.11	0.1	0	0	0	0	0

Grain Size	SL2	SL3	SL4	SL5	SL6	SL7	SL8	SL9	SL10
Very Coarse Sand	0.12	0.77	3.03	0.63	0.04	0.06	0.17	0	0.22
Coarse Sand	0.29	4.96	11.43	1.05	1.15	0.21	1.5	0.17	0.51
Medium Sand	1.67	9.24	16.55	2.53	5.58	0.65	9.15	6.14	0.49
Fine Sand	34.63	28.57	21.65	7.25	47.47	1.73	19.9	11.16	5.57
Very Fine Sand	53.76	44.94	41.06	54.05	44.03	45.07	57.03	37.65	89.08
Coarse Silt	3.66	4.61	2.64	2.16	0.97	2.94	5.25	1.56	1.85
Medium Silt	2.25	2.81	1.61	1.35	0.38	0.87	2.60	0.57	0.62
Fine Silt	1.8	1.7	1.1	30.1	0.3	47.48	2.21	0.41	0.51
Coarse Clay	1.03	1.36	0.28	0.19	0.07	0.45	1.09	0.26	0.08
Fine Clay	0.79	1.04	0.54	0.59	0	0.55	1.11	42.07	1.08

Table 2. Grain size (%) of beach sediments collected from the coast of Carita Beach.

Grain Size	SP1	SP2	SP3	SP4	SP5	SP6
Gravel	5.63	5.03	0.23	2.30	0.56	2.38
Very Coarse Sand	6.36	5.15	0.39	5.44	1.15	3.88
Coarse Sand	9.69	5.59	1.28	16.82	4.32	14.72
Medium Sand	44.08	34.46	39.67	54.57	68.39	52.93
Fine Sand	31.88	47	38.57	19.81	23.11	17.56
Very Fine Sand	1.35	1.54	1.12	0.36	1.13	6.93
Coarse Silt	0	0.13	0.27	0.13	0.26	0.54
Medium Silt	0	0.31	0.25	0	0.23	0.32
Fine Silt	0.45	0.22	18	0.15	0.30	0.23
Coarse Clay	0.36	0.285	0	0.14	0.28	0.15
Fine Clay	0.20	0.41	0.22	0.27	0.27	0.35

To determine sediment texture, percentage data are required from the results of grain size analysis in the form of percentages of gravel, sand, silt, and clay. The dominant texture at the foreshore station is muddy sand, whereas that at the beach station is dominated by slightly gravely sand (**Fig. 3** and **Fig. 4**). The difference in the sediment fraction size is related to the origin of the sediment source. This explains why sediments sourced from the sea experience transportation and deposition at each location. If the sediment source comes from the direction of the bay, its size will be finer. The coarser fraction size can be assumed to be the result of the wave transport process, which is carried to the coastal area [2]. In general, sediments in coastal areas are finer rocks, and the remains of plants are commonly referred to as thermogenic deposits [23].

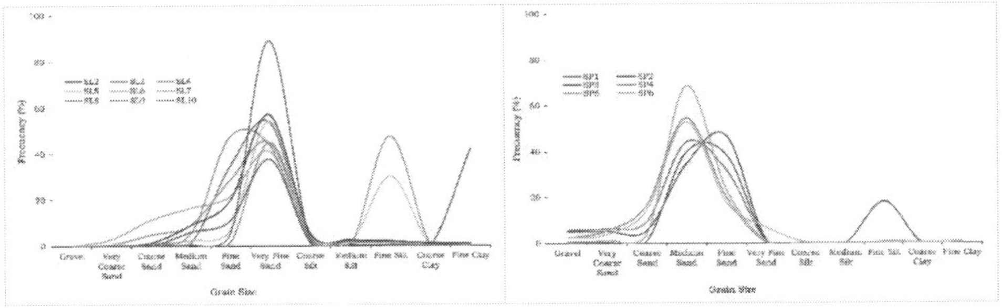

Fig. 3. Sediment particle types frequency of foreshore (left) and beach (right) at the coast of Carita Beach

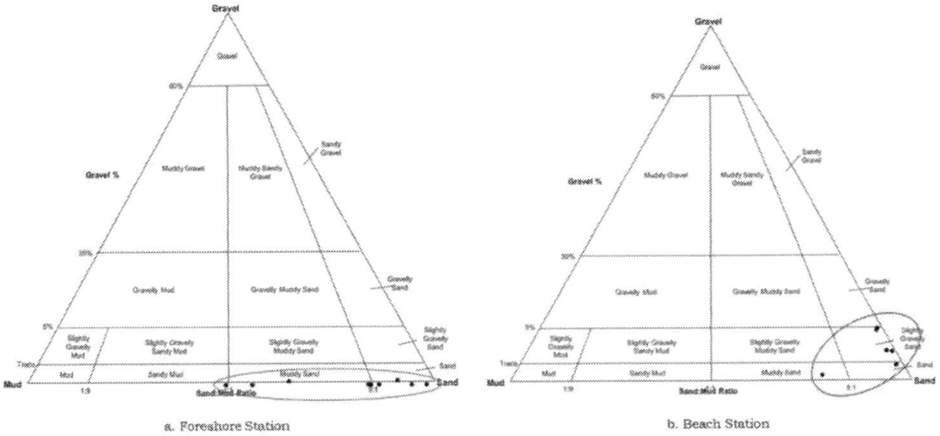

Fig. 4. Sediment aggregation of foreshore (a) and beach (b) at the coast of Carita Beach

3.2 Organic matter in sediments

The grain size composition is important, leading to potential interactions with the binding of organic matter (OM) [24]. The total OM in sediment is an essential indicator for assessing the quality of the aquatic environment [25]. OM describes the amount of organic material stored in sediment, which comes from the remains of living organisms and other organic materials, so that it can provide information about the level of pollution and biological activity that occurs in an aquatic ecosystem [26]. The results of the calculation of the total organic content at each sampling point are shown in **Fig. 5.** The organic content in the foreshore sediments ranged from low to moderate, whereas all beach sediments had very low values. The finer sediment grain size, the higher OM content [27]. The results were below the OM value from sediment of Angke Kapuk Protected Forest, Jakarta, with a value of 14.6% to 29.43% [28].

As the OM characteristics, the total organic carbon (TOC) value also shows a higher value in the foreshore sediment and lower in the beach sediment (**Fig. 6**). The highest TOC was at SL3 and the lowest was at SP1. The TOC of all sediments collected on the beach was either low or very low. In contrast, the characteristics of the foreshore sediments ranged from moderate to very high. The high TOC values at SL3 and SL4 (**Fig. 6**) were due to the high total OM value (**Fig. 5**). Increasing organic carbon content is positively correlated with increasing organic matter [29]. This is because the organic carbon content represents the fraction of organic matter that escapes remineralization during sedimentation [30]. The TOC from this study is still lower than from sediments of Muara Sungai Banger, Pekalongan, ranging from 5.1% to 20.46% [31].

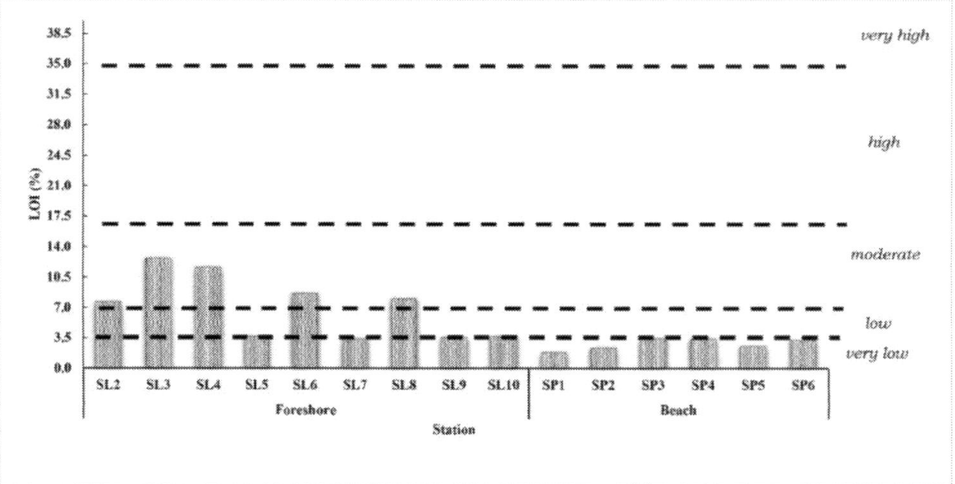

Fig. 5. Total organic matter (%) in the sediment collected from the beach and foreshore of Carita Beach.

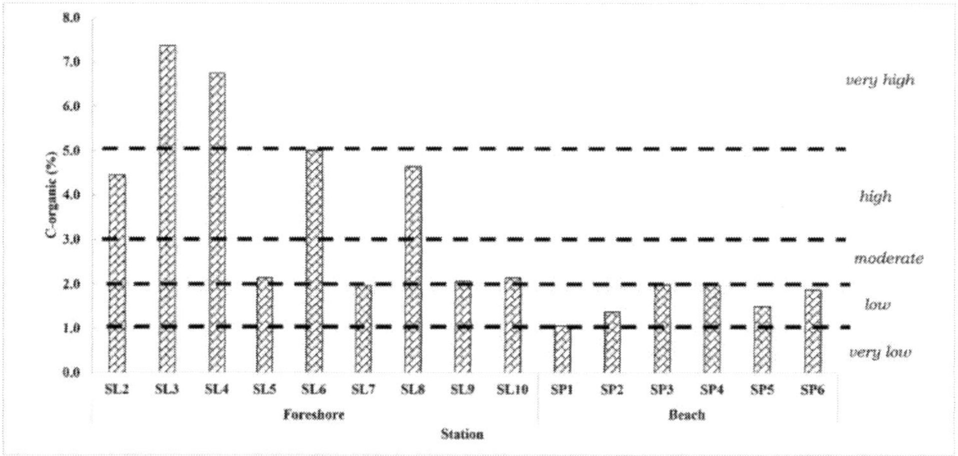

Fig. 6. Carbon organic content (%) in the sediment collected from the beach and foreshore at Carita Beach.

Regression analysis was used to determine the relationship between the OM content and sediment (mud fraction) (**Fig.7**). The closeness of the relationships between the variables can be observed from the determination value (R). The determination is negative if R is negative, no relationship if r is 0 (zero), and positive or very strong if R is plus. The closer the value is to 1, the stronger the relationship. The regression results showed that correlation between OM and mud was 41.46 % in the foreshore sediments and 21.99% in the beach sediments. The organic matter content increases with increasing mud content in the sediment [32, 33].

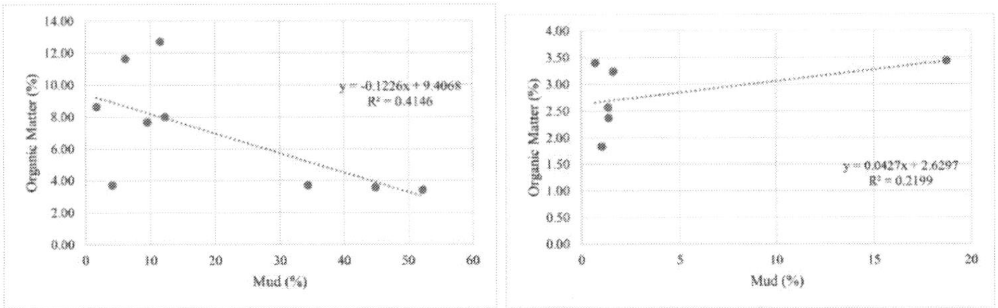

Fig. 7. Relation of sediment and mud of foreshore (left) and beach (right) at Carita Beach.

3.3 Statistic of sediments

Average Sediment Grain Size (Mean)

The average grain size of sediment on Carita Beach ranges from 25.83 – 362.0 µm. The sediment was classified as medium sand or very fine sand, and only a small amount was classified as fine sand. If a location is dominated by sediment dominated by coarse sediment grains, then this indicates that the strength of the flow that transports the sediment is quite large, whereas if it is dominated by fine sediment, it indicates the weak strength or energy that transports the sediment [15].

Sediment Grain Sorting (Sorting)

Sorting describes the level of uniformity of the sediment grains. A smaller sorting value indicates that the sediment is more well-sorted, so that the sediment grain size is more uniform. Based on the results of the analysis, most of the sediment in the Carita coastal waters is classified as poorly sorted or well-sorted. This shows that the strength of the current in these waters is unstable, meaning that under certain conditions, currents with high strength occur and can change at certain times [21].

Dominance of Sediment Grain Size (Skewness)

Skewness describes the dominant direction of the grain size of the population, perhaps symmetry, leaning towards coarse-grained sediments or fine-grained sediments [15]. Therefore, skewness can be used to determine sedimentation dynamics. Based on the results of the statistical analysis, bottom sediments in Carita coastal waters vary greatly, from symmetrical to very coarse skewed, and some have positive and negative values. This shows that the sediment underwent a transportation process and was deposited in the water area. This is in accordance with Nugroho and Basit [23], who stated that sediment transport in this area is influenced by fluctuating currents so that they are able to sort fine sediment grains of every size.

Sediment Grain Size Distribution (Kurtosis)

Most of the grain sizes of the bottom sediments in the study area are very fine sand. The gradation of the bottom sediments was classified as good (there were variations in grain size). The average sediment grain size is between 25.83 – 362.0 µm. All the sediments in the study area were classified as leptokurtic – platykurtic. This shows that there is a certain sediment grain size that dominates the sediment in the study area, namely the grain size of very fine sand sediment.

The statistical analysis of sediments is generally characterized by four parameters: mean, sorting, skewness, and kurtosis. To determine the grain size parameters, statistical calculations were performed, and the results are listed in **Table 3**. The results of the grain size parameter analysis show that Carita Beach is dominated by very fine sand - coarse sand, with a grain size range of 72.11µ - 378.7 µ, while at stations SL5, SL7, and SL9 Coarse Silt (coarse mud) is dominated with a grain size range of 19.17 µ- 10.67 µ. This indicates that the sediment underwent deposition. The deposition process is characterized by the deposition of fine coarse sediments [23].

The sediment sortation value at the Carita Beach location is moderately well-sorted and very poorly sorted (**Table 3**). This is because of the influence of currents in several water areas, such as small or large river estuaries, and the influence of activities and environmental conditions, which result in large sediment grains. The precipitate was then randomly mixed. The skewness value at different research locations varies, starting from the very fine skewed - Very Coarse Skewed category. SL5, SL9, and SPK3 were classified as Very Fine Skewed, with dominance in the range of skewness values (-0.640) – (-0.751). SL2, SL7, and SPK 5 are included in the fine-skewed category, with dominance in the skewness value range (-0.158) to (-0.240). SL3, SL6, and SPK 6 were included in the symmetrical classification category with dominance in the skewness value range (0.016) – (0.036). SL4, SL10, SPK1, and SPK4 are classified as Coarse Skewed, with dominance in the skewness value range of 0.116 – 0.264. Meanwhile, SPK2 is classified as very coarse skewed with a dominant skewness value of 0.334, which is included in the asymmetric to large-size category. This shows that the sediment underwent a transportation process and was deposited in that area. Based on the statistical calculations above, it can be shown that sediment transport in this area is influenced by fluctuating currents, so that it is able to sort every size of sediment grain from fine to muddy.

Table 3. Statistic characteristics of beach and foreshore sediments at the coast of Carita Beach

Sampling Point	Mean (µ)	Median (D_{50}) (µm)	Classification	Sorting	Classification	Skewness	Classification	Kurtosis	Classification
SL2	111.0	105.5	Very Fine Sand	2.033	Poorly Sorted	-0.158	Fine Skewed	1.704	Very Leptokurtic
SL3	123.0	113.3	Very Fine Sand	2.631	Poorly Sorted	-0.031	Symmetrical	1.880	Very Leptokurtic
SL4	168.4	136.6	Fine Sand	2.562	Poorly Sorted	0.264	Course Skewed	0.981	Mesokurtic
SL5	30.67	76.79	Coarse Silt	4.983	Very Poorly Sorted	-0.640	Very Fine Skewed	0.588	Very Platykurtic
SL6	131.7	133.0	Fine Sand	1.632	Moderately Sorted	0.036	Symmetrical	0.846	Platykurtic
SL7	19.17	25.83	Coarse Silt	4.573	Very Poorly Sorted	-0.240	Fine Skewed	0.495	Very Platykurtic
SL8	110.9	99.15	Very Fine Sand	2.460	Poorly Sorted	-0.016	Symmetrical	2.263	Very Platykurtic
SL9	16.43	69.16	Coarse Silt	11.00	Very Poorly Sorted	-0.658	Very Fine Skewed	0.487	Very Platykurtic
SL10	89.65	89.65	Very Fine Sand	1.306	Well Sorted	0.116	Coarse Skewed	0.960	Mesokurtic
SPK1	343.2	320.3	Medium Sand	2.168	Poorly Sorted	0.234	Coarse Skewed	1.284	Leptokurtic
SPK2	266.5	250.9	Medium Sand	2.038	Poorly Sorted	0.334	Very Coarse Skewed	1.285	Leptokurtic
SPK3	72.11	214.9	Very Fine Sand	6.749	Very Poorly Sorted	-0.715	Very Fine Skewed	2.405	Very Leptokurtic
SPK4	378.7	362.0	Medium Sand	1.918	Moderately Sorted	0.152	Coarse Skewed	1.478	Leptokurtic
SPK5	300.5	320.2	Medium Sand	1.558	Moderately Well Sorted	-0.194	Fine Skewed	1.167	Leptokurtic
SPK6	331.1	341.9	Medium Sand	2.082	Poorly Sorted	-0.047	Symmetrical	1.591	Very Leptokurtic

4 Conclusions

The results of the Carita Coastal sediment grain size analysis were dominated by very fine sand at the foreshore station and fine sand at the beach. These characteristics are supported by the lower water content and organic matter in coastal sediments compared to foreshore sediments. Statistically, the sediment distribution shows good sorting, high variation in skewness, and leptokurtic dominance.

This study is the final field trip output of the Chemical Oceanography course, Master of Marine Science Program, Dept. ITK-FPIK IPB University, in collaboration with LPSPL Serang, Ministry of Marine Affairs and Fisheries.

References

1. N. Bujan, C. Ronadh, M. Gerd, *Mar. Geol.* **417**, 106012 (2019) https://doi.org/10.1016/j.margeo.2019.106012
2. W.A. Gemilang, U.J. Wisha G.A. Rahmawan, R. Dhiauddin, J. Kel. Nas. **13**, 2: 65-74 (2018)
3. A.M. Yasin, E. Sukiyah, V. Isnaniawardhani, IJSR, 2139-7064p (2016)
4. Y. Wang, W. Ouyang, C. Lin, W. Zhu, A. Critto, M. Tysklind, X. Wang, M. He, B. Wang, H. Wu, *Environ. Sci. Technol.* **55**, 10: 6783-6790 (2021) https://doi.org/10.1021/acs.est.1c00135
5. S. Kim, D.S. Yang, Y.S. Kim, *Environ Monit Assess.* **192**, 502: 1-15 (2020) https://doi.org/10.1007/s10661-020-08475-z
6. J. Xu, G.D. Bland, Y. Gu, H. Ziaei, X. Xiao, A. Deonarine, D. Reible, P. Bireta, T.P. Hoelen, G.V. Lowry, *Environ. Sci. Technol.* **55**, 18: 12393–12402 (2021) https://doi.org/10.1021/acs.est.1c03572
7. [ISO] International Organization for Standarization, 5667-12 (2017)
8. [USACE] US Army Corps of Engineers, (Washington DC. Department of The Army, US Army Corps of Engineers, 2003)
9. B.A. Schumacher, US Environ. Protect. Agency, Environ. Sci. Div. Nat, Exposure Res. Lab. (2002)
10. A. Mudroch. *Manual of physico-chemical analysis of aquatic sediments 1st ed* (Routledge, 1997) https://doi.org/10.1201/9780203748176
11. Blott, S.J. and K. Pye. 2001, Earth. Surf. Processes. Landforms. **26,** 1237–1248 (2001)
12. T.D.T. Oyedotun, *Interdiscip. Environ. Rev,* **22**, 1: 43-66 (2022) https://doi.org/10.1504/IER.2022.122993
13. S. Sharma, D.N. Singh, *Mar. Georesour. Geotechnol.* **33,** 447-465 (2015) https://doi.org/10.1080/1064119X.2014.953232
14. F. Listiani, H.R. Mahardhika, N.A Prayogo, *Omni-Akuatika.* **14**, 20: 63-68 (2015)
15. G. Varga, G. Ujvari, J. Kovacs, *Quatenary. Intern.* **502**, 60-70 (2019) https://doi.org/10.1016/j.quaint.2017.09.021
16. Y. Ping, W. Houjie, W. Xiao, B.I. Naishuang, *J. Ocean. Univ. China.* **19**, 3: 589-600 (2020). https://doi.org/10.1007/s11802-020-4221-y
17. R. Akbarimansyah, F. Fauziyah, E.N. Ningsih, F. Agustriani, F. Supriyadi, A.A.P. Febrianti, *JITKT.* **15**, 2: 251-264 (2023)
18. I. Nuraini, D.B. Wiyanto, J. *Ilmiah Kel. Perik*, **2**, 3: 243-254 (2021)

19. R.R. Anggraini, U. Yanuhar, Y. Risjani, *JITKT.* **12**, 1: 235-246 (2020) http://doi.org/10.29244/jitkt.v12i1.28705
20. S. Schmutz, J. Sendzimir, (Vienna, BOKU – University of Natural Resources and Life Sciences, 2018)
21. M. Joensuu, C.A. Pilditch, R. Harris, S. Hietanen, H. Pettersson, A. Norkko, *Limnol. Oceano.* **63,** 173-186 (2018) https://doi.org/10.1002/lno.10622
22. S. Ouillon, *Water.* **10**, 390 (2018) https://doi.org/10.3390/w10040390
23. D.D. Truong, D.Q. Tri, N.C. Don, *Civil. Engine.* **7**, 10: 1634-1649 (2021) http://dx.doi.org/10.28991/cej-2021-03091749.
24. Q. Liu, P. Yang, Z. Hu, Q. Shu, Y. Chen, *Ecol. Indic.* **137**, 108716 (2022) https://doi.org/10.1016/j.ecolind.2022.108716
25. A. Baran, M. Mierzwa-Hersztek, K. Gondek, M. Tarnawski, M. Szara, O. Gorczyca, T. Koniarz, *Environ. Geochem. Health.* **41**. 2893-2910 (2019) https://doi.org/10.1007/s10653-019-00359-7
26. S. Dubois, N. Savoye, A. Grémare, M. Plus, K. Charlier, A. Beltoise, H. Blanchet, *J. Mar. Syst.* **94**, 64-73 (2012).
27. T. Prartono, N.M.N. Natih, A.S. Atmadipoera, S. Susanti, R. Afifah, D.S. Yolanda, Lestari, *BIO. Web. Conf.* **106**, 03006 (2024)
28. T. Herawati, D. C. Januaristy, *BIO. Web. Conf.* **74**, 05007 (2023)
29. A. A. Sidi, I. A. Okunlola, N. M. Waziri, *J. Geosci. Environ. Prot.* **11**, 10: 218-226 (2023)
30. A. Dianto, I. Ridwansyah, L. Subehi, *IOP. Conf. Ser: Earth. Environ. Sci.* **535**, 1: 012030 (2020)
31. M. F. Lazuardi, L. Maslukah, S. Widada, *Indones. J. Oceanogr.* **4**, 3: 86-95 (2022)
32. A. R. N. Kinasih, P. W. Purnomo, Ruswahyudi, *Manage. Aquat, Resour.* **4**, 3: 99-107 (2015)
33. O. Serrano, P. S. Lavery, C. M. Duarte, G. A. Kendrick, A. Calafat, P. H. York, A. Steven, P. I. Macreadie, *Biogeosci.* **13,** 4915-4926 (2016)

Microplastics in the digestive tract of sunu grouper *(Plectropomus leopardus* Lacepede, 1802) from coastal waters of North Maluku

Najamuddin [1,2*], *Abdurrachman* Baksir[1,2], *Halikuddin* Umasangaji[1,2], and *Erlena* Umanahu[1]

[1]Marine Science Study Program, Fisheries and Marine Science Faculty, Khairun University, Indonesia
[2]Magister of Marine Science Study Program, Graduate Studies Program, Khairun University, Indonesia

Abstract. Microplastic pollution threatens marine fish resources as it contaminates various types of fish. This study aimed to analyze the characteristics and abundance of microplastics in the digestive tract of sunu groupers. Data were collected at the Ternate Archipelago Fisheries Port and Ternate Hygienic Fish Market. The data collected included the fish's length and weight, stomach, and microplastics from 30 sunu groupers samples. The results of this study showed that all samples of sunu grouper tested contained microplastics in their digestive tract. Microplastics consist of fibers, films, and fragments, consisting of black, grey, red, white, green, brown, blue, purple, and yellow. The abundance of microplastics in the grouper's digestive tract ranges from to 6-32 particles/individual. The presence of microplastics in marine biota has a high potential to pose a threat to marine food sources. Therefore, it is very important to conduct a massive awareness campaign about the threat of microplastics to the environment and marine food sources so that mitigation efforts are urgent and carried out simultaneously and jointly by all parties.

1 Introduction

Sunu grouper (*Plectropomus leopardus*) is one of the fish that has high economic value and is a leading commodity in the fisheries sector is the sunu grouper (*Plectropomus leopardus*). Several studies have found the benefits of grouper fish, such as being able to reduce the risk of coronary heart disease, containing omega 3 with a content of around 200 mg, containing vitamin A, lowering high cholesterol levels, anti-inflammatory effect, preventing Alzheimer's disease, containing amino acids, vitamin B complex, taurine and selenium, increasing body immunity, caring for the digestive system and being an antioxidant [1]. However, marine fish resources, including grouper fish are currently threatened because they are contaminated by dangerous chemicals such as heavy metals and microplastics.

Microplastics are the result of fragmentation of macro-sized plastic waste through various processes, such as photolysis by ultra violet light, fragmentation by hydro-

* Corresponding author: najamuddin313@gmail.com

oceanographic activities such as waves and currents, or other activities. Fragmentation of macroplastics which are classified as microplastics, are plastic particles with a size of <5 mm and greater than 330 μm [2,3].

Microplastics consist of primary and secondary microplastics. Primary microplastics are microplastics that are directly produced or produced for certain purposes by processing raw and completely unprocessed polymer materials that are generally found in soap, detergent, cosmetics and clothing. Secondary microplastics originate from the fragmentation or decomposition of large plastics. Microplastics are very dangerous because they persist in the environment for a long period of time. The presence of microplastics in the environment is increasing because they enter the food chain and bioaccumulate in top predators, including humans.

The very small size of microplastic particles means that fish can ingest microplastics either directly or indirectly by eating zooplankton that have accumulated microplastics [4,5]. Microplastics ingested by fish translocate into the digestive tract and gills and enter the circulatory system. Microplastics that enter the blood circulation have the potential to enter fish organs [6,7]. When microplastic particles accumulate in large quantities in the fish body, they can block the fish's digestive tract [8].

Several research results in Indonesian waters have also proven the presence of microplastics in several types of marine fish such as [9-12]. The above study was carried out on the digestive tract of several types of marine fish such as mackerel, flying fish, herring, Carangidae rabbitfish, lemuru, snapper and grouper. Research conducted in various locations has proven the presence of microplastics in fish which are widely consumed by the community. What about the fish in the waters of Ternate Island that have also been contaminated with microplastics? To complete and strengthen the existing research result, it is necessary to carry out research regarding the microplastic content in the digestive tract of sunu groupers in Ternate City, which is one of the coastal cities where people are highly dependent on marine fish. The aim of this study was to analyze the content and abundance of microplastics in the digestive tract of the sunu grouper (*Plectropomus leopardus*).

2 Materials and methods

Samples of sunu grouper (*Plectropomus leopardus*) were collected from fishermen's catches at the Ternate Nusantara Fisheries Harbor (Station 1) and the Ternate Hygienic Fish Market (Station 2) each with 15 fish from each location so that the total samples collected were 30 fishes. The data collected included the length and weight of the fish, the length and weight of the fish digestive tract, and the characteristics of the microplastics. To avoid contamination from the surrounding environment, all equipment used was protected by cleansing with distilled water. Fish samples were dissected to determine the contents of their digestive tracts. Fish sample dissection was based on previously published guidelines [13]. Fish samples were dissected along the gastrointestinal tract from the anterior part (esophagus) to the posterior part (anus).

The microplastic analysis procedures were based on [14]. The weight and length of the digestive tract were determined, after the tract was removed from the body of the fish, placed in a test tube and oven-dried at 80 ºC for 24 h. The samples were dried placed in a solution of 2 ml 30% H_2O_2 and 6 g of NaCl/20 ml left to stand for 24 h and then incubated in a water bath at 80 ºC. This is the process of organic material destruction. The use of a 30% H_2O_2 solution eliminates a large amount of biogenic organic materials on the surface of microplastics (biofilms and macrofouling organisms), and the use of NaCl is recommended to separate microplastics from other materials. The sample was then filtered using a vacuum pump and Whatman filter paper number 42 with a size 2.5 μm. The type of microplastic was determined using a digital microscope connected to laptop integrated

 https://doi.org/10.1051/bioconf/202516801011

cooling tech software with 100-fold magnification.

Types of microplastics are plastic fragments (compact color), pellets (compact color), filaments/fibers (compact color), and plastic films (transparent color). The process of observing microplastics has been standardized to avoid selection errors. The characteristics of the microplastics used as standards are 1) no visible cellular or organic structure; 2) the fibers are of the same thickness throughout their entire length and must not taper at the ends; 3) colored particles have a homogeneous color; 4) the fibers are not segmented, or appear to be bent on the ribbon; and 5) the particles are not shiny [14]. Most plastic pieces are flexible and do not break when produced. Plastic pieces often bounce or spring when produced. If a piece broke when touched, it was not counted as plastic. The concentration of microplastics in the fish samples was measured as particles per individual.

The distribution of microplastics in the digestive tract of grouper fish at the two stations was analyzed using one-way ANOVA and the relationship between fish biometrics consisting of fish length and weight and fish digestive tract length-weight with the abundance of microplastics in the digestive tract of sunu grouper fish at the two stations used correlation analysis.

3 Results and discussion

3.1 Biometrics of sunu grouper (*Plectropomus leopardus)*

The results of biometric measurements of 30 samples of sunu grouper (*Plectropomus leopardus*) consisting of the length and weight of the fish, and the length-weight of the fish's digestive tract are presented in Fig 1 and 2. The results of biometric measurements of the sunu grouper at the first location of the Nusantara Fisheries Port have a length ranging from 29.5 cm – 50.3 cm with an average of 41.9 cm, a weight ranges between 1.12 kg – 2.14 kg with an average of 1.4 kg, a length of the digestive tract ranging between 4 cm – 10.5 cm with an average of 7.0 cm and a weight of the digestive tract ranges from 2.93 g – 47.01 g with an average of 12.8 g.

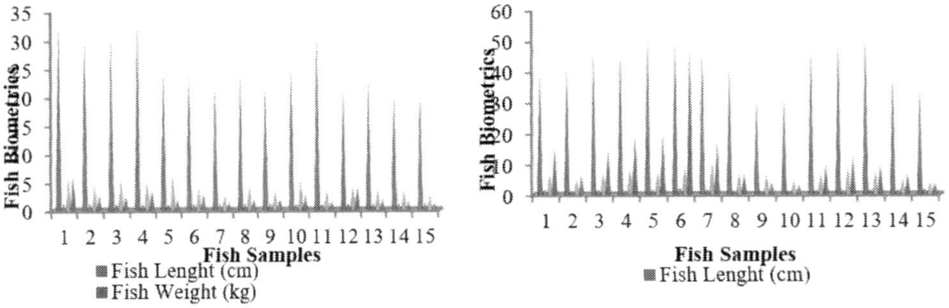

Fig 1. Biometric samples of sunu grouper (*Plectropomus leopardus*) at Station 1 (left side) and Station 2 (right side)

The results of this study are similar to [15], who found that the length range of sunu grouper (*Plectropomus leopardus*) caught in the waters around the Spermonde Islands, South Sulawesi ranged from 15.0-83.0 cm with a weight ranging from 100-4,610 g. Likewise [16] stated that the average length of sunu grouper is around 40 cm. Several studies have found that the length of the fish's digestive tract ranges from <1 cm to 20 times its body length. The highest length of the fish digestive tract generally occurs in herbivorous and detrivorous species, whereas the lowest is found in carnivorous and predatory species [17].

https://doi.org/10.1051/bioconf/202516801011

The dimensions of the size of the fish digestive tract are important for analysis as a morphological indicator of trophic levels in the ecological system. Several factors influence the dimensions of the fish's digestive tract, including food factors, fish size (mass and length), fish body shape, feeding history, ontogeny and phylogeny [18].

3.2 Characteristics and abundance of microplastics in the digestive tract of sunu grouper (*Plectropomus leopardus)*

There are three types of microplastics found in the digestive tract of sunu grouper consist of 3 types, namely fiber, film and fragments (Fig 2). The microplastics found in the digestive tract of sunu grouper consist of nine colors, namely black, grey, red, white, green, brown, blue, purple and yellow. The diverse colors of microplastics reflect the diverse of sources of plastic pollution in the environment [19]. The percentage of microplastic types and microplastic color found in the digestive tract of sunu grouper fish at the two sampling locations are presented in Fig 3 and 4.

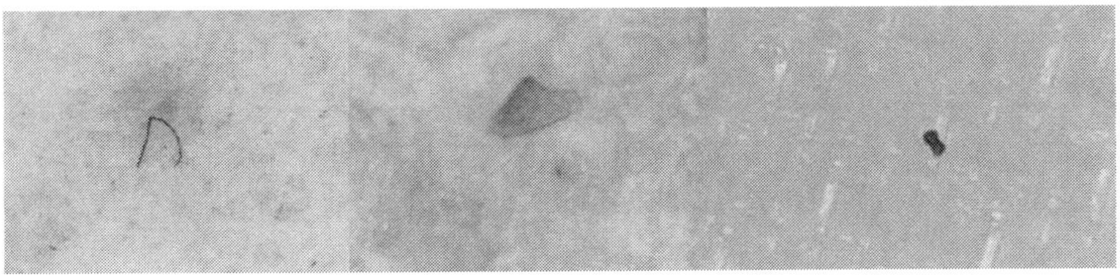

Fig 2. Samples of microplastics found in the digestive tract of sunu grouper fish (a) fiber, (b) film, (c) fragment

The types of microplastics include plastic fragments (compact color), plastic fibers (compact color), and plastic films (transparent color). The fragments are microplastic shaped like fragments which generally come from the fragmentation of hard plastic such as plastic food boxes, plastic bottles and other hard plastic products. The fibers are microplastic-shaped like threads and generally come from the fragmentation of fabric, clothing, nets, ropes, sacks and textile waste. The films are microplastic shaped like a thin sheet, which generally comes from the fragmentation of thin plastic materials, such as plastic packaging and plastic bags.

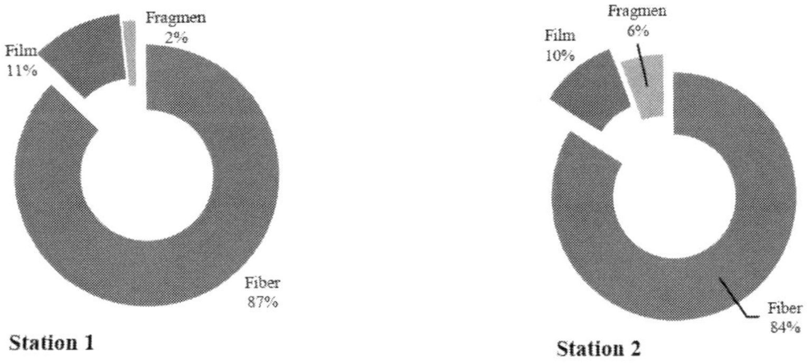

Fig 3. Percentage of microplastics in the digestive tract of sunu grouper

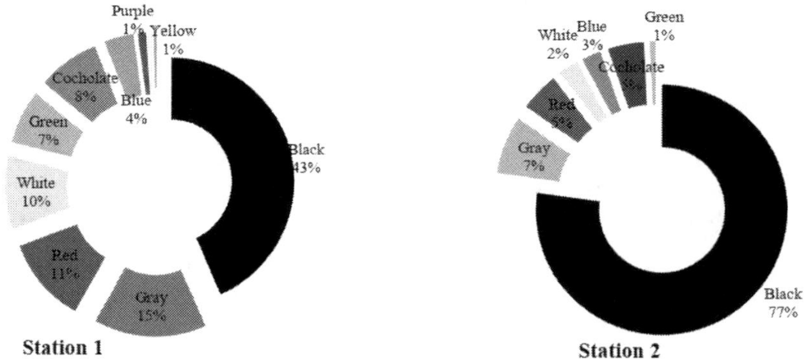

Fig 4. Percentage of microplastic colors in the digestive tract of sunu grouper

The dominant color of the microplastics was black that is 43% at sampling location one and 77% at sampling location 2. The results of this study are similar to [16] in research conducted in Tuban Sea Waters, East Java which found black microplastics as the dominant color in pelagic fish at (82%) and in demersal fish at (69%). Similarly, [20,21] reported that the color variations of microplastic particles found in marine biota were dominated by black (45%). Research by [22] also stated that the accumulation of microplastics in seafood was dominated by black (70%).

The dominant black color was found because it is similar to the color of fish food and some common prey [23]. The ingestion of colorful microplastics by fish is considered food by marine organisms because of their limited ability to recognize food selectivity [24]. Likewise [6,25,26] suggested that microplastics with various attractive colors might be ingested by biota because of their attractiveness and similarity to real prey or food. Meanwhile, the very small size of microplastics and their high buoyancy mean that microplastics are always available in the water column so that the potential for ingestion by demersal and pelagic fish is very high.

The color characteristics of microplastics have also been used as an index of photodegradation and the length of time the plastic remains on the sea [27]. The black color found in microplastics can change because of weathering and transportation in water. When microplastics are in the environment, they undergo degradation simultaneously. This mechanism does not only depend on environmental conditions, but is also influenced by the physical and chemical characteristics of the polymer materials that make up microplastics [28]. Black microplastics are predominantly found in the gastrointestinal tract of fish and in the environment, indicating that these microplastics have been in the tissues and environment for a long time, thereby absorbing a number of pollutants and causing the color to change to black [29,30].

The presence of other colors such as yellow, blue, and red indicates that the original color of the microplastic source and the change in the presence of these colors is due to polyethylene polymer compounds that change color because they contain or have absorbed PCB compounds. The color change process is an indication of long exposure times in seawater, resulting in the polymer oxidation [27].

The results of the microplastics abundance analysis in the digestive tract of sunu grouper at the two sampling locations are presented in Figure 5. The abundance of microplastics at Station 1 of the Nusantara Fisheries Harbor ranged between 6-32 particles/individual and at Station 2 of the Ternate Hygienic Fish Market ranged from 11-29 to individual. The abundance of microplastics obtained in this study was similar to that of [31] of 14-20 particles/ individual in the digestive tract of pelagic fish in the southwest

Plymouth Beach.

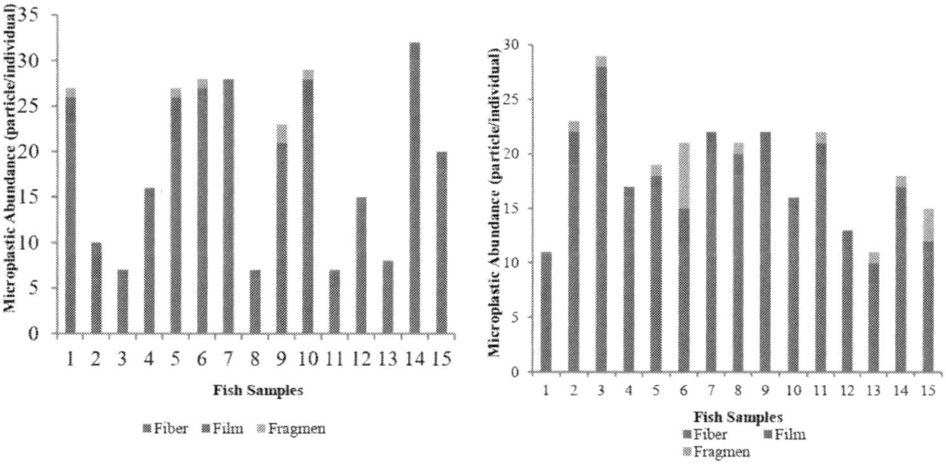

Fig 5. Abundance of microplastics in the digestive tract of sunu grouper at Station 1(left side) and Station 2 (right side).

The types of microplastics found in this studi also strengthen the results of previous research such as [16], which found three types of microplastics, namely fibers, fragments, and pellets in the gastrointestinal tract of pelagic and demersal fish in the waters of the Tuban Sea. The dominant types of microplastics were fiber (74%) in pelagic fish and fiber (84%) in demersal fish. Several other studies such as [22, 31-36] also reported the fiber form as the most dominant microplastic in the gastrointestinal tract of fish samples.

The types of microplastic particles found in this study also strengthen the results of previous research such as [16], which found three types of microplastics, namely fiber, fragment and pellet, in the gastrointestinal tract of pelagic and demersal fish in the waters of the Tuban Sea with the dominant type of microplastic being fine fiber in the pelagic and demersal fish. Several other studies have also reported fiber-type microplastics to be the most dominant microplastics found in the gastrointestinal tract.

The main source of fiber microplastics is household waste as a result of washing clothes where these products are made from polyethylene and polyamide polymers. The fabric washing process can release 1900-700,000 fibers per 6 kg into the environment [37-41]. In addition, microplastic fibers can also come from nets or ropes containing polyamide polymers which are often used in fishing gear by the fishing industry [42-44].

Fiber type microplastics are often found in the gastrointestinal tract of pelagic fish because of their thin shape and size and many float on the surface of water [3,4]. In addition, fibers released from wastewater treatment plant drainage pipes can enter the water surface and eventually accumulate in sediment
[26] and fiber microplastics have also been found in demersal fish species. Fiber type microplastics are generally made from acrylic, polyester or polyamide polymers.

In addition, the presence of microplastic fragments in the gastrointestinal tract of fish is secondary microplastic originating from the degradation of larger plastic debris through natural weathering processes in accordance with [45,46]. Fragment types are secondary microplastics whose presence in the environment and biota is closely related to the population size of an area [47,48]. The city of Ternate is a coastal city where most people carry out activities in coastal areas so that it has the potential to produce macro and microplastic waste.

Fragments type microplastic generally consist of polyethylene, which is commonly used in consumer products (e.g., plastic bags, bottles, hats, and containers [6]. In addition, they can originate from daily use of food packaging materials, containers, and other items is also the main source of fragments [49].

Pellet microplastics are primary microplastics whose presence in the environment and biota is direct waste from microplastic products and not the result of fragmentation of macroplastics. However, if the shape is round and the edges are sharp and irregular, macroplastic fragmentation is indicated. In demersal species, microplastics exist in the form of pellets, which are generally considered primary microplastics [50]. Pellets as a primary source are usually directly produced by raw material factories for making plastic products [51], such as industrial applications including medicine and the medical device plastic industry which are produced on a micro scale including those used to form larger plastic products (resin pellets pre-production) [27].

The type of microplastics in the marine environment and biota depends on the source and fragmentation process of the microplastics as well as the length of time the microplastics remain in the environment. Variations in the type of microplastics are the result of visualization of the source and fragmentation process of macroplastics which then enter the digestive tract because they are preyed upon by fish owing to their shape resembling prey [52,53].

The microplastic fragmentation process can be the result of degradation or erosion on the surface of the particles resulting from biological processes, photodegradation, chemical weathering, or physical influences (wave action, wind, weather and ultraviolet radiation). This result in various forms of microplastic particles [40].

[26,27] reported that the type of plastic polymer most often found in marine environments and fish organisms in the sea is polyethylene, which can reach 88.4%. Meanwhile [3,4,20,21], polyethylene polymer is most widely produced by the plastics industry because it is used on a large scale such as daily household needs, in cosmetics, detergents, cleaning and maintenance products, packaging, shopping bags, food containers and drinking water bottles [54]. Polyethylen is the most dangerous plastic polymer in aquatic environment because of its durability and is mostly used in packaging [55]. Polyethylene is a low density polymer often found in the digestive tract of pelagic fish. Polyethylene production is high because it has a density below one, so there is a high probability that this type of polymer will be abundant [56].

The distribution of microplastics in the digestive tract of fish is strongly influenced by suitable habitat conditions [7]. In addition, there are other factors such as fish eating patterns and habits, sources of microplastics, and oceanographic dynamics. For distribution of microplastics in the digestive tract of grouper fish in both locations, the results of one-way ANOVA showed an F count of 2.079, F table of 4.20, and a significant value of 0.101. Because the significant value obtained was greater than 0.05 and the F count $<$ F table, it can be concluded that the abundance of microplastics in the digestive tract of sunu grouper fish at the two locations was not significantly different.

3.3 Relationship between fish biometrics and the abundance of microplastics in the digestive tract of sunu grouper (*Plectropomus leopardus*)

The results of the analysis of the relationship between fish biometrics consisting of fish length and weight and fish digestive tract length weight with the abundance of microplastics in the digestive tract of sunu grouper fish at two sampling locations are presented in Fig. 6 and 7.

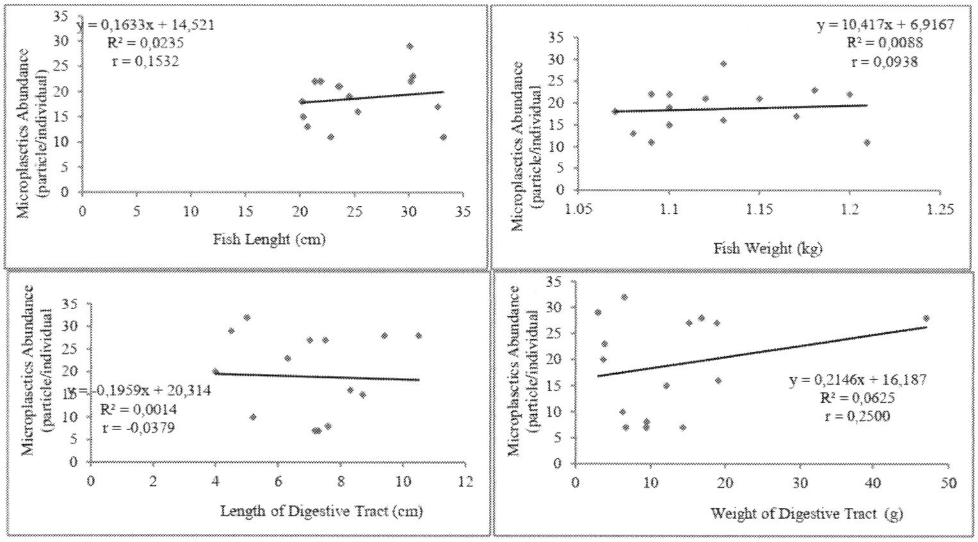

Fig 6. Correlation between microplastic abundance and fish biometrics at Station 1.

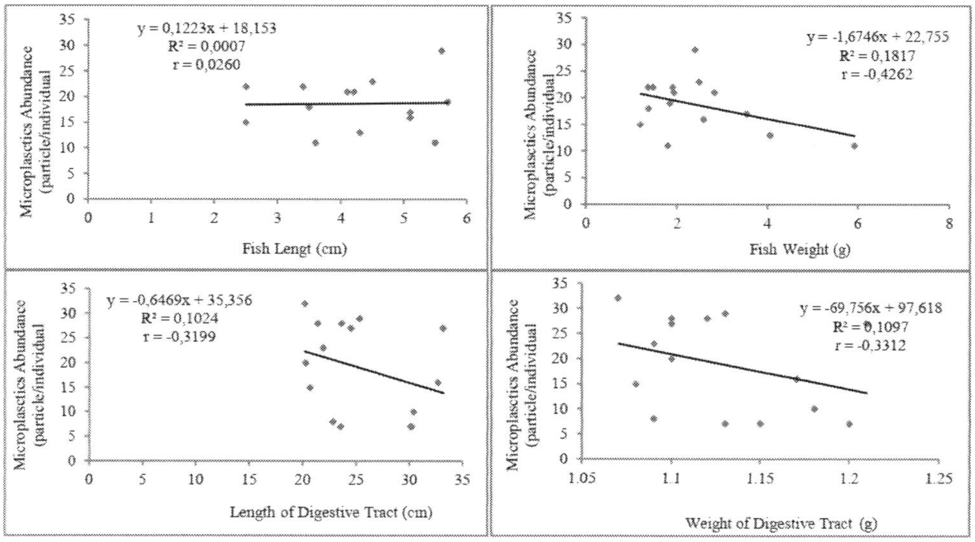

Fig 7. Correlation between microplastic abundance and fish biometrics at Station 2

The results of the analysis of the relationship between fish biometrics and the abundance of microplastics in the digestive tract of sunu grouper fish in general at sampling location one showed a positive relationship, meaning that if there was an increase in the biometric size of the fish, the abundance of microplastics also increased. However, based on the regression coefficient value which was <0.25, there was a very weak correlation. At the second sampling location, a negative correlation was obtained, which means that if there is an increase in the biometric size of the fish, there will be a decrease in the abundance of microplastics in the digestive tract of the sunu grouper. The correlation coefficient was <0.4 indicates a weak correlation. This shows that other factors that influence the abundance of microplastics in the fish digestive tract have a greater influence than fish biometrics. The influencing factors were feeding patterns, habits, habitat conditions, and oceanographic dynamics.

4 Conclusion

The presence of microplastics in marine biota has a high potential to pose a threat to marine food sources. All sunu grouper samples tested contained microplastics found in their digestive tracts. The microplastics found consisted of three types, namely fiber, film and fragments and consisted of nine colors, namely black, gray, red, white, green, brown, blue, purple and yellow. The abundance of microplastics in the grouper's digestive tract ranges from to 6-32 particles/individual. Analysis of the relationship between fish biometrics and abundance of microplastics in the fish digestive tract showed a very weak relationship. The recommendation suggested from this research is the importance of a massive awareness campaign about the threat of microplastics to the environment and marine food sources so that mitigation efforts are urgent and carried out simultaneously and jointly by all parties.

References

1. Ernaningsih, Grouper fish resources management. Al Zikra Press. Jogjakarta-Makassar (2018).
2. F.R. Storck, S.A.E. Kools, S. Rinck-Pfeiffer, Microplastic in fresh water resources. Science Brief. Global Water Research Coalition, Australia, (2015).
3. GESAMP (Group of Experts on the Scientific Aspects of Marine Environmental Protection), Sources, fate and effects of microplastic in the marine ocean : Part 2 of a global assessment. A report to inform the Second United Nations Environment Assembly GESAMP Working Group 40 2nd phase (2015).
4. M. Cole, P. Lindeque, C. Halsband, and T.S. Galloway, Mar. Pollut. Bulletin, *62*, 2588–2597 (2011).
5. O. Guven, K. Gokdag, B. Jovanovic, A.E. Kideys, Environ. Pollut. 223: 1-9 (2017).
6. S.L. Wright, R.C. Thompson, T.S. Galloway, Environ. Pollut. 178, 483– 492 (2013).
7. B.M. Rao, Indian Journal Fish 66 (1): 142-152 (2019).
8. C. Chih-Feng, J. Yun-Ru, Lima YC, C. Chiu-Wen, D. Cheng-Di, Sci. of the Total Environ. 781 (2021).
9. C.M. Rochman, The complex mixture, fate and toxicity of chemicals associated with plastic debris in the marine environment. In: Bergmann, M., Gutow, L. and Klages, M., Eds., Marine anthropogenic litter, Springer, Cham, 117-140 (2015).
10. N.Y. Ramadhani, B.M. Wibawa and T. Laksamana, Jurnal Sains and Seni. 6(1): 22-26 (2017).
11. C.I. Yudhantari, I.G. Hendrawan, and N.L.P.R. Puspitha, Journal of Marine Research and Technology 2(2): 48-52 (2018).
12. S.W. Sandra, and A.D. Radityaningrum, Journal of Environmental Science, 19(3), 638-648. (2021).
13. The Civic Laboratory for Environmental Action Research, Marine Plastics in Fish a citizen science dissection & analysis protocol. http://civiclaboratory.nl
14. N.R. Buwono, Y. Risjani, A. Soegianto, Emerging Contaminants 7 (2021): 172-178.
15. Ernaningsih, Budimawan, N. Nessa and Sudirman, Aust. J. Basic & Appl. Sci., 8(18): 438-444 (2014).
16. Fauziyah, Characteristics of microplastics in the gastrointestinal tract pelagic and demersal fish in Tuban Sea Waters, Java East. Thesis. Institut Teknologi Sepuluh

Nopember, Indonesia (2020).

17. K.A. Ray, and E. Ringo, The gastrointestinal tract of fish in aquaculture nutrition: Gut health, probiotics and prebiotics, first edition. West Sussex : John Wiley & Sons, Ltd. (2014).
18. J.M. Wilson, and L.F.C. Castro, Fish Physiology. 30 (2011). India: Academic Press.
19. J. Bellas, J. Martínez-Armental, A. Martínez- Cámara, V. Besada, and C. Martínez-Gómez, Marine Pollution Bulletin. 109 (1), 55-60 (2016).
20. C. Lopes, J. Raimundo, M. Caetano, and S. Garrido, Limnology and Oceanography Letters. 5 (1): 103 112 (2020).
21. A.L. Lusher, M. Mchugh, and R.C. Thompson, Marine Pollution Bulletin. 67 (1-2): 94- 99 (2013).
22. R. Akhbarizadeh, F. Moore, and B. Keshavarzi, Food Additives and Contaminants: Part A. 36 (11): 1696-1708 (2019).
23. N.C. Ory, C. Gallardo, M. Lenz, and M. Thiel, Environmental Pollution. 240: 566-573 (2018).
24. F. Murphy, M. Russell, C. Ewins, and B. Quinn, Marine Pollution Bulletin. 122 (1-2): 353-359 (2017).
25. L. Van-Cauwenberghe, M. Claessens, M. Vandegehuchte, and C.R. Janssen, Environmental Pollution, 199 (2015):10- 17.
26. M. Firdaus, Y. Trihadiningrum, and P. Lestari, Marine Pollution Bulletin. 150 (2020): 1-9.
27. L. Gallitelli, P. Girard, U. Andriolo, C. Roebroek, M. Liro, G. Suaria, Science of the Total Environment, 2024.
28. Vera N. de Ruijtera, Paula E. Redondo-Hasselerharmb, Koelmans AA, Environmental Pollution, 2025.
29. H. Hiwari, P.P. Noir, N.I. Yudin, P.S.Y. Lintang, and G.M. Putri, National Seminar Proceeding, Indonesian Biodiversity Society. 5 (2): 165- 171 (2019).
30. N. Suwartiningsih, I. Setyowati, and R. Astuti, Jurnal Biodjati. 5 (1): 33-49 (2020).
31. Z. Yang, A. Isobe , J. Zhang, V. Viyakarn, H. Arakawa, Marine Pollution Bulletin 212 (2025).
32. K. Jabeen, L. Su, J. Li, D. Yang, C. Tong, J. Mu, and H. Shi, Environmental Pollution. 221 (2016): 1-9.
33. S.M.O. Azad, P. Towatana, S. Pradit, B.G. Patricia, and H.T. Hue, International Journal Technology. Vol. 14 (7): 1017-1032 (2018).
34. M.R. Cordova, A.I.S. Purwiyanto, and Y. Suteja, Marine pollution bulletin. Vol. 142 (2019): 183-188.
35. S. Karbalaei, A. Golieskardi, H.B. Hamzah, S. Abdulwahid, P. Hanachi, T.R. Walker, and A. Karami, Marine pollution bulletin. Vol. 148(2019):5-15.
36. W. Klangnurak, and S. Chunniyom, Environmental Science and Pollution Research International. 2020.
37. P.M. Silva, and M.A. Nanny, Water. 12 (4): 1-16. 2020.
38. E. Hernandez, B. Nowack, D.M. Mitrano, Environ. Sci. Technol. 51 (2017): 7036 7046.
39. X. Peng, M. Chen, S. Chen, S. Dasgupta, H. Xu, K. Ta, M. Du, J. Li, Z. Gio, and S. Bai, Geochemical Perspectives Letters. 9 (2018): 1-5.

40. X. Wen, C. Du, P. Xu, G. Zeng, D. Huang, L. Yin, Q. Yin, L. Hu, J. Wan, J. Zhang, S. Tan, and R. Deng, Marine pollution bulletin. 136 (2018): 414-423.
41. I.E. Napper, A. Bakir, S.J. Rowland, and R.C. Thompson, Marine Pollution Bulletin, 99(1–2), 178–185 (2015).
42. S.E. Nelms, T.S. Galloway, B.J. Godley, D. Jarvis, and P.K. Lindeque, Environmental Pollution. 238 (2018): 999-1007.
43. E. Hansen, N.H. Nilsson, D. Lithner, C. Lassen, Hazardous substances in plastic materials. Denmark: COWI & Danish Technological Institute. 2013.
44. M.R. Ismail, M.W. Lewaru, and D.J. Prihadi, An International Sientific Journal. 23 (2019): 173- 181.
45. M.A. Asadi, Y.A. Ritonga, D. Yona, A. Hertika, Nature Environment and Pollution Technology. 18 (4): 1169-1176. 2019.
46. K. James, K. Vasant, S. Padua, V. Gopinath, K.S. Abilash, R. Jeyabaskaran, A. Babu, and S. John, Marine Pollution Bulletin. 154 (2020): 1-12 (111027)
47. M.A. Browne, J.N. Stewart, S.G. Tamara, J.R. Steve, and C.T. Richard, Current Biology. 23 (2013): 2388-2392.
48. J.P.W. Desforges, M. Galbraith, N. Dangerfield, and P.S. Ross, Marine pollution bulletin. 79 (1-2): 94-99. 2014.
49. W. Wang, and W. Jung, Trends in analytical chemistry. 108 (2018): 195-202.
50. L. Zhu, H. Wang, B. Chen, X. Sun, K. Qu, and B. Xia, Science of The Total Environment. 677 (2019): 493-501.
51. J. Kingfisher, Micro-plastic debris accumulation on puget sound beaches. Port Townsend Marine Science Center. 2011
52. H.S. Carson, M.S. Nerheim, K.A. Carroll, and M. Eriksen, Marine Pollution Bulletin. 75 (1-2): 126-132. 2013.
53. A. Rodriguez-Seijo, and R. Pereira, Comprehensive analytical chemistry. 75 (2017).
54. R.A. Naik, Rowles III, L.S. Hossain, A.I. Yen, M. Aldossary, R.M. Apul, O.G.J. Conkle, and N.B. Saleh, Sci. of the Tot. Environ. 139690 (2020).
55. E.G. Cassola, V. Zadjelovic, M.I. Gibson, and J.A. Christie-Oleza, Journalof hazardous materials. Vol. 369 (2019): 691-698.
56. C. Lassen, S.F. Hansen, K. Magnusson, N.B. Hartmann, P.R. Jensen, T.G. Nielsen, and A. Brinch, Microplastics: occurrence, effects and sources of releases to the environment in Denmark. Denmark: The Danish Environmental Protection Agency. 2015.

The pollution status of Banten Bay waters based on physical and chemical parameters using CCME index

Ameliani Wardania Putri[1], *Sulistiono* Sulistiono[1*], *Ali* Mashar[1], and *Aliati* Iswantari[1]

[1]Aquatic Resources Management Department, Faculty of Fisheries and Marine Science, IPB University, Bogor, Indonesia, 16680

Abstract. Banten Bay are located in northern part of Cilegon City, Banten Province. The water quality of Banten Bay is influenced by both environmental and anthropogenic factors, which if not properly managed can lead to pollution. This study aimed to assess the pollution status of Banten Bay waters using the Canadian Council of Ministers of the Environment (CCME) Water Quality Index, both temporally and spatially. The observed parameters were temperature, turbidity, TSS, pH, dissolved oxygen, ammonia, nitrate, nitrite, and total phosphate. The results indicate that most parameters are within the seawater quality standards for marine life, with the CCME index classifying the pollution status as fairly good. However, nutrient enrichment from land-based activities and sediment input from river discharge have been identified as key factors influencing water quality. Potential pollution sources include domestic wastewater, agricultural runoff, and industrial effluents, which, if unmanaged, can deteriorate the water quality. This study highlights the need for sustainable waste management practices to maintain the ecological health of the bay.

1 Introduction

The Banten Bay, northern part of Cilegon City, Banten Province. Banten Bay has an area of 150 km^2 and relatively shallow water depth of approximately 7 m [1]. Various activities such as urban areas, fish farming, shrimp farming, sea transportation, recreation, and industry have been developed [2]. These activities (anthropogenic) can produce a variety of both liquid and solid waste discharges into the marine environment, causing a decrease in water quality and possible water pollution [3], due to appropriate environmental management [4]. Water pollution describes a decrease in water quality as a result of input from organisms, components, and materials which act as pollutant and results in water defunction [5].

Water quality is an indicator used to assess the extent of management measures [6]. The assessment includes several physical parameters (transparency, turbidity, temperature, total suspended solid, and total dissolved solid), chemical parameters (pH, salinity, dissolved oxygen, biochemical oxygen demand, chemical oxygen demand, ammonia, nitrate, nitrite, and total phosphate), and biological parameters (total coliform and plankton) [7]. Other

* Corresponding author: onosulistiono@gmail.com

factors affecting water quality include ocean currents, rainfall, and anthropogenic activities[8].

The pollution status of Banten Bay has been determined mostly in coastal areas such as the waters of Bojonegara Beach [9], Cengkok Beach [10], Kronjo coast of Tangerang Regency [11], water from Tanjung Pasir Coast [12], and water of Karangantu Coast [13]. Research related to the status of water pollution in the middle of the sea in the waters of Banten Bay in determining the status of water pollution has not been carried out. This study aimed to investigate the pollution status of Banten Bay based on physical and chemical parameters using the CCME WQI method temporally and spatially.

2 Materials and methods

2.1 Time and location

The research was carried out for seven months from July 2022 to January 2023 in the waters of Banten Bay, Banten Province. Observations and sampling of water samples were carried out at six locations representing the area starting from the outermost line of Karangantu Beach, Banten Province, to the waters of Panjang Island, Banten Province. Pollution status monitoring was conducted on different dates and months based on the season. A map of the research locations for the Banten Bay waters is presented in Figure 1.

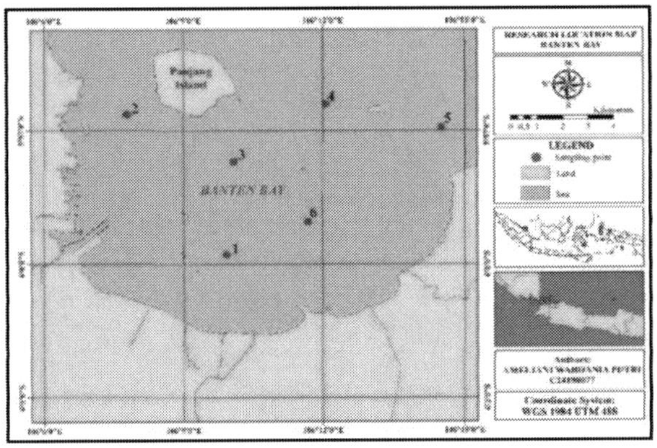

Fig. 1. Map of research locations in the waters of Banten Bay, Banten Province.

2.2 Data Collection

No ethics approval required in collecting data, since water quality data was collected directly from water in the field using purposive sampling to determine the location points for sampling. The parameters observed in situ were dissolved oxygen, pH and temperature were measured in situ. The samples were collected from each location using a vertical water sampler. The water sample was placed in a sample plastic bottle (250 ml in size) without preservatives and a sample plastic bottle (100 ml in size) with sulfuric acid (H_2SO_4) preservative was taken. Both types of plastic bottles are marked by its sampling point code and then put into a coolbox. Then, the water samples are taken to laboratory. In the meantime, water samples were put into a refrigerator so that the water content in the sample plastic bottle does not change before the analysis is done.

Unpreserved samples 250 ml were used to measure turbidity, Total Suspended Solid (TSS), nitrite, and total phosphate parameters, while 100 ml H_2SO_4 preserved samples were used to measure ammonia and nitrate parameters in the laboratory for ex situ observation using the APHA standard [14]. The samples were analyzed at the Macro Biology Laboratory, Department of Aquatic Resources Management and Environmental Laboratory, Department of Aquaculture, Faculty of Fisheries and Marine Sciences, Bogor Agricultural University.

Secondary data collection was carried out using a licensing procedure from IPB University to Meteorology, Climatology and Geophysics Agency (BMKG). The data collected was rainfall data and ocean currents from July 2022 to January 2023 from six research locations based on the Meteorology, Climatology and Geophysics Agency (BMKG) at Serang Station, Banten Province. The water quality parameters are listed in Table 1.

Table 1. Water quality parameters along with measuring instruments and analytic methods used.

Parameter	Units	Measuring instruments	Analytic methods
Physics			
Turbidity	NTU	Turbidity meter	Nephelometric method
Temperature	°C	Thermometer	Visual
Total Suspended Solid	mg/L	Balance sheet	Gravimetry
Chemical			
pH	-	pH meter	Electrometric method
Dissolved Oxygen	mg/L	DO meter	Visual
Ammonia (NH_3)	mg/L	Spectrophotometry	Phenate method
Total Phospate	mg/L	Spectrophotometry	Ascorbic acid method
Nitrate (NO_3)	mg/L	Spectrophotometry	Cadmium reduction method
Nitrite (NO_2)	mg/L	Spectrophotometry	Sulfanilamide method

2.3 Data Analysis

The data obtained were compared descriptively to those of seawater quality standards for Indonesia's marine biota using PPRI No.22 of 2021 in Appendix VIII [15]. The Canadian Council of Ministers of the Environment Water Quality Index (WQI) was used to determine the status of pollution in Banten Bay waters. This method was used to compare the water quality between the physical and chemical parameters of the waters and their pollution status in the research results. The pollution status of the Banten Bay waters was assessed using the CCME WQI method according to [16]. Data were processed using Microsoft Excel.

The Canadian Council of Minister of the Environment Water Quality Index (CCME WQI) method is one of the most effective methods for determining water quality status. It was created by the British Columbia Ministry of Environment, Lands and Parks and then developed through the Alberta Environment. In the CCME method, the determination of the types of parameters, quality standards, and time periods depends on the actual conditions that occur in each water area. This method evaluates the dynamics of changes in water quality at certain locations based on time-series data and can compare the overall water quality index at each location using water quality standards and the same variables [17].

The CCME WQI calculation consists of several formulas and variables that determine the status of water pollution in an area. These formulas and variables use values F1, F2, and F3. The value of each of these calculations has a different function, as explained below.

1. F1 (*Scope*) functions shows the percentage of parameters that do not meet quality standards (a parameter does not meet standards if at least in one test it does not reach the quality standard).

$$F_1 = \left(\frac{Number\ of\ failed\ parameters}{Total\ number\ of\ parameters}\right) \times 100 \quad (1)$$

2. F2 (*Frequency*) functions shows the percentage of parameters that do not meet quality standard in every tests (failed test).

$$F_2 = \left(\frac{\textit{Number o failed tests}}{\textit{Total number of tests}}\right) \times 100 \qquad (2)$$

3. F3 (*Amplitude*) functions show degree of difference of the test value to the quality standards. It only calculates parameters of each tests that do not comply with the standard. Results can be acquired by these steps:
   a) For parameters indicating increasing pollution during elevated value (value exceeds quality standard), establish excursion by this formula:

$$\text{Excursion} = \left(\frac{\textit{Deviant test values}}{\textit{Quality standards}}\right) - 1 \qquad (3)$$

   b) For parameters indicating increasing pollution during reduced value (value below quality standard, for example DO parameter), establish excursion by this formula:

$$\text{Excursion} = \left(\frac{\textit{Quality standards}}{\textit{Deviant test values}}\right) - 1 \qquad (4)$$

   c) Establish normalization of deviations value from the test results for each parameter that do not comply with the standard (non sampling error or nse) by this formula:

$$\text{nse} = \left(\frac{\sum_{i=1}^{n} \textit{Excursion}_i}{\textit{Number of test values}}\right) \qquad (5)$$

Determine the degree of difference of the test value to the quality standards to specify the inappropriateness of the water to the quality standards. This can be done as follows:

$$F_3 = \left(\frac{nse}{0{,}01nse + 0{,}01}\right) \qquad (6)$$

The level of water pollution can be calculated as the accumulated values of F1, F2, and F3, which were previously calculated using the following formula:

$$\text{CCME WQI} = 100 - \left(\frac{\sqrt{F1^2 + F2^2 + F3^2}}{1{,}732}\right) \qquad (7)$$

The water criteria used in the CCME WQI method consist of five class classifications. Each class had a certain range of values to determine the status of water pollution as presented in Table 2.

Table 2. Criteria for waters based on the CCME WQI.

CCME WQI Score	Water Criteria
95-100	Excellent
80-94	Good
60-79	Fair
45-59	Marginal
0-44	Poor

Source : Canadian Council of Minister of the Environment (CCME) (2001)

3 Results and discussion

3.1 Water Quality in Banten Bay

The results of the water quality parameter tests that have been analyzed both physically and chemically in the temporal waters are presented in Table 3.

Table 3. Minimum, maximum, average, and standard deviation values of physical and chemical parameters of water quality in Banten Bay during July 2022 – January 2023.

Parameters	Units	Quality standard	Month						
			Jul-22	Aug	Sep	Oct	Nov	Dec	Jan-23
Physics									
Turbidity	NTU	5	3,71-8,06*	4,48-9,17*	4,58-9,17*	4,89-7,16*	4,73-8,34*	4,24-9,03*	3,85-9,53*
			5,68±1,81	6,44±1,92	6,46±1,89	5,91±1,01	6,24±1,46	6,34±2,05	6,33±2,33
Temperature	°C	28-30	27,8-29*	29,3-30,8*	29-30	29,4-31*	28-30	27-28,8*	26,7-28,2*
			28,62±0,42	30,3±0,55	29,5±0,55	30,05±0,66	28,83±0,98	28,29±0,68	27,4±0,55
TSS	mg/L	20	5,5-10	7,8-13,4	9,1-10,3	9,2-12,4	8,1-11,3	9,5-12,9	8,2-13,8
			8,17±1,94	10,64±2,2	9,73±0,46	10,6±1,3	9,62±1,06	10,62±1,3	11,03±1,9
Chemical									
pH	-	7-8,5	7-7	7,51-7,84	7-7,57	7-7	7-7,32	7-7	7,11-7,19
			7±0	7,65±0,11	7,33±0,22	7±0	7,09±0,14	7±0	7,15±0,03
Dissolved Oxygen	mg/L	> 5	7,05-7,3	6,35-7,95	4,5-5,4*	5,6-8	7-8	6,9-7,8	7,6-9,1
			7,15±0,12	7,13±0,66	5,03±0,33	7,11±0,89	7,5±0,55	7,4±0,33	8,3±0,57
Ammonia (NH_3)	mg/L	0,3	0,24-0,43*	0,22-0,36*	0,12-0,2	0,07-0,1	0,03-0,09	0,03-0,12	0,02-0,06
			0,32±0,08	0,3±0,05	0,16±0,03	0,09±0,01	0,05±0,02	0,08±0,03	0,04±0,02
Total Phospate	mg/L	-	0,015-0,02	0,01-0,025	0,002-0,01	0,001-0,008	0-0,006	0,009-0,02	0,004-0,02
			0,02±0,003	0,02±0,01	0,01±0,002	0,004±0,003	0,002±0,002	0,02±0,004	0,01±0,005
Nitrate (NO_3)	mg/L	0,06	0,03-0,05	0,03-0,07*	0,01-0,06	0,02-0,05	0,02-0,05	0,012-0,05	0,021-0,04
			0,04±0,006	0,04±0,01	0,04±0,017	0,035±0,013	0,033±0,011	0,03±0,015	0,03±0,007
Nitrite (NO_2)	mg/L	-	0,03-0,08	0,03-0,12	0,05-0,25	0,03-0,2	0-0,15	0,013-0,05	0-0,1
			0,05±0,02	0,09±0,04	0,14±0,08	0,12±0,07	0,07±0,06	0,03±0,01	0,03±0,04

* Parameter values are not in accordance with the quality standard

Most of the physical and chemical parameters per month are not within the range of the quality standards, such as turbidity, temperature, ammonia, nitrate, and dissolved oxygen. The values vary: turbidity and TSS parameters were at the highest mostly during the rainy season (October 2022 – January 2023), while other parameters mostly during the dry season (July-September 2022). The highest average of overall water quality parameters values was found in the temperature parameter (by 30.3°C) during August 2022, while the lowest value was found in the total phosphate parameter (by 0.002 mg/L) during November 2022.

The minimum, maximum, average and standard deviation of physical and chemical parameters and their quality standards for marine biota is spatially presented in Table 4.

Table 4. Minimum, maximum, average, and standard deviation values of physical and chemical parameters of water quality in Banten Bay at each observation location.

Parameters	Units	Quality standard	Station					
			1	2	3	4	5	6
Physics								
Turbidity	NTU	5	7,16-9,17*	6,11-7,17*	3,96-4,98	3,71-4,89	4,83-5,31*	6,12-9,53*
			8,4±0,84	6,88±0,38	4,56±0,35	4,42±0,49	5,09±0,2	7,83±1,07
Temperature	°C	28-30	26,7-31*	27,6-30,1*	27,7-30,7*	26,9-30,3*	28-30,6*	27,3-30,8*
			28,82±1,67	29,2±0,96	28,97±1,1	28,79±1,11	29,04±0,94	29,17±1,24
TSS	mg/L	20	6,5-13,4	7,5-10,3	9-12,6	5,5-11,3	9,47-11,7	8,1-13,8
			10,21±2,15	9,12±0,97	10,61±1,38	9,19±1,98	10,3±0,79	10,91±2,09
Chemical								
pH	-	7-8,5	7-7,51	7-7,84	7-7,62	7-7,63	7-7,64	7-7,63
			7,16±0,19	7,17±0,31	7,19±0,27	7,16±0,25	7,2±0,29	7,16±0,25
Dissolved Oxygen	mg/L	> 5	5,2-7,95	5,4-9,1	4,5-8,1*	5,1-7,8	5,2-8,5	4,8-8,7*
			7,11±0,91	7,11±1,11	7,06±1,27	6,55±0,93	7,34±1,07	7,36±1,23
Ammonia (NH_3)	mg/L	0,3	0,03-0,31*	0,03-0,28	0,03-0,43*	0,02-0,36*	0,06-0,29	0,03-0,38*
			0,13±0,11	0,14±0,1	0,18±0,16	0,15±0,13	0,14±0,09	0,15±0,14
Total Phospate	mg/L	-	0-0,025	0,001-0,019	0,001-0,019	0,003-0,022	0-0,018	0,003-0,022
			0,012±0,009	0,009±0,007	0,011±0,007	0,012±0,006	0,008±0,007	0,013±0,009
Nitrate (NO_3)	mg/L	0,06	0,03-0,07*	0,014-0,05	0,02-0,04	0,01-0,05	0,031-0,05	0,02-0,05
			0,05±0,01	0,03±0,01	0,03±0,01	0,03±0,01	0,04±0,01	0,03±0,01
Nitrite (NO_2)	mg/L	-	0,05-0,25	0,005-0,12	0,01-0,17	0-0,21	0-0,05	0-0,1
			0,14±0,07	0,07±0,04	0,09±0,06	0,09±0,08	0,02±0,02	0,05±0,04

* Parameter values are not in accordance with the quality standard

The analysis of water quality parameter reveals spatial variations across the six sampling stations. Turbidity is notably higher at Stations 1 and 6, exceeding the quality standard of 5 NTU. This suggests increased sedimentation activity in these areas, likely influenced by river inflows or anthropogenic factors. Water temperature remains relatively consistent across all stations, ranging from 26.7°C to 31°C, which falls within the acceptable range of 28-30°C. Total Suspended Solids (TSS) fluctuate between stations, with the highest concentrations observed at Station 5 and 6. Dissolved Oxygen (DO) levels show a decline at certain stations, particularly Station 3 and 5, where values fall below the minimum standard of 5 mg/L. This indicates a potential risk of hypoxic conditions, which could negatively impacts aquatic life. pH levels remain stable across all stations, ranging from 7 to 7.84, within the acceptable range. Nitrate (NO_3) and nitrite (NO_2) concentrations are within permissible limits across all stations, although minor variations are observed. In contrast, ammonia (NH_3) levels are elevated at certain stations, suggesting potential contamination from domestic or industrial sources. Overall, water quality appears to deteriorate in some stations, particularly in terms of DO and turbidity, while better conditions are found at stations farther from pollution sources or river inflows. These spatial variations highlight the influence of environmental factors such as depth, anthropogenic activities, and nutrient inputs on water quality in the study area.

3.2 Water Pollution Status

The status of water pollution from July 2022 to January 2023 can be determined using the Canadian Council of Minister of the Environment Water Quality Index (CCME WQI) method. Analysis using the CCME method in Banten Bay waters from July 2022 to January 2023 showed varying criteria. The results of the overall analysis using the monthly CCME method are shown in Figure 2.

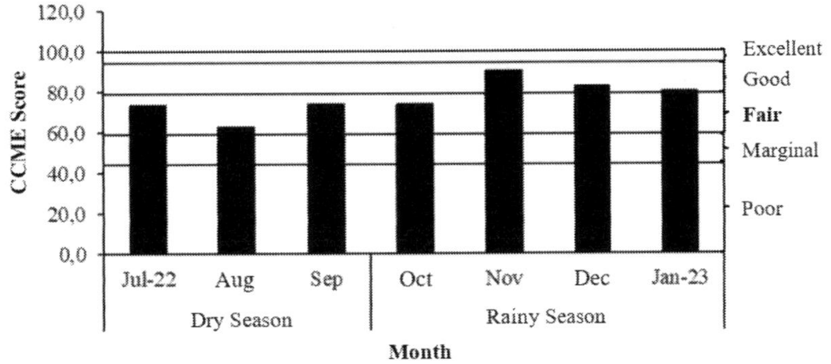

Fig. 2. Banten Bay water pollution status per month.

The highest pollution level in Banten Bay waters was in August 2022 during the dry season at 62.34 categorized as fairly good, while the lowest was in November 2022 during the rainy season at 89.88 classified as good status. During the rainy season, water pollution tend to be good due to the flushing of pollution loads in the sea because of the currents and tides that form are quite strong in this season. The results of the overall analysis of the CCME method from July 2022 to January 2023 show that the condition status of the Banten Bay waters is in the criteria of sufficiently good, although there is tendency to more adequate status during rainy season.

The overall results of the analysis of the pollution status data in Banten Bay waters using the CCME method for each observation location (Stations 1 – 6) are presented in Figure 3.

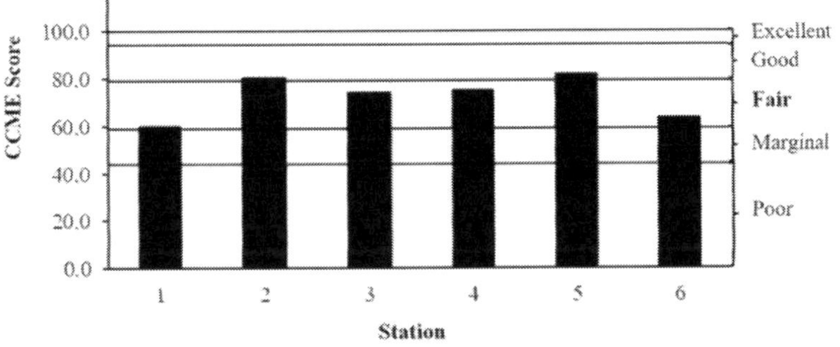

Fig. 3. Banten Bay water pollution status per station.

Spatial analysis shows that the highest pollution level occurred at Station 1 with a score of 59.95, categorized as fairly good status, while the lowest occurred at Station 5 with a score of 81.56, categorized as good status. Station 1 and 6 fall in the fairly good category because those locations are close to the estuaries of the Karangantu and Kemayungan-Linduk rivers, where pollution load from the river easily accumulates in these areas. Meanwhile, station 2-5 is in the middle of the sea, hence smaller pollution load since it was thinned by the distribution of currents and ocean waves from the Banten Bay to the high seas of Northern Java. The results of the overall analysis of the CCME method at Stations 1 – 6 show that the condition status of Banten Bay waters is sufficiently good.

3.3 Discussion

All analyzed data are acquired directly on the sampling location and thus available within the paper. Based on temporal and spatial analysis, the overall water quality of Banten Bay is sufficiently good. Some of the parameters such as turbidity, temperature, ammonia, dissolved oxygen, and nitrate are not within quality standards. According to Juniardi *et al.* [18], in 2019 the conditions of Banten Bay waters, which were analyzed using the CCME method, showed that these waters were still in the fair good category, because several water quality parameters studied were still in accordance with the quality standards (safe category). This is also supported by research from Wardani *et al.* [19] which shows the same results as research on the pollution status of Banten Bay waters from July 2022 to January 2023, where the pollution status of Banten Bay waters is in the fair good category. The input of nutrients, organic matter, and heavy metals in Banten Bay waters is still relatively low compared to Jakarta Bay and its surroundings; therefore, the pollution status of the waters in the bay is still in the fair good category [20].

Research using the CCME WQI method shows that there is consistency in the results of the water pollution status from time to time at each station. This is supported by research from Eljaiek-Urzola *et al.* [21] in the Caribbean Sea and El-Nemr *et al.* [22] in the waters of Red Sea Coast, Egypt which states the pollution status results are quite the same and fluctuate slightly every year. According to previous research from Akhtar *et al.* [23], the CCME method is quite effective and specific for determining the status of water pollution in an area.

Pollution level differs between months and locations. During the dry season (July-September 2022), higher levels of pollution was detected since higher amounts of industry, agricultural, and domestic runoff were dumped into sea. However, a slower pace of agitation and rinsing occurred during the season since low sea current (0,04-0,28 m/2) prompted the pollution burden to settle at the location. Whereas during the rainy season (October 2022-January 2023), a faster pace of agitation and rinsing occurred with sea current of 0,28-0,37 m/s which caused the pollution burden to be easily diluted so as to reduce pollution level to low category [24]. Spatially, station 1 and 6 fall into fairly good category, which are higher in terms of pollution compared to other stations since its location is adjacent to the estuaries of Karangantu and Kemayungan-Linduk rivers which results in higher accumulation of pollution burden by industry, agricultural, and also domestic organic and inorganic waste carried by river runoff at the stations.

Overall, pollution level at Banten Bay waters is influenced the most by turbidity. Turbidity values that exceed the quality standard can be caused by high rainfall, large amount of organic matter, substrate runoff from rivers and land into the water column, high sedimentation rates, wind direction and speed, as well as patterns of currents and tides. Increasing turbidity levels in waters can also cause a decrease in the concentration of dissolved oxygen [25]. During the rainy season, turbidity and TSS values tended to increase compared to those observed during the dry season [26].

However, ammonia and nitrate may also have its effects to increased water pollution. Levels of ammonia (NH_3) that do not reach quality standards are caused by influx of several sources of ammonia, such as the breakdown of organic nitrogen (urea and protein), decomposing inorganic nitrogen in waters, and the decomposition of organic matter by microbes and fungi from plants and biota or dead aquatic animals [27]. Nitrate (NO_3) levels that do not reach quality standards are caused by several factors, namely temperature and dissolved oxygen. Higher temperatures cause lower levels of nitrate in waters [28]. Nitrates also play an important role in marine biogeochemical cycles, and their presence must still be controlled according to quality standard thresholds [29]. Dissolved oxygen levels lower than the quality standard can be caused by increasing concentrations of ammonia and turbidity levels in the oceans, as well as the process of decomposing organic matter in the region [30].

It is clear that several water quality parameters do not comply with quality standards. This could be caused by several factors, such as abrasion, dynamics of sediment transport, conversion of mangrove land into ponds, development of industrial areas, increase in beach construction by years, and illegal sand mining by local residents and private parties surrounding the bay [24]. Pollution in water bodies can occur because of the input of waste and other foreign substances into water bodies, causing changes in water quality [31]. The greatest influence of water pollution in Banten Bay is from anthropogenic factors related to domestic and industrial activities. Allowing this to continue, and it can cause an increase of pollution status in Banten Bay which is dangerous for the survival of humans and aquatic biota around the aquatic ecosystem. Biota contaminated from these waters can poison the human body if consumed continuously [32].

Therefore, there is a need for recommendations to sustainable management of Banten Bay waters based on their pollution status. Recommendations for the management of Banten Bay waters can be carried out in various ways, including having an appropriate wastewater treatment installation system for anthropogenic activities around Banten Bay waters to control the input of waste into these waters. This effort aims to reduce the levels of ammonia, nitrate, nitrite, and total phosphate, which are not in accordance with the quality standards, and can improve the status of pollution in Banten Bay waters. In addition, it is necessary to coordinate, monitor, and evaluate activities between related stakeholders so that the management of Banten Bay waters can be carried out optimally and create a sustainable Banten Bay waters ecosystem [33].

Monitoring and managing the input of garbage and waste originating from the waters of Banten Bay is necessary to maintain the quality of the waters in good condition, especially for the parameters of temperature, dissolved oxygen, and pH, which are often used in the survival of marine biota [34]. The management of these areas can also be carried out by conducting reviews and making appropriate policies related to development efforts around coastal areas (especially settlements and industries) so that the levels of turbidity, TSS, and other organic pollution loads that accumulate in the area can always be controlled so that they are always in good condition (according to quality standards) to create a sustainable Banten Bay aquatic ecosystem [35].

4 Conclusion

The water quality status of Banten Bay, as determined by average concentration, was generally suitable for marine biota. However, temporal and spatial analysis revealed instances where specific parameters fell into categories unsuitable for marine life. The pollution status of Banten Bay waters assessed using the CCME method shows that the water conditions are fairly good or in lightly polluted category. Overall, the primary parameters identified to contributing the pollution status of Banten Bay is turbidity. It is caused by various activities surrounding the bay such as industrial, agricultural, and domestic activities that can contribute to increased sedimentation and pollutant loads influenced by river inflows. Seasonal variations and local oceanographic dynamics, such as currents and tides, also affect the distribution of pollution in this area. Pollution control in Banten Bay requires continuos water quality monitoring, strict regulations on industrial and domestic waste disposal, as well as conservation and coastal ecosystem rehabilitation strategies. Collaboration between the government, local communities, and the industrial sector is essential to ensure sustainable water management and maintain the ecological balance of Banten Bay.

Thank you to all those who helped author in completing this research. Thank you to IPB University for being co-funder of this research. Thank you to the technician from the Environmental Laboratory and also Ecobiology and Conservation Laboratory, Faculty of Fisheries and Marine Science, IPB University for providing a facility for this research. The authors would like to express their sincere gratitude to the anonymous reviewer for insightful comment and constructive feedback in order to improve the manuscript.

References

1. U.J. Wisha, S. Husrin, J. Prihantono, Ilmu Kelautan : Indonesian Journal of Marine Sciences, **20**, 2 (2015)
2. A. Bayhaqi, U.J. Wisha, M.Y. Iswari, Trends in Sciences, **19**, 4 (2022)
3. D.R. Pratama, M. Yusuf, M. Helmi, Jurnal Oseanografi, **5**, 4 (2016)
4. M.G. Uddin, S. Nash, A. Rahman, A.I. Olbert, Science of the Total Environment, **868** (2023)
5. M.M.M. Syeed, M.S. Hossain, M.R. Karim, M.F. Uddin, M. Hasan, R.H. Khan, Environmental and Sustainability Indicators, **18**, 100247 (2023)
6. K.K. Vadde, J. Wang, L. Cao, T. Yuan, A.J. McCarthy, R. Sekar, Water, **10**, 183 (2018)
7. M.G. Mahmoud, E.A. El-Khir, M.H. Ebeid, L.A. Mohamed, M.A. Fahmy, K.S. Shaban, Journal of Environmental Protection, **11**, 1 (2020)
8. G. Wang, X. Feng, J. Zhang, Z. Huang, Y. Bai, W. Song, H. Su, Frontiers in Marine Science, **10**, 1 (2023)
9. W. Berkademi, K. Mizuno, T.E.B. Soesilo, International Journal on Advanced Science Engineering Informatioin Technology, **13**, 4 (2023)
10. Sugiarti, A. Rahmadya, D. Rohaningsih, R. Novianti, A. Waluyo, S. Aisyah, Rosidah, Limnotek Perairan Darat Tropis di Indonesia, **6**, 2 (2024)
11. R.Y. Kartika, Tingkat pencemaran pesisir Kronjo Kabupaten Tangerang, Banten, Undergraduate Thesis, (Institut Pertanian Bogor, Bogor, 2014)
12. A. Sahidin, Zahidah, H. Herawati, Y. Wardiatno, I. Setyobudiandi, R. Partasasmita, Biodiversitas, **19**, 3 (2018)
13. M. Limbong, D. Simbolon, A. Purbayanto, R. Yusfiandayani, K. Amri, Depik Jurnal Ilmu-Ilmu Perairan, Pesisir, dan Perikanan, **12**, 3 (2023)
14. APHA, Standard methods for the examination of water and waste water 23rd edition, (AWWA (American Water Works Association) and WEF (Water Environment Federation), Washington D.C., 2017)
15. PPRI, Peraturan Pemerintah Republik Indonesia Nomor 22 Tahun 2021 tentang Penyelenggaraan Perlindungan dan Pengelolaan Lingkungan Hidup, Appendix VIII, (PPRI, Jakarta, 2021)
16. CCME, Canadian water quality guidelines for the protection of aquatic life: CCME water quality index 1.0, user's manual, (Canadian Council of Ministers of the Environment, Winnipeg, 2001)

17. Y. Romdania, A. Herison, G. E. Susilo, E. Novilyansa, Jurnal Spatial Wahana Komunikasi dan Informasi Geografi, **18**, 1 (2018)
18. E. Juniardi, Sulistiono, S. Hariyadi, IOP Conf. Series: Earth and Environmental Science, **950**, 012052 (2022)
19. N.K. Wardani, T. Prartono, Sulistiono, Jurnal Pendidikan IPA Indonesia. **9**, 4 (2020)
20. Sugiarti, S. Hariyadi, S.H. Nasution. Limnotek Perairan Darat Tropis di Indonesia, **23**, 1 (2016)
21. M. Eljaiek-Urzola, N. Romero-Sierra, L. Segrera-Cabarcas, D. Valdelamar-Martinez, E. Quinones-Bolanos, Water, **11**, 856 (2019)
22. A. El-Nemr, G.F. El-Said, S. Ragab, A. Khaled, A. El-Sikaily, Chemosphere, **165** (2016)
23. N. Akhtar, M.I.S. Ishak, M.I. Ahmad, K. Umar, M.S.M. Yusuf, M.T. Anees, A. Qadir, Y.K.A. Al-Manasir, Water, **13**, 905 (2021)
24. T. Solihuddin, J. Prihantono, E. Mustikasari, S. Husrin, Jurnal Geologi Kelautan, **18**, 2 (2020)
25. N.H. Quang, J. Sasaki, H. Higa, N.H. Huan, Water, **9**, 570 (2017)
26. B.B. Sinaga, Y. Suteja, I.G.B.S. Dharma, Journal of Marine and Aquatic Sciences, **6**, 2 (2020)
27. D.N. Handiani, A. Heriati, Jurnal Ilmu Lingkungan. **18**, 2 (2020)
28. S.I. Patty, H. Arfah, M.S. Abdul, Jurnal Pesisir dan Laut Tropis, **1**, 1 (2015)
29. W. Niu, H. Li, X. Guo, J. Chen, X. Shi, Y. Zhu, Frontiers in Marine Science, **10**, 2 (2023)
30. S.I. Patty, Jurnal Ilmiah Platax, **6**, 1 (2018)
31. T.A. Kurniawan, E.R. Bandala, M.H.D. Othman, H.H. Goh, A. Anouzla, K.W. Chew, F. Aziz, H.E. Al-Hazmi, A.N. Khoir, Water Supply, **24**, 2 (2024)
32. G. Prayoga, S. Hariyadi, Sulistiono, H. Effendi, IOP Conf. Series: Earth and Environmental Science, **744**, 012055 (2021) (32)
33. D. Adyasari, M.A. Pratama, N.A. Teguh, A. Sabdaningsih, M.A. Kusumaningtyas, N. Dimova, Marine Pollution Bulletin, **171**, 112689 (2021)
34. H. Darmawan, A. Masduqi, Jurnal Teknik Pomits, **3**, 1 (2014)
35. V. Liyubayina, Jurnal Planesa, **9**, 1 (2018)

Characteristics microplastics in gills and digestive tract of fish at seagrass ecosystem on Bintan Island, Indonesia

Fadhliyah Idris[1], *Try* Febrianto[1], *Aditya* Hikmat Nugraha[1], *Arief* Pratomo[1], *Chandra* Joei Koenawan[1], *Falmi* Yandri[1], *Rika* Anggraini[1], *Wiwin*[1], *Puspa* Ayu[1], *Jelita* Rahma Hidayati[1*]

[1]Department of Marine Science, Faculty of Marine Sciences and Fisheries, Raja Ali Haji Maritime University, Indonesia

Abstract. Plastic waste is still present. Plastics are still required in food and drink containers because of their practical use and light, sturdy, and durable nature. Plastic waste can naturally decompose into small particles of less than 5 mm, called microplastics. Microplastics are currently of concern because of their impact on marine biota, including fish, in the seagrass ecosystem. This study aimed to determine the characteristics of microplastics found in fish organs, particularly in the digestive tract and gills, in the waters of Bintan Island and its surroundings. Sampling was performed on fish collected from fish trapped in gill nets. Ten fish samples were collected for each species, and the microplastics were analyzed using optical microscopy in the laboratory. The results showed that five forms of microplastics were found in the digestive tract and intestines of the fish, with abundances ranges between 49.67 – 122.38 particles/gram dry weight. The highest abundance of colors was found in both transparent colors in the digestive tract and black colors in the gills.

1. Introduction

The Seagrass ecosystem is a shallow water ecosystem consisting of seagrass, the only flowering plant that can live below the sea surface. This ecosystem is an essential nursery and feeding ground for various types of marine biota, including fish [1]. Therefore, the presence of seagrass in the marine environment is important for the sustainability of the diversity of fish species that use seagrass as a habitat to find food [2].

Plastic waste is a pollutant that negatively impacts marine biota [3]. Plastic waste is one of the most common types of waste both on land and at sea. It is estimated that approximately 10% of all newly produced plastic waste is dumped through rivers and ends up in the sea. Almost all types of plastic float or drift in water bodies, causing plastic waste to disintegrate and degrade by sunlight, oxidation, mechanical abrasion, and the formation of plastic particles.

Plastic particles measuring ≤5 mm are known as microplastics [4]. Microplastics that accumulate in the body can cause physical and chemical damage, such as damage to internal

* Corresponding author: jelitarahmahidayati@umrah.ac.id

organs, blockage of the digestive tract, carcinogenic properties, and endocrine disorders. The impact of microplastics shows that they are dangerous for organisms in marine waters.

Dompak is an area that has the potential to contribute microplastics because of human activities, such as capturing fisheries and beach tourism. Pengudang Village, located in the Bintan Regency, has abundant natural resources, including seagrass ecosystems. Pengudang is also included in the Regional Marine Conservation Area and Seagrass Protection Area (DPPL), where the seagrass area is a habitat for marine biota. Seagrass, as a habitat for marine biota, causes the abundance and diversity of marine biota to be higher in the seagrass ecosystem than in areas without vegetation. Plastic waste in marine waters can affect the ecosystem through the relationship of marine biota to the environment, such as plastic consumption by marine animals, zooplankton, and large predators [5]. Large plastics can potentially entangle marine animals, causing injuries, difficulty swimming, and even death [6].

Microplastics in the sea have the potential to be eaten by marine biota, ranging from small organisms such as zooplankton to higher trophic levels [7]. Zooplankton can be accidentally exposed to microplastics because of their nature in searching for food by filtering (filter feeding). The impact of microplastics entering zooplankton reduces their physiological performance. Their feces contains microplastics and allows the transfer of contaminants to predators [15]. Microplastics can affect the food chain and disrupt the structure and function of the ecosystem as a whole. Organisms exposed to microplastics can transfer particles to their predators, which can then accumulate in the food web and reach higher concentrations at higher trophic levels [16]. For example, a previous study [8] showed that microplastics are found in the digestive tract of sea cucumbers living in seagrass. Microplastics have also been found in *Strombus* spp. and sediments [9]. Based on the explanation above, further studies are needed on the characteristics of microplastics in fish, especially in the digestive tract and gills, which are the entry and accumulation sites for microplastics and seagrass in the habitat of fish.

2. Materials and method

2.1. Sampling area

This study was conducted on Bintan Island (Dompak and Pengudang). The location of the fish sampling was determined based on the characteristics of the area and the presence of seagrass as a fish habitat. The research locations are shown in Figure 1.

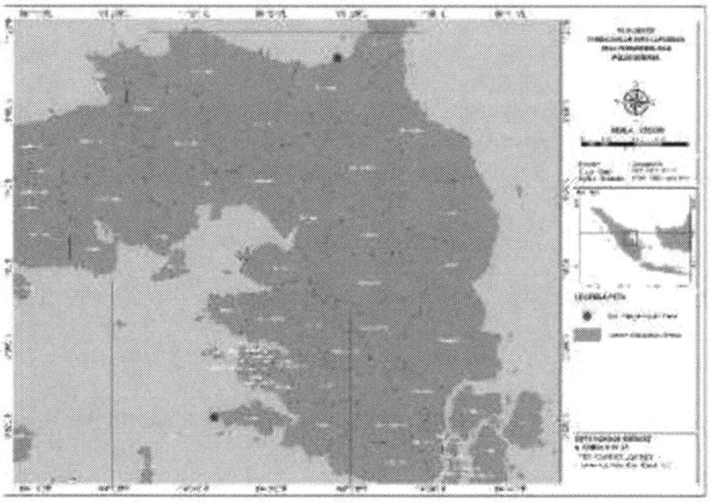

Fig 1. Research location map

2.2. Sampling Collection

Sampling was conducted at two locations determined during the preliminary survey, Dompak and Pengudang. Fish samples were collected using a net measuring 100 × 2 m with a mesh size of 2 in. The net was set horizontally parallel to the coastline in a seagrass meadow area of 100 m. The fish samples trapped in the net were then taken from 10 fish of each species to the laboratory for microplastic content analysis.

2.3. Sampling Analysis

The collected fish samples were dissected and the intestines and gills were collected for analysis. Several stages were conducted during the analysis. The first stage involved the removal of organic matter by adding a 30% H_2O_2 solution and incubating for 24 h at room temperature [7]. Furthermore, in the second stage, the microplastic particles were separated from other particles using $ZnCl_2$ through a floating system for 24 h until the two deposits were [10]. The last stage is observation, which uses a microscope at 100x magnification to determine the amount, shape, and color of the microplastics.

2.4. Data analysis

Microplastic abundance calculations were performed to determine the presence of microplastics in each sample. The abundance of microplastics was calculated based on the number of microplastic particles per dry weight of the sample.

$$abundance\ of\ microplastic = \frac{microplastic\ particle}{dry\ weight\ of\ the\ sample}$$

3. Results and discussion

3.1. Type of Fish

Seagrass is an essential habitat for the survival of various fish species in which certain fish live in groups or associations. The interactions between these fish can involve multiple activities, such as foraging, protection from predators, or reproduction. Fish associations in seagrass habitats can also provide significant ecological benefits, such as maintaining ecosystem balance and sustainability of fish populations.

Paraplotosus sp. are caught in Dompak and Pengudang, *Siganus* sp. are only found in Dompak waters, and *Lethrinus* sp. are only found in Pengudang waters. This is likely due to the characteristics and high adaptability of *Paraplotosus* sp. fish. These three types of fish have a relatively high economic value and are often hunted by local fishermen. The figure of fish types are presented in below (Fig.2).

A **B** **C**

Fig 2. The fish are caught (A; *Paraplotosus* sp. B; *Lethrinus* sp. C; *Siganus* sp.)

3.2. Microplastic abundance in fish

The results of this study showed variations in the accumulation of microplastics from the three types of fish caught, with the highest abundance value found in the *Lethrinus* sp. type of fish (122.38 particles/gram dry weight), followed by the *Paraplotosus* sp. type of fish (63.86 particles/gram dry weight), and the lowest abundance of microplastics found in the *Siganus* sp. type of 49.67 particles/gram dry weight. The abundance of microplastic in each type of fish is shown in Figure 3.

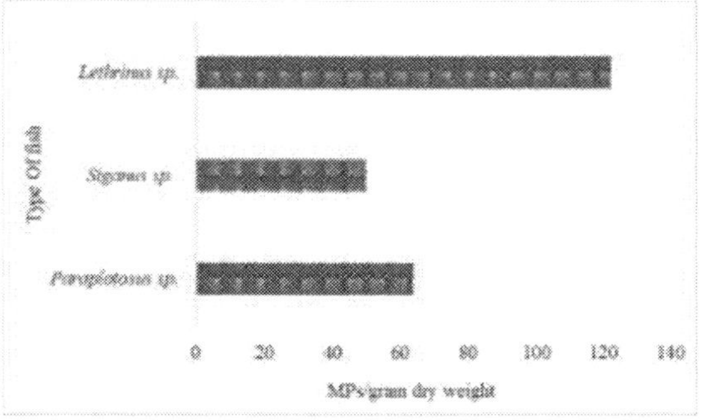

Fig 3. Microplastics abundance in fish

The high abundance of microplastics in *Lethrinus* sp. is thought to be due to several factors, including eating habits. *Lethrinus* sp. are predatory fish that eat crustaceans, mollusks, and small fish. According to [11], the type of food for *Lethrinus* sp. fish is crabs and shellfish because they live at the bottom of the water, so *Lethrinus* sp. fish quickly eat crabs and shellfish. Crabs and shellfish are filter feeders and deposit feeder biota that search for food by eating sediments and organic materials. Based on a previous study [12], microplastics were found in the sediment and *Strombus* sp.

3.3. Composition type and color of Microplastic

Microplastics found in the results of observations of intestines and gills consist of fiber, fragment, film, foam, and pellet types (Figure 4).

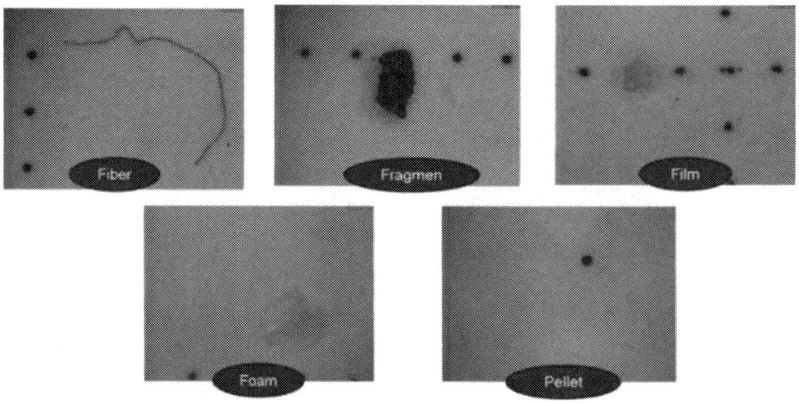

Fig 4. Microplastics types in gills and digestive tract

The characteristics of microplastics in dompak waters according to the color of the intestines of *Paraplotosus* sp. fish are transparent, brown, blue, white, black, and red. In contrast, the gills were red, black, brown, transparent, and dominated by black. The intestines of *Siganus* sp. fish include brown, blue, red, transparent, and black, whereas the gills consist of brown, blue, red, transparent, and black, which are dominated by black (Figure 4A).

Meanwhile, the characteristics of the abundance of microplastics in Pengudang waters in terms of color in the intestines of *Paraplotosus* sp. fish included transparent, white, brown, blue, yellow, black, and green. At the same time, the gills consist of transparent, brown, blue, red, black, and green colors, which are dominated by transparent colors in the intestines and black in the gills. The intestines of *Lethrinus* sp. fish consist of transparent, brown, blue, yellow, red, green, black, and white, while the gills consist of transparent, brown, red, blue, black, and green, which are dominated by transparent color in the intestines and black in the gills (Figure 4B).

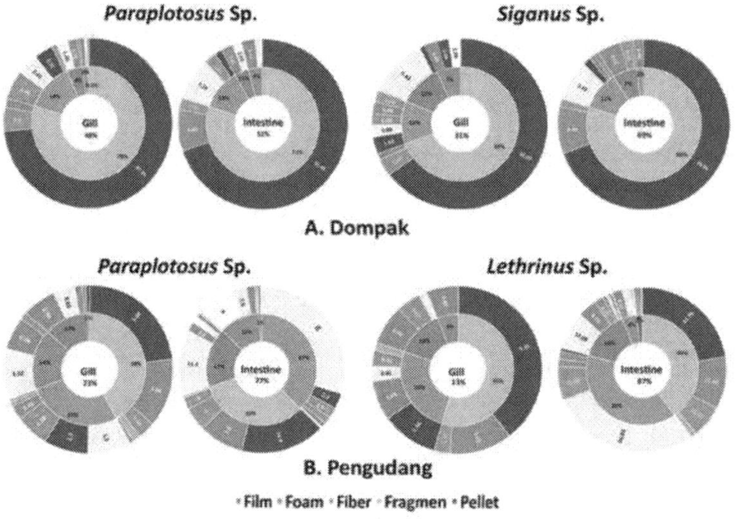

Fig 5. Composition type and color of microplastic

Figure 5 shows the sunburst diagram of the composition of the shape and color of microplastic particles in fish species and parts of the internal organs found in the seagrass habitat of Dompak (A) and Pengudang (B). The inner ring shows the composition of the microplastic particles. The outer ring shows the composition of the colored microplastics expressed as the number of particles per gram.

The results in Figure 5 show that the dominant type of microplastic was fragments, with an average percentage of 51%. The high type of fragment is possible because it is a primary source of high in water due to the fragmentation of hard plastics such as plastic bottles and packaging. Fish consider microplastics as their food, so there is an accumulation of microplastics in their bodies [17]. According to [13], fragments result from the cutting of plastic products with solid synthetic polymers. Film-type microplastics originate from plastic bags and food packaging. Film is a secondary plastic polymer that originates from the fragmentation of plastic bags or plastic packaging and has a low density. Other types of microplastics are elongated plastic fibers originating from the fragmentation of fishing net monofilaments, ropes, and synthetic fabrics. These plastic fibers can come from high fishing activities around the area, contributing debris to the water [14]. Pellets are primary microplastics directly produced by factories as raw materials for making plastic products.

4. Conclusion

Microplastics were found in every type of fish, both in the gills and digestive tract, with a characteristic black color. Fragments were the types of microplastics found. The highest abundance of microplastics was found in *Lethrinus* sp. (122.38 particles/gram dry weight). This indicates that the fish found in the Pengudang and Dompak waters were contaminated with microplastics.

This research was financially supported by an internal grant from the Raja Ali Haji Maritime University. Therefore, the author would like to express gratitude and appreciation to all parties involved in this research.

References

1. E.W. Koch, D.M. Booth, C. Palinkas, (Instituto Biodiversida de Marinha, Rio deJaneiro, 2012)
2. A. Syukur, Y. Wardianto, I. Muchsin, M. M. Kamal, J. Akuatik **1** (1): 50-60 (2011)
3. M.R. Cordova, Oseana, **42** (3): 21 – 30 (2017)
4. M. D. S. Wirawan, F. Dhafir, I. M. Budiarsa, G. B. N. Shamdas, Media Eksakta, **17** (2) : 73-78 (2021)
5. D. Zhang, X. Liu, W. Huang, J. Li, C. Wang, D. Zhang, C. Zhang, Environmental Pollution, **25** (9): 1-48 (2020)
6. L.G.A. Barboza, A. Cozar, B.C. G. Gimenez, T.L. Barros, P.J. Kershaw, L. Guilhermino, Academic Press, **30** (5):305-328 (2019)
7. D. Yona, F.A.D. Prikah,, M.A. As'adi, J. Ilmu Lingkungan, **18** (2):375–383 (2020)
8. F. Idris, T. Febrianto, J.R. Hidayati, Rajb, A.H. Nugraha, *Microplastic abundance in sea cucumber at seagrass ecosystem of Bintan Island and surrounding area, Indonesia*, in Proceedings of the International Conference on Integrated Coastal Management & Marine Biotechnology, ICMMBT, Bogor, Indonesia (2022)
9. F. Idris, Rajb, J.R. Hidayati, T. Febrianto, A.H. Nugraha, *Microplastic contamination in*

sediment and Strombus sp. *on Bintan Island, Indonesia*, in Proceedings of the International Seminar On Marine Science and Sustainability, ISOSS, Bogor, Indonesia (2023)

10. D.B. Makrima, J. Suprijanto, B. Yulianto, J Marine Research, **11** (3) : 467-474 (2022)
11. Kafabihi, F. Idris, R.D. Putra, A.H. Nugraha, T. Apriadi, A.D. Syakti, *Microplastic contamination extent on Strombus* sp. *in North Bintan Waters*, in Proceedings of the International Conference on Integrated Coastal Management & Marine Biotechnology, ICMMBT, Bogor, Indonesia (2022)
12. Y.H. Restiangsih, N. J. Ikhtio Indonesia, **19** (1) : 115- 126 (2019)
13. J. Kingfisher, *Microplastic debris accumulation on puget sound beaches. port townsend marine science center* (2011)
14. S. Katsanevakis, A. Katsarou, Water, Air, and Soil Pollution. **15** (9):325-337 (2004)
15. A.A. Amanu, A.P. Zahrani, F.A. Ristaati, A.R. Ardillah, D.O. Radianto. Manufaktur: Publikasi Sub Rumpun Ilmu Keteknikan Industri. **2**(2): 12-24 (2024).
16. Mardiyana, A. Kristingsih. JPPL. 2(1) **(2020).**
17. F.A.F. Ashuri, Ghitarina, AbdunnurJurnal Aquarine. 10(1): 29-34 **(2023).**

Antioxidant and DPP-IV inhibitory activities of *Sargassum cristaefolium* extract: A therapeutic potential agent for type-2 diabetes

Agung Giri Samudra[1,2], *Agung* Endro Nugroho[3*], *Retno* Murwanti[3,4]

[1]Faculty of Pharmacy, Universitas Gadjah Mada, Yogyakarta 55281, Indonesia
[2]Department of Pharmacy, Faculty of Mathematics and Natural Sciences, Bengkulu University. Bengkulu 38371, Indonesia
[3]Department of Pharmacology and Clinical Pharmacy, Faculty of Pharmacy, Universitas Gadjah Mada, Yogyakarta 55281, Indonesia
[4]Medicinal Plants and Natural Products Research Center, Faculty of Pharmacy, Universitas Gadjah Mada, 55281 Yogyakarta, Indonesia

Abstract. Diabetes mellitus (DM) is a metabolic disorder characterized by elevated blood glucose levels due to impaired insulin function. Dipeptidyl peptidase 4 (DPP-IV) inhibitors are potential agents for managing type-2 DM. This study evaluates the total phenolic and flavonoid content, antioxidant activity, and DPP-IV inhibitory potential of the ethanolic extract and ethyl acetate fraction of *Sargassum cristaefolium*. Antioxidant activity was assessed using the DPPH assay, while DPP-IV inhibition was compared with sitagliptin. Gas Chromatography-Mass Spectrometry (GC-MS) was used to identify active compounds. The total phenolic content in the ethanol extract and ethyl acetate fraction was 1.04% and 7.33% GAE, respectively, while total flavonoid content was 1111.70 and 3529.01 µgQE/g. Antioxidant activity increased in a concentration-dependent manner, with IC_{50} values of 1.38 mg/mL (ethanol extract) and 0.71 mg/mL (ethyl acetate fraction). The ethanolic extract (500 ppm) and ethyl acetate fraction (62.5 ppm) exhibited DPP-IV inhibitory activity of 92.07% and 95.87%, respectively. GC-MS identified hexadecenoic acid, methyl ester, and trans-13-octadecenoic acid, methyl ester as dominant compounds. *S. cristaefolium* exhibits promising antioxidant and DPP-IV inhibitory activity, supporting its potential role in type-2 DM management.

1 Introduction

According to the International Diabetes Federation (IDF) Diabetes Atlas, it is projected that by 2030, approximately 643 million adults aged 20 to 79 years, representing an estimated prevalence of 11.3%, will live with diabetes. By 2045, this number is predicted to increase to approximately 783 million adults, with a projected prevalence of 12.2% [1]. Diabetes mellitus refers to a variety of metabolic illnesses that are generally defined as persistently elevated blood sugar levels [2]. This syndrome is caused by a decrease in the production of

* Corresponding author: nugroho_ae@ugm.ac.id

insulin, a decrease in the effectiveness of insulin, or a combination of both mechanisms [3]. Elevated blood glucose levels in diabetes lead to increased production of harmful free radicals, contributing to tissue damage and complications, such as nephropathy, neuropathy, retinopathy, and cognitive impairment [4].

Oxidative stress is recognized as a major contributor to the development and progression of diabetes [5]. It occurs because of an imbalance between the excessive production of reactive oxygen species (ROS) and deficiencies in the antioxidant defense system. This imbalance is strongly associated with the onset of various diseases, including diabetes and its complications [6, 7]. Bioactive compounds, such as alkaloids, phenolic acids, flavonoids, steroids, saponins, and glycosides, have shown potential in diabetes therapy because of their dual function as DPP-IV inhibitors and antioxidants. DPP-IV inhibitors alleviate hyperglycemic conditions by extending the half-life of incretin hormones including glucagon-like peptide-1 (GLP-1) and glucose-dependent insulinotropic polypeptide (GIP). These incretins play critical roles in enhancing insulin secretion and maintaining β-cell mass. Concurrently, antioxidants combat oxidative stress, a key factor in diabetes pathogenesis, by scavenging free radicals, thereby protecting pancreatic β-cells from oxidative damage [8, 9].

The exploration of marine plants provides an alternative approach to drug discovery. Accordingly, research on drug development and discovery uses materials from extraction processes, fractions, sub-fractions, and the isolation of active compounds from plants, either separately or in combination [10, 11]. Marine macroalgae, which are abundant producers of biologically active compounds in marine environments, exhibit significant potential for medical applications [12]. *Sargassum*, a genus of brown algae belonging to the family Sargassaceae, encompasses approximately 400 species worldwide. Throughout generations, *Sargassum* species, including *S.fusiforme*, have been integral to traditional diets and folk health practices in Korea and China [13]. Known for its pharmacological properties, such as hypoglycemic, anticancer, immunomodulatory, anti-obesity, and anti-inflammatory activities [14], *S. cristaefolium* was selected for this study to analyze its bioactive components, specifically phenolics and flavonoids, using gas chromatography-mass spectrophotometry (GC-MS). These bioactive compounds are of particular interest because of their potential roles in diabetes therapy, including their ability to inhibit the DPP-IV enzyme, which prolongs incretin activity and enhances insulin secretion, and their antioxidant properties, which mitigate oxidative stress and protect the pancreas. Additionally, this study aimed to evaluate the antioxidant and DPP-IV inhibitory activities of the ethanol extract and ethyl acetate fraction of *S.cristaefolium.*

2 Research Method

2.1 Collection of brown marine algae

Brown marine algae (*S. cristaefolium*) were collected from the coastal area of Gunung Kidul Beach in Yogyakarta, Indonesia (8°08'11.2″ S, 110°33'52.0″ E). Marine algae were collected for sample identification at the Department of Pharmaceutical Biology, Faculty of Pharmacy, Universitas Gadjah Mada, Indonesia.

2.2 Preparation of marine algae extracts

S. cristaefolium was washed with running water to remove dirt, coral, and sand. It was then cut into small pieces, dried in the sun, and covered with a black cloth for three days. Dried *S.cristaefolium* was ground to obtain a dry powder. *S. cristaefolium* extract was obtained by maceration by soaking 100 g of dry powder in 400mL of ethanol solvent for one week at

37°C. It was then sieved through filter paper and concentrated in a rotary vacuum evaporator to yield a concentrated ethanol extract [15]. The fractionation was performed using the liquid-liquid method with an ethyl acetate (semi-polar) solvent. Fractionation was performed using the separating funnel method (liquid-liquid extraction) by suspending the extract in 100 mL of distilled water, followed by the addition of 100 mL of ethyl acetate, shaking, and leaving it until homogeneous and separated into two phases. The ethyl acetate phase was separated from the insoluble ethyl acetate phase and collected in different containers. The mixture was then evaporated using a rotary vacuum evaporator. The concentrated ethyl acetate fraction was stored in an airtight bottle at ± 4°C [16].

2.3 Determination of total phenolics

Total phenolic content was determined using a UV-Vis spectrophotometer (Model UV-1800, Shimadzu). Twenty milligrams of the sample was weighed carefully, and 1 mL of distilled water was added. Folin-Ciocalteu reagent (0.5 mL) was added to the water phase, followed by the addition of 5 mL of aquabidest. The solution was allowed to stand for 10 min at ambient temperature, after which sodium carbonate solution (1.5 mL) at a concentration of 20% was added. Aquabidest was added to a volume of 10 ml and the concentration was adjusted as required. It was then transferred to a cuvette for spectrophotometric analysis, with absorbance maintained at a wavelength of 760 nm [17].

2.4 Determination of total flavonoids

Total flavonoid content was determined using a UV-Vis spectrophotometer (Model UV-1800, Shimadzu). Twenty milligrams of each sample were weighed and placed in a test tube. Then, 0.3 mL of 5% sodium nitrite was added and the mixture was allowed to react for 5 min. Then, 0.6 mL of 10% aluminium chloride was added and left for 5 min. Next, 2 mL of 1 M NaOH was added to the mixture. The total volume was adjusted to 10 mL using distilled water in a measuring flask. The solution was then diluted, as required. The prepared solution was transferred to a cuvette and absorbance was measured at a wavelength of 510 nm [17].

2.5 DPPH radical scavenging activity

The DPPH radical scavenging activity was determined using a UV-Vis spectrophotometer (Model UV-1800, Shimadzu). A total of 50 μL of test sample at varying concentrations, corresponding to IC_{50} values (the concentrations of extract or fraction yielding 50% antioxidant activity relative to the control as determined by linear regression), was combined with 1.0 mL of 0.4 mM DPPH and 3.950 mL of ethanol. The mixture was then completely blended and left undisturbed for 30 min. After incubation, the absorbance of the solution was measured at 517 nm. A blank solution, consisting of 50 μL of the extract and 4.950 mL of ethanol was used as a reference. Absorbance measurements were conducted on a control solution consisting of 1.0 mL and ethanol (4.0 mL) [17].

2.6 GC-MS chemical profiling

Chemical profiling determination via GC-MS was conducted using a Thermo Scientific Trace 1310 Gas Chromatograph, coupled with a Thermo Scientific ISQ LT Single Quadrupole Mass Spectrometer. The system was equipped with a TriPlus RSH auto-injector for precise sample handling and was controlled using the Chromeleon software for data acquisition and analysis. A sample solution of 1μL was injected into a GC-MS system equipped with a capillary column measuring 30 mm in length, 0.25 mm in diameter, and 0.25 µm in thickness. Helium was used as the carrier gas at a rate of 1 mL/min, with a split ratio of 1:50. The initial oven temperature was set to 150°C and held isothermally for five minutes. Subsequently, the temperature was increased at a rate of 10°C/min until it reached 250°C, which was maintained for an additional five minutes [18].

2.7 DPP-IV assay

An Elabscience® DPP-IV Inhibitor Screening Assay Kit (Catalog No: E-BC-D007) was used for measurements using a fluorescence microplate reader. For blank wells, 20 µL of buffer solution was added. The control wells received 20 µL of enzyme working solution, and the sample wells received 20 µL of enzyme working solution. Next, 30 µL of sample solvent was added to both the blank and control wells, and 30 µL of the sample was added to the sample wells. The contents were thoroughly mixed using a microplate reader for three seconds and then incubated at 37 °C for 10 min. Following this, 170 µL of the reaction working solution was added to each well, and the plate was further incubated at 37 °C for 30 min. Subsequently, the fluorescence intensity of each well was measured at excitation and emission wavelengths of 360 and 460 nm, respectively [19].
The inhibition rate (%) is calculated using the formula:

$$\text{Inhibition Rate (\%)} = \frac{(\text{Fcontrol} - \text{Fsample})}{(\text{Fcontrol} - \text{Fblank})} x\ 100\%$$

Fblank: The fluorescence intensity of blank well.
Fsample: The fluorescence intensity of sample well.
Fcontrol: The fluorescence intensity of control well.

3 Results and Discussion

3.1 Determination of yields, total phenolic content (TPC), total flavonoid content (TFC)

The extraction results of 250 g of dried *S. cristaefolium* simplicial with ethanol solvent using the maceration method are shown in Table 1. The extraction yield of *S. cristaefolium* varied significantly depending on the solvent used. Ethanol extraction resulted in a sticky extract weight of 8.19 ± 0.92 grams, corresponding to a yield of 3.27 ± 0.37%. In contrast, the ethyl acetate fraction produced a sticky extract weight of only 1.89 ± 0.38 grams, with a much lower yield of 0.75 ± 0.15%.

Based on the literature, *Sargassum*.sp contains various chemical compounds, including terpenoids, saponins, phlorotannin, cardiac glycosides, phenols, and flavonoids [24]. This difference in yield can be attributed to the polarity of the solvent used. Ethanol, a polar solvent, is capable of extracting a broad range of compounds, including polar and semi-polar constituents, such as polyphenols, flavonoids, and other hydrophilic bioactive compounds commonly found in Sargassum species. These compounds were highly soluble in ethanol, resulting in higher extraction yields. In contrast, ethyl acetate is a semi-polar solvent with limited solubility for highly polar compounds. It is more selective for extracting less polar or semi-polar constituents, such as terpenoids or certain lipophilic compounds. The lower yield observed with the ethyl acetate fraction is consistent with the selective nature of this solvent, which targets a narrower range of bioactive compounds [21].

Table 1. Yield of Ethanol Extract and Ethyl Acetate Fraction from *S.cristaefolium* Sample Extract

Sample Extract	Dry weight of simplicial (g)	Weight of sticky extract ± SEM (g)	Yield ± SEM (%)
Ethanol	250	8.19 ± 0.92	3.27 ± 0.37
Ethyl Acetate Fraction	250	1.89 ± 0.38	0.75 ± 0.15

The results showed that the total phenolic content (TPC) and total flavonoid content (TFC) varied between the ethanol extract and ethyl acetate fraction of *S. cristaefolium* (shown in Table 2). The ethanol extract exhibited lower TPC (1.04 ± 0.02% GAE) and TFC (1111.70 ± 48.91 μg QE/g) than the ethyl acetate fraction, which had higher TPC (7.33 ± 0.04% GAE) and TFC (3529.01 ± 54.32 μg QE/g).

Table 2. Total Phenolic Contents (Gallic Acid Equivalent) and Total Flavonoid Contents (Quercetin Equivalent) of Ethanol Extract and Ethyl Acetate Fraction

Sample Extract	Average TPCs ± SEM (% GAE)	Average TFCs ± SEM (μgQE/g)
Ethanol	1.04 ± 0.02	1111.70 ± 48.91
Ethyl Acetate Fraction	7.33 ± 0.04	3529.01 ± 54.32

The results reflect differences in the content of chemical compounds between the ethanol extract and ethyl acetate fraction. This disparity can be attributed to the polarity of the solvents used and their efficacy in the extraction of phenolic and flavonoid compounds. Ethanol is a polar solvent that is proficient in extracting a broad spectrum of bioactive compounds, including both polar and semi-polar constituents. However, ethyl acetate, a semi-polar solvent, is more selective and efficient for extracting phenolic and flavonoid compounds, particularly those with moderate polarity [22, 23]. Studies have indicated that ethyl acetate is effective for extracting phenolic compounds because of its intermediate polarity, which aligns well with the polarity of many phenolic compounds.

Previous research on *Sargassum polycystum* explains that extraction with ethyl acetate solvent results in higher total phenolic and total flavonoid contents than extraction using ethanol solvent [28]. A comparative study between the ethanol extract and ethyl acetate fraction of green tea leaves also indicated that the ethyl acetate fraction contained the most significant amount of flavonoid compounds [25]. Polyphenols and flavonoids were found in all extracts and fractions of *S.cristaefolium*, indicating that this plant is rich in antioxidant compounds. These phenolic and flavonoid compounds play an essential role in protecting cells from damage caused by free radicals, because they have an ideal chemical structure that allows them to effectively counteract the harmful reactivity of free radicals [26].

3.2 DPPH radical scavenging activity

Antioxidants are believed to act by donating hydrogen, which is why they are effective at scavenging DPPH compounds. An indicator of antioxidant activity is the concentration of antioxidants required to reduce the initial DPPH concentration by 50% (IC_{50}). A lower IC_{50} value corresponds to higher antioxidant power. The ethyl acetate fraction exhibited a lower IC_{50} value of 0.469 ± 0.0003 mg/mL, indicating stronger antioxidant activity than the ethanol extract, which had an IC_{50} value of 1.389 ± 0.0021 mg/mL (Tables 3 and 4).

This enhanced antioxidant activity in the ethyl acetate fraction was correlated with its higher TPC and TFC. Phenolic and flavonoid compounds are known for their antioxidant properties, primarily because of their ability to donate hydrogen atoms or electrons to neutralize free radicals. The higher concentrations of these compounds in the ethyl acetate fraction likely contributed to their superior radical-scavenging activity [27]. Additionally, the specific phenolic profile present in the ethyl acetate fraction may play a role in its potent antioxidant activity. Ethyl acetate tends to extract phenolic compounds with moderate polarity, which are associated with strong antioxidant activity [28].

Table 3. Antioxidant (DPPH) of the Ethanol Extract of *S.cristaefolium*

Concentration (mg/mL)	Average Free Radical Scavenging ± SEM (%)	IC_{50} ± SEM (mg/mL)
0.288	12.283 ± 0.038	1.390 ± 0.0021
0.576	25.813 ± 0.038	
1.153	44.407 ± 0.100	
1.729	62.396 ± 0.136	
2.306	76.757 ± 0.236	

Table 4. Antioxidant (DPPH) of the Ethyl Acetate Fraction of *S.cristaefolium*

Concentration (mg/mL)	Average Free Radical Scavenging ± SEM (%)	IC_{50} ± SEM (mg/mL)
0.113	27.322 ± 0.000	0.469 ± 0.0003
0.226	35.993 ± 0.037	
0.452	49.654 ± 0.037	
0.678	65.027 ± 0.000	
0.904	73.734 ± 0.036	

In the ethyl acetate fraction of *S.cristaefolium*, more significant activity in the sequestration of free radicals was observed. This activity increased with increasing concentrations. Flavonoid compounds contribute to the antioxidant activity. The higher the number of flavonoid compounds, the higher is the antioxidant ability of the ethanol and ethyl acetate fractions. The ability of the ethyl acetate fraction to contain more flavonoids may indicate that this fraction has significant potential to function as an effective antioxidant agent in protecting cells from damage caused by free radicals.

3.3 GC/MS chemical profiling

GC-MS analysis allows for the identification of major non-volatile compounds, particularly with high accuracy when compounds are derivatized (Figures 1, 2, Table 5 and 6) [29]. In the ethanol extract and ethyl acetate fraction, prominent peaks were observed, indicating the presence of major compounds, such as hexadecanoic acid methyl ester (RT 17.46) and trans-13-octadecenoic acid methyl ester (RT 19.14). The chemical structures of the main bioactive compounds are shown in Figure 3. These compounds exhibited higher peak intensities in the ethyl acetate fraction, highlighting their predominance in this fraction.

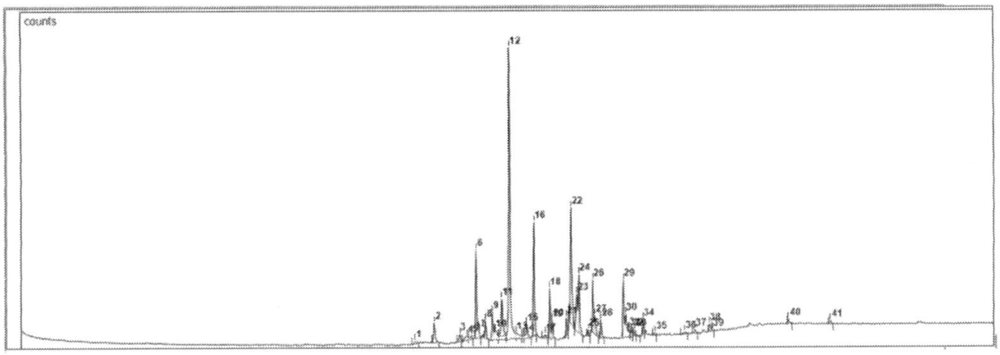

Fig. 1. GC-MS Chromatogram of the Ethanol Extract of *S.cristaefolium*

Table 5. Compounds Identified in the Ethanol Extract of *S.cristaefolium* by GC-MS

No.	Ret.Time (min)	Compound Name	Mol. Weight	Area %
1	14,88	[1,1'-Bicyclopropyl]-2-octanoic acid, 2'-hexyl-, methyl ester	322	0,43
2	15,40	Methyl tetradecanoate	242	2,20
3	16,13	Tetradecanoic acid, ethyl ester	256	0,65
4	16,33	6-Hydroxy-4,4,7a-trimethyl-5,6,7,7a-tetrahydrobenzofuran-2(4H)-one	196	0,92
5	16,45	Cyclopropanebutanoic acid, 2-[[2-[[2-[(2-pentylcyclopropyl)methyl]cyclopropyl]methyl]cyclopropyl]methyl]-, methyl ester	374	0,56
6	16,57	Neophytadiene	278	6,47
7	16,66	[1,1'-Bicyclopropyl]-2-octanoic acid, 2'-hexyl-, methyl ester	322	0,52
8	16,83	3,7,11,15-Tetramethyl-2-hexadecen-1-ol	296	1,24
9	17,02	3,7,11,15-Tetramethyl-2-hexadecen-1-ol	296	2,25
10	17,09	2-Methyl-E,E-3,13-octadecadien-1-ol	280	1,45
11	17,27	(Z)-Methyl hexadec-11-enoate	268	3,55
12	17,46	Hexadecanoic acid, methyl ester	270	26,19
13	17,65	Cyclopropanebutanoic acid, 2-[[2-[[2-[(2-pentylcyclopropyl)methyl]cyclopropyl]methyl]cyclopropyl]methyl]-, methyl ester	374	1,24
14	17,83	Cyclopropanebutanoic acid, 2-[[2-[[2-[(2-pentylcyclopropyl)methyl]cyclopropyl]methyl]cyclopropyl]methyl]-, methyl ester	374	0,35
15	17,93	Ethyl 9-hexadecenoate	282	1,57
16	18,13	Hexadecanoic acid, ethyl ester	284	7,44
17	18,43	Cyclopropanedodecanoic acid, 2-octyl-, methyl ester	366	0,44
18	18,56	Arachidonic acid	304	2,97
19	18,63	cis-5,8,11,14,17-Eicosapentaenoic acid	302	0,01
20	18,63	cis-5,8,11,14,17-Eicosapentaenoic acid	302	1,37
21	19,02	Methyl stearidonate	290	1,20
22	19,14	trans-13-Octadecenoic acid, methyl ester	296	10,08
23	19,31	Phytol	296	3,87
24	19,37	Heptadecanoic acid, 16-methyl-, methyl ester	298	4,20
25	19,62	9,12,15-Octadecatrienoic acid, 2,3-dihydroxypropyl ester, (Z,Z,Z)-	352	0,37
26	19,74	(E)-9-Octadecenoic acid ethyl ester	310	3,38
27	19,80	(E)-9-Octadecenoic acid ethyl ester	310	1,48
28	19,97	Heptadecanoic acid, 15-methyl-, ethyl ester	312	1,19
29	20,58	5,8,11,14-Eicosatetraenoic acid, methyl ester, (all-Z)-	318	4,20
30	20,65	cis-5,8,11,14,17-Eicosapentaenoic acid	302	4,00
31	20,73	6,9,12,15-Docosatetraenoic acid, methyl ester	346	0,15
32	20,80	6,9,12,15-Docosatetraenoic acid, methyl ester	346	0,21
33	20,89	cis-13-Eicosenoic acid	310	0,26
34	21,11	5,8,11,14-Eicosatetraenoic acid, ethyl ester, (all-Z)-	332	0,95
35	21,43	Ethyl iso-allocholate	436	0,31
36	22,25	Spirost-8-en-11-one, 3-hydroxy-, (3ß,5a,14ß,20ß,22ß,25R)-	428	0,32
37	22,52	Ethyl iso-allocholate	436	0,36
38	22,92	Hexa-t-butylselenatrisiletane	506	0,78
39	23,01	Ethyl iso-allocholate	436	0,04
40	25,06	Ethyl iso-allocholate	436	0,49
41	26,20	Octasiloxane, 1,1,3,3,5,5,7,7,9,9,11,11,13,13,15,15-hexadecamethyl-	578	0,37

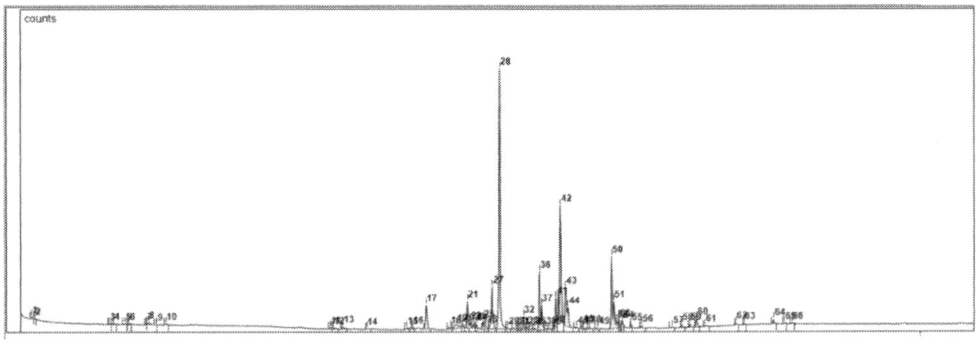

Fig. 2. GC-MS chromatogram of the Ethyl Acetate Fraction of *S.cristaefolium*

Table 6. Compounds Identified in the Ethyl Acetate Fraction of *S.cristaefolium* by GC-MS.

No	Ret.Time (min)	Compound Name	Mol. Weight	Area %
1	4,31	Isoxazolidine-3,5-dicarboxylic acid, dimethyl ester	189	0,40
2	4,39	Isoxazolidine-3,5-dicarboxylic acid, dimethyl ester	189	0,14
3	6,46	9-Octadecenoic acid (Z)-, hexyl ester	366	0,13
4	6,58	9-Octadecenoic acid, (2-phenyl-1,3-dioxolan-4-yl)methyl ester, cis-	444	0,25
5	6,91	9-Octadecenoic acid (Z)-, 2-(acetyloxy)-1-[(acetyloxy)methyl]ethyl ester	440	0,14
6	7,01	9-Octadecenoic acid (Z)-, 2-(acetyloxy)-1-[(acetyloxy)methyl]ethyl ester	440	0,17
7	7,50	9-Octadecenoic acid (Z)-, 2-(acetyloxy)-1-[(acetyloxy)methyl]ethyl ester	440	0,15
8	7,56	trans-Verbenyl caprate	306	0,62
9	7,79	9-Octadecenoic acid (Z)-, hexyl ester	366	0,03
10	8,06	9-Octadecenoic acid, (2-phenyl-1,3-dioxolan-4-yl)methyl ester, cis-	444	0,19
11	12,68	9-Octadecenoic acid (Z)-, 2-(acetyloxy)-1-[(acetyloxy)methyl]ethyl ester	440	0,29
12	12,76	Hexadecane, 1,1-bis(dodecyloxy)-	594	0,27
13	13,04	2,4-Di-tert-butylphenol	206	0,43
14	13,73	Estra-1,3,5(10)-trien-17ß-ol	256	0,16
15	14,86	Methyl 3-hydroxydodecanoate	230	0,26
16	15,02	1-Hexadecanol, 2-methyl-	256	0,14
17	15,39	Methyl tetradecanoate	242	3,37
18	16,10	1-Hexadecanol, 2-methyl-	256	0,51
19	16,26	6-Hydroxy-4,4,7a-trimethyl-5,6,7,7a-tetrahydrobenzofuran-2(4H)-one	196	1,76
20	16,45	Tetradecanoic acid, 12-methyl-, methyl ester	256	0,40
21	16,57	Neophytadiene	278	2,63
22	16,66	2-Pentadecanone, 6,10,14-trimethyl-	268	0,85
23	16,75	1,3-di-iso-propylnaphthalene	212	0,44
24	16,83	3,7,11,15-Tetramethyl-2-hexadecen-1-ol	296	0,33
25	17,02	3,7,11,15-Tetramethyl-2-hexadecen-1-ol	296	0,77
26	17,09	cis-13-Eicosenoic acid	310	0,67
27	17,26	9-Hexadecenoic acid, methyl ester, (Z)-	268	4,96
28	17,46	Hexadecanoic acid, methyl ester	270	27,98

No	Ret.Time (min)	Compound Name	Mol. Weight	Area %
29	17,69	Cyclopropanebutanoic acid, 2-[[2-[[2-[(2-pentylcyclopropyl)methyl]cyclopropyl]methyl]cyclopropyl]methyl]-, methyl ester	374	0,33
30	17,92	Cyclopropanebutanoic acid, 2-[[2-[[2-[(2-pentylcyclopropyl)methyl]cyclopropyl]methyl]cyclopropyl]methyl]-, methyl ester	374	0,37
31	17,98	12,15-Octadecadiynoic acid, methyl ester	290	0,16
32	18,12	Hexadecanoic acid, ethyl ester	284	1,32
33	18,22	cis-10-Heptadecenoic acid, methyl ester	282	0,37
34	18,31	Cyclopropanedodecanoic acid, 2-octyl-, methyl ester	366	0,13
35	18,44	Hexadecanoic acid, 14-methyl-, methyl ester	284	0,25
36	18,56	5,8,11,14-Eicosatetraenoic acid, methyl ester, (all-Z)-	318	4,93
37	18,63	Doconexent	328	2,40
38	18,73	cis-5,8,11,14,17-Eicosapentaenoic acid	302	0,16
39	18,88	Cyclopropanebutanoic acid, 2-[[2-[[2-[(2-pentylcyclopropyl)methyl]cyclopropyl]methyl]cyclopropyl]methyl]-, methyl ester	374	0,16
40	18,95	6,9,12-Octadecatrienoic acid, methyl ester	292	0,25
41	19,02	Methyl stearidonate	290	2,36
42	19,14	trans-13-Octadecenoic acid, methyl ester	296	12,90
43	19,30	Phytol	296	5,66
44	19,36	Heptadecanoic acid, 16-methyl-, methyl ester	298	2,21
45	19,61	6,9,12,15-Docosatetraenoic acid, methyl ester	346	0,32
46	19,74	trans-9-Octadecenoic acid, pentyl ester	352	0,31
47	19,80	9-Octadecenoic acid (Z)-, 2-hydroxy-1-(hydroxymethyl)ethyl ester	356	0,17
48	19,96	Ethyl 14-methyl-hexadecanoate	298	0,36
49	20,19	Ethyl iso-allocholate	436	0,15
50	20,57	5,8,11,14-Eicosatetraenoic acid, methyl ester, (all-Z)-	318	6,79
51	20,64	5,8,11,14,17-Eicosapentaenoic acid, methyl ester, (all-Z)-	316	3,84
52	20,72	6,9,12,15-Docosatetraenoic acid, methyl ester	346	0,42
53	20,79	Methyl 8,11,14,17-eicosatetraenoate	318	0,52
54	20,88	Methyl 9-eicosenoate	324	0,53
55	21,11	Cyclopropanebutanoic acid, 2-[[2-[[2-[(2-pentylcyclopropyl)methyl]cyclopropyl]methyl]cyclopropyl]methyl]-, methyl ester	374	0,32
56	21,40	Ethyl iso-allocholate	436	0,34
57	22,25	Spirost-8-en-11-one, 3-hydroxy-, (3ß,5a,14ß,20ß,22ß,25R)-	428	0,17
58	22,52	Methyl 11-docosenoate	352	0,38
59	22,74	Docosanoic acid, methyl ester	354	0,37
60	22,92	Diisooctyl phthalate	390	0,94
61	23,17	Methyl glycocholate, 3TMS derivative	695	0,21
62	24,02	Spirost-8-en-11-one, 3-hydroxy-, (3ß,5a,14ß,20ß,22ß,25R)-	428	0,31
63	24,25	Methyl glycocholate, 3TMS derivative	695	0,20
64	25,06	Ethyl iso-allocholate	436	0,54
65	25,36	Octasiloxane, 1,1,3,3,5,5,7,7,9,9,11,11,13,13,15,15-hexadecamethyl-	578	0,17
66	25,58	Methyl glycocholate, 3TMS derivative	695	0,21

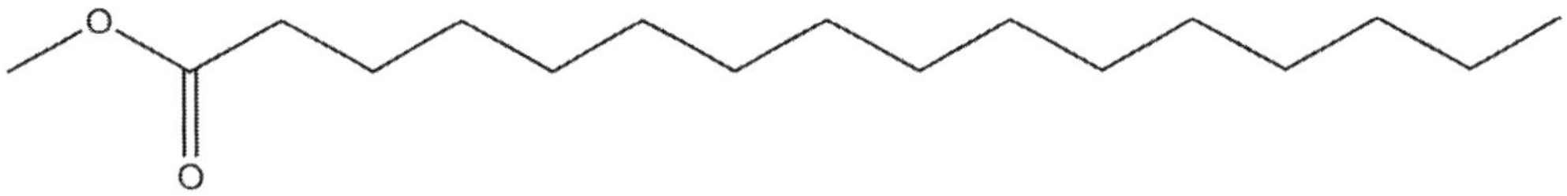

hexadecanoic acid, methyl ester [30]

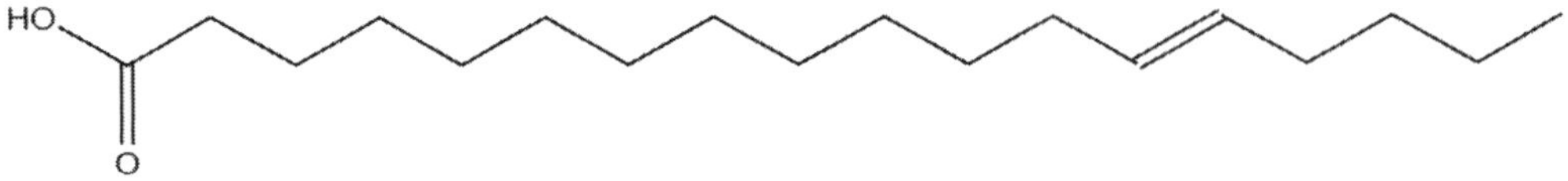

trans-13-octadecenoic acid, methyl ester [31]

Fig. 3. Major bioactive compounds

Hexadecanoic acid, known as palmitic acid, possesses antioxidant, anti-inflammatory, hypocholesterolemic, and 5-α reductase inhibitory properties, preserves beta-cell function, and inhibits α-amylase and intestinal α-glucosidase activity in vitro [32, 33, 34]. Trans-13-octadecenoic acid has anti-inflammatory effects [35]. Given the significant role of reactive oxygen species (ROS) in the pathogenesis of type-2 diabetes mellitus (T2DM), understanding the interplay among diabetes mellitus, aging, inflammation, ROS, and oxidative stress is important [36]. Many flavonoids and phenolic compounds, renowned for their anti-inflammatory and antioxidant actions, mitigate ROS effects, thereby potentially alleviating diabetes-related complications over the long term [32].

3.4 DPP-IV assay

In the in-vitro DPP-IV inhibition assay, sitagliptin was employed as the positive control alongside two samples consisting of the ethanol extract and ethyl acetate fraction of *S.cristaefolium*, detailed results are presented in Table 7. Sitagliptin functions as an inhibitor of dipeptidyl peptidase-IV (DPP-IV, also known as CD26), which is a clinically approved treatment for diabetes mellitus. Sitagliptin exhibited antioxidant properties and specifically inhibited DPP-IV [37]. Inhibiting DPP-IV enzyme is an effective strategy to maintain incretin hormones in the bloodstream, thereby enhancing insulin-dependent glucose disposal from the blood [38].

Table 7. DPP-IV inhibitory activity

Sample	Concentration (ppm)	Inhibition (%)
Sitagliptin	40	85.55 ± 1.64
Ethanol Extract	500	92.07 ±1.47
Ethyl Acetate Fraction	62.5	95,87 ± 2,54

These findings indicated that the ethyl acetate fraction exhibited superior DPP-IV inhibitory activity compared to the ethanol extract (Table 7). This aligns with previous observations, highlighting the high antioxidant capacity of the ethyl acetate fraction. Additionally, GC-MS analysis revealed increased concentrations of hexadecanoic acid methyl ester and trans-13-octadecenoic acid methyl ester in the ethyl acetate fraction. These compounds may contribute to the observed DPP-IV inhibitory activity, possibly enhanced by the antioxidant, anti-inflammatory, and antidiabetic properties of the sample.

4 Conclusions

This study shows that the *Sargassum cristaefolium* extract contains beneficial compounds, including antioxidants, total phenols, and flavonoids, which may help manage diabetes. The research highlights the potential of marine algae as natural inhibitors of the DPP-IV enzyme, which plays a role in blood sugar regulation. The ethyl acetate fraction demonstrated stronger antioxidant activity than the ethanolic extract. Additionally, chemical analysis (GC-MS) identified key compounds, such as hexadecanoic acid, methyl ester, and trans-13-octadecenoic acid, methyl ester, in the ethyl acetate fraction. These compounds may contribute to its ability to inhibit DPP-IV activity, which, combined with its antioxidant effects, suggests potential antidiabetic benefits.

This work was supported by an Indonesian Education Scholarship and the Ministry of Education, Culture, Research, and Technology of the Republic of Indonesia for providing funding with BPI Identification Number 202209090572.

References

[1] H. Sun, P. Saeedi, S. Karuranga, M. Pinkepank, K. Ogurtsova, B.B. Duncan, et al. Diabetes Res Clin Pract. **183**, 109119 (2022).

[2] W. Setyani, R. Murwanti, T.N.S. Sulaiman, T. Hertiani. J Adv Pharm Technol Res.**14**, 4 (2023).

[3] A. Petersmann, D. Müller-Wieland, U.A. Müller, R. Landgraf, M. Nauck, G. Freckmann, et al.**127**,S 01,S1–7, (2019).

[4] A. M. Abd-Elkader, S. Labib, T.F. Taha, F. Althobaiti, A. Aldhahrani, H. M. Salem, et al. Saudi J Biol Sci.**29**,3,1428–33 (2022).

[5] S. Bhattacharya, P.C. Sil. React Oxyg species. **5**,13,15–34 (2018).

[6] H. Ramos, P. Bogdanov, J. Huerta, A. Deàs-Just, C. Hernández, R. Simó. Antioxidants. **11** (2022).

[7] J.O. Unuofin, S.L. Lebelo. Oxid Med Cell Longev.**1**,1356893 (2020).

[8] M.A. Nauck, J.J. Meier. Diabetes, Obes Metab.**20**:5–21 (2018).

[9] A.K. Singh, D. Yadav, N. Sharma, & J.O. Jin. Pharmaceuticals, **14**(6), 586 (2021).

[10] A.E. Nugroho, G. Kusumaramdani, A. Widyaninggar, D.P. Anggoro, S. Pramono. Int Food Res J. **21**,2,785 (2014).

[11] S. Pramono, A.E. Nugroho. Int Food Res J. **22**,4 (2015).

[12] R.R. Remya, A.V. Samrot, S.S. Kumar, V. Mohanavel, A. Karthick, V.K. Chinnaiyan, et al. Adsorpt Sci Technol. 9104835(2022).

[13] S. Lim, A-H. Choi, M. Kwon, E-J. Joung, T. Shin, S-G. Lee, et al. Food Chem.**278**:178–84 (2019).

[14] J. Liu, S. Luthuli, Y. Yang, Y. Cheng, Y. Zhang, M. Wu, et al. T. Food Science and Nutrition. **8 (**2020).

[15] K. Hira, H. Farhat, N. Sohail, M. Ansari, J. Ara, S. Ehteshamul-Haque. Clin Phytoscience.**7**,1,40 (2021).

[16] W. Ko, H. Lee, N. Kim, H.G. Jo, E-R. Woo, K. Lee, et al. Antioxidants. **10** (2021).

[17] I. Isnindar, S. Luliana. Maj Obat Tradis. **25**,3,140–5(2020).

[18] E. Lukitaningsih, R. Rumiyati. Borneo J Pharm.**4**,1,29–35(2021).

[19] H.S. Parmar, P. Jain, D.S. Chauhan, M.K. Bhinchar, V. Munjal, M. Yusuf, K. Choube, A. Tawani, V. Tiwari, E. Manivannan, A. Kumar. Diabetes Research and Clinical Practice. 1;**97**(1):105-11 (2012).

[20] M.K.A. Sobuj, Islam MA, Haque MA, Islam MM, Alam MJ, Rafiquzzaman SM. J

Food Meas Charact.**15**,1653–65(2021).

[21] E. Edison, A. Dirahmi, N.M. Ariani, M. Ilza. Jurnal Pengolahan Hasil Perikanan Indonesia. **30**;23(1):58-66(2020).

[22] A. Altemimi, N. Lakhssassi, A. Baharlouei, D.G. Watson, & D.A. Lightfoot. Plants, **6**(4), 42 (2017).

[23] T. Lefebvre, E. Destandau, E. Lesellier. Journal of Chromatography A. Jan **4**;1635:461770 (2021).

[24] A. Arsianti, A. Bahtiar, V.K. Wangsaputra, N.N. Azizah, W. Fachri, L.D. Nadapdap, et al. Pharmacogn J.**12**,1 (2020).

[25] K. Indarti, E.F. Apriani, A.E. Wibowo, P. Simanjuntak. Pharmacogn J.**11**,4 (2019).

[26] D. Bešlo, N. Golubić, V. Rastija, D. Agić, M. Karnaš, D. Šubarić, et al. Antioxidants.**12**,6:1141 (2023).

[27] M.D. Catarino, R. Silva-Reis, A. Chouh, S. Silva, S. S. Braga, A. M. S. Silva, & S. M. Cardoso. Marine drugs, **21**(3), 172 (2023).

[28] S. Mariem, F. Hanen, J. Inès, S. Mejdi, K. Riadh. South African Journal of Botany. 1;**94**:114-21 (2014).

[29] A. A. Shalini, M.S. Ali, V. Anuradha, N. Yogananth, P. Bhuvana. Biocatal Agric Biotechnol.**19**,101148(2019).

[30] National Institute of Standards and Technology (NIST). (n.d.). Chemistry WebBook. Retrieved January 26, 2025, from https://webbook.nist.gov/cgi/cbook.cgi?ID=C112390&Mask=8

[31] National Institute of Standards and Technology (NIST). (n.d.). Chemistry WebBook. Retrieved January 26, 2025, from https://webbook.nist.gov/cgi/inchi?ID=U333613&Mask=200

[32] O.R. Olasehinde, O.R. Afolabi, O.V. Owolabi, A.B. Akawa, O.B. Omiyale. Sci African.**16**,e01178 (2022).

[33] R. Agada, D. Thagriki, D. E. Lydia, A. Khusro, J. Alkahtani, M.M. Al-Shaqha, et al. J King Saud Univ - Sci.**33**,2,101342 (2021).

[34] D.C. Tan, N.K. Kassim, I.S. Ismail, M. Hamid, M.S.A. Bustamam. Biomed Res Int.**1**, 7603125(2019).

[35] M.S. Khan, M. Uzair, M. Hanif, M. Jabeen, B.A. Chaudhary. Pak J Pharm Sci.**35**,6 (2022).

[36] M. Halim, A. Halim. Diabetes Metab Syndr Clin Res Rev.**13**,2,1165–72(2019).

[37] X. Wang, J. Ke, Y. Zhu, B. Cao, R. Yin, Y. Wang, et al. Cell Death Discov. **7**,1:236 (2021).

[38] R. Mahmood, W.K. Kayani, T. Ahmed, F. Malik, S. Hussain, M. Ashfaq, et al. BMC Complement Med Ther.**20**,1–17 (2020).

A review prospective study: Exploring bounty marine biota bioprospecting

Neviaty Putri Zamani[1*], *Alda Awayan* Banjarsari[2], *Jean Claudia* De Soysa[2], *Jane Cita Syafitri* Sudrajat[2], *Arta* Silaban[2]

[1]Center for Transdisciplinary and Sustainability Science. IPB University. Kampus IPB Baranang Siang, Bogor 16144. Indonesia
[2]Department of Marine Science and Technology, Faculty of Fisheries and Marine Sciences, IPB University, Bogor, Indonesia

Abstract. The exploration of secondary metabolite compounds from marine biota has become an important research subject, as these resources are diverse. The purpose of this article is to provide an overview of bioprospection research on secondary metabolites in marine biota exploration and clinical trials from 2014 to 2024 in Indonesia and around the world. This study used a Systematic Literature Review (SLR). Based on the 341 articles, 126 articles on Porifera were the most researched, while the least studied phylum was Arthropoda in one article. The most conducted bioprospection stage was compound exploration (304 articles), while there were two articles about clinical trials as the least researched. The most studied compound was an alkaloid in 49 articles, and the most studied potential was antibacterial in 60 articles. The results of this article review contain an overview of the extent of bioprospection of secondary metabolites in marine biota over the last ten years.

1 Introduction

The ocean is home to a wide variety of unique species, some of which remain unexploited [1][2]. The diversity of marine animals and plant species has attracted the attention of researchers to explore the great potential of marine ecosystems [2]. These species have unique biological functions that require the production of secondary metabolites. Secondary metabolites from various species can be used by humans as medicinal ingredients, food additives, and various other purposes [3][1]. These benefits can be discovered through marine bioprospection research.

Bioprospection is the exploration of marine organisms to identify and utilize biota for human use. The bioprospection stage includes exploration, identification, and characterization of natural compounds that can be used for scientific and commercial purposes. [4] This process is very important in scientific research as it enables the discovery of new compounds that can be used to develop useful health products, food, and raw materials for industrial products[5]. The role of bioprospection in the discovery of natural compounds,

* Corresponding author: neviaty@apps.ipb.ac.id

such as secondary metabolites, in potential species is very important [6]. Bioprospected products can increase knowledge and economic development owing to their high added value [7].

Research on secondary metabolites in marine bioprospection has increased in recent years, particularly in the context of drug and natural resource development [8]. Indonesia has great potential for this research because of its large species diversity and strategic location for studying marine ecosystems. This article demonstrates the economic and medical value of marine biodiversity, which can play a role in environmental conservation and preservation. This can increase awareness and support conservation efforts and help improve the management of marine resources. This study aimed to determine the secondary metabolites from marine biota to increase the knowledge of the biotechnological potential associated with marine biota species.

2 Materials and Methods

The data used were articles published from 2014 to 2024 and taken from Google Scholar. The search was conducted until April 24, 2024. There were 121 keywords used to search the articles in English and Indonesian of sea cucumber extract, sea star bioactive, sea urchin bioactive, sea star anti, sea urchin extract, echinoderm secondary metabolite, secondary metabolites coral, secondary metabolites jellyfish, hydrozoan secondary metabolism, crustacean bioactive compounds, mollusca secondary metabolites, cuttlefish bioactive compounds, perna viridis bioactive compounds, echinoderm anticancer, echinoderm antibacterial, echinoderm anti-inflammatory, starfish anti, sea cucumber against mice, cnidaria phytochemistry, cnidaria secondary metabolites, soft coral secondary metabolites, jellyfish compounds, secondary metabolite compounds in mangroves, gastropod secondary metabolites, bivalvia secondary metabolites, marine secondary metabolites, marine bioactive compounds, bioactive compounds, bioactive phytochemicals, bioactive echinodermata, bioactive cnidaria, bioactive angiospermae, bioactive mollusca, bioactive tracheopyta, and bioactive ochro@@ The data were collected using Google Sheets and further processed using Microsoft Excel to perform calculations, tabulations, and visualization of the research results. Visualization of cooperation between countries was performed using Flourish (https://app.flourish.studio).

3 Results and Discussion

A total of 341 articles were found during 2014–2024 on secondary metabolites of marine biota in Indonesia and from around the world (Fig. 1). Research on secondary metabolites can be said to be increasing despite a decrease in publications in 2020, 2021 and 2022. This may be due to a decrease in activities during the pandemic.
The stages used in this study were divided into four groups based on field exploration, compound exploration, animal testing, and clinical trials (Fig. 2). Compound exploration is the most commonly used method. This shows that the next stages of bioprospection research, such as animal testing and clinical trials, can be conducted. Funding and cooperation between experts in different fields, such as marine biology, pharmacy, doctors, law, and others, must be provided.

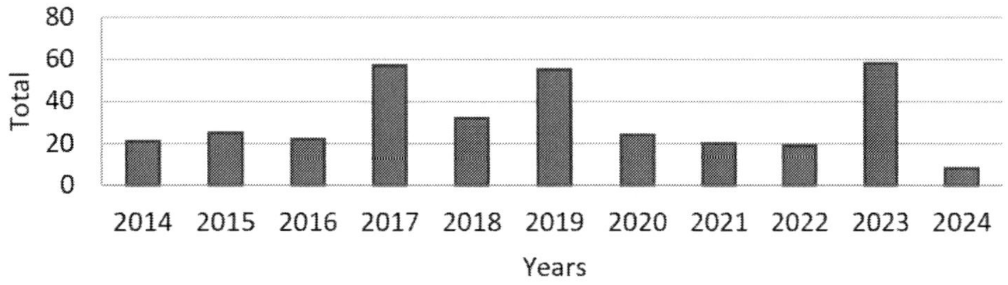

Fig. 1. Total of articles published from 2014 to 2024

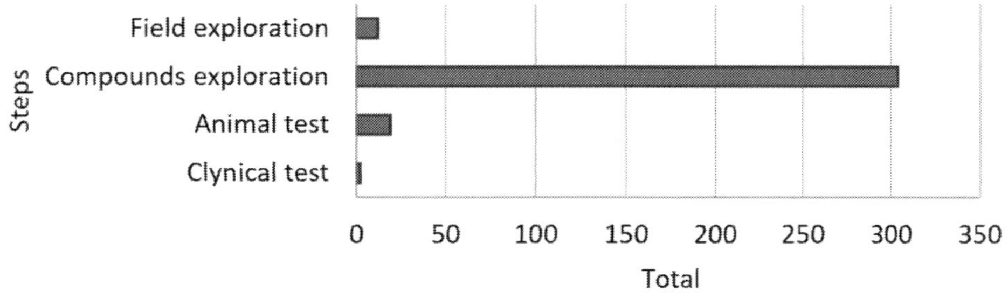

Fig. 2. Total of articles based on the stages of research conducted

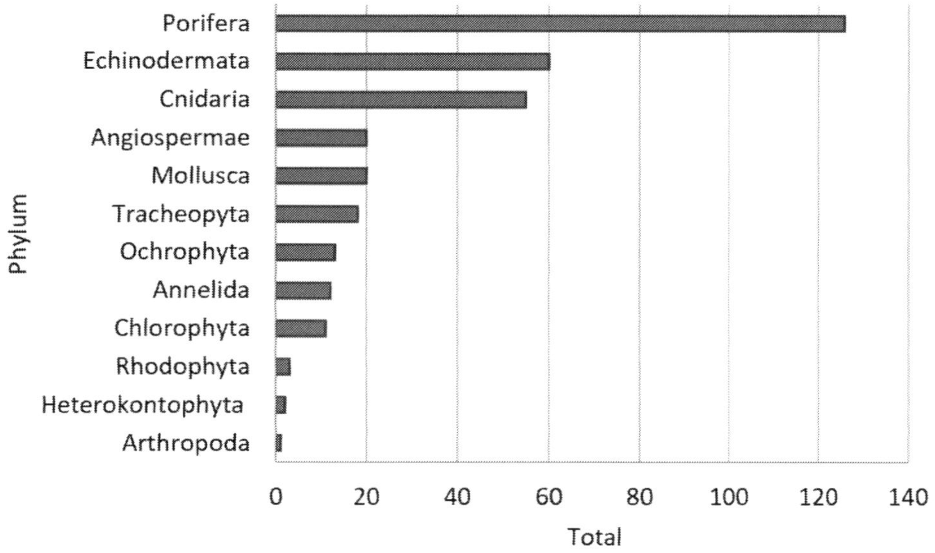

Fig. 3. Total of articles published based on the phylum studied

Porifera was the most studied phylum (Fig. 3). There are 23 genera of porifera or sponges, most of which are from the Demospongiae class. The Demospongiae class includes more than 90% sponge species that live in marine and freshwater environments [9]. Therefore, species from this class are often found in nature and have high abundance and diversity compared to other classes, such as Xestospongia sp. The most commonly found sponge species was Xestospongia sp.. This genus is the most often studied because of its diversity and potential secondary metabolite content. The abundance of the genus Xestospongia in various habitats, particularly in tropical and subtropical areas, has been widely studied [10]. The second most common phylum was Echinodermata. The genus found in Echinodermata comprises 17 species. The species studied in the search was Holothuria scabra. Holothuria scabra is one of the most studied holothuroideas and has been discussed in literature since the early 19th century [11]. The potential abundance of this species provides opportunities and makes it easier to use.

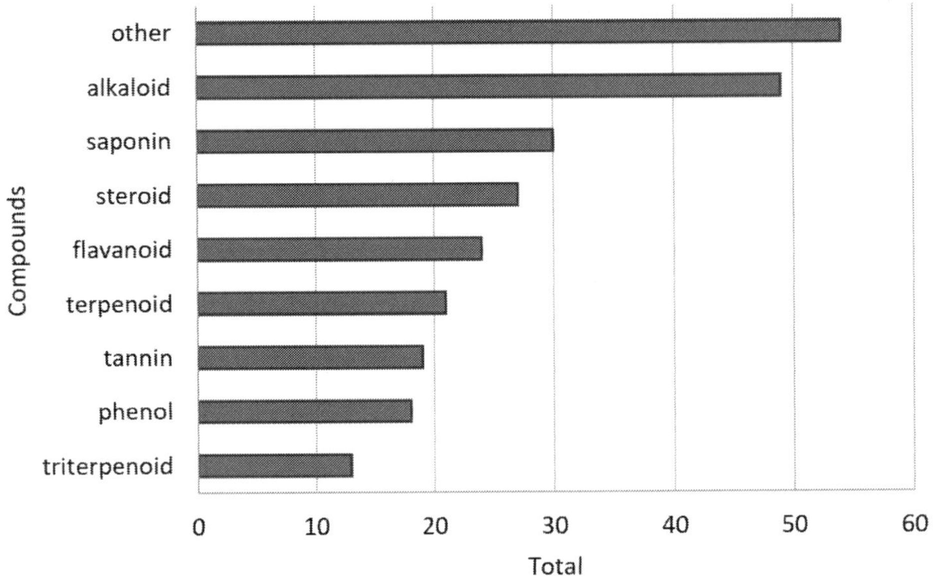

Fig. 4. Total of articles published based on the compounds studied

The most studied compounds in the bioprospection of secondary metabolites in marine life are alkaloids. Marine life produces various secondary metabolites that play important roles in the fields of biology and pharmacy. The secondary metabolites obtained were six commonly found compounds: alkaloids, saponins, steroids, flavonoids, trepenoids, tannins, phenols, and triterpenoids. Alkaloids are a class of nitrogen compounds with biological activities such as anticancer, antitumor, antibacterial, and antiviral activities.

Secondary metabolites are chemical compounds produced as a part of secondary metabolism [12]. These compounds are often not essential for the basic growth and development of organisms, but provide adaptive advantages such as protection from predators, competition with other organisms, and defense against pathogens. Secondary metabolites produced in biota are useful for self-defense against predator attacks and for competition with other species [13]. The content is also utilized by organisms for antibacterial, antiviral, and anticancer activities, which can inhibit the growth of pathogenic bacteria, viruses, and cancer cells [14].

The activities observed in bioprospection research on marine life include antibacterial, cytotoxic, antioxidant, anticancer, anti-inflammatory, antimicrobial, antitumor, and antiviral activities (Fig. 5). The most bioactive compounds secreted by marine biota are antibacterial, cytotoxic, and antioxidant compounds. This is in line with a review conducted by [15], which found that most groups of marine animals and plants produce antioxidant and antibacterial compounds. The antibacterial and antioxidant activities found in high amounts are also an indication that the methods used to identify these compounds are much cheaper and easier than other compounds. Cytotoxicity is the biological property of a chemical compound that can damage and destroy mammalian cells [16]. Bioactive compounds with cytotoxic activity can be used as drugs for various diseases such as cancer and leukemia [17][14].

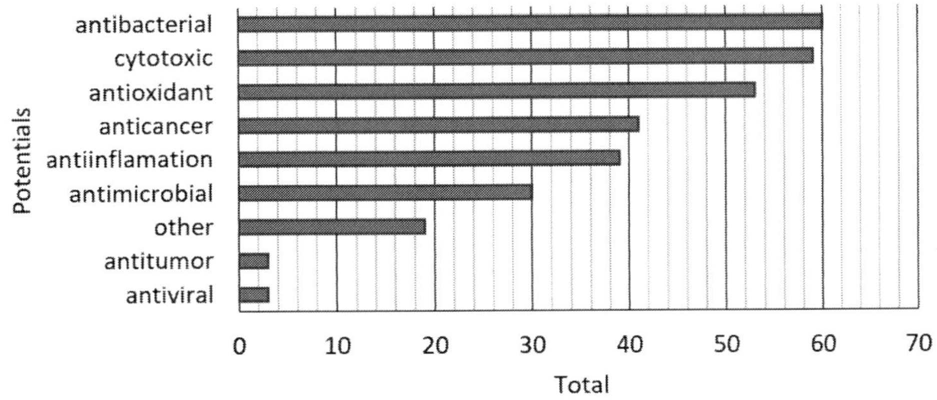

Fig. 5. Total of articles by potential studied

The research connections between countries come from the country of origin of the author mentioned in the article (Fig. 6). Research on secondary metabolites in marine biota has mostly been conducted in Indonesia. This can occur because Indonesia is an archipelago with abundant natural marine resources. The highest level of cooperation between author countries is Saudi Arabia, with eight authors from different countries. The most cooperative relationship was carried out by the authors from Malaysia and Indonesia in six links.

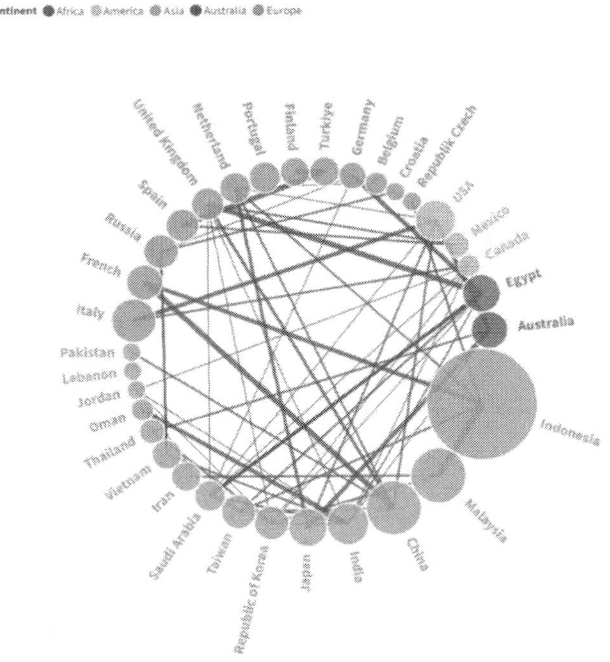

Fig. 6. Connections Cooperation of researchers between countries

Technological advances in marine bioprospection have further increased secondary metabolite research, despite a decline in 2020, 2021, and 2023, which may be due to decreased activity caused by the pandemic. This increase has occurred because technological developments, such as DNA sequencing, spectrometry, and other methods, can identify and characterize compounds more efficiently. Increased research on secondary metabolism has also occurred in other biota, such as plants and fungi [18][19][20].

Bioprospecting marine life to utilize secondary metabolites plays an important role in commercial and medical benefits, including the conservation of marine ecosystems [21]. This is because it allows scientists to identify exploitation-prone species, understand key species, and understand ecological interactions that are important for maintaining ecosystem balance. The discovery of bioactive compounds from the ocean can also reduce the pressure on already dwindling terrestrial resources, thereby encouraging more sustainable management of natural resources, not only on land but also in the ocean.

The utilization of marine biota resources also has an economic impact. The discovery of new compounds encourages commercial industries to develop innovative products that add value to biota [22]. The utilization and management of marine biota can also create jobs from research to industry at national and international levels [23]. This supports the welfare of the country and its local communities.

Bioprospecting of marine biota also faces various challenges, such as the sustainability of marine resource exploitation [24]. The overexploitation of resources, such as overfishing, threatens species that are vulnerable to extinction. Clear regulations that facilitate sustainable management and access to marine biota resources are obstacles that must be overcome through national regulations and cooperation between countries [25].

References

1. Shubina LK, Makarieva TN, Denisenko VA, Popov RS, Dyshlovoy SA, Grebnev BB, Dmitrenok PS, von Amsberg G, Stonik VA. Gracilosulfates A–G. Marine drugs. **18**, (2020)
2. Varijakzhan D, Loh J, Yap W, Yusoff K, Seboussi R, Lim S, Lai K, Chong C. **19**, (2021).
3. Khodzori FA, Mazlan NB, Chong WS, Ong KH, Palaniveloo K, Shah MD. **13**, (2023)
4. Dixit S, Shukla A, Singh V, Upadhyay SK. **19,** (2021)
5. Ashraf M, Hussain N, Baqar Z, Kumar A, Ferreira LF, Iqbal HM. InMicrobial biomolecules Jan 1 (pp. 301-324). Academic Press. (2023)
6. Arunachalam K, Yang X, Sasidharan SP, editors. Bioprospecting of tropical medicinal plants. Springer Nature; (2023)
7. de Oliveira I, Albuquerque B, Ueda JM, Alves MJ, Ferreira IC, Barros L, Heleno SA. Natural Secondary Metabolites: From Nature, Through Science, to Industry. **22**, (2023)
8. Hosseini H, Al-Jabri HM, Moheimani NR, Siddiqui SA, Saadaoui I. **62,** (2022).
9. Marzuki I. Eksplorasi spons indonesia: seputar kepulauan spermonde. 2018.
10. Khodzori FA, Mazlan NB, Chong WS, Ong KH, Palaniveloo K, Shah MD. Biomolecules. **13**, (2023)
11. Hamel, J.F., Eeckhaut, I., Conand, C., Sun, J., Caulier, G. and Mercier, A., *Advances in marine biology*, *91*, pp.1-286. (2022)
12. Saroinsong MS, Kandou FE, Papu A, Singkoh MF. Jurnal MIPA. **3,** (2014)
13. Hafizah I, Muliati FF, Sulastrianah S. Medula: Jurnal Ilmiah Fakultas Kedokteran Universitas Halu Oleo. **4,** (2016).

14. Rijai L. Jurnal Sains dan Kesehatan. **2,** (2019)
15. Zamani NP, Rahman L, Rosada RL, Tirtama W. *Overview of bioactivity studies on marine natural products.* InIOP Conference Series: Earth and Environmental Science Dec 1 (Vol. 944, No. 1, p. 012029). IOP Publishing. (2021)
16. Marliza H, Nurliyasman N, Haryani R, Lestari V. Jurnal Katalisator. **7,** (2022).
17. Hartmann T. Phytochemistry. **68,** (2007)
18. Wurtzel ET, Kutchan TM. Science. **353,** (2016)
19. Keller NP, Turner G, Bennett JW. Nature reviews microbiology. **3,** (2005).
20. Verpoorte, R., Contin, A. and Memelink, J., 2002. *Phytochemistry reviews, 1,* pp.13-25.
21. Carmona-Antoñanzas G, Tocher DR, Martinez-Rubio L, Leaver MJ. Gene. **534,** (2014)
22. Thirumurugan D, Cholarajan A, Raja SS, Vijayakumar R. Secondary metabolites-sources and applications. 1, (2018)
23. Pott DM, Osorio S, Vallarino JG. Frontiers in plant science. **10,** (2019)
24. Pushpangadan P, George V, Ijinu TP, Chithra MA. Journal of Traditional Medicine & Clinical Naturopathy. 7, (2018).
25. Flemsæter F. Regulating marine bioprospecting. Exploring the establishment of new regulatory regimes in the blue bioeconomy. Ocean & coastal management. **194,** (2020)

Determining potential location for development planning of ocean current energy for electricity resources in Bengkulu

Rauzatul Nazzla[1*], *Indra* Jaya[2], *Donwill* Panggabean[3], *and Ayi* Rahmat[2]

[1]Postgraduate of Marine Technology, Department of Marine Science and Technology, Faculty of Fisheries and Marine Science, Jl. Agatis Gedung Marine Center Babakan Dramaga Bogor
[2]Department of Marine Science and Technology, Faculty of Fisheries and Marine Science, Jl. Agatis Gedung Marine Center Babakan Dramaga Bogor
[3]Department of Postgraduate School, Jl. Pondok Cabe, Tangerang Selatan-Banten

Abstract. This study aims to determine the potential locations for alternative energy sources from waves and currents. Located in the Indian Ocean, the hydrodynamics potential of Bengkulu waters is quite high. In this study, we used data obtained from the OSCAR satellite series and bathymetric data obtained from Dishidrosal. The data series for currents are 5-year from 2019 to 2023 and were analyzed to classify the distribution values of ocean currents and bathymetry to generate the seabed topography profile. The method used in this study employ Inverse Distance Weighting and Fuzzy Logic. The sea surface current velocity is represented by the distribution of the average current speed (cm/s), which is divided into three classes slow (3.08–3.50), medium (3.5-7.84), and fast (7.84 – 12.65). The fuzzy analysis results show the estimation of suitable sites using defuzzification results at approximately 12 m. The classes for sea depth (m) were shallow (0.13-5.0), medium (5.0-20), and deep (20-315.35). The potential location is in the northern part of the province, specifically in North Bengkulu, Central Bengkulu, Bengkulu City, Seluma, and South Bengkulu, which topographically allows energy accumulation. These three districts can be designated as locations for the development of alternative electrical energy using ocean waves and currents.

1 Introduction

Energy is one of the most crucial factor of social development, economic growth, and human well-being [1]. It is predicted that by 2050, the world's energy consumption will rise by 50% over its current level [2]. Ocean energy is currently among the world's top exploitable renewable energy resources, along with solar and wind energies [3]. The energy policy target in indonesia, produced from various power plants and ocean electricity called (PLTAL). Ocean currents are driven by wind, water density differences, and tides [4]. The wave velocity is fast but unstable, whereas the ocean current energy has the characteristics of slow flow velocity but stable electrical energy transmission and a small impact on the power grid

* Corresponding author: rauzatulnazzla@apps.ipb.ac.id

[5]. Compared with tidal range plants, current plants have a lower environmental impact and are potentially less disruptive to marine ecosystems. Both types of energy utilize the motion of water to generate power, and tidal energy depends on predictable tidal cycles, whereas current energy captures energy from ocean currents, which are driven by a broader set of forces, including the Earth's rotation and temperature gradients in the water. While reviewed from a geographical perspective, ocean currents can be constructed in open ocean environments and often in deeper waters. Current energy can be captured in many coastal regions, even when tides are not particularly strong. This gives the current energy a potentially wider geographic range than tidal energy. Location selection is an important issue in contradictory criteria, or so-called complex multi-criteria decision-making [6][7].

Located on the coast of the Indian Ocean, Bengkulu Waters surface currents range from 0,2 -0,5 m/s and in 20 m depth range from 0,2 - 0,4 m/s [8]. What differentiates the concept of current and tide is the movement of water to propel the turbine. The generator converts the kinetic energy into electrical energy [9]. However, physical characteristics and hydrodynamic factors make the study area potentially vulnerable to disaster navigational risks. Therefore, it is essential to determine locations that are topographically safe, allow energy accumulation, and pay attention to shipping traffic safety.

Feasibility studies and power plant installations are expensive. An alternative approach can be conducted using available data, such as the data provided by the OSCAR. The raw data are computed using a simplified physical model for geostrophy, Ekman, and thermal wind dynamics from satellite-sensed sea surface height gradients, ocean vector winds, and sea surface temperature gradients. A global 0.25° × 0.25°-grid is used to display the daily averaged surface currents across an estimated well-mixed top 30 meters.

The satellite data approach has the advantage of providing a wide coverage. Previous research [10] has shown the potential of ORE (Ocean Renewable Energy) throughout Indonesia, including in Bengkulu Province, using satellite data and GIS analysis. This study intends to further examine one of the previously studies energy sources. Therefore, the purpose of this study is to produce a map of depth and ocean current distribution, and to produce a map of potential locations based on suitability values using Fuzzy Logic.

2 Methods

In this study, two variable inputs contributed to the determination of potential sites for power plant installation development planning. The bathymetry profile was obtained from the Dishidrosal and current velocities from satellite data. Satellite altimetry is an essential remote sensing technique for monitoring dynamic ocean conditions, including major ocean currents. Ocean currents are determined by satellite synthetic aperture radar (SAR). The data processing mechanism is shown in Figure 1, and an explanation is provided below.

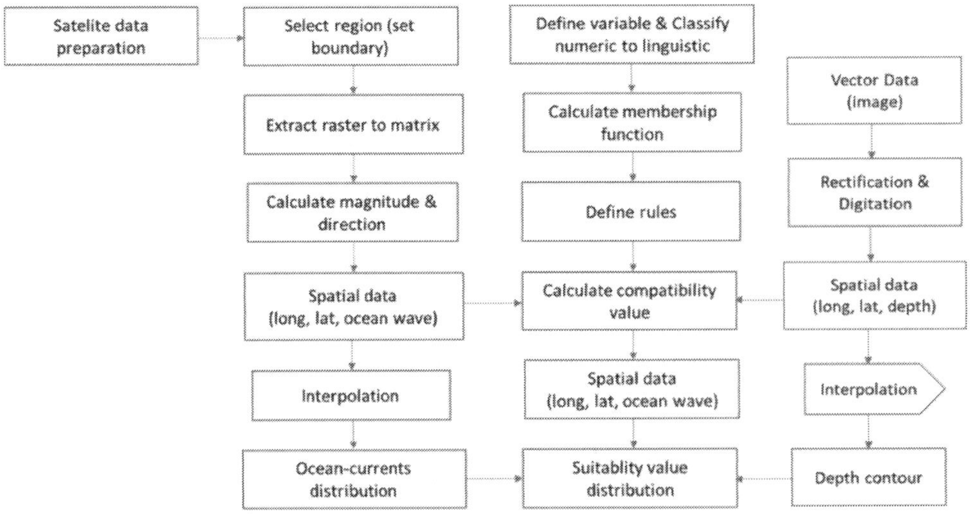

Figure 1. Research Workflow

2.1. Depth

Most available bathymetric charts are processed from echosounder data through digitizing and plotting to provide batchymetry chart contain with coordinate data and depth values. This value later proceeds with the fuzzy-logic process. The spatial distribution was interpolated using Inverse Distance Weighting (IDW).

2.2. Sea-surface Currents

Ocean surface currents are driven by global wind systems, which draw energy from the sun. The patterns of these surface currents are influenced by the wind direction, the Coriolis effect resulting from the Earth's rotation, and the geographical features that interact with the currents [4].

Vector represents the current velocity. With the north serving as a zero and clockwise measurement, its direction refers to the flow direction and uses degrees (°) as a unit. The water flowing through the columns moved more slowly than at the surface. Pressure and density are the two physical characteristics of water that affect it. Currently the u and v components have data. Two vectors were computed using Eq. (1) to determine the magnitude:

$$|v| = \sqrt{(u^2 + v^2)} \tag{1a}$$

$$\Theta = \left(\frac{v}{u}\right) \Big/ 180 * \pi \tag{1b}$$

$|v|$ is the magnitude representing the speed of the current or vector's length in the horizontal plane.
u is an east-west component of the current-velocity; the value indicates flow direction, positive means east, and negative means west.
v is a north-south component of the current velocity, the value indicates flow direction, positive means north, and negative means south

Θ is direction represented as degree of angle
π is a mathematical constant representing the ratio of the circumference of a circle' to its diameter.

2.3. Fuzzy logic

Fuzzy set theory is among the most preferred theories in decision making introduced by Zadeh (1965) for dealing with uncertainty and imprecision information associated with another. There are two potential sources for generating a fuzzy membership function, expert knowledge and real data [11]. The three steps of a fuzzy-logic system are Rule Evaluation, Defuzzification, and Fuzzification. Membership functions are used in the fuzzification and defuzzification processes to transfer non-fuzzy input values to fuzzy linguistic concepts and vice versa. A function called the membership degree indicates how much an input fits into a set [12][13][14]. A trapezoidal membership function for two variables represents fuzzy sets in a two-dimensional space. This function is useful in fuzzy logic systems for defining the degrees of membership of various inputs. The formula for trapezoidal membership is as follow:

$$\mu \text{ membership } [x, y] = \begin{cases} 0; & 0; \\ \frac{x-a}{b-a}; & \frac{y-a}{b-a}; \\ 1; & 1; \\ \frac{d-x}{d-c}; & \frac{d-y}{d-c}; \end{cases} \tag{2a}$$

α = smallest domain value with a zero membership level
b = smallest domain value that has a membership degree of one
c = value of the largest domain with a degree of membership of one
d = the value of the largest domain that has zero membership degree

As a result of normalization, all input and output variable values fell within the range [0,1]. The most popular methods are (i) min–max normalization, (ii) Z-score normalization, and (iii) decimal scaling normalization [15].

2.4. Spatial Analysis

Currents can be synoptically detected by satellite remote sensors [16] processed and visualized by geographic information systems (GIS). Calculating the statistical features and performing geoprocessing operations included how interpolation was performed. This method, known as Inverse Distance Weighting (IDW) interpolation, uses deterministic mathematical functions to assume a closer value that is more closely related than a further value. The IDW formula can be expressed as follows:

$$FZ_x = \sum_{i-1}^{n} W_i Z_i \tag{3a}$$

$$W_i = \frac{\left(\frac{1}{d_i}\right)^p =}{\sum_{j=1}^{n}\left(\frac{1}{d_j}\right)^p} \tag{3b}$$

$$\sum_{i=1}^{n} W_i = 1, p > 1 \tag{3c}$$

The IDW process in this study can be classified as a preprocessing step to fill in missing data gaps and increase the spatial resolution. The processed parameters include the depth and current components (u and v). After the empty values are filled in, the magnitude of the current velocity and direction of the current at each point are calculated as show in Table 1.

Table 1. Matrix format for parameters processed by IDW and example

Longitude (x)	Latitude (y)	Depth (m)	u	v	Magnitude ($|v|$)	direction (Θ)
102.7895	-4.4420	3.92746	5.4245	11.4339	12.6554	25.3939
101.7005	-3.3022	4.00431	-0.4001	-0.7638	0.8622	-152.4291

The idw process was conducted for each year from 2019 to 2023. A composite process must be carried out to determine the average of the 5-year data.

3 Result and Discussion

There are 2 (two) main results in the form of spatial information and graphs. Spatial information contained in the bathymetry chart, sea surface currents, and suitable value distribution. Graph as a result of data analysis to interpret the physical properties.

3.1 Depth Distribution

Most available bathymetric charts are processed from echo-sounder data using a digitizing and plotting procedure to produce digital data-sets [22]. Depth is an important factor for determining the optimal location for generator installation. If it is too shallow, it is affected by sedimentation, which can reduce the water capacity and damage the turbine [18] [19]. The depth distribution represents in seven districts and divided in 3 classes as shown in Table 2.

Table 2. Depth data 2019-2023

	Muko muko	**North Bkl**	**Central Bkl**	**Bkl City**	**Seluma**	**South Bkl**	**Kaur**
Minimum	1.76	1.31	3.65	3.33	3.34	0.13	0.32
Maximum	41.97	17.50	19.67	44.54	39.53	29.35	315.34
Mean	10.25	9.69	10.56	19.03	16.30	13.11	89.37

*Bkl: Bengkulu

This depth distribution provides information on the topographic conditions of a body of water. This information is directed at considerations of navigation safety and the protection of the power plant itself. The depth distribution and bathymetric chart are shown in Figure 1.

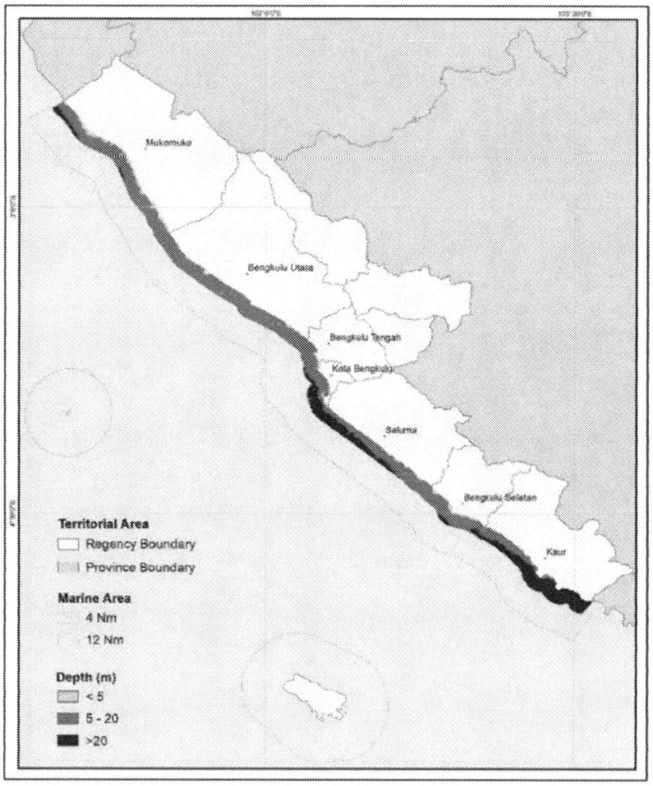

Figure 1. Depth Distribution

Shallow waters, especially near the coast, tend to have predictable ocean current patterns, making energy production more reliable. Shallow waters also allow the use of fixed bottom structures (turbines mounted directly on the seabed), which are simpler and more stable than floating systems. This can lead to more efficient energy extraction, as the velocity of the currents tends to be faster in shallow coastal areas. However, in areas with very shallow depths, the flow speed of the currents may be reduced, thereby limiting the energy extraction potential. In such locations, energy generation may be less efficient than in deeper water areas, where currents are faster.

3.2 Sea-surface Currents Distribution

Water currents initially involve at least two dimensions. A current measurement can be expressed as the direction of the current's movement or as its flow speed [20]. Sea surface current distribution is represented by the magnitude obtained from vector data.

Table 2. Sea-surface currents velocity (cm/s) for year 2019-2023

	2019	2020	2021	2022	2023	2019-2023
Minimum	1.00	1.89	0.65	1.28	0.60	0.60
Maximum	8.54	2.18	1.29	2.88	2.32	8.54
Mean	4.12	1.99	1.04	2.26	1.11	4.57

Sea surface currents distribution 5 years range from 1- 8.54 cm/s for 2019, 1.89 - 2.18 cm/s for 2020, 0.65 - 1.29 cm/s for 2021, 1.28 - 2.88 cm/s for 2022, and 0.60 - 2.32 cm/s for

2023. This means that from 2019-2023 to the current velocity range from 0.60 - 8.54 cm/s. Ocean currents are the culmulative of both local and distant influences, such as tides, winds, buoyancy fluxes, and various wave types; these forces have a significant impact on ocean currents [21] as shown on Figure 2.

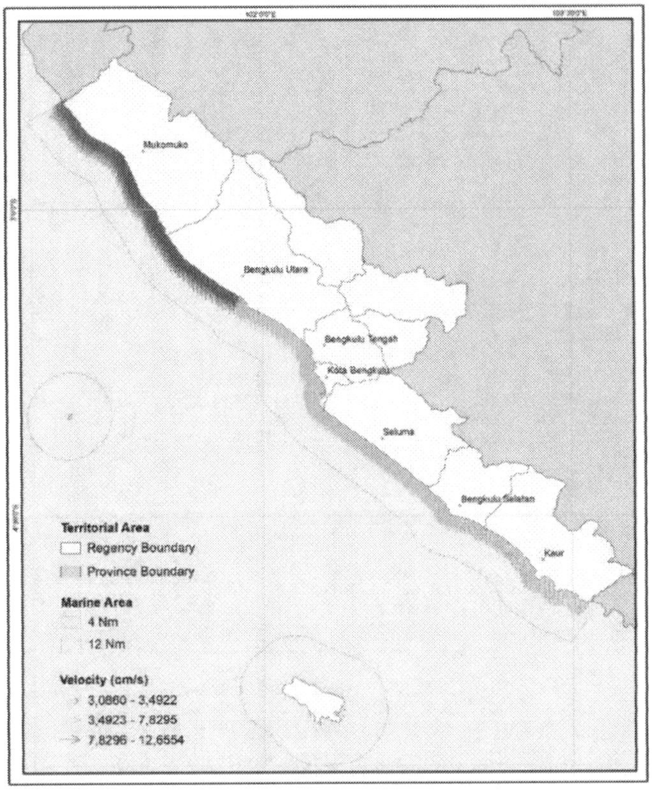

Figure 2. Sea-surface Current Distribution

The power of the current is proportional to the cubed velocity and flux. Minimum values of ocean energy source for ORE (Ocean Renewable Energy) ideally have to reach a value of 0.5 m/s [10]. The higher the speed, the more valuable is the ocean current [24]. The faster the current is, the more energy that can be extracted. Ocean currents must sufficiently fast to be economically viable for power plants. However, in places where ocean currents are generally slow but constant, weak-current power plants can still be implemented. Weak current power plants can also be useful in remote coastal regions, islands, and off-grid communities, where other forms of energy generation may be too costly or impractical.

3.3 Determining the Suitable Site

Building mathematical models while considering the element of uncertainty in decision -making process can enhance the adequacy, thereby increasing the validity and factual efficacy of the conclusions drawn from their analysis [22]. When determining a suitable site, it is possible to identify two main situations requiring a multi-criteria approach using the fuzzy logic system [23]. A fuzzy inference system (FIS) implements a nonlinear mapping from an input space to an output space [15]. In this process, we must define the membership function applied to the current velocity and depth. Both variables were divided into 3 groups, as shown in Table 3.

Table 3. Data range and membership

Currents		Depth	
range (cm/s)	**class**	**range (m)**	**class**
3.08 –3.50	slow	0.13 - 5.0	shallow
3.50 –7.84	medium	5.0 - 20.0	medium
7.84 – 12.65	fast	20.0 - 315.35	deep

Currents variable range from 3.08 – 12.65 cm/s and depth range from 0.13 – 315.35 m. Membership for currents consists of 3 class that is slow, medium, and fast. The depth consists of shallow, medium, and deep layers. In this case, a suitable membership function is the trapezoidal function. The trapezoidal membership function formula for the current and depth variables is as follows:

$$\mu \text{ membership } [x, y] = \begin{cases} 0; & 0; \\ \frac{x-0.308}{0.35-0.308}; & \frac{x-0.13}{5-0.308}; \\ 1; & 1; \\ \frac{0.784-x}{0.784-0.783}; & \frac{315.35-x}{315.35-19}; \end{cases}$$

A Trapezoidal is a type of membership function to describe values that fall within a certain range, with varying degrees of membership. In fuzzy inference systems, trapezoidal membership functions help evaluate how strongly an input value belongs to a particular fuzzy set, which is then used in rules to make decisions or predictions. After normalizing the available data of the case study and determining the linguistic variable, the next step is to contrive the rules that define the relationship between the input variables and output. The use of linguistic variables while formulating the 'if-then' rules to implement the knowledge base in the system [15]. The linguistic variable from the case study data consists of velocity members that should be associated with depth members and then combined with rule results to form a fuzzy output that represents the overall decision made by the fuzzy system. At this stage, fuzzy rules are applied to the fuzzified inputs, the degree of satisfaction of each rule is calculated, and the results are combined to create a final fuzzy output set as shown in Table 4.

Table 4. Relationship between velocity and depth variable with decision

Velocity\ Depth (input)	shallow	medium	deep	Decision/output
slow	NR	NR	NR	NR = Not Recommended
medium	NR	R	R	R = Recommended
fast	NR	VR	R	VR = Very Recommended

In defuzzification, fuzzy output is converted to a crisp value as the decision to generate a precise value from a fuzzy set. Figure 3 illustrates the relationship between the input and output applied to the graph of the trapezoidal function. The output is the decision of the site represented by a specific depth.

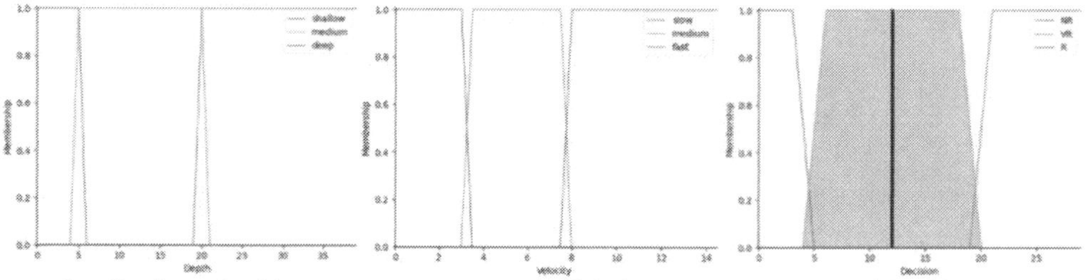

a) Depth membership b) Velocity c) Decision

Figure 3. Graph of membership function

The sea-surface current velocity is represented by the distribution of the average current speed, which is divided into three classes: slow, medium, and fast. Respectively 3.08 –3.50 cm/s, 3.5 - 7.84 cm/s, and 7.84 – 12.65 cm/s. The classes for sea depth are shallow, medium, and deep. Respectively is 0.13 - 5.0 m, 5.0 - 20 m, and 20 - 315.35 m. A suitable site can be estimated using the value of the defuzzification results obtained. At a depth of approximately 12 m.

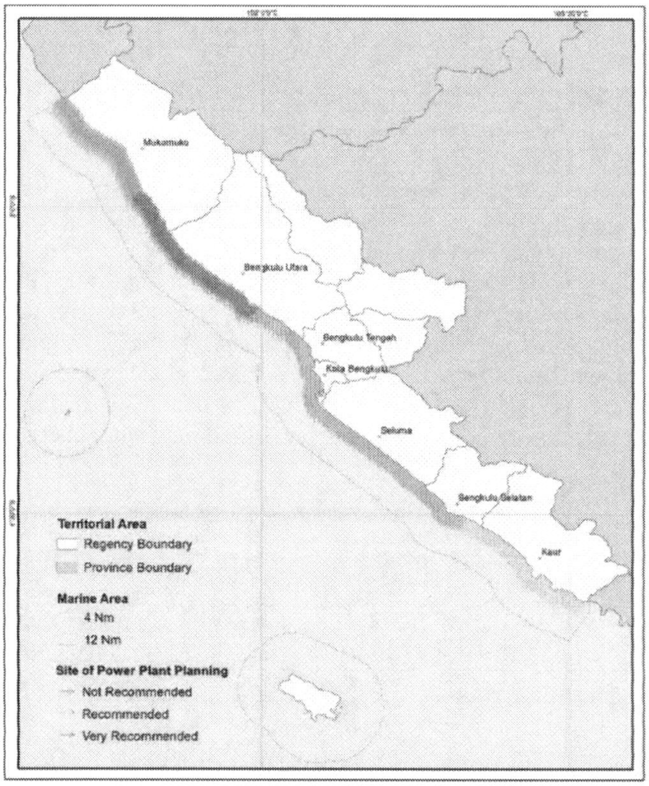

Figure 3. Site of Power-plant Planning

Mukomuko has the fastest velocity, but approximately 75% of the area classified as not recommended. It is considered that a strong current will affect the performance and increase the need for high maintenance for power-plant installation. However, this is approximately 25% of area that is recommended with approximately 75% of North Bengkulu Regency. Meanwhile, 25% of remaining area is classified as highly recommended along with Central Bengkulu, Bengkulu City, Seluma, and South Bengkulu Regency. Meanwhile, the Kaur Regency has the slowest velocity but a deeper slope than the others. Slow currents but large water masses can still produce sufficient energy to drive rotors and generate electricity, although the design and technology must be adapted to maximize efficiency under these conditions.

4 Discussion and Conclusion

The value of this ocean current distribution is the result of incoming and returning currents for one year. In coastal regions, incoming currents can be driven by ocean gyres (large circular ocean current systems), bringing water into the area. Returning currents can be driven by the Coriolis effect or tidal changes, where water returns to the open ocean. The resultant currents would describe the overall movement of the water within the coastal zone. Using this information, we can estimate the energy that can rotate the turbine through the mass of water that arrives and returns. The greater the speed of the ocean current, the greater the speed of the turbine to rotate and the greater the power obtained [24].

In shallow waters, currents tend to be more easily disrupted by the seafloor and may lose energy faster, especially near coastlines. Current energy plants are often deployed at depths ranging from a few meters to 50 m. Currently, the recommended depth for this case study is 5-20 m.

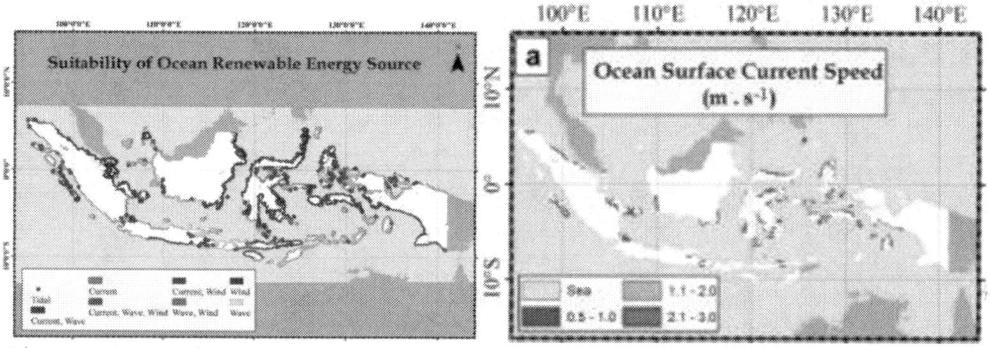

Figure 4. 1) Map of potential location of whole ORE sources, 2) Map of suitable location for current source [13]

The western part of Indonesia has only one location with a strong current in the range of 1.1 m/s until 2 m/s, and is located in the Malacca Strait [12] as shown in Figure 4. The Sea Surface Current (SSC) and waves had no significant value difference with the change in season. The average SSC for the whole of Indonesia is still below 0.5 m/s, but there is also a maximum value of up to 3.6 m/s in the west and second transition seasons. To verify the results study, we can compare with some expected data values or historical data from an established power plant. For example, computing using the MAE formula $= \frac{1}{n}\sum_{i=1}^{n}|y_i - \hat{y}_i|$.

The distribution of value provides a sufficient illustration of the energy potential in the area but seems vague. This caused by the nature of the data, which have a low spatial resolution (big pixel: 4 km). This limitation has to make the dataset goes through interpolation process to fill in the data gaps and increase the spatial resolution.

It is recommended that in situ measurements be taken to calibrate and cross-validate satellite data. To be more rigid, it is necessary to collect field data by increasing spatial resolution. The steps are to add the number of data collection stations or collect data to cover the empty pixel areas. Future work will involve examining waves and tidal currents to compare the energy sources that can produce optimum power for prospective renewable electricity sources.

References

1. R. Cullen. "Evaluating renewable energy policies". Australian Journal of Agricultural and Resource Economics. 61,1–18. DOI: 10.1111/1467-8489.12175. (2017)
2. U. S. E. I. Administration, "International Energy Outlook 2019," (2019).
3. F. Amjad, E. B. Agyekum, L. A. Shah, A. Abbas. "Site location and allocation decision for onshore wind farms, using spatial multi-criteria analysis and density-based clustering. A techno-economic-environmental assessment, Ghana". Sustainable Energy Technologies and Assessments. **47**, 101503 (2021)
4. NOAA. "What is a current?". https://oceanservice.noaa.gov/facts/current.html. (2024)
5. L. Chen, F. Zhang, X. Lin. Design of Water Deflection Mechanism for Ocean Current Energy Generation Device. IOP Conf. Series: Journal of Physics: Conf. Series 1064 012063. doi:10.1088/1742-6596/1064/1/012063. (2018)
6. Y. Wu, J. Zhang. J. Yuan, S. Geng, H. Zhang. Study of decision framework of offshore wind power station site selection based on ELECTRE-III under intuitionistic fuzzy environment: A case of China. Energy Conversion and Management, **113**, 66-81. (2016)
7. M. C. Toklu, Ö. Uygun. Location selection for wind plant using AHP and axiomatic design in fuzzy environment. Periodicals of Engineering and Natural Sciences (PEN), 6(2), 120-128. (2018).
8. K. Fahmi, E. Indrayanti, WB. Setyawan. Kajian Arus dan Batimetri di Perairan Pesisir Bengkulu. JURNAL OSEANOGRAFI. **3**, 4: 549 – 559. (2014)
9. EBTKE. Pengembangan Energi Arus Laut. https://ebtke.esdm.go.id/post/2011/04/25/138/pengembangan.energi.arus.laut [July 4, 2024] (2011)
10. N.P. Purba, J. Kelvin, R. Sandro, S. Gibran, R.A.I. Permata, F. Maulida. & M. K. Martasuganda. Suitable Locations of Ocean Renewable Energy (ORE) in Indonesia Region – GIS Approached. Conference and Exhibition Indonesia - New, Renewable Energy and Energy Conservation (The 3rd Indo-EBTKE ConEx 2014). Energy Procedia **65**. 230–238. 2015
11. I. Derbel, N. Hachani, H. Ounelli. M2embership Functions Generation Based on Density Function. Conference: 2008 International Conference on Computational Intelligence and Security, CIS. Suzhou, China. DOI:10.1109/CIS.2008.211. (2008)
12. G. Wang, L. Wu, Y. Liu, X. Ye. A review on fuzzy preference modeling methods for group decision-making. J. Intell. Fuzzy Syst. **40**, 10645–10660. (2021)
13. P. Perny, M. Roubens. Fuzzy preference modelling. In Fuzzy Sets in Decision Analysis, Operations Research and Statistics; Springer: Boston, MA, USA. pp. 3–30. (1998)
14. L. Liu & R. Vuillemot. A Generic Interactive Membership Function for Categorization of Quantities. Hal Open Science. (2023)
15. K. Pujaru, S. Adak, T. K. Kar, S. Patra, S. Jana. A Mamdani fuzzy inference system with trapezoidal membership functions for investigating fishery production. Decision Analytics Journal. **11**, June 2024, 100481. (2024)
16. V. Klemas. Remote Sensing of Coastal and Ocean Currents: An Overview. Journal of Coastal Research **28**, 3:576-586. DOI:10.2112/JCOASTRES-D-11-00197.1. (2012)
17. R. H. Stewart. Introduction To Physical Oceanography. Texas A & M University. 2000.

18. B.S.Thapa., O. G., Dahlhaug., & B Thapa. Sediment erosion in hydro turbines and its effect on the flow around guide vanes of Francis turbine. Renewable and Sustainable Energy Reviews. **49**, September 2015, 1100-1113
19. D. Felix, I. Albayrak, A. Abgottspon, and R. M. Boes. Optimization of hydropower plants with respect to fine sediment focusing on turbine switch-offs during floods. IOP Conf. Series: Earth and Environmental Science. 28th IAHR symposium on Hydraulic Machinery and Systems (IAHR2016). **49**, 122011. IOP Publishing. (2016)
20. VIMS. Measuring Currents.https://www.vims.edu/research/units/labgroups/tc_tutorial/current_measure.php. [15 August 2024]
21. Y. Ashkenazy, H. Gildor. on the Probability and Spatial Distribution of Ocean Surface Currents. Journal Of Physical Oceanograph. American Meteorological Society. **41**. (2011)
22. M. A. D. de O. Ferreira, L. C. Ribeiro, H. S. Schuffner, M. P. Libório, P. I. Ekel. Fuzzy-Set-Based Multi-Attribute Decision-Making, Its Computing Implementation, and Applications. Axioms. **13**, 142. https://doi.org/10.3390/axioms13030142. (2024)
23. C. Kahraman, S. C. Onar, B. Oztaysi. "Fuzzy multicriteria decision-making: a literature review". International journal of computational intelligence systems. **8**, 4, 637-666. (2015).
24. H. Rahmi L., B. Dharmala S., A. Gediana, A. Yusup , W. Septria. Optimasi Pembangkit Listrik Tenaga Arus Laut Menggunakan Sistem Turbin Savonius Termodifikasi. Berkala Fisika. **18**, 2, April 2015, p 75 – 82. ISSN: 1410 – 9662. (2015)

Ecologically benign parameters: setting Bangka Island, East Likupang for a nature ecotourism destination

Christin R. M. Malinggas[1,2*], *Anik M.* Hariati[3], *Luchman* Hakim[4], *Dewa Gede Raka* Wiadnya[5], and *Riski Agung* Lestariadi[6]

[1]Post Graduate Programme of Fisheries and Marine Science, Universitas Brawijaya Malang, Indonesia
[2]Department of Research and Development of North Sulawesi Province, Indonesia
[3]Department of Aquaculture, Faculty of Fisheries and Marine Science, Universitas Brawijaya Malang, Indonesia
[4]Department of Biology, Faculty of Mathematics and Natural Sciences, Universitas Brawijaya Malang, Indonesia
[5]Department of Fisheries and Marine Resources Utilization, Faculty of Fisheries and Marine Science, Universitas Brawijaya Malang, Indonesia
[6]Department of Fisheries Agrobusiness, Faculty of Fisheries and Marine Science, Universitas Brawijaya Malang, Indonesia

Abstract. Likupang was established as a new super-priority tourism destination through the ministerial regulation of Indonesia. This study aims to analyze the tourism benign performance of Bangka Island, a small island targeted as the main underwater attraction within Likupang. Six dive sites were selected from 21 sites available for dive tourism. Tourism Sustainability Index (SI) was estimated based on eight attributes: coral health, coral life-forms, number of fish species, fish diversity index, fish dominance index, water current, water visibility, and water depth. Coral health and life-forms were measured using a point intercept transect of 50 m with three replicates. Fish diversity was measured based on a line-imagery transect of 50×10 m^2 with three repetitions, and species richness was based on total species count. Fish diversity index was calculated based on the presence of each species at each sampling site. Analysis resulted in an average life-coral cover of 37.7±5.7 %, indicating *moderate* health status. There were 227 fish species found in all dive sites, with a high diversity index of 3.28, and a low dominance index of 0.07. With water depths range between 4-10 m, coral reefs were visible from the surface. These all attributes resulted a sustainability index of 32.4, categorized as *almost sustainable*. Fish diversity is considered to be the most sensitive attribute affecting the sustainability of a destination. As tourism creates more jobs, local tourism management may involve local fishermen. This new compensation may reduce fish and coral reef extraction.

* Corresponding author: malinggas_ch78@student.ub.ac.id; dgr_wiadnya@ub.ac.id

1 Introduction, local tourism management

In recent decades, the tourism sector has become one of the driving sectors of a country's economic growth, affecting the global economy [1]. Based on data from World Travel and Tourism [2], the growth of the Gross Domestic Product (GDP) of the tourism sector globally exceeded economic growth from 2011 to 2018. In Indonesia, the tourism sector is one of the largest contributors to foreign exchange after oil and gas [3]. To develop tourism in Indonesia, Government Regulation Number 50 of 2011 was issued, concerning the National Tourism Development Master Plan 2010-2025. The regulation includes ten priority tourism destinations to be developed, and Likupang is one of the priority tourism destinations in North Sulawesi. More specifically, Likupang is also part of five super priority tourism destinations according to the Ministry of Tourism and Creative Economy (MTCE) (National Plan of Action). White sand beaches, panoramic views, coral gardens, and cultural heritage sites are the main attractions of Likupang [4]. The most attractive site, yet less explored, is Bangka Island within Likuang. It has at least 21 spectacular dive sites on its island.

As a super priority tourism development, Likupang, including Bangka Island, is predicted to receive a tremendous increase in tourism arrivals due to progressive promotion and a diversion from the already overcrowded visit of Bunaken National Park, a nearby area [5]. Consequently, Likupang may ultimately turn into mass tourism, a negative risk that should be prevented. The marine waters of Bangka Island have been recently declared as a Marine Protected Area (Ministerial Regulation of Marine Affairs and Fisheries, MMAF, No. 128/2023). Tourism arrivals to Bangka Island will soon be regulated to ensure the sustainable use of natural resources within and surrounding protected areas.

Bangka Island covers a total land area of approximately 4,788 ha and is inhabited by 2,749 local populations. Eight resorts were present to support underwater tourism, each with a dive center, a dive instructor, and some local dive guides. The Island is surrounded by 1,062 ha of coral reef ecosystems, which are attractive for tourism. This study aims to analyze coral health, fish diversity, and other attributes that determine tourism sustainability. This information is important for determining management strategies that are in line with sustainable tourism principles.

2 Material and Methods

2.1 Research Location and Data Collection

This study was conducted between April and December of 2023. Six dive sites were selected from 21 dive sites currently available for tourism; two dive sites were located inside the marine protected area (ST4 and ST5), and four of the remaining sites were selected randomly from 19 dive sites (Figure 1). The selection was based on the fact that only 20 % of coral reefs located inside the protected area.

The supporting policy documents were based on Government Regulation No. 50/2011, Ministerial Regulation of MTCE No. 11/2022, and MMAF No. 128/2023. Tourism amenities were investigated through resorts and local accommodations on the island. Coral reef examination uses the point intercept transect (PIT) method [6], and fish diversity through underwater visual censuses [7]. These were completed by two experienced dive instructors (MSDT-424341 and PRO-SSI-80816) who were certified in coral reef and fish examinations.

2.2 Parameters and attributes

Coral reef examinations were conducted using coral-life forms and detailed forms of Hard Coral Life. A line transects of 50 m was deployed at each dive site parallel to the coastline, with three repetitions (total transect line of 150 m). A total of 12 coral life-form categories were identified in every 50 cm of point in the transect line: Hard Coral Life (HCL), Soft Coral (SC), Others (OT), Sponge (SP), Zoanthids (ZO), Rubble (R), Sand (S), Rock (RC), Dead Coral (DC), Dead Coral with algae (DCA), Halimeda (HA), and Turf Algae (TA) [8]. This summarized a total of 300 life-form data points within the site. Hard Coral Life (HCL) was further separated into 10 categories: Acropora Branching (ACB), Acropora Submassive (ACS), Acropora Tabulate (ACT), Acropora Digitate (ACD), Coral Branching (CB), Coral Massive (CM), Coral Submassive (CS), Coral Encrusting (CE), Coral Mushroom (CMR), and Coral Heliophore (CHL) [9]. All parameters were examined via scuba diving following the transect line. Each life form was recorded using a special form attached to a slate.

Fish diversity was based on species identification within the transect line with an additional imagery transect of 5 m to the left and 5 m to the right of the transect line (total imagery area of 150 × 10 m^2) [7]. Species confirmation was performed by direct identification and video scanning of the transect line. Divers identified and recorded the species in the slate. The second diver video scanned all the fish along the transect line. The species name was determined following species correction from the scanned video.

Substrate depth (m) was measured using a dive watch (gauge) during coral examination by the divers. The data consisted of minimum and maximum depths along the transect line. The water current was measured using a Flowatch FL-03 current meter (cm sec^{-1}). The water visibility was measured using a Secchi Disk. This indicates transparency of coral reefs, which can be seen from surface water, when 100 % of visibility.

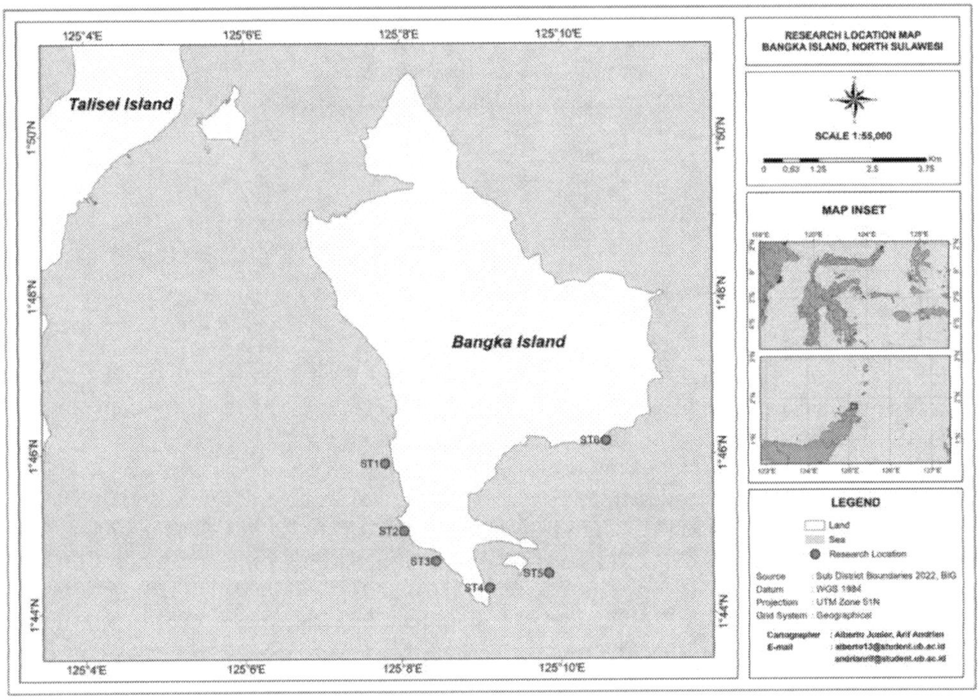

Fig. 1. Bangka Island and dive site locations selected for this research. ST1 = Linggua, ST2 = Coral Eye, ST3 = Busabora, ST6 = Areng Kambing, ST4 = Tanjung Husi, ST5 = Sahaung. ST4 and ST5 are dive sites located inside the protected area (Ministerial Decree of MMAF No. 128/2023).

2.3 Data Analysis

Coral health status was calculated based on the percentage ratio between total HCL divided by total life form in each 100 points of transect, using the following formula 1 [8]:

$$CH = \frac{\sum_{i=1}^{100} HCL}{\sum_{i=1}^{100} TLF} * 100\ \% \ldots\ldots\ldots\ldots (1)$$

where CH = Coral Health, HCL = Hard Coral Life, and TLF = total life form (100).

Coral Health (CH) is divided into four categories [10]: (1) CH < 25 % is categorized as bad, (2) CH < 50 % is Moderate, CH < 75 % is good, and (4) CH ≥ 75 % is excellent [10]. The Hard Coral Life Form was calculated based on the total number of HCL forms found within the 50 m transect. The categories are (Table 1): (1) life forms < 4, are categorized as bad; (2) life forms < 6, moderate; (3) life forms < 8, good; and life forms ≥ 8, as excellence.

The Fish Diversity Index (FDI) is calculated using Shannon and Weaver [11] based on the following formula 2:

$$H' = \sum_{i=1}^{s} - P_i \ln P_i \,;\, where\ P_i = \frac{n_i}{N} \ldots\ldots\ldots\ldots (2)$$

where H' is Fish Diversity Index, P_i is the probability of the presence of the ith species, n_i is the number of ith species, S is the number of species, and N is the total number of individual species.

The Fish Dominance Index (FDoI) describes the extent to which a species dominates the community. The formula to calculate the FDoI value uses Simpson [11] based on the following equation 3.

$$FDoI = \sum_{i=1}^{n} p_i^2 = \sum_{i=1}^{n} \left(\frac{n_i}{N}\right)^2 \ldots\ldots\ldots\ldots (3)$$

where FDoI = Dominance Index, P_i is the probability of the presence of the ith species, n_i is the number of ith species, and N is the total individual number of all species. The total fish species were summed from all six sampled dive sites. This value was used to score the sustainability index, as shown in Table 1.

The sustainability status of the six dive sites on Bangka Island, Likupang, was analyzed using the Rapfish method with a Multi-Dimensional Scaling (MDS) approach [12]. All the attribute criteria, scores, and indicators used to determine the categories are presented in Table 1.

Table 1. Criteria (ecological attributes), scores, indicators, and measures used to assess sustainability index of in Bangka Island for dive site

No	Attribute	Score	Indicator / Measure	References
1	Coral Health	1 - 4	Percent life-coral cover < 25 % = 1, bad < 50 % = 2, moderate < 75 % = 3, good ≥ 75%, = 4, excellence	[13]
2	Coral-life form	1 - 4	Number of life form < 4 life form = 1, bad < 6 life form = 2, moderate < 8 life form = 3, good ≥ 8 life form = 4, excellence	[14]
3	Fish Diversity Index (FDI)	1 - 4	FDI < 1, score = 1, is bad FDI < 2, score = 2, is moderate	[14]

No	Attribute	Score	Indicator / Measure	References
			FDI < 3, score = 3, is good FDI ≥ 3, score = 4, excellence	
4	Number of fish species	1 - 4	≤ 20 fish species = 1, bad < 50 fish species = 2, moderate ≤ 100 fish species = 3, good ≥ 100 fish species = 4, excellence	[14,15]
5	Fish Dominance Index (FDoI)	1 - 4	≤ 1 dominance index = 1, bad ≤ 0.7 dominance index = 2, moderate ≤ 0.5 dominance index = 3, good ≤ 0.3 dominance index = 4, excellence	[14]
6	Water Current (cm sec^{-1})	1 - 4	> 50 cm sec^{-1} = 1, bad > 30 cm sec^{-1} = 2, moderate > 15 cm sec^{-1} = 3, good ≤ 15 cm sec^{-1} = 4, excellence	[13]
7	Water visibility (%)	1 - 4	< 20 % = 1, bad < 50 % = 2, moderate < 80 % = 3, good ≥80 % = 4, excellence	[16]
8	Water depth (m)	1 - 4	≥ 30 m = 1, bad < 30 m = 2, moderate < 20 m = 3, good < 10 m = 4, excellence	[13]

The (SI) was calculated as the average value from all eight attributes. The field values from all eight attributes are fed into Table 1. It should match one of the following four categories: 1 (bad), 2 (moderate), 3 (good), or 4 (excellence). The score value of each attribute was calculated as the percentage ratio to the highest score value, which was 4. This value means sustainability index of a specific attribute. The Sustainability Index (SI) in this context is the average sustainability index for all eight attributes. The SI was divided into four categories, as presented in Table 2 [14].

Table 2. Range of Sustainability Index (SI) and meanings per category index

No	Sustainability Index (SI)	Category
1	0.0 – 25.0	not sustainable
2	26.0 – 50.0	almost sustainable
3	51.0 – 75.0	simply sustainable
4	76.0 – 100.0	very sustainable

The sustainability index for each attribute was then fed into the RapFish Excel WorkBook model [12]. Through RapFish modeling, leverage analysis can be performed to identify sensitive attributes that can increase or decrease the sustainability index. The leverage value was calculated as the percentage of the Root Mean Square (RMS) of each attribute. The Leverage values would vary between 0 – 10. The higher the RMS value, the more sensitive the attribute that affects the sustainability index.

3 Results and Discussion

Bangka is an island with a total land area of approximately 4,788 ha. The island is inhabited by 2,749 a population. The coastal area is surrounded by coral gardens and seagrass beds. The clear water, diverse fish species, and beach panorama all provided visitor satisfaction with a unique experience. As tourism grows, eight resorts are equipped with rooms and villas. Each resort anchors a speed boat to pick reserved visitors from Likupang and accompany dives or snorkeling within 21 available dive sites. Back packers can use regular boats at low cost. The island is now included in the super-priority tourism destination part of Likupang (MTCE Regulation No. 11/2022). The increase in tourism visits is mainly due to progressive promotion from the local government and overflow visitors from Bunaken National Park, which is already overcrowded. On the other hand, part of the coastal marine area has been declared a marine protected area (MMAF Regulation No. 128/2023). To maintain the sustainability of Bangka Island as a tourist destination, visitors must be regulated.

3.1 Coral Health

Data on the percentage coral coverage and other physical parameters are presented in Table 3. Line transects were deployed at 4-10 m water depth. Almost all the coral reefs can be seen in the surface water (visibility of 100 %).

The highest coral cover was found at Linggua dive site (56.7±5.5 %). This was significantly higher than the coral cover at Tanjung Husi and Sahaung (Figure 1), both of which are located inside the protected area (Table 3). The establishment of the area as a marine protected area may increase coral cover in the future. The lowest coral cover was found at Site6, Areng Kambing (Figure 1), categorized as *Bad* (Ministerial Decree of Environment No. 4/2001). Most coral cover was categorized as *Moderate* (37.7 %) (Table 3). The average percentage of coral cover at 6 dive sites on Bangka Island was higher than the percentage of coral cover in Bunaken National Park (32.9 %) [17].

Most coral reefs in Indonesia are threatened by destructive overfishing activities [18]. The community on Bangka Island is mainly dependent on marine resources, especially fish, as a source of food and livelihood [19]. Destructive fishing is the most effective and cheapest method to harvest fish. This is probably the reason for the bad and moderate life coral cover surrounding the island. The presence of eight resorts with dive centers may create new alternative jobs for locals. More frequent visits to dive sites may indicate the presence of management. This may prevent local fishermen from using destructive fishing equipment. Hence, more tourism activities within dive sites can increase coral coverage [20]. The coral cover surrounding the island that was investigated 10 years ago [21] was significantly lower than that found in this study.

Table 3. Hard coral coverage (%-HCL) together with coral life forms and other physical parameters measured at all the selected dive sites

Location	Depth (m)	Current (cm sec^{-1})	Water visibility (%)	Total Life Form	Hard Coral Coverage (%)	Coral Health categories
ST 1	5-7	20.69	100	8	56.7 ± 5.5^{a}	good
ST 2	4-6	10.39	100	10	46.7 ± 3.0^{ab}	moderate
ST 3	5-7	12.18	100	8	32.3 ± 7.9^{b}	moderate
ST 4	5-7	13.89	100	9	34.3 ± 1.9^{b}	moderate
ST 5	7-10	21.49	83	8	40.3 ± 4.8^{b}	moderate
ST 6	5-7	13.58	100	8	16.0 ± 1.5^{c}	bad

Remark: ST1, Linggua dive site, ST2 = Coral Eye, ST3 = Busabora, ST4 = Tanjung Husi, ST5 = Sahaung, ST6 = Areng Kambing

Ten categories of coral life forms were measured in this study. At least eight life forms were found at almost all dive sites. All ten life forms were found specifically at the Coral Eye (Table 3). This indicates good coral diversity within the area. Bangka Island is located at the heart of the coral triangle [22] within the Sulu Sulawesi Eco-Region [23]. This area is recognized as having the most diverse marine biodiversity. Specific to Bangka Island, a new sponge species (*Dercitus bangkae*) associated with mangroves was recently identified [24].

Coral life forms were distributed unevenly between and among dive sites (Table 4). Branching coral (CB), especially Acropora, together with massive coral (CM), were the most dominant in all dive sites. Acropora branching (ACB) can easily crack into rubble under strong currents [25]. The average current in the area was considered moderate (Table 3) and favorable for the growth of branching corals. However, Coral Massive (CM), which can survive in strong currents [26] also dominates the dive site. Acropora digitata (ACD) and mushroom coral (CMR) were rarely found at all dive sites. Mushroom corals are typical of solitary corals. However, its presence may never be dominant in the area. Generally, under favorable environmental conditions, every life form competes with each other, and there is an opportunity for every life form to dominate in the system.

Table 4. Distribution of coral life form (%) among dive sites

Dive Site	ACB	ACS	ACT	ACD	CB	CM	CS	CE	CMR	CHL
ST 1	5.9	26.5	2.9	0.0	29.4	25.9	2.9	5.3	1.2	0.0
ST2	23.6	8.6	1.4	2.1	15.7	17.9	7.1	5.0	1.4	17.1
ST 3	6.2	41.2	0.0	1.0	5.2	35.1	5.2	1.0	0.0	5.2
ST 4	12.6	21.4	3.9	1.9	20.4	24.3	0.0	11.7	1.0	2.9
ST 5	15.7	15.7	3.3	3.3	21.5	9.9	0.0	20.7	0.0	10.0
ST 6	50.0	2.1	6.3	2.1	16.7	18.8	0.0	2.1	2.1	0.0
Average	19.0^{a}	19.3^{a}	3.0^{b}	1.7^{b}	18.2^{a}	22.0^{a}	2.5^{b}	7.6^{b}	1.0^{b}	5.9^{b}
Error	16.5	13.9	2.2	1.1	8.0	8.6	3.7	7.4	0.8	6.7

3.2 Reef fish

Based on records from underwater visual censuses and confirmed with video scanning, 227 reef fish species were found in all dive sites. It consisted of 36 families. The family with the highest number of species was Pomacentridae (45 species), followed by Labridae (36 species) and Chaetodontidae (21 species). The total reef fish species in Indonesia may account for more than 500 species [27]. Indonesia has exported more than 600 ornamental reef fishes [28]. However, under a total dive coverage of approximately 9,000 m^2, this finding can be considered rich [29]. High coral diversity with colorful ornamental fish is the most attractive for underwater tourism, leading the area to become a favorite destination after Bunaken National Park.

Total reef fish species between dive sites varied between 88 - 124 species, and between 21 - 30 families (Table 5). The Fish Diversity Index was excellent (Table 2) for all sites. Most species belong to the family Pomacentridae, commonly called Damselfish and Clownfish. Both groups are very common in the aquarium trade for ornamental reef species [28]. Wrasses are a common name for the family Labridae. It is popular in aquarium trade fishes, mainly because of its abundance. In contrast, members of the family Chaetodontidae (butterfly fishes) are recognized for their unique and attractive body shape and behavior.

Species of the family Chaetodontidae are well-known as indicator species for coral health [30]. The more species found in an area, the healthier the reef [31].

Table 5. Diversity index, dominance index, total species count and total family of reef fishes in each dive site

Parameters	Dive Site					
	ST 1	ST 2	ST 3	ST 4	ST 5	ST 6
Diversity index	3.08	3.37	3.43	3.11	3.06	3.65
Dominance index	0.08	0.07	0.07	0.09	0.09	0.04
Total species	88	124	120	92	93	97
Total family	21	30	24	22	26	22

The diversity index at each dive site ranged from 3.06 to 3.65, with an average value of 3.28, and the dominance index ranged from 0.04 to 0.09, with an average of 0.07 (Table 5). The ecological index analysis indicated that the coral reef fish diversity level on Bangka Island, Likupang, is high, with an uneven distribution of individuals among species, resulting in a stable community. This was demonstrated by the average value of the fish dominance index < 0.09. These findings suggest the presence of an ecosystem balance and equitable distribution of different fish species [10]. The high fish diversity in Bangka Island's coral reef ecosystem is the main attraction for underwater ecotourism [29].

3.3 Sustainability status

The Sustainability Index (SI) analysis resulted in an average score of 32.4. This value matched the category of *almost sustainable* (Table2) [13][14][15][16]. Sustainable tourism is defined as tourism together with associated infrastructures that both now and in the future operate within natural capacities for the regeneration and future productivity of natural resources; recognize the contribution that people and communities, customs, and lifestyles make to the tourism experience; and accept that these people must have an equitable share in the economic benefits of local people and communities in the host areas [32]. Tourism infrastructure includes, attractions among others. The eight parameters currently used to measure the sustainability index for Bangka Island are only for underwater attraction. Therefore, it only matches the underwater activities.

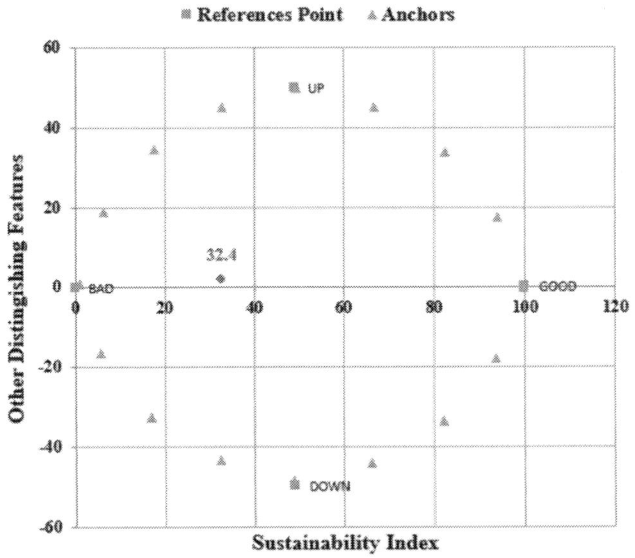

Fig 2. Sustainable Index (SI) of tourism at Bangka Island based on RapFish analysis [15]

Leverage is one or more sensitive attributes, resulting in a negative or positive impact on the sustainability index. It is calculated as the Root Mean Square (RMS) of each attribute [15]. The values vary between 0 and 10. The leverage results are shown in Figure 3. Two attributes, the number of fish species and fish dominance index, were recognized as the most sensitive indices that determine underwater tourism arrivals. This means that a high number of fish species and low fish dominance (high fish diversity) are the main motivations for tourists to visit Bangka Island [33]. To maintain underwater activities that are attractive to tourism, local tourism management should keep reef fish rich in species. Consequently, fishing on ornamental reef fish should be limited or prohibited.

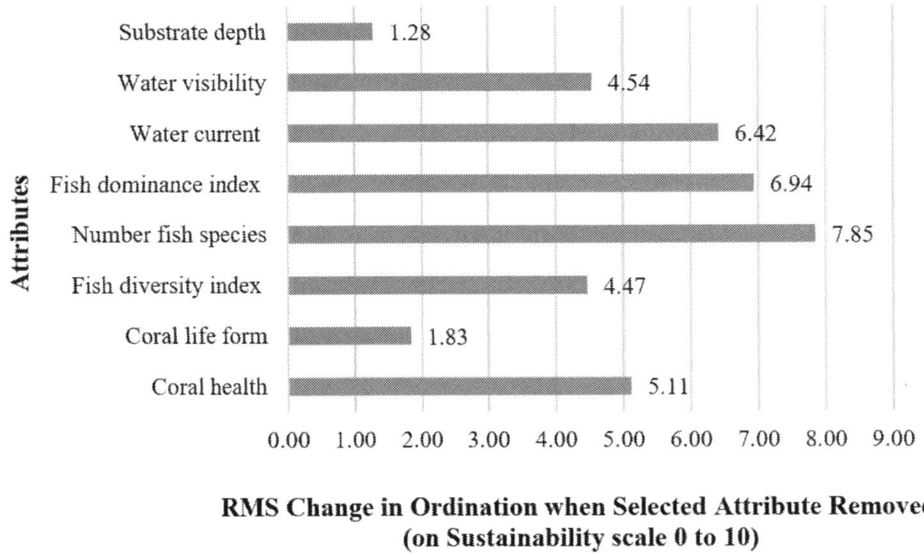

Fig 3. Leverage attributes that sensitive to the sustainability of Bangka Island as tourism destination

Through MMAF Regulation No. 128/2023, approximately 20 % of marine waters of Bangka Island were set aside as protected areas. This actually covers two dive sites (Tanjung Husi and Sahaung) where fishing or any extraction will soon be prohibited. However, the other 19 dive sites were still open for extraction by local ornamental reef fishermen. To reduce further extraction by fishermen, the government or local tourism management should introduce a compensation path into a new alternative livelihood. As the number of tourism visits increases, this can be achieved through their involvement in tourism activities. Resorts, accommodations, restaurants, and local transport and services create jobs that can be fulfilled by locals [34]. To meet all the criteria of a specific job, local tourism management must conduct skill and capacity training.

4 Conclusion

Based on these eight attributes, it can be concluded that tourism in Bangka Island was categorized as almost sustainable with a sustainability index of 32.4. Fish species richness and diversity were the most sensitive attributes that could maintain tourism. Setting aside part of Bangka Island's surrounding waters as protected area can affect only two dive sites (Tanjung Husi and Sahaung). As tourism creates many job opportunities for locals, local tourism management and the government must conduct skill and capacity training to meet the criteria required for the job.

I would like to express my acknowledgment to The North Sulawesi Provincial Government for the scholarship during my doctoral studies, the Faculty of Fisheries and Marine Science, Universitas Brawijaya, Malang, for providing the opportunity to gain knowledge, as well as dive instructors (Efra Wanta and Pratama Diffi Samuel) who have provided support in this research and Bogor Agricultural University for providing the opportunity to participate in the International Seminar on Ocean Sciences and Sustainability 2024.

References

[1] N. Comerio and F. Strozzi, "Tourism and its economic impact: A literature review using bibliometric tools," *Tour. Econ.*, vol. 25, no. 1, pp. 109–131, 2019, doi: 10.1177/1354816618793762.

[2] World Travel and Tourism Council, "Travel and Tourism: Economic Impact 2021." [Online]. Available: https://wttc.org/Portals/0/Documents/EIR/EIR2021 Global Infographic.pdf?ver=2021-04-06-170951-897

[3] A. Wiratha and N. Kemalasari, "Strategi Inovasi Pariwisata Indonesia Untuk Bertahan Pada Masa Pandemi Covid 19," *J. Multidisiplin Madani*, vol. 2, no. 5, pp. 2335–2356, 2022, doi: 10.55927/mudima.v2i5.378.

[4] K. S. Andaria, R. H. E. Sendouw, A. L. Lonto, X. E. Lobja, and N. J. Sindua, "Identification of Potential Tourism in East Likupang Subdistrict Bangka Island North Minahasa," in *International Conference on Social Sciences*, 2018, pp. 742–746. doi: 10.2991/icss-18.2018.151.

[5] M. F. Farhum, J. Jompa, M. Restu, and D. Salman, "Carrying capacity analysis in Bunaken National Park to support marine tourism activity (dive and snorkelling tourism)," *IOP Conf. Ser. Earth Environ. Sci.*, vol. 763, no. 1, 2021, doi: 10.1088/1755-1315/763/1/012069.

[6] M. Facon *et al.*, "A comparative study of the accuracy and effectiveness of Line and Point Intercept Transect methods for coral reef monitoring in the southwestern Indian Ocean islands," *Ecol. Indic.*, vol. 60, pp. 1045–1055, 2016, doi:

10.1016/j.ecolind.2015.09.005.

[7] F. Tony, D. Gede, R. Wiadnya, and L. Hakim, "Diversity of reef fish in Halang Melingkau Island , South Kalimantan , Indonesia," *Biodiversitas*, vol. 21, no. 10, pp. 4804–4812, 2020, doi: 10.13057/biodiv/d211046.

[8] R. Kurniawan, R. D. Putra, and L. P. Aji, "Method for Calculating the Coral Health Index (Case Study: Mapur Island, Bintan, Indonesia)," *BIO Web Conf.*, vol. 134, 2024, doi: 10.1051/bioconf/202413406022.

[9] F. Tony, S. Soemarno, D. G. K. Wiadnya, and L. Hakim, "Habitat Biodiversity as a Determinant of Fish Community Structure on Coral Reefs in Halang Melingkau Island, Kotabaru, South Kalimantan, Indonesia," *Egypt. J. Aquat. Biol. Fish.*, vol. 25, no. 1, pp. 351–370, 2021.

[10] D. Oetama, L. Hakim, T. D. Lelono, and M. Musa, "Health status of coral reef at Moramo Bay Marine Conservation Area, Southeast Sulawesi, Indonesia," *Biodiversitas*, vol. 25, no. 4, pp. 1454–1464, 2024, doi: 10.13057/biodiv/d250413.

[11] L. Lianthuamluaia *et al.*, "Understanding spatial and temporal patterns of fish diversity and assemblage structure vis-a-vis environmental parameters in a tropical Indian reservoir," *Environ. Sci. Pollut. Res.*, vol. 26, no. 9, pp. 9089–9098, 2019, doi: 10.1007/s11356-019-04305-z.

[12] A. A. A. Putri, R. Purwaningsih, and S. Hartini, "A comparison of rap-tourism method and multi attribute aggregation in sustainability assessment of tourist destination," in *IOP Conference Series: Earth and Environmental Science*, 2022. doi: 10.1088/1755-1315/1098/1/012046.

[13] F. F. Muhsoni and M. Efendy, "Coral Reefs Eco Tourism Sustainability Management In The Gili Labak Island, Using Rapfish Method," *J. Kelaut.*, vol. 10, no. 2, pp. 192–204, 2017, doi: :10.21107/jk.v10i2.3235.

[14] D. B. Wiyanto, N. Harahab, Rudianto, and A. Sartimbul, "Cultural Heritage Conservation of 'the United State Army Transport (Usat) Liberty' Shipwreck Site As a Sustainable Scuba Diving Ecotourism," *Int. J. Conserv. Sci.*, vol. 11, no. 4, pp. 931–944, 2020.

[15] D. Ario, M. Karuniasa, M. P. Patria, and K. A. Roeroe, "Assessment of Bunaken Island for sustainable tourism destination using the rapid appraisal for fisheries," in *IOP Conference Series: Earth and Environmental Science*, 2021, pp. 0–12. doi: 10.1088/1755-1315/716/1/012094.

[16] M. Fattah, C. A. Intyas, H. Ruminar, and M. E. Arif, "Evaluation of the Sustainability Status of the Bangsring Underwater With Indonesia'S Coral Reef Ecosystem," *Geoj. Tour. Geosites*, vol. 49, no. 3, pp. 858–865, 2023, doi: 10.30892/gtg.49302-1086.

[17] J. N. W. Schaduw *et al.*, "Data On Percentage Coral Reef Cover In Small Islands Bunaken National Park," *Data Br.*, vol. 31, p. 105713, 2020, doi: 10.1016/j.dib.2020.105713.

[18] G. G. R. Bernard, A. L. Kellam, and S. Szereday, "Differential size frequency distribution of hard coral colonies across physical reef health gradients in Northeast Peninsula Malaysia," *Reg. Stud. Mar. Sci.*, vol. 61, p. 102872, 2023, doi: 10.1016/j.rsma.2023.102872.

[19] F. Ferrigno, C. N. Bianchi, R. Lasagna, C. Morri, G. F. Russo, and R. Sandulli, "Corals in high diversity reefs resist human impact," *Ecol. Indic.*, vol. 70, pp. 106–113, 2016, doi: 10.1016/j.ecolind.2016.05.050.

[20] A. Diedrich, "The impacts of tourism on coral reef conservation awareness and

support in coastal communities in Belize," *Coral Reefs*, vol. 26, no. 4, pp. 985–996, 2007, doi: 10.1007/s00338-007-0224-z.

[21] M. Ponti, F. Fratangeli, N. Dondi, M. S. Reinach, C. Serra, and M. J. Sweet, "Baseline reef health surveys at Bangka Island (North Sulawesi, Indonesia) reveal new threats," *PeerJ*, no. 10, 2016, doi: 10.7717/peerj.2614.

[22] P. P. Linggi and A. Burhanuddin, "The role of coral triangle initiative on coral reefs, fisheries, and food securities in Indonesia's environmental conservation," *IOP Conf. Ser. Earth Environ. Sci.*, vol. 343, no. 1, 2019, doi: 10.1088/1755-1315/343/1/012092.

[23] R. Ambo-Rappe and A. M. Moore, *Sulawesi seas, Indonesia*, Second Edi. Elsevier Ltd., 2018. doi: 10.1016/B978-0-08-100853-9.00032-4.

[24] B. Calcinai, A. Bastari, D. M. Makapedua, and C. Cerrano, "Mangrove sponges from Bangka Island (North Sulawesi, Indonesia) with the description of a new species," *J. Mar. Biol. Assoc. United Kingdom*, vol. 97, no. 6, pp. 1417–1422, 2017, doi: 10.1017/S0025315416000710.

[25] A. Kesuma, L. I. Bernawis, B. Subhan, Muhidin, and N. M. Sangadji, "Distribution of Coral Life Forms Based on Sea Current Velocity in Kei Besar Island, Maluku, and Mare Island, North Maluku," *IOP Conf. Ser. Earth Environ. Sci.*, vol. 1350, no. 1, pp. 0–19, 2024, doi: 10.1088/1755-1315/1350/1/012017.

[26] Y. Cheng, C. Chin, D. Lin, and C. Wang, "The Probability of an Unrecoverable Coral Community in Dongsha Atoll Marine National Park Due to Recurrent Disturbances," *Sustainability*, vol. 12, no. 21, 2020, [Online]. Available: https://doi.org/10.3390/su12219052

[27] G. Bintoro *et al.*, "Reef Fish Monitoring as a Coral Reef Resilience Indicator in the Sempu Strait, South of East Java, Indonesia," *Biodiversitas J. Biol. Divers.*, vol. 24 Number, pp. 4950–4959, 2023, doi: 10.13057/biodiv/d240938.

[28] D. G. R. Wiadnya *et al.*, "DNA barcoding of the most common marine ornamental fish species spilled over from a small-sized marine protected area, Bali Barat National Park, Indonesia," *Biodiversitas*, vol. 24, no. 1, pp. 47–54, 2023, doi: 10.13057/biodiv/d240107.

[29] F. Yulianda, *Ekowisata perairan suatu konsep kesesuaian dan daya dukung wisata bahari dan wisata air tawar*. Pt Penerbit Ipb Press, 2019.

[30] R. D. Putra *et al.*, "the Diversity and Abundance of Chaetodontidae Family and Its Relationship With Live Coral Cover in Spermonde Islands, Makassar," *J. Sustain. Sci. Manag.*, vol. 19, no. 1, pp. 113–126, 2024, doi: 10.46754/jssm.2024.01.010.

[31] E. S. Darling, N. A. J. Graham, F. A. Januchowski-Hartley, K. L. Nash, M. S. Pratchett, and S. K. Wilson, "Relationships Between Structural Complexity, Coral Traits, and Reef Fish Assemblages," *Coral Reefs*, vol. 36, no. 2, pp. 561–575, 2017, doi: 10.1007/s00338-017-1539-z.

[32] G. Miller and A. Torres-Delgado, "Measuring sustainable tourism: a state of the art review of sustainable tourism indicators," *J. Sustain. Tour.*, vol. 31, no. 7, pp. 1483–1496, 2023, doi: 10.1080/09669582.2023.2213859.

[33] N. H. Mohd Salleh, N. Mohd Rozar, N. A. Osnin, M. S. Che Din, N. A. Mohd Azli, and S. N. R. Syed Muhammad, "Dive Site Attractiveness Index (Dsai) As an Innovation in Scuba Diving Tourism: a Case of Perhentian Islands," no. November, pp. 11–13, 2021, doi: 10.46754/gtc.2021.11.004.

[34] A. Thommandru, M. Espinoza-Maguiña, E. Ramirez-Asis, S. Ray, M. Naved, and M. Guzman-Avalos, "Role of tourism and hospitality business in economic

development," *Mater. Today Proc.*, vol. 80, pp. 2901–2904, 2023, doi: 10.1016/j.matpr.2021.07.059.

Fish community structure in the seagrass ecosystem of Tidung Besar Island, Seribu Islands

Firyal Kintania Fadhila[1], *Sulistiono* Sulistiono[1*], *Niken T.M.* Pratiwi[1]

[1]Department of Aquatic Resources Management, Faculty of Fisheries and Marine Sciences, Bogor Agricultural Institute, Bogor, 16680

Abstract. Seagrass plays an important role as a habitat for spawning, nurturing, protecting, and foraging marine biota such as fish. However, the fish community structure in the seagrass ecosystem of Tidung Besar Island has not been studied. This study aimed to analyze the structure of fish communities in the seagrass ecosystem of Tidung Besar Island during high- and low-tide conditions. Fish sampling was carried out for three months from October to December 2023 using gillnets with four different mesh sizes, lengths, and widths. The fish samples were analyzed in the laboratory. During the study, 1144 individuals were collected, consisting of 12 orders, 30 families, and 70 species. The fish composition was dominated by *Dischistodus chrysopoecilus* (white-spot damsel) at 612 Ind/m^2. The abundance and dominance of species were higher at high tide, whereas species richness, diversity, and evenness were higher at low tide. Fish diversity on Tidung Besar Island is included in the medium category, with conditions that tend to be stable.

1. Introduction

Tidung Island is one of the islands inhabited by the Seribu Islands. The Island is divided into two islands, namely, Tidung Besar Island and Tidung Kecil Island, which are connected by the Love Bridge. Tidung Besar Island has an area of 50.13 Ha, while Tidung Kecil Island is 17.40 Ha. Tidung Besar Island is used as a residential area, whereas Tidung Kecil Island is used as a conservation area [1].

The Tidung Besar Island has great tourism potential because of its diverse biota. This potential needs to be studied further to obtain the correct management steps. The strategic location of Tidung Island has its own advantages, especially for the utilization of shallow seawater areas, because the island is easy to reach. In addition to having a strategic location,

* Corresponding author: onosulistiono@gmail.com

the island has three types of coastal ecosystems with high biodiversity: coral reefs, seagrass, and mangroves. The high species biodiversity on the island can be utilized for fishery activities [2].

Seagrass ecosystems play an important role in marine ecosystems. Seagrass plays an important role as a primary producer, sediment trapper, and nutrient recycler [3]. The seagrass ecosystem also plays an important role as a shelter, spawning, nursery ground, and a place to find food for the biota who live on it [1]. Various biota are associated with seagrass, such as herbivorous fish, reef fish, and various species of macrozoobenthos from the classes Gastropoda, Bivalvia, Echinoidea, and Holothuroidea [4].

Community structure can be interpreted as a concept that discusses the arrangement or composition of species and their abundance in a community [5]. Community structure refers to an ecological term that studies the types and numbers of species found in an ecosystem. Community structure consists of ecological indices that reflect the dominance, evenness, and richness of an organism [6]. However, there has been no research related to the structure of fish communities in the seagrass ecosystem of Tidung Besar Island, so this research needs to be carried out on fish structure communities so that the abundance, richness, diversity, and evenness of fish in the seagrass ecosystem of Tidung Besar Island can be known so that it can be used as a basis for sustainable fisheries management.

2. Materials and Methods

2.1 Time and Location

The study was conducted from October to December 2023, with a sampling interval of one month, in the seagrass ecosystem of Tidung Besar Island, Seribu Islands (Figure 1).

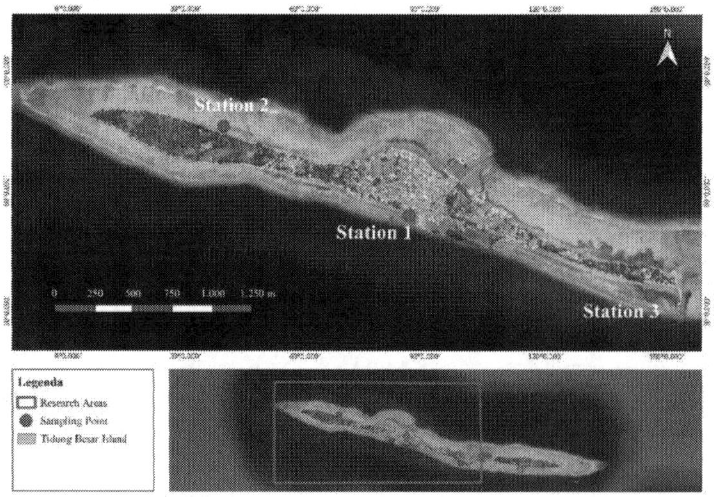

Fig. 1. Research areas at Tidung Besar Island of Seribu Island, Jakarta

The determination of sampling points was based on the percentage of different seagrass density levels and different regional conditions. The seagrass density at Station 1 was 8,9394 stands/m^2. Station 1 is a residential area that is widely used as a homestay for tourists. The seagrass density at Station 2 is 18,6364 stands/m^2. Station 2 is also a residential area, but in the northern area, people are still less aware of marine cleanliness, so there is still a lot of garbage and many people still dispose of their waste directly into the sea. Most of the people in the northern part were fishermen. This caused them to anchor their ships or boats directly in front of their houses. The seagrass density at Station 3 was 8,9394 stands/m^2. Station 3 is an area that is not used in any place and there is mangrove planting in the area.

2.2 Data Collection

Samples of fish caught came from three stations on Tidung Besar Island during high- and low-tide conditions. It was conducted three times at each station in October, November, and December 2023. Based on rainfall data from the Indonesian Agency for Meteorological, Climatological, and Geophysics, October and December were included in the dry season, while November was included in the rainy season. The catchment area is determined by a line transect 50 – 100 m long and 5 – 10 m wide. Fish samples were caught using gillnets with a total catchment area of 500 m^2. The catch time ranged from two to three hours per station for each condition at high and low tides.

The gillnet has a float attached to the top so that the net can float and an anchor located at the bottom to maintain the position of the fishing gear [7]. The nets used in this study consisted of four types of different mesh sizes, lengths, and widths. First gillnet has a mesh of 0,5 inches with a length of 20 m and a width of 1,5 m, the second gillnet has a mesh of 1,75 inches with a length of 40 m and a width of 1 m, the third gillnet has a mesh of 1,5 inches with a length of 40 m and a width of 1 m, and the fourth gillnet has a mesh of 2 inches with a length of 30 m and a width of 1 m.

Samples of the caught fish were collected in the box. The fish were sorted based on their shape and color, which tended to be the same. After sorting, the fish samples were placed in medium – sized plastic bags that had been perforated in several parts so that the fish samples could be completely submerged. The fish samples were placed in a coolbox. Two coolboxes were used: a coolbox containing 20 liters of 10% formalin and a coolbox containing 20 liters of 96% alcohol. The first fish was preserved in formalin for 24 – 48 h. After 24 – 48 h, the fish samples were cleaned using clean water, transferred into a coolbox containing 96% alcohol, and soaked again until they were identified.

The preserved fish samples were identified by class and species. Identification was carried out using several reference books (Allen, 1997); Carpenter and Niem (2001); Allen *et al.* (2003); Allen and Erdmann (2012). Each identified species was counted as number of individuals. Furthermore, the length of the fish body was measured using a scale and the weight of the fish was measured using a balance. The identification results are listed in a table that also contains data related to the order, family, species, local name, number, length distribution, weight distribution, station where the fish were found, resident status, economic value, and IUCN status.

The density of seagrass was determined by stretching a 100-meter line transect. Within a 100-meter distance, the density was measured using a quadrant transect divided into four boxes. Each box was calculated for seagrass stands, and each type of seagrass was identified. After the data were collected, the density of seagrass at each sampling point was analyzed.

Water quality parameters were measured using several tools. A pH meter was used to measure salinity, a refractometer was used to measure dissolved oxygen, a DO meter was used to measure temperature, a thermometer was used to measure depth, and a graduated stick was used to measure depth. For the pH, salinity, temperature, and dissolved oxygen parameters, three repetitions were performed to obtain the average.

2.3 Data Analysis

2.3.1 Species Abundance

Species abundance was calculated as the number of individuals found per unit area of the observation area. Species abundance was calculated using the following formula [12]:

$$K_i = \frac{n_i}{A}$$

Description:
K_i : Species abundance (Ind/m^2)
n_i : Number of individuals of species (Ind)
A : Observation area (m^2)

2.3.2 Richness Index (D)

Fish species richness can be calculated using the Margalef species richness index model [13].

$$\left[D = \frac{S-1}{\ln N}\right]$$

Description:
S: Amount of species
N: Amount of individuals of all species
The following are the criteria for assessing the species richness index [13]:
$D < 3.50$: Low species richness
$3.50 \leq D \leq 5.00$: Medium species richness
$D > 5.00$: High species richness

2.3.3 Diversity Index (H')

According to [14], the diversity index is used to measure the relative abundance based on the number of species and individuals of each species. There is a Shannon-Wiener diversity index formula is as follows [15]:

$$\left[H' = \sum_{i=1}^{n} p_i \ln p_i\right]$$

$$p_i = \frac{n_i}{N}$$

Description:

p_i: Proportion of the amount of individuals
n_i: Amount of individuals for each species
According to [12], there are criteria for the diversity index that include
$H' < 1$: Low diversity
$1 \leq H' \leq 3$: Medium diversity
$H' > 3$: High diversity

2.3.4 Evenness Index (E)

The magnitude of similarity in the distribution of the number of individuals of each type can be calculated using a method called the uniformity index according to [15]. The formula used is as follows:

$$\left[E = \frac{H'}{\ln S}\right]$$

Based on [16], the uniformity index criteria are as follows:
$0 \leq E \leq 0.5$: The community is in a depressed condition
$0.5 \leq E \leq 0.75$: The community is in an unstable condition
$0.75 \leq E \leq 1$: The community is in a stable condition

2.3.5 Dominance Index (C)

The dominance index describes whether a dominant species exists in an ecosystem. The dominance index formula is as follows [17].

$$\left[C = \sum_{i=1}^{n} {p_i}^2\right]$$

According to [18], the dominance index criteria are as follows:
$0.00 < C \leq 0.50$: Low dominance
$0.50 < C \leq 0.75$: Medium dominance
$0.75 < C \leq 1.00$: High dominance

2.3.6 Seagrass Density (Ki)

Seagrass density was used to calculate the number of seagrass species in a given area [19]. According to [20], the formula for the density of seagrass species is as follows:

$$K_i = \frac{\Sigma n_i}{A}$$

Description:
K_i: Density of the i-th species (stand/m^2)
Σn_i: Number of individuals or stands in the i-th transect
A: Sampling area (m^2)

2.3.7 Similarity Analaysis

One of the analyses that can be used for similarity is the Bray–Curtis similarity analysis. It was used to determine the similarities between research stations based on the number of individuals and the number of species at each station using a grouping system. The grouping principle is carried out by determining the correlation value between stations, which is continued by compiling a matrix called the Bray Curtis similarity matrix [21]. Bray Curtis similarity analysis can be performed using the PAST (Paleontological Statistics) software [22]. The Bray Curtis similarity analysis was based on two variables: the number of individuals and the number of species found at each station. The results are presented as a dendrogram that forms several clusters. Based on the clusters formed, further analysis was carried out to determine the relationship between the influencing variables and the influenced variables.

2.3.8 Principal Component Analysis (PCA)

Principal Component Analysis (PCA) is an analysis that can provide information on the relationship between environmental parameters such as pH, temperature, dissolved oxygen that can affect the abundance and diversity of fish [23]. Principal Component Analysis was performed using the XLSTAT software, which can be connected to Microsoft Excel. This software can assist with statistical data processing, such as identifying several characteristics and classifying variability [24].

3. Results

3.1 Fish Composition

The fish samples collected were 1144 individuals, consisting of 393 individuals in October, 252 individuals in November, and 499 individuals in December. Individuals were caught in the Actinopterygii class (consisting of 12 orders, 30 families, and 70 species). The order Perciformes is the order with the highest number of families and species, consisting of 11 families and 46 species.

Table 1. Composition of fish catches

Order/Family	Species	Common Name	Amount	Weight Range (gr)	Long Range (cm)	Station			Resident Status	Economic Values			IUCN Status
						1	2	3		C	F	D	
ACANTHURIFORMES													
Acanthuridae	*Acanthurus xanthopterus*	Yellowfin surgeonfish	1	87,8835	15,7			✓	OV	✓			LC
Chaetodontidae	*Chaetodon octofasciatus*	Eightband butterflyfish	3	14,1748 - 28,2495	7,5 - 9,6	✓			OV	✓		✓	LC
	Chaetodon trifasciatus	Melon butterflyfish	2	36,8544 - 59,5340	10,9 - 12,1			✓	OV	✓		✓	LC
Siganidae	*Siganus fuscescens*	Mottled spinefoot	12	42,5243 - 192,7766	14,1 - 22,2	✓	✓	✓	TR	✓			LC
	Siganus virgatus	Barhead spinefoot	17	22,6796 - 96,3883	11,7 - 18,0	✓	✓	✓	TR	✓			LC

Order/Family	Species	Common Name	Amount	Weight Range (gr)	Long Range (cm)	Station			Resident Status	Economic Values			IUCN Status
						1	2	3		C	F	D	
ATHERINIFORMES													
Atherinidae	*Atherinomorus endrachtensis*	Endracht hardyhead	7	2,8350 - 2,840	6,9 - 7,6	✓			RV		✓		N/A
BELONIFORMES													
Belonidae	*Tylosurus crocodilus*	Crocodilian longtom	1	323,1843	58,7	✓			OV	✓			LC
Hemiramphidae	*Hemiramphus far*	Barred garfish	1	221,1261	32,7	✓			OV	✓		✓	N/A
Zenarchopteridae	*Zenarchopterus buffonis*	Buffon's river-garfish	2	2,8349 - 2,8540	12,2 - 13,6	✓			RV	✓		✓	LC
CARANGIFORMES													
Carangidae	*Carangoides coeruleopinnatus*	Coastal trevally	1	87,8835	18,3			✓	OV	✓			DD
	Decapterus macarellus	Mackerel scad	1	53,8641	18,1			✓	OV	✓			LC
	Selaroides leptolepis	Yellowstripe scad	2	25,5146 - 70,8738	13- 17,5			✓	OV	✓			LC
CENTRARCHIFORMES													
Kyphosidae	*Kyphosus vaigiensis*	Brassy chub	1	62,3689	15,3			✓	OV	✓			DD
GOBIIFORMES													
Gobiidae	*Acentrogobius nebulosus*	Shadow goby	1	1,5025	4,9		✓		RV	✓	✓	✓	LC
	Amblygobius bynoensis	Freckled goby	1	8,5049	6,5		✓		RV	✓	✓		LC
HOLOCENTRIFORMES													
Holocentridae	*Sargocentron rubrum*	Red squirrelfish	1	79,3786	16,4	✓			OV	✓			LC
KURTIFORMES													
Apogonidae	*Cheilodipterus quinquelineatus*	Five-lined cardinalfish	1	17,0097	10,9	✓			RV		✓	✓	LC
	Fibramia lateralis	Humpback cardinal	6	2,1262 - 5,6699	5,4 - 6,6	✓			RV		✓		N/A
PERCIFORMES													
Epinephelidae	*Epinephelus quoyanus*	Longfin grouper	8	42,5243 - 68,0388	13,9 - 16,8	✓	✓	✓	OV	✓			LC
Gerreidae	*Gerres erythrourus*	Deepbody silverbiddy	1	73,7087	13	✓			TR	✓	✓		LC
	Gerres oyena	Common silver-biddy	202	22,6796 - 36,8544	11,1 - 13,2	✓	✓	✓	TR	✓	✓		LC
Labridae	*Cheilinus oxycephalus*	Snooty wrasse	5	34,0194 - 45,3592	13,0 - 13,8	✓	✓	✓	OV	✓			LC
	Cheilinus trilobatus	Tripletail wrasse	1	56,699	13,7	✓			OV	✓		✓	LC
	Choerodon anchorago	Orange-dotted tuskfish	89	1,1340 - 121,9029	5,3 - 17,9	✓	✓	✓	RV	✓			LC
	Halichoeres argus	Argus wrasse	3	48,1942 - 56,6990	14,3 - 15,9	✓		✓	RV	✓		✓	LC

Order/Family	Species	Common Name	Amount	Weight Range (gr)	Long Range (cm)	Station			Resident Status	Economic Values			IUCN Status
						1	2	3		C	F	D	
	Halichoeres chloropterus	Pastel-green wrasse	19	2,8350 - 14,1748	5,8 - 10,5	✓	✓		RV	✓	✓		LC
	Halichoeres richmondi	Richmond's wrasse	3	2,8350 - 2,8401	6,8 - 7,4		✓		RV			✓	LC
	Halichoeres scapularis	Zigzag wrasse	4	2,8350 - 45,3592	7,2 - 14,2		✓		RV	✓			LC
	Hemigymnus melapterus	Thick-lipped wrasse	19	39,6893 - 167,2671	12,9 - 21,5	✓	✓	✓	PR	✓			LC
	Stethojulis trilineata	Three-lined rainbowfish	5	36,8544 - 42,5243	12,4 - 13,0		✓	✓	OV			✓	LC
	Lethrinus atkinsoni	Pacific yellowtail emperor	1	25,5146	11,1	✓			OV	✓	✓		LC
Lethrinidae	*Lethrinus harak*	Thumbprint emperor	4	39,6893 - 62,3689	13,9 - 16,4	✓	✓	✓	TR	✓	✓		LC
	Lethrinus lentjan	Pink ear emperor	36	31,1845 - 51,0291	11,4 - 15,2	✓	✓	✓	TR	✓	✓		LC
	Lutjanus carponotatus	Spanish flag snapper	12	62,3689 - 141,7475	15,0 - 20,3	✓	✓	✓	TR	✓	✓		LC
Lutjanidae	*Lutjanus decussatus*	Checkered snapper	4	56,6990 - 68,0388	14,9 - 15,8	✓		✓	PR	✓			LC
	Lutjanus fulviflamma	Dory snapper	2	48,19415 - 56,6990	14,0 - 15,1		✓		OV	✓	✓		LC
	Nemipterus furcosus	Rosy threadfin bream	1	73,7087	17,9	✓			OV	✓			LC
	Pentapodus trivittatus	Three-striped whiptail	40	11,3398 - 133,2427	9,8 - 21,0	✓	✓	✓	RV	✓			LC
	Scolopsis aurata	Yellowstripe monocle bream	4	34,0194 - 87,8835	12,7 - 17,3	✓		✓	OV	✓			LC
Nemipteridae	*Scolopsis ciliata*	Whitestreak monocle bream	14	28,3495 - 53,8641	12,9 - 15,9		✓	✓	OV	✓			LC
	Scolopsis lineata	Striped monocle bream	19	17,0097 - 113,3980	10,4 - 19,3	✓	✓	✓	OV	✓			LC
	Scolopsis margaritifera	Pearly monocle bream	5	34,0194 - 53,8641	13,4 - 14,8		✓	✓	OV	✓			LC
Pomacanthidae	*Pomacanthus sexstriatus*	Sixbar angelfish	1	34,0194	10,1			✓	OV			✓	LC
	Abudefduf bengalensis	Narrow-banded sergeant major	14	17,0097 - 73,7087	10,3 - 14,3	✓	✓	✓	OV			✓	LC
Pomacentridae	*Abudefduf sexfasciatus*	Scissortail sergeant	5	19,8445 - 62,3689	10,5 - 14,2	✓	✓		OV			✓	LC
	Abudefduf sordidus	Blackspot sergeant	1	39,6893	11	✓			OV	✓	✓		LC

Order/Family	Species	Common Name	Amou nt	Weight Range (gr)	Long Range (cm)	Station			Reside nt Status	Economic Values			IUC N Stat us
						1	2	3		C	F	D	
	Abudefduf vaigiensis	Sergeant-major	16	19,8445 - 138,9126	9,7 - 12,0	✓	✓		TR			✓	VU
	Dischistodus chrysopoecilus	White-spot damsel	306	1,9845 - 87,8835	4,1 - 15,7	✓	✓	✓	PR		✓		LC
	Dischistodus fasciatus	Banded damsel	1	28,3495	10,3	✓			OV	✓	✓		N/A
	Dischistodus perspicillatus	White damsel	90	2,83495 - 93,5534	6,4 - 16,2	✓	✓	✓	OV		✓		LC
	Dischistodus prosopotaenia	Honey-head damsel	1	28,3495	11,5		✓		OV	✓	✓		LC
	Pomacentrus javanicus	Java damsel	1	76,54365	14,7	✓			OV	✓	✓		N/A
	Pomacentrus littoralis	Smoky damsel	9	22,6796 - 82,2136	9,5 - 14,7	✓		✓	TR		✓		N/A
	Stegastes nigricans	Dusky farmerfish	1	85,0485	16,2			✓	OV		✓		LC
	Calotomus spinidens	Spinytooth parrotfish	3	48,1942 - 59,5340	12,8- 15,0	✓	✓		OV	✓			LC
	Hipposcarus harid	Candelamoa parrotfish	2	59,5339 - 79,8738	15,7 - 16,9	✓		✓	PR	✓			LC
	Scarus ghobban	Blue-barred parrotfish	38	8,50485 - 136,0776	7,9- 18,5	✓	✓	✓	TR	✓			EN
Scaridae	*Scarus psittacus*	Palenose parrotfish	45	36,85435 - 85,0485	12,5 - 16,3	✓	✓	✓	OV	✓			DD
	Scarus quoyi	Quoy's parrotfish	19	45,3592 - 70,8738	12,3- 14,6	✓	✓	✓	OV	✓			LC
	Scarus rivulatus	Surf parrotfish	6	65,2039 - 153,0873	14,1 - 20,6		✓	✓	OV	✓			DD
	Scarus spinus	Greensnout parrotfish	1	56,699	14		✓		OV	✓			LC
Scorpaenidae	*Scorpaenopsis diabolus*	False stonefish	1	249,4756	20,2			✓	OV	x	x	x	N/A
Sphyraenidae	*Sphyraena jello*	Pickhandle barracuda	1	269,3203	37		✓		OV	✓			N/A
Synanceiidae	*Synanceia horrida*	Estuarine stonefish	1	464,9318	16,6			✓	OV	x	x	x	LC
SCOMBRIFORMES													
Scombridae	*Rastrelliger kanagurta*	Indian mackerel	1	45,3592	16,4			✓	OV	✓			LC
SYNGNATHIFORMES													
Centriscidae	*Aeoliscus strigatus*	Razorfish	2	0,6804	12,9			✓	PR				DD
TETRAODONTIFORMES													
Balistidae	*Balistoides viridescens*	Titan triggerfish	1	48,1942	12		✓		OV	✓			LC
Monacanthidae	*Acreichthys tomentosus*	Bristle-tail file-fish	6	8,5049 - 170,0970	7,4 - 9,3	✓	✓		RV		✓		LC

Order/Family	Species	Common Name	Amount	Weight Range (gr)	Long Range (cm)	Station			Resident Status	Economic Values			IUCN Status
						1	2	3		C	F	D	
Ostraciidae	*Ostracion cubicum*	Yellow boxfish	7	36,8544 - 215,4562	9,2 - 19,0	✓	✓	✓	PR	✓		✓	N/A
Tetraodontidae	*Arothron mappa*	Map puffer	2	291,9998 - 1312,5819	18,8 - 34,1			✓	TR			✓	LC

**Note: PR: permanent resident, TR: temporary resident, RV: regular visitors, OV: occasional visitors, C: consumption, F: feed, D: decorative, LC: least concern, DD: data deficient, VU: vulnerable, EN: endangered, N/A: no data found on IUCN.*

3.2 Abundance and Richness Index

Species abundance showed varying results at each station. In October – November, fish species abundance was not significantly different, whereas in December, a significant difference was observed. The highest and lowest fish abundances were found at Station 3 in December under high- and low-tide conditions, respectively.

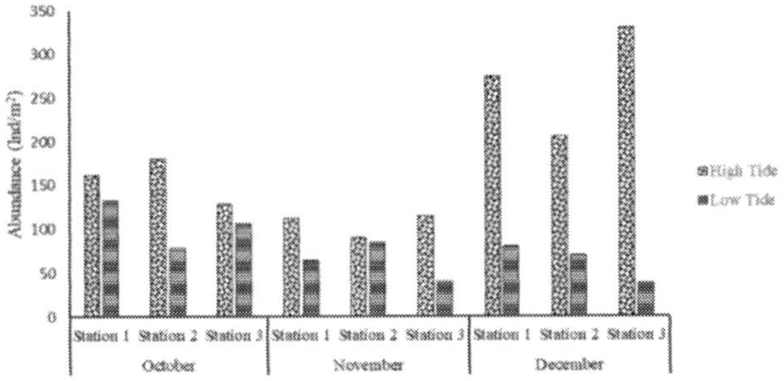

Fig. 2. Fish abundance during high and low tide conditions in Tidung Besar Island, Seribu Islands

The comparison between the richness values based on the amount and biomass of fish showed an insignificant difference. The highest richness of fish species was found at Station 2 in October during low-tide conditions, which were included in the high category. The lowest richness of fish species was found at Station 2 in November during high-tide conditions, which were included in the low category.

(a)

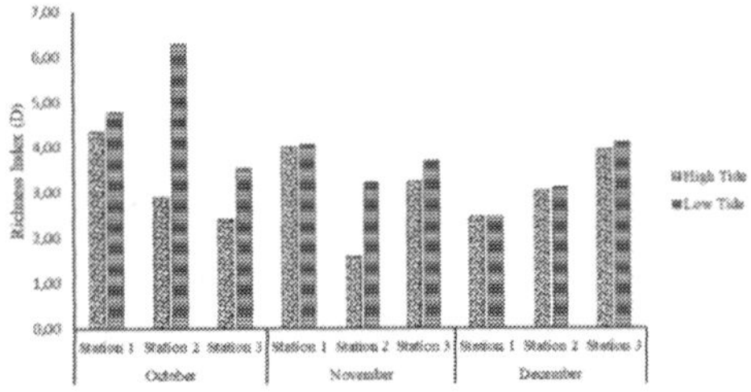

(b)

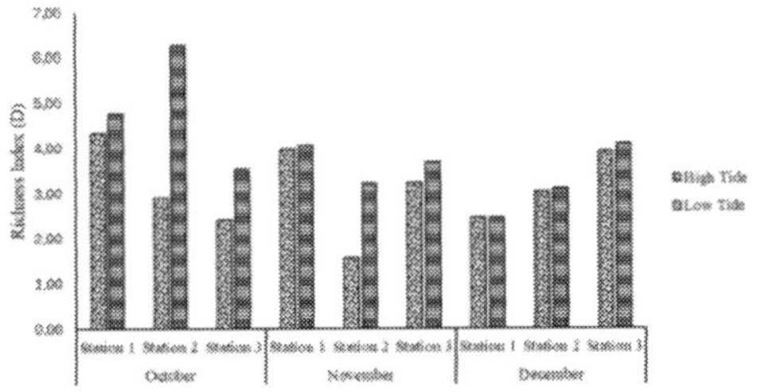

Fig. 3. Fish richness based on number (a) and biomass (b) in Tidung Besar Island, Seribu Islands

3.3 Diversity, Evenness, and Dominance Index

Fish diversity based on the amount and biomass was highest at Station 2 in October during low tide. The lowest diversity of fish species was observed at Station 1 in December during high tide. The highest and lowest diversity of fish species were included in the moderate category. Differences were observed in the values at Station 1 in October and Station 3 in December. Based on the number of fish, Station 1 in October and Station 3 in December had higher diversity values during low tide, while based on fish biomass, the diversity value of fish was higher during high tide.

(a) (b)

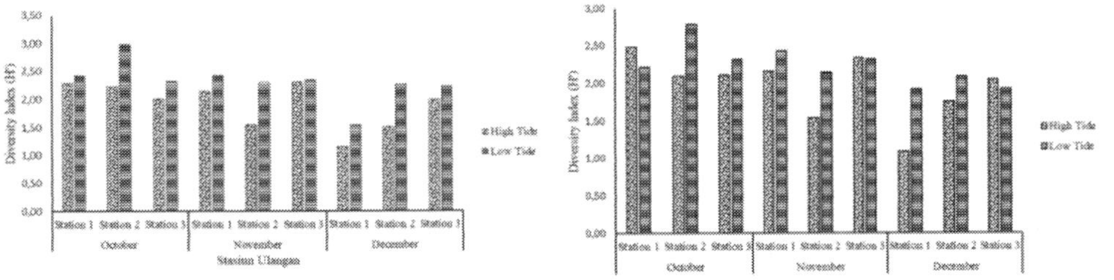

Fig. 4. Fish diversity based on number (a) and biomass (b) in Tidung Besar Island, Seribu Islands

The highest fish evenness was found at Station 3 in November during low-tide conditions, which were included in the stable category. The lowest evenness of fish species was found at Station 1 in December during high-tide conditions, which were included in the depressed category. There were differences in the values at stations 1 and 3 in October. Based on the number of fish, Stations 1 and 3 in October had higher evenness values during low tide conditions, while based on fish biomass, during high tide conditions the evenness values of fish were higher.

(a) (b)

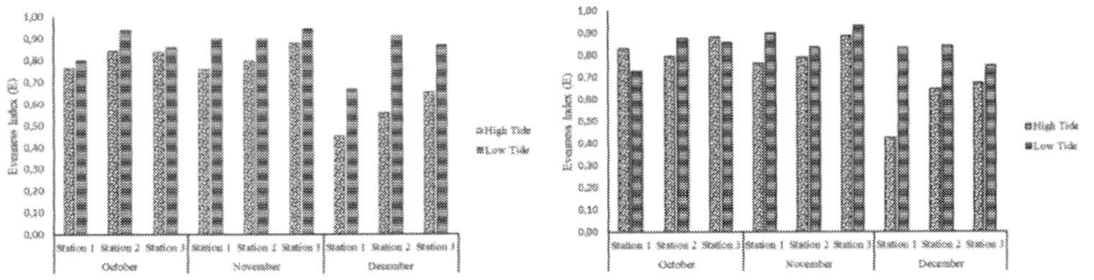

Fig. 5. Fish evenness based on number (a) and biomass (b) in Tidung Besar Island, Seribu Islands

The dominance values at Stations 1, 2, and 3 in October – November showed no very different results between high and low tide conditions. In contrast to October and November, at Stations 1, 2, and 3 in December, the difference in dominance values was significant between high and low tides. The highest dominance was observed at Station 1 in December, when high tide conditions were included in the moderate category. A comparison between dominance values based on the amount and biomass of fish showed significant differences. Based on the number of fish, Station 1 in October had a higher dominance value during high tide conditions, whereas based on fish biomass, the fish dominance value was higher during low tide conditions.

(a) (b)

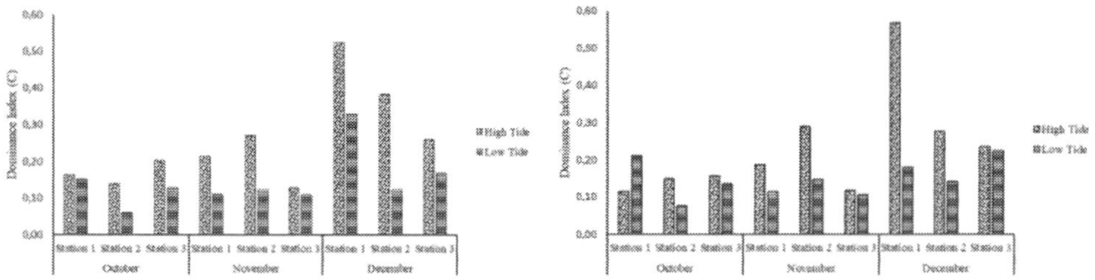

Fig. 6. Fish dominance based on number (a) and biomass (b) in Tidung Besar Island, Seribu Islands

3.4 Types and Density of Seagrass

There were four types of seagrass found in the observation area. These four species were *Enhalus acoroides*, *Thalassia hemprichii*, *Cymodocea rotundata*, and *Halophila ovalis*. The dominant seagrass type on Tidung Besar Island was *T. hemprichii* with a species density value of 93,58 stands/m^2. The highest total density was observed at Station 3, which was 19,15 stands/m^2.

Table 2. Types and density of seagrass of Tidung Besar Island, Seribu Islands

Types of Seagrass	**Density (Ind/m^2)**			
	Station			**Density of type**
	1	**2**	**3**	
Enhalus acoroides	3,52	3,52	57,70	64,73
Thalassia hemprichi	31,39	43,27	18,91	93,58
Cymodocea rotundata	0,00	16,24	0,00	16,24
Halophila ovalis	0,85	11,52	0,00	12,36
Total	35,76	74,55	76,61	
Average	8,94	18,64	19,15	

3.5 Water Quality Parameters

The results of the measurements of physicochemical parameters, including temperature, depth, pH, salinity, and dissolved oxygen, are presented in Table 3. The physicochemical parameters of water vary both spatially and temporally.

Table 3. Water quality parameters at Tidung Besar Island, Seribu Islands

Condition	**Water Quality Parameters**				
	Physical Parameters		**Chemical Parameters**		
	Temperature	**Depth**	**pH**	**Salinity**	**Dissolved Oxygen**
High Tide	29,6 - 32,3	70,5 - 126,4	6,76 - 8,08	34 - 36	4,5 - 6,4
Low Tide	29,9 - 33,2	48,2 - 90,4	6,93 - 8,05		4,1 - 5,1

3.6 Similarity Analysis

A similarity analysis was conducted based on data on the number of individuals and the number of species found at each station. This is because the t-test that has been conducted showed that the differences in seasons in October, November, and December were not significant, so it can be categorized into one season. The clustering results based on Bray-Curtis analysis showed that the number of individuals and species on Tidung Besar Island were divided into four clusters. This grouping was based on 80 value of the data.

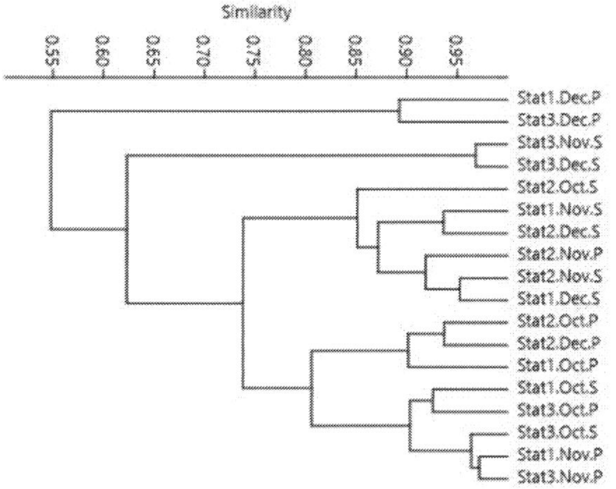

Fig. 7. Similarity of number of individuals and species between stations in Tidung Besar Island, Seribu Islands

3.7 Principal Component Analysis

The relationship between the abundance of fish species and diversity and environmental parameters is shown in Figure 7. Results of the principal component analysis were divided into four clusters. F1 and F2 in each biplot are factorial axes that represent linear combinations of the original variables. The total percentages of F1 and F2 illustrate the magnitude of the influence of environmental parameters on the abundance and diversity of fish species.

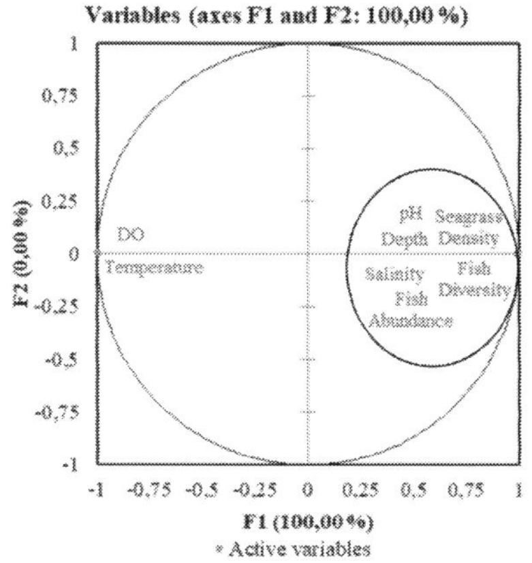

Cluster 1

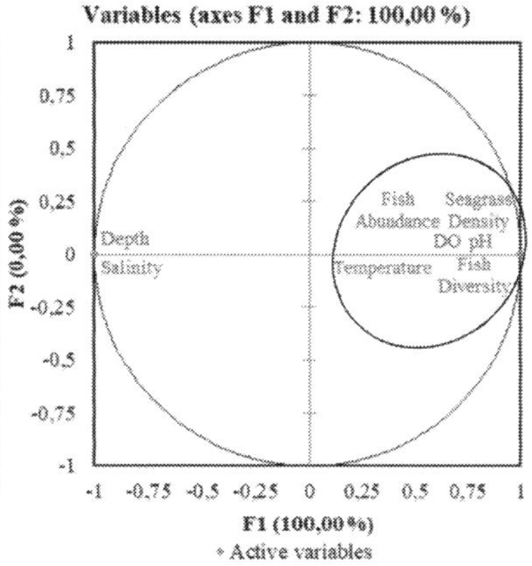

Cluster 2

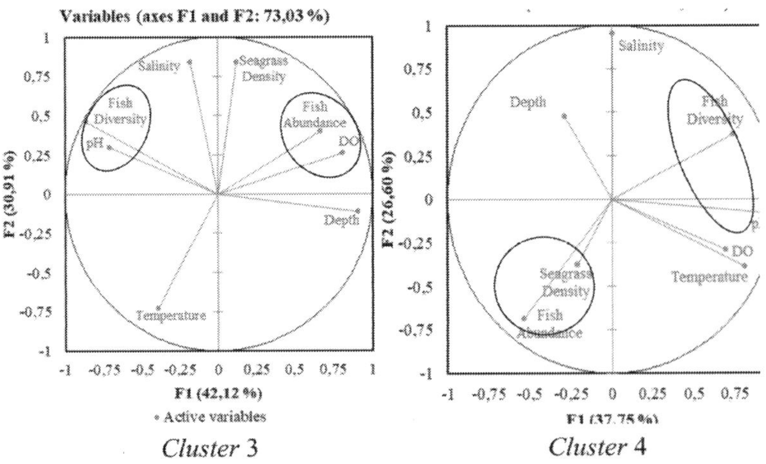

Fig. 8. Relationship between environmental factors, species abundance, and fish diversity at Tidung Besar Island, Seribu Islands

4. Discussion

The composition of fish species included the number of individuals, species, families, and orders recorded during the study period. Perciformes had the highest number of families and species compared to the other orders. This is supported by the statement in [25], where Perciformes is the order with the highest number of species in the fish world. This causes Perciformes to be found in many aquatic ecosystems and has cosmopolitan properties. Other orders found in the seagrass ecosystem of Tidung Besar Island include Acanthuriformes, Beloniformes, Carangiformes, Tetraodontiformes, Kurtiformes, Gobiiformes, Athiriniformes, Centrachiformes, Holocentriformes, Scombriformes, and Syngnathiformes. One of the families in Perciformes is the Pomacentridae family, which has the highest number of species. Compared to other families that have only one to nine species, Pomacentridae consists of 12 species. This is supported by research conducted by [26], which states that Pomacentridae has a worldwide distribution. Pomacentridae prefers tropical habitats both on the edges of waters such as seagrass and mangrove ecosystems, as well as around coral reefs or lagoon ecosystems. The species with the highest number of individuals obtained during the study was *Dischistodus chrysopoecilus* (white-spot damsel), which came from the Pomacentridae, with a total of 306 individuals. In addition to *Dischistodus chrysopoecilus*, *Gerres oyena* (common silver-biddy), which belongs to the order Gerreidae, also obtained a high number of individuals, namely 202 individuals. *Dischistodus chrysopoecilus* and *Gerres oyena* were obtained from Perciformes. This is supported by the statement of Kholis *et al.* that the types of fish found in the seagrass ecosystem generally come from the families Lutjanidae, Gerreidae, Haemulidae, Pomacentridae, Labridae, Sparidae, Atherinidae and Leiognathidae [27].

Based on the data obtained, *Dischistodus chrysopoecilus* and *Gerres oyena* were found at almost every station during high and low tides. This is because of the characteristics of fish from the Perciformes (*Dischistodus chrysopoecilus* belongs to the Pomacentridae family and *Gerres oyena* belongs to the family Gerreidae). This correlates with the seagrass ecosystem, which acts as a shelter, feeding ground, spawning ground, and nursery ground for marine biota [28]. The Pomacentridae is included in the temporary residents (TR) of the seagrass ecosystem, which only occupies the seagrass ecosystem in the juvenile phase [29], whereas the Gerreidae family uses the seagrass ecosystem as a spawning ground. In general, the spawning period for fish of the Gerreidae family occurs throughout the year with two peak spawning periods, namely in March and October – December [30].

The highest abundance was found at Station 3 in December during high tide, with a value of 330 Ind/m^2. This was evidenced by the number of catches at Station 3 in December during high tide, which reached 165 individuals. One of the factors that can cause high fish abundance in an ecosystem is temperature. Beroperai *et al.* stated that if a body of water has a temperature that tends to be low and higher in nutrient availability, the abundance value will be higher when compared to waters that have a higher temperature and insufficient food availability [31]. According to Latuconsina and Rappe, the high and low abundances of fish in an aquatic ecosystem can be influenced by high and low oceanographic parameters such as turbidity, temperature, salinity, pH, and dissolved oxygen [31]. In addition to high and low oceanographic parameters, fish abundance can also be influenced by human activities, such as overfishing, the use of environmentally unfriendly fishing gear, and no size selection in fishing activities [32]. As shown in Figure 3, fish abundance is higher at high tide than at low tide, according to Hutami *et al.*, because at high tide, the nutrients in the waters tend to be higher than at low tide [33].

The richness index was positively correlated with the diversity index, which caused diversity at Station 2 (in October) when the low tide was also high. This is evidenced by the number of diverse species of 24 species from 39 individuals obtained, where the richness index value at this station was 6,2780, which was included in the high category, and the diversity index value was 2,9817, which was included in the medium category. High and low diversity cannot be separated from the influence of environmental conditions and human activities such as fishing, presence of invasive species, and habitat destruction [34]. The water quality in the seagrass ecosystem of Tidung Besar Island is included in the good category; however, the high fishing activity of local fishermen is the cause of the diversity value of all observed stations being included in the medium category.

The highest evenness value was found at Station 3 in November when the low tide had a value of 0,9440 which means that the fish community at that station was in a stable condition. This is in accordance with the catch results, which only contained 20 individuals but had a total of 12 species. This high evenness value in a community indicates that the waters have a large community balance; conversely, if the evenness value is low, then there are fish species that have a relatively large number of individuals, whereas other species have a relatively small number of individuals. This can be seen at Station 1 (in December) during high tide conditions where the number of catches was 137, but was dominated by the *Dischistodus chrysopoecilus* species of 97 individuals. There were differences in the structure of the fish community at high tide and low tide. During high tide, abundance and dominance were higher, while during low tide, diversity and evenness were higher. This is because of the different daily rhythms of fish and differences in fish orientation to find food and shelter [35]. The dominance index was used to determine the most commonly found species from a given location [36].

The highest dominance was found at Station 1 in December when the tide conditions had a value of 0,5233 which was included in the medium dominance category. This is evidenced by the high catch of 137 individuals, but only 13 species. According to Yohanista, if a community has a high dominance value, it means that the dominance in the community is only centered on one species, whereas if the dominance value is low, the dominance is centered on several species [37]. This proves the dominance at Station 1 (in December) when the tide conditions were centered on the species *Dischistodus chrysopoecilus*. This dominance index has a negative correlation with diversity; the higher the dominance value, the lower the diversity value in a community. In accordance with the results shown in Figures 4 and 6, the highest dominance value and the lowest diversity were found at Station 1 in December when the tide conditions were.

Fish abundance obtained during the study cannot be separated from the ecosystem factor itself, namely, the seagrass ecosystem. The seagrass ecosystem is one of the ecosystems that is a habitat for fish found in the sea. Station 1 had the physical characteristics of sediments dominated by sandy substrates mixed with dead coral. Three types of seagrass vegetation were found (*Enhalus acoroides*, *Thalassia hemprichii*, and *Halophila ovalis*). Station 2 had the physical characteristics of sediment dominated by muddy sand substrates mixed with dead coral. Four types of seagrass vegetation were found (*Enhalus acoroides*, *Thalassia hemprichii*, *Cymodocea rotundata*, and *Halophila ovalis*). Station 3 had the physical characteristics of sediments dominated by sandy substrates mixed with dead coral. Two types of seagrass vegetation were identified (*Enhalus acoroides* and *Thalassia hemprichii*).

Fish associated with the seagrass species *Enhalus acoroides* generally come from the families Pomacentridae, Siganidae, Sphyraenidae, Nemipteridae, Acanthuridae, and Labridae. The identification results showed that the fish from Pomacentridae had the largest number. This may indicate that fish of the Pomacentridae family are closely associated with seagrass. Pomacentridae use seagrass as a shelter from predators during the juvenile phase. The types of fish from Siganidae depend on the seagrass species *Enhalus acoroides* for their livelihood [38]. This is supported by research by Latuconsina *et al.*, who found that fragments of *Enhalus acoroides* seagrass dominated the stomach contents of fish from the Siganidae family [32].

Acanthuridae, Kyphosidae, Hemiramphidae, and Siganidae are fish families strongly associated with *Thalassia hemprichii* seagrass. Fish from these families use *Thallassia hemprichii* as a food source. This was evidenced by the discovery of this type of seagrass in the stomachs of these fish in a study conducted by Adrim [39]. Fish of Apogonidae are one of the fish that spend most of their lives in seagrass, either for spawning, foraging, or sheltering [40]. Fish of Siganidae prefer *Thallassia hemprichii* and *Enhalus acoroides* because they tend to be larger and wider, respectively. With this shape, these two types of seagrass can provide an ecological niche for fish as a nursery area.

Bray-Curtis clustering divides clustering into four main clusters. This clustering was based on data on the number of individuals and species at each station. The formation of these four clusters was due to the high similarity between the stations in each cluster. This means that between stations in each cluster, there is high similarity based on the number of individuals and the number of species at each station. The grouping into four clusters indicated that there was a relatively low similarity between clusters.

The environment plays an important role in the marine biota. Environmental parameters, such as seagrass density, pH, salinity, and depth, had a strong effect on the abundance of fish species and diversity (Cluster 1). Environmental parameters, such as seagrass density, dissolved oxygen, temperature, and pH, have a strong effect on the abundance of fish species and diversity (Cluster 2). In contrast to clusters 1 and 2, DO and seagrass density in clusters 3 and 4 had a strong effect on the abundance of fish species, while pH had a strong effect on the value of fish diversity.

The abundance of species (Ki) and diversity of fish (H') can be influenced by factors such as temperature, pH, salinity, dissolved oxygen, and water depth. As shown in Figure 7, the abundance of species and the diversity of fish in this study were greatly influenced by factors such as temperature, salinity, pH, depth, and dissolved oxygen. The magnitudes of temperature, salinity, pH, and dissolved oxygen obtained during the study were included in the good category, which is suitable for the life of marine biota in accordance with seawater quality standards in accordance with Government Regulation of the Republic of Indonesia Number 22 of 2021 concerning the Implementation of Environmental Protection and Management. This is in accordance with the function of seagrass as a place for spawning, shelter, foraging, and nurturing fish, which causes the diversity of fish in the seagrass ecosystem of Tidung Besar Island to be included in the moderate category. The parameters of water depth and dissolved oxygen are interrelated, which means that as the depth increases, it will affect the entry of light into the water, which will affect seagrass photosynthesis and the level of dissolved oxygen present [41]. The depth level in the seagrass ecosystem of Tidung Besar Island has a range of values of 48,2 cm at low tide – 137,6 cm at high tide. This value is neither too low nor too high, which means that the seagrass ecosystem of Tidung Besar Island can support the life of biota.

Based on the data obtained, Tidung Besar Island experienced two different types of tides. In October – November, there is a mixed tide prevailing semi-diurnally, while in December there is a diurnal tide. Mixed tide prevailing semi-dirunal is a type of tide that occurs twice in one day and twice in one day with different wave heights, whereas diurnal tide is a type of tide that only occurs once in one day [42]. Sea tides influence the presence of fish in the seagrass ecosystem of Tidung Besar Island. Generally, seagrass is located in shallow waters, which causes water quality, movement, and depth to be influenced by the rhythm of the tides within a certain time span. In addition, food availability, shelter quality, and other environmental factors also occur during the tidal cycle.

These tidal changes can affect animal migration patterns in one ecosystem. During high tide, there will generally be more fish entering shallower waters, such as seagrass. Likewise, when the water starts to recede, fish generally leave shallower waters to avoid the increase in temperature or salinity that occurs and the risk of predation. For example, *Siganus fuscences* and *Siganus virgatus* were found at Stations 1, 2, and 3 to find food because these are herbivorous fish that eat seagrass leaves. These fish, which belong to the Siganidae family, come to seagrass during high tide.

Community structure in an ecosystem is not only influenced by factors such as relationships between species, but also by the relative number of organisms of the species, so that the relative abundance of a species will affect the function of the community, the distribution of individuals in a community, the balance of a community, and the stability of the community [43]. The abundance, richness, diversity, evenness, and dominance indicate different values every month because they are caused by changes in the environmental conditions of the water. The physicochemical parameters of the waters in the seagrass ecosystem of Tidung Besar Island change every month, although the changes are not significant. This is caused by changes in weather and season. According to Patty, seasonal changes can affect changes in temperature and salinity distribution patterns [44]. This is in accordance with data from the Indonesian Agency for Meteorological, Climatological, and Geophysics, which states that October and December 2023 are the dry seasons, while November 2023 is the rainy season. However, the results of the t – test showed that there was no significant difference between the rainy and dry seasons; therefore, the three months can be categorized as transitional seasons.

The fish community structure in the seagrass ecosystem of Tidung Besar Island had varying values for each station and month. Based on these results, this variation is caused by factors such as the environmental conditions of the ecosystem and the correlation between ecosystems. Stations 2 and 3 have seagrass ecosystems and coral reef ecosystems that are close together so that the two ecosystems have strong connectivity, whereas Station 1 has seagrass, mangrove, and coral reef ecosystems that are close together to create connectivity between habitats. This is supported by Adrim's research showing that the diversity of fish species in the seagrass ecosystem is supported by the correlation between ecosystems [39].

Stations 1 and 2 have seagrass beds close to coral reefs. According to Latuconsina *et al.*, there are families of fish typical of coral reefs that can be found in seagrass beds, such as Labridae, Serranidae, Scaridae, Siganidae, Lutjanidae, Lethrinidae, and Pomacentridae [35]. This is in accordance with the findings at Stations 1 and 2, namely that there are fish species such as *C. oxycephalus*, *C. trilobatus* (Labridae), *S. ghobban* and *S. quoyi* (Scaridae), *L. carponotatus* and *L. decussatus* (Lutjanidae), *L. harak* and *L. lentjan* (Lethrinidae), *A. bengalensis* and *A. sordidus* (Pomacentridae). The discovery of typical coral fish at stations 1 and 2 proves that there is habitat connectivity between seagrass and coral reefs. Typical coral reef fish found in seagrass tend to use seagrass beds as a place to find food and as a nursery and protection area as found by Latuconsina *et al.* in the seagrass ecosystem of the Ambon Dalam Bay waters [32].

Station 3 had a seagrass bed located close to the coral reefs and mangroves. According to Latuconsina *et al.*, several typical mangrove families found in seagrass are Carangidae, Sphyraenidae, and Gerreidae [35]. This is in accordance with the findings at Station 3, namely that there are fish species, such as *C. coeruleopinnatus* (Carangidae), *S. jello* (Sphyraenidae), and *G. oyena* (Gerreidae). According to Kimirei *et al.*, the presence of typical coral reefs and mangrove fish in seagrass proves the high connectivity found in the three ecosystems [45]. This is related to the ontogenetic distribution patterns of these fish. Generally, ontogenetic habitat patterns shift from shallower waters, such as seagrass and mangroves, to deeper waters, such as coral reefs.

The presence of typical fish from coral reefs and mangrove ecosystems proves that the richness of fish species in the seagrass ecosystem on Tidung Besar Island is influenced by the presence of coral reefs and mangrove ecosystems that are correlated with each other. It is important to maintain the stability of each habitat in order to create stable water conditions. The seagrass ecosystem on Tidung Besar Island can be categorized as quite good when viewed from the perspective of the diversity of existing fish. Good environmental conditions support the presence of existing fishes. Human activities, such as throwing garbage into the sea and anchoring ships in areas where there is seagrass, need to be avoided to reduce the risk of habitat damage. Fishermen's fishing activities must also be monitored. This is because there are still many fishermen who carry out fishing activities with endangered status, namely, *Scarus ghobban* (based on IUCN data).

5. Conclusion

The diversity of fish on Tidung Besar Island is included in the moderate category, with conditions that tend to be stable. The abundance of species and dominance mostly had higher values during the high tides. Richness, diversity, and dominance mostly had higher values during the low tides. The structure of the fish community in the seagrass ecosystem of Tidung Besar Island is influenced by the type and density of the seagrass.

References

1. R. Nurmasari, A.H.A. Putri, S. Rosmaida, U. Nurkhalifah, F. Ramadhan, Jurnal Teknologi Perikanan dan Kelautan. **14**, 1 (2023)
2. R.M. Wirakusuma, Jurnal Manajemen Resort & Leisure. **11**, 1 (2014)
3. S.N. Sari, E. Nurfaizi, Y. Anjeli, M. Fawwaz, A. Topano, Ghaista: Islamic Education Journal. **4**, 2 (2023)
4. S.W. Ningsih, W.A. Setyani, N.T. Spj, Journal of Marine Research. **9**, 3 (2020)
5. T. Herawati, R.A.R Sidik, A. Sahidin, H. Herawati, Jurnal Perikanan Universitas Gadjah Mada. **22**, 2 (2020)
6. F.N. Shabrina, D. Saptarini, E. Setiawan, Jurnal Sains dan Seni ITS. **9**, 2 (2020)
7. A.D.P. Fitri, H. Boesono, B.B. Jayanto, T.D. Hapsari, F. Adiyanto, Aquaculture, Aquarium, Conservation & Legislation. **14**, 1 (2021)
8. G. Allen, A field guide for anglers and divers: marine fishes of Southeast Asia, (Western Australian Museum, Perth, 1997)
9. K.E. Carpenter, V.H. Niem, FAO Species Identification Guide for Fishery Purposes: The Living Marine Resources of the Western Central Pacific: Volume 6: Bony Fishes Part 4 (Labridae to Latimeriidae), Estuarine Crocodiles, Sea Turtles, Sea Snakes, and Marine Mammals (FAO, Rome, 2001)
10. G. Allen, R. Steene, P. Humann, N. Deloach, Reef fish identification tropical pacific, (New World Publications, Jacksonville, 2003)
11. G.R. Allen, M.V. Erdmann, Reef fishes of the East Indies – volume I, (Tropical Reef Research, Perth, 2012)
12. M.F. Fachrul, Metode sampling ekobiologi, (PT Bumi Aksara, Jakarta, 2007)
13. A.E. Magurran, Measuring biological diversity (Blakwell Publishing, London, 2004)
14. M. Horinouchi, Laguna. **12** (2005)
15. C.J. Krebs, Ecological methodology second edition (An Imprint of Addison Wesley Longman, Menlo Park, 1999)
16. C.J. Krebs, Ecological methodology (NY Harper and Row Publisher, New York, 1989)

17. E.P. Odum, Fundamentals of ecology third edition (W.B. Saunders Company, Philadelphia, 1993)
18. P. Labrosse, Underwater visual fish census surveys (Secretariat of the Pacific Community Noumea, New Caledonia, 2002)
19. F. Feryatun, B. Hendarto, N. Widyorini, Journal of Management of Aquatic Resources. **1**, 1 (2012)
20. I.P. Putra, Kelautan dan Perikanan. **3**, 12 (2014)
21. L. Virta, J. Soininen, A. Norkko, Microbial Ecology. **81**, 1 (2020)
22. W. Handayani, A. Winara, Jurnal Agroforestri Indonesia. **3**, 2 (2020)
23. S. Nolan, M. Ramli, Bahtiar, Jurnal Biologis Tropis. **19**, 2 (2019)
24. F. Laila, U. Zanetta, A. Karuniawan, B. Waluyo, IOP Conference Series: Earth and Environmental Sciene. **1**, 1 (2021)
25. N. Mote, E. Indrayani, Jurnal Ruaya. **11**, 1 (2023)
26. A. Sen, C.R. Sreeraj, C. Raghunathan, K. Chandra, International Journal of Fisheries and Aquatic Studies. **9**, 1 (2020)
27. N. Kholis, M.P. Patria, T. Soedjiarti, Research Article. **1**, 1 (2017)
28. I. Sinaga, N.S. Sihombing, L. Sitinjak, T.S. Siregar, Jurnal Penelitian Terapan Perikanan dan Kelautan. **4**, 2 (2022)
29. D. Lirman, Journal of Experimental Marine Biology and Ecology. **180**, 1 (1994)
30. H. Lamtane, H. Pratap, S. Ndaro, Western Indian Ocean Journal of Marine Science. **6**, 1 (2009)
31. R.A. Beroperai, R. Ayorbaba, H.M. Bonai, R.M. Rahanra, Unes Journal of Scientech Research. **6**, 2 (2021)
32. H. Latuconsina, R.A. Rappe, Iktiologi Indonesia. **13**, 1 (2013)
33. G.H. Hutami, M.R. Muskanafola, B. Sulardiono, Journal of Maquares. **6**, 3 (2017)
34. Romini, Royandi, A.H. Yanti, Life Science. **12**, 1 (2023)
35. H. Latuconsina, A. Padang, A.M. Ena, Jurnal Agribisnis Perikanan. **12**, 1 (2019)
36. Supriadi, A. Romadhon, A. Farid, Jurnal Kelautan. **8**, 1 (2015)
37. M. Yohanista, M.I. Rume, M.H.D. Nela, Jurnal Ilmu Kelautan dan Perikanan. **5**, 2 (2023)
38. A.H. Nugraha, S. Wahyuni, F. Idris, A. Zahid, Authentic Research of Global Fisheries Application Journal (Aurelia Journal). **5**, 1 (2023)
39. M. Adrim, Oseana. **31**, 4 (2006)
40. N.A. Ardhiani, D.S. Ardyanti, A. Suryanda, Jurnal Ekologi, Masyarakat dan Sains. **1**, 2 (2020)
41. A.G. Putra, Ruswahyuni, N. Widyorini, Diponegoro Journal of Maquares Management of Aquatic Resources. **4**, 2 (2015)
42. E. Supriyadi, Siswanto, W.S. Pranowo, Jurnal Metereologi dan Geofisika. **19**, 1 (2019)
43. H. Latuconsina, M. Sangadji, L. Sarfan, Jurnal Ilmiah Agribisnis dan Perikanan. **6**, 3 (2014)
44. S.I. Patty, Jurnal Ilmiah Platax. **1**, 3 (2013)
45. I.A. Kimirei, I. Nagelkerken, B. Griffioen, C. Wagner, Y.D. Mgaya, Estuarine, Coastal and Shelf Science. **92**, 1 (2011)

Sustainable management of marine resources through customary practices: The implementation of sasi on sea cucumbers in Werka Village, Maluku Province

Siti Hajar Suryawati[1*], *Radityo* Pramoda[2], *Hakim* Miftakhul Huda[2], *Adriani* Sri Nastiti[3], *Lismining* Pujiyani Astuti[3], *Andrias* Steward Samusamu[3], *Arip* Rahman[3], *Raymon* Rahmanov Zedta[3], *Reny* Puspasari[3], *Masayu* Rahmia Anwar Putri[3]

[1]Research Center for Industrial, Service and Trade Economics National Research and Innovation Agency (Jakarta-Indonesia)
[2]Research Center for Behavioral and Circular Economics, National Research and Innovation Agency (Jakarta-Indonesia)
[3]Research Center for Conservation of Marine and Inland Water Resources National Research and Innovation Agency (Bogor-Indonesia)

Abstract: Natural resource management based on local wisdom is a key approach to effective and sustainable biodiversity conservation outside protected areas. One example is the **sasi** practice for sea cucumbers in Werka Village, Southeast Maluku Regency, Maluku Province. This study examines the application of Traditional Ecological Knowledge (TEK) in **sasi** practices to sustain sea cucumber populations. The research aims to analyze the implementation of **sasi** in Werka Village, the role of indigenous peoples, and the challenges they face in resource conservation. Data were collected through interviews and analyzed using qualitative scoring and a descriptive case study approach. The findings indicate that the success of **sasi** in Werka Village is driven by the integration of economic, social, and institutional factors, with strong support from indigenous communities and local authorities.

1. Introduction

The challenge in resource conservation is the increasingly limited avaibility of natural areas for protection. Indonesia must adopt natural resource conservation management based on local wisdom to promote sustainable use [1]. In-situ natural resource management based on local wisdom has become an increasingly important topic in efforts to balance human use with ecosystem conservation. Local wisdom is defined as 'noble values that apply in the community's life system to, among other things, protect and manage the environment sustainably' [2]. One approach that draws on indigenous peoples' understanding of their relationship with the natural environment is Traditional Ecological Knowledge (TEK)

* Corresponding author: sitisuryawati@gmail.com

This understanding encompasses not only the management of natural resources but also the ways in which these relationships are established and passed down as part of local cultural traditions [3]. Indigenous peoples in various parts of the world, including Indonesia, have long developed natural resource management practices based on TEK to maintain the ecological and social balance in their regions. An example of TEK implementation in Indonesia is the *sasi* custom in Werka Village, Kei Besar District, Southeast Maluku Regency, Maluku Province. The vast waters of Maluku make the marine and fisheries sector crucial as a primary driver of economic development on the Maluku Islands. If managed properly, the abundance of marine resources can bring significant benefits to local communities, which are the rightful stewards of these waters. However, these benefits may diminish if resource availability declines or becomes extinct while community needs continue to grow [4].

Sasi in Werka Village is a customary practice that temporarily prohibit the use of natural resources for a specific period. This tradition has been passed down for generations and is an integral part of local community life. The marine resource included in the sasi practice in Werka Village are sea cucumbers. The market demand for sea cucumbers in Southeast Maluku is high [5] because of their promising economic value. However, this global demand places significant pressure on the sea cucumber population in Werka Village. Sasi practice is becoming increasingly relevant as a means to help maintain resource sustainability in the face of rising demand.

The indigenous people of Werka Village play a key role in the implementation of *sasi*, not only as direct managers of marine resources but also as the primary custodians of traditional practices. Traditional leaders have the authority to determine when *sasi* is enforced and lifted, based on observations of resource conditions and ecosystem cycles. The community's active involvement in each stage of decision-making fosters a strong sense of ownership of resource sustainability in their waters. The practice of *sasi* on sea cucumbers in Werka Village has been ongoing for many years and is an integral part of local community life.

The practice of *sasi* on sea cucumbers in Werka Village faces challenges that require attention of various stakeholders. Economic pressures from outside can lead to violations of *sasi* rules, posing a serious threat to both the preservation of sea cucumbers and the sustainability of the *sasi* practice itself. Additionally, modernization and social changes may influence the younger generation's adherence to and appreciation of traditional values passed down by their ancestors. Government efforts, particularly by the Ministry of Marine Affairs and Fisheries (MMAF), to support management based on local wisdom could make Werka Village a valuable study area for assessing the effectiveness of *sasi* in preserving sea cucumber populations. Thus, this study aimed to examine the application of Traditional Ecological Knowledge (TEK) in the practice of *sasi* in Werka Village, the role of indigenous communities in its implementation, and the challenges encountered in sustaining sea cucumber conservation.

2. Data and Methods

2.1 Time and Location

This study was conducted in 2024 in Werka, Maluku Province (Figure 1). Werka was chosen as the research location because it is a coastal village that has practiced a *sasi* culture for sea cucumber resources since ancient times.

Source: Primary Data (processed)

Fig. 1. Location Map of Werka Village, Southeast Maluku Regency, Maluku Province

Figure 1 shows that the name 'Werka' consists of the words Wer (meaning 'to pull') and Ka (meaning 'a rope stronger than wire'). The Ka rope, according to local indigenous beliefs, is symbolized as an heirloom. The village of Werka is traditionally led by a king (rat) who operates independently, not under the jurisdiction of Ursiw and Loor Lim, the larger kingdoms on Kei Besar Island.

2.2 Data and Analysis

This study used both the primary and secondary data. Primary data were gathered through interviews with key stakeholders familiar with the sasi customary practice in Werka Village, including (1) pastors, (2) fishermen, (3) traditional leaders, and (4) the village head and secretary. A total of 29 respondents were selected using topic guides and structured questionnaires to obtain insights into (1) sea cucumber sasi practices, (2) challenges and opportunities, and (3) economic, social, and institutional factors.

To analyze the interview data, a scoring technique was applied to assess respondents' perceptions. The findings were then pdescriptively presented using a qualitative case study approach, providing a comprehensive understanding of the research results. Purposive sampling was employed to, ensure that the respondents were directly involved in sasi practices, thereby making their perspectives highly relevant to the study's objectives. The ideal score was calculated using the formula *Criterion Score = Scale Value x Number of Respondents*, with the assigned scores based on the respondents' answers, as shown in Table 1.

Table 1. Scoring Criteria for Respondent Answers

Score	Criterion
4	Very positive
3	Positive
2	Negative
1	Very negative

Referring to Table 1, after scoring, the data were categorized within the rating scale to determine overall perception results. The percentage of responses was calculated using the formula p = (f/n) × 100%, where *f* represents the frequency of responses and *n* denotes the ideal score.

To ensure methodological transparency, several measures were employed to assess reliability and validity.

Reliability

- Source triangulation: Information was cross-checked across different respondent groups (fishermen, traditional leaders, village officials, and pastors) to minimize bias.
- Method triangulation: Interviews were supplemented with field observations and document analyses to enhance data consistency.
- Response consistency check: Key questions were phrased in different ways to verify that the respondents provided stable and non-contradictory answers.

Validity

- Content validity: Research instruments (questionnaires and interview guides) were developed based on the relevant literature and reviewed by experts and traditional leaders before data collection.
- Construct validity: The scoring method used to assess respondents' perceptions was based on established indicators from previous studies of customary conservation practices and marine resource sustainability.
- Member checking: Key respondents reviewed the interpretations and analyses of their responses to ensure accuracy.

Additionally, secondary data were collected through literature review, drawing from books, scientific articles, national regulations, and other relevant sources related to the research topic.

3. Results and Discussions

3.1 Traditional Ecological Knowledge (TEK) and Sasi Practices of Werka Village

The development of the marine and fisheries sectors in Indonesia faces various challenges, one of which is overfishing [1]. This challenge requires a utilization approach that considers aspects of environmental sustainability, such as the concept of TEK. This concept encompasses a variety of practices, habits, and understandings that have been passed down from generation to generation by indigenous people in relation to the conservation of natural resources. TEK, which is realized through the practice of sasi customs, serves as a legal system that helps to restore the population [6]. Resource conservation in the sea cucumber sasi custom, as a conservation area in Werka Village, is an important tool for managing its potential, which makes it vulnerable to overexploitation. The mechanism used to maintain the potential of sea cucumbers focuses on optimizing their growth, considering factors such as life cycle and ecosystem conditions.

The perception of the people of Werka Village regarding the benefits of conservation areas in sasi customs for the environment and their relationship with daily life are shown in Figure 2.

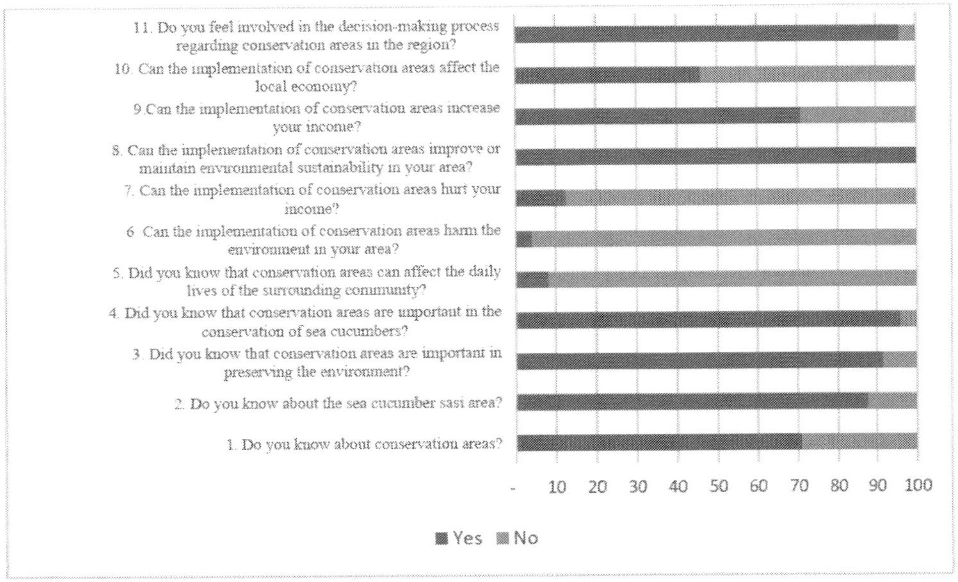

Source: Primary Data (processed)

Fig. 2. Perceptions of Traditional Sasi Practices in Werka Village

Based on Figure 2, the majority of respondents in Werka Village had a high awareness of conservation areas (71%) and the importance of conservation for the environment (92%). Only 8% of respondents felt that conservation areas affected their daily lives. This indicates that most people believe that sasi practices do not directly impact their economies. Most respondents considered conservation areas for sasi practices to be harmless to the environment (96%) and believed that these practices can increase their income (71%). They also stated that they are always involved in decision-making related to the regulation of conservation areas (96%). In general, people of Werka have a positive view of the environmental benefits of conservation areas related to the practice of sasi, while their perceptions of the economic impact and its relationship with daily life are mixed. The respondents' perceptions of the economic impact of sea cucumber sasi practices are explained in Figure 3.

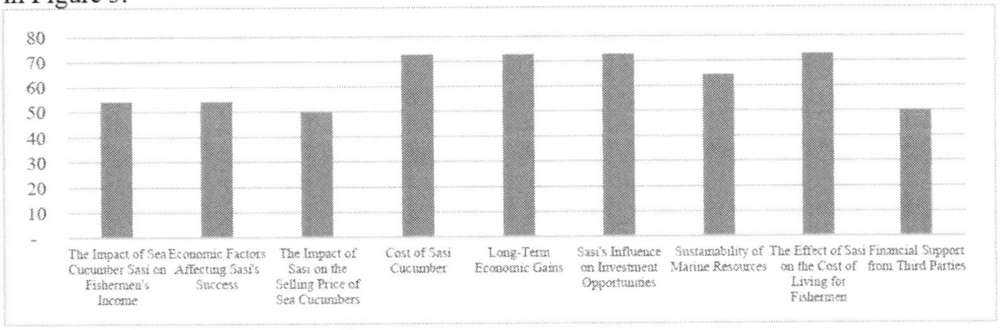

Source: Primary Data (processed)

Fig. 3. Perception of the Economic Impact of Sasi Traditional Practices in Werka Village

According to respondents, sasi teripang had a moderate influence on fishermen's income (54%) and selling prices (50%). This indicates that the direct economic contribution of sales is not significant. Respondents assessed the costs of implementing the sasi, its long-term economic benefits, and investment opportunities, giving them a fairly high score (73%). This reflects the long-term economic benefits experienced by the community in attracting investment. The practice of sasi is also viewed by the community as contributing to conservation areas for in-situ conservation, as it has a positive impact on reducing the cost of living (73%). Some respondents (50%) expressed a need for external support in the practice of sasi, particularly in the form of financial assistance from third parties. Overall, the information provided by respondents indicates that sea cucumber sasi has a long-term positive impact on the economy, although challenges related to fishermen's direct income and the need for external financial support remain.

The basis for deciding on the implementation of sasi in the waters of Werka Village was determined deliberatively, considering both ecological and social needs. These deliberations ensure that decision-making is not only based on scientific data but also takes into account the local wisdom that has existed for a long time. The concept of TEK in the practice of sasi creates a strong sense of ownership of natural resources by involving all the elements of society. The prayer ritual that accompanies the opening and closing of the sasi holds deep meaning, serving both as a form of respect for nature and as a moral reminder for the community to act responsibly toward the preservation of natural resources [7].

The prayer ritual conducted in Werka Village serves to strengthen social and spiritual ties between the community and the environment, as well as to increase compliance with established customary rules. This phenomenon fosters a greater sense of community responsibility for maintaining the in-situ area they possess. The combination of TEK and custom-based management makes natural resource management more effectivein conservation areas. Custom-based management also maintains a balance between human needs and ecosystems in the context of current global environmental changes. The implementation of sea cucumber sasi in Werka Village is not only limited to the application of customary rules but also involves the active participation of indigenous peoples and support from customary institutions. This ensures that the practice of sasi is conducted consistently with local values, allowing it to contribute to the preservation of sea cucumbers and the environment.

The integration of the concepts of TEK and the Sasi management system in an era facing environmental challenges offers a relevant contextual approach to natural resource management. The active role of the people of Werka Village in maintaining the sustainability of sea cucumber resources by considering social, spiritual, and ecological aspects can serve as an example in other areas with similar conditions. The practice of sasi in Werka Village emphasizes the importance of local wisdom in maintaining an environmental balance and the sustainability of community life. Sasi, a form of traditional wisdom interpreted through special symbols that signify prohibition, is believed to maintain a balance between humans and their environment [8].

3.2 The Practice of Sasi for Sea Cucumbers in Werka Village

Werka Village implemented delineation provisions for the practice of regional and type sasi. The boundary for regional sasi extends from the coastline to the area bordering deep waters (blue water), whereas the boundary for sasi jenis (sea cucumber sasi) encompasses all waters (Figure 4). The sea cucumbers types found in the waters of Werka Village include *Holothuria scabra*, *Thelenota ananas*, *Holothuria leucospilota*, *Holothuria coluber*, *Holothuria edulis*, *Holothuria atra*, *Bohadschia argus*, *Bohadschia vitiensis*, *Bohadschia marmorata*, and *Actinopyga lecanora*.

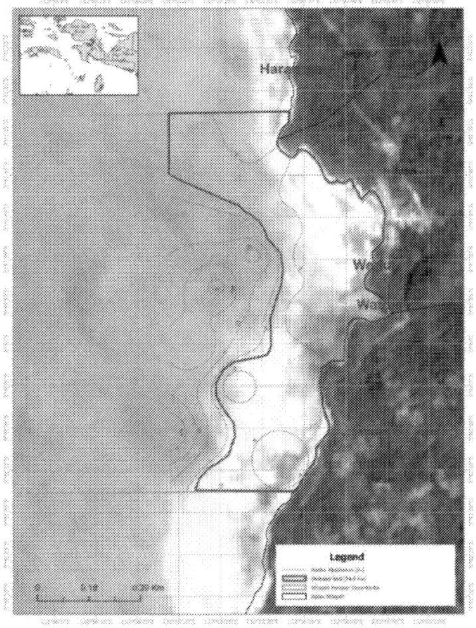

Primary Data (processed)

Fig. 4. Map of the Boundary of the Sasi Area in Werka Village

Figure 4 show that the area of the sasi region, based on spatial analysis, is approximately 54.51 hectares, while the area of waters designated as a type of sasi region is about 158.84 hectares.

The effectiveness of sea cucumber sasi in Werka Village is highly dependent on the active participation of indigenous peoples and the support of traditional leaders. Conservation areas managed by indigenous peoples tend to be more successful when supported by strong local institutions and community involvement in preserving natural resources [9]. The mapping of social impacts in Werka Village (Figure 5) illustrates that each individual plays an important role in the sustainability of sasi practices.

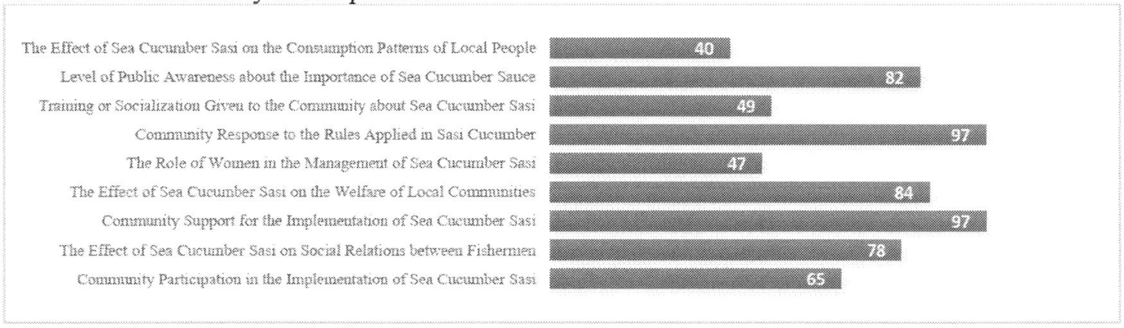

Source: Primary Data (processed)

Fig. 5. Perceptions of the Social Impact of Traditional Sasi Practices in Werka Village

Figure 5 shows that the social impact related to participation (65%) and community support (97%) for the implementation of sea cucumber sasi in this region was quite high. This demonstrates that the practice of sea cucumber sasi in Werka Village has a positive influence on social relations among fishermen (78%) and on the welfare of the local community (84%). The level of public awareness about the importance of sea cucumber sasi was also quite high (82%), although the training or socialization provided was still low (49%).

Based on the information from the respondents, women's involvement in the management of sea cucumber sasi is still relatively low (47%). This low participation rate is due to the prevailing belief of the community that women should focus on household responsibilities and child care. The sea cucumber sasi custom in Werka Village emphasizes conservation and resource management more than changes in consumption habits. This is evident from the opinions of 40% of respondents, who stated that the sea cucumber sasi custom does not affect their diet. The strong community support of 97% for the implemented sasi rule reflects a positive response, indicating that this customary practice is effective in conserving sea cucumber resources.

Traditional leaders are responsible for determining when a sasi is opened and closed, including the implementation of rituals related to respect for nature [10]. The decision to close or break the sasi is made by the king and the traditional elders. Opening and closing ceremonies are usually marked by a traditional ceremony, which is intended to inform the public that the sasi will be opened or closed. When the sasi is closed, the entire community is prohibited from carrying out activities that could harm the aquatic ecosystem (such as using poisons like bores or tuba roots to catch fish). A ban on these toxins is enforced because they can damage seagrass and coral habitats. This prohibition applies to all people living inside and outside Werka Village. Resources that are prohibited from being caught include biota within the sasi area, including sea cucumbers and organisms in coral reef areas [11].

The timing of the sasi opening was determined based on information about the conditions of sea cucumbers that were ready to be harvested, as reported by fishermen. All types of fish can be caught by local residents or outsiders at the time of opening the sasi; however, sea cucumbers must not be caught. This restriction occurred because the auction winner purchased all types of sea cucumbers in the sasi area. The individual rights of the indigenous people of Werka Village are compromised when the harvest is handed over to the auction winner. This situation provides flexibility for auction winners to exploit natural resources without regard for sustainability in pursuit of economic benefits [12]. The auction price for the sasi area and the selection of the winner fall under the authority of the king and the customary elders (the auction price at the time of this study was 5 million rupiahs).

Provisions related to the size of harvested sea cucumbers have not yet been established. This situation can jeopardize the sustainability of fishery resources, as the focus of sasi practices is primarily on maximizing economic benefits. Catching sea cucumbers at the time of opening the sasi still employs traditional, environmentally friendly methods, such as diving and hand collection. This method has proven to be more sustainable than modern techniques, which often destroy sea cucumber habitats and lead to population decline [13]. The traditional methods practiced by the people of Werka Village not only help preserve the sea cucumber population but also maintain the marine ecosystem as a whole.

The form of sanctions and their application to violators during the closing of the sasi in Werka Village are shown in Table 2.

Table 2. Procedure for the Implementation of Customary Sanctions in Werka Village

Procedure	Explanation
Reporting	- Reporting is carried out by anyone who witnesses a violation firsthand - If the report is proven to be true, the reporter will receive an incentive of 5 million rupiah, which will be obtained from customary fines

Procedure	Explanation
Sanction	- Every perpetrator proven to have committed a violation will be subject to customary sanctions in the form of a fine of 1 lela/cannon (approximately 20 million rupiah) - Customary fines collected from violators are handed over to the village government as a form of compensation
Implementation	The application of sanctions against violators is carried out based on the decisions of customary institutions
Supervision	Supervision is carried out by the community, the apparatus, and customary institutions

Source: Primary Data (processed)

Table 2 shows that the provision of sanctions to violators can have a deterrent effect, preventing them from repeating their offenses. Strict sanctions for violators serve as a disincentive to prevent overexploitation [14]. Community-based supervision is a potential solution for increasing compliance with sasi rules. Community involvement fosters a stronger sense of ownership over managed resources and enhances awareness of the importance of sasi practices

One of the key aspects of the efficiency of sasi practices in Werka Village is the synergy between the TEK and community-based conservation practices. The inherited knowledge about sea cucumber resource behavior, its life cycle, and the optimal environmental conditions for its growth provides the basis for management decisions [15]. Involving the community in the decision-making process to determine the opening and closing times of the month can strengthen their sense of ownership of managed resources. This is closely linked to the compliance of all people in Werka Village with the rules established by the traditional elders, community leaders, and religious leaders.

The moral support of customary institutions also affects the success of sasi practice in Werka Village. The function of these institutions is not only to supervise but also to mediate and maintain communication among community members. Rituals and traditions related to sasi provide deeper meaning for the community, making them aware of the importance of protecting natural resources and ecosystems [16]. The prayer ritual before and after the month is not only symbolic but also serves to strengthen the community's commitment to environmental sustainability. The local wisdom approach applied in Werka Village demonstrated that traditional practices can contribute to the long-term preservation of marine ecosystems. This shows that managing sea cucumber resources through traditional methods can be an effective solution for addressing the challenges of in-situ conservation in future conservation areas.

The effectiveness of sea cucumber sasi in Werka Village is a clear example of how the integration of local knowledge with sustainable management can benefit both the community and environment. The management of sea cucumber sasi is not only an effort to conserve marine resources but also a manifestation of local wisdom that can serve as a model for other regions facing similar challenges. The practice of sasi in Werka Village provides valuable insights into sustainable natural resource management through the active participation of the community and the support of indigenous institutions.

3.3 Challenges and Opportunities

Technological advances and globalization are increasingly eroding the nation's culture and identity. Additionally, sasi faces challenges and weaknesses in various aspects that are feared to undermine the rules that have been implemented so far. The influence of external culture on people in coastal areas has led to a decline in awareness of the sasi tradition. This foreign culture, resulting from globalization, causes certain segments of society, particularly the younger generation, to take local rules that are spiritual or traditional rituals less seriously. Local cultural values are considered very important because they represent the identity of a society and reflect how people address issues, such as environmental sustainability and the weakening of social institutions [1].

The practice of sea cucumber sasi in Werka Village can be said to preserve marine resources; however, there are still challenges in ensuring that this method provides sustainability. The main challenge that will be faced in the future is external economic pressure, which encourages some individuals to violate Sasi rules to meet the demands of the global market. High demand for sea cucumbers is the primary driver of overexploitation. This economic pressure has the potential to undermine the success of the sasi practice if it is not balanced by adequate protective measures [6].

Another challenge that needs to be addressed is optimizing sea cucumber resources so that they can provide benefits and become the main source of income for the community in Werka Village. This situation arises because the public is unaware that certain types of sea cucumbers have a high selling value. As a result, the selling price of sea cucumbers in the waters of Werka Village remains low. The involvement of the government, particularly the Ministry of Marine Affairs and Fisheries (MMAF) and local authorities, is urgently required to overcome this challenge. The government can also work to provide economic incentives to the community as a reward for efforts to preserve sea cucumbers. These incentives can be in the form of welfare programs or alternative income opportunities related to environmental conservation. Such efforts will allow people to experience the direct benefits of conservation practices, which, in turn, will enhance their commitment to sea cucumber sasi.

Climate change and ecosystem damage are also threats to the sustainability of sea cucumbers today [17]. In light of this, education and increased public awareness of the importance of maintaining the marine ecosystem sustainably are essential. Training and socialization programs that focus on the long-term benefits of conservation practices can help communities to better understand the value of natural resource sustainability. The synergy of the government, local authorities, customary institutions, and village officials is urgently needed to create rules that strengthen the formal management of sea cucumber sasi customary practices in Werka Village. The purpose of establishing these regulations is to encourage effective implementation of sasi in the long term.

The practice of sasi within the framework of TEK provides an understanding that conservation efforts carried out by traditional communities can serve as a reference for policy-making related to natural resource conservation. Management rooted in local wisdom deserves the attention of central and regional authorities to protect the potential of natural resources [18]. Overall, although the challenges faced are quite complex, there are opportunities to improve the implementation of sea cucumber sasi practices. The synergy of all government elements and the community in Werka Village is believed to enable the ongoing development of sasi as a form of conservation of in-situ natural resources based on sustainable local wisdom.

3.4 Spiritual and Institutional Integration

The concept of TEK refers to an understanding of a community's traditional practices in managing and preserving the surrounding ecosystem. The practice of sea cucumber sasi in Werka Village reflects the typical integration of TEK from social, spiritual, and ecological aspects. TEK not only encompasses technical knowledge about natural resources but also includes ethical values, spirituality, and belief systems that are passed down from generation to generation. This reflects one of the main principles of TEK, which is to maintain a harmonious relationship with nature through a holistic approach that considers the long-term well-being of an ecosystem [19].

The spiritual integration seen in the practice of sasi in Werka Village confirms the role of spirituality in the TEK. The ritual of prayer and respect for the sea in the management of sea cucumber sasi is an integral part of the opening and closing of the sea cucumber season. This shows that the management of natural resources is not only based on rational calculations but also relates to nature as a spiritual entity. In the context of TEK, ecosystems are often considered living beings that possess spirit or power, which must be respected through rituals and actions that are in harmony with nature [20]. This phenomenon reinforces the awareness that humans are part of a larger network of life, making ecosystem damage not only harmful but also disruptive to cosmological balance.

Emphasis on the spiritual aspect of resource management also strengthens the collective moral values embraced by the community. Community awareness of protecting sea cucumber resources through sasi is a spiritual responsibility that legitimazes customary provisions. The perspective of TEK in relation to the relationship between humans and nature, which is often considered reciprocal, can be interpreted as nature providing welfare while humans are obliged to maintain this balance [21]. The practice of sasi does not depend solely on written rules but also on high moral awareness within society, which is reinforced by the values upheld by customary and religious institutions [7].

The role of customary institutions is crucial in the practice of sea cucumber sasi in Werka Village. Indigenous institutions have strong social legitimacy to establish and ensure compliance with sasi rules. Customary institutions act as supervisors and law enforcers, and ensure that the practice of sasi is carried out in accordance with established traditions. This approach creates public awareness that non-compliance with the provisions that have been set not only has an impact on individuals, but can also damage the balance of the existing ecosystem. Customary institutions in the management of sasi, act as a liaison between the older generation who inherit knowledge about the ecosystem and the younger generation who will continue this tradition. TEK provides a foundation for sustainable resource management, where traditional knowledge and spiritual values are inherited and maintained by the community [22].

The impact of institutional roles on the implementation of sea cucumber sasi custom in Werka Village is presented in Figure 6.

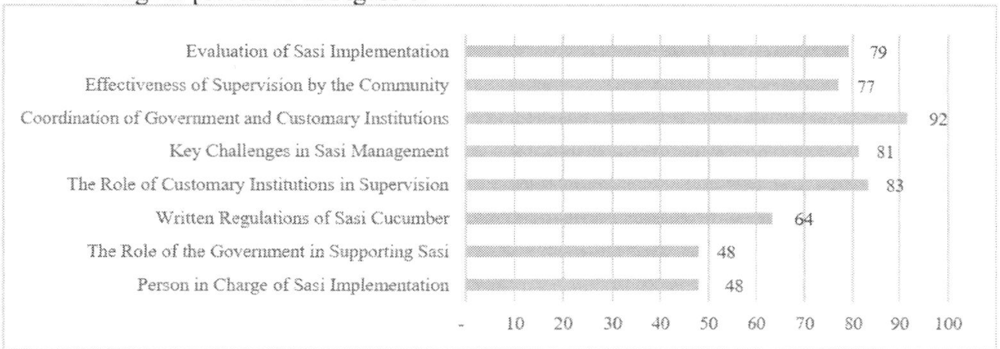

Source: Primary Data (processed)

Fig. 6. Perceptions of the Institutional Impact on Sasi Practices in Werka Village

Figure 6 illustrates that the responsibility for implementing sasi and the government's role in supporting it are still relatively low, with each receiving a score of 48%. This indicates the need to clarify the government's role as the primary entity responsible for the implementation of sea cucumber sasi. According to 64% of respondents, a written regulation on sasi practice may exist but has not been legally established or optimally implemented. The role of customary institutions in supervision was perceived as strong by 83% of the public. This perception indicates that customary institutions in Werka Village play an important role in maintaining compliance with the rules of the sasi.

The main challenge of sea cucumber sasi management, according to respondents, is facing significant obstacles (81%), while coordination between the government and indigenous institutions is reported to be very good (92%). This situation indicates that although institutional coordination is effective, actions must still be taken to address the obstacles encountered. Supervision carried out by the community (77%) and evaluation of the implementation of sea cucumber sasi (79%) show that the implementation has functioned well. According to the results of the study, the efforts needed to improve the shortcomings of sea cucumber sasi practice involve enhancing government responsibility and formalizing written regulations.

The merging of social, spiritual, and ecological dimensions in the management of natural resources can create a sustainable and effective system. The support of customary and religious institutions ensures that the people of Werka Village are not only bound by written laws but also possess deep spiritual and ethical values [23]. This encourages them to care for the environment and preserve sea cucumber resources as integral parts of their culture and spirituality. The practice of rituals and ceremonies to honor the sea as a form of gratitude for the resources obtained demonstrates the close bond between spiritual aspects and resource management. Natural resource management based on local wisdom in Werka Village, which unites spiritual and institutional aspects, provides a holistic solution to all challenges regarding ownership of its potential. This situation reflects an environment imbued with local wisdom values, which are integral to the identity of the people of Werka Village.

4. Conclusion

The management of natural resources based on local wisdom in Werka Village, through the traditional practice of sea cucumber sasi, demonstrates that TEK plays a central role in preserving the marine ecosystem. The active participation of the community, involvement of traditional leaders, and support from the government are key factors in the effective management of these resources. The purpose of implementing the sea cucumber sasi custom is to provide recovery time for the population while simultaneously preserving the ecosystem as a whole. Despite its implementation challenges, the practice of sea cucumber sasi in Werka Village has provided a custom-based management model that can be applied to other regions in Indonesia.

Community involvement at every stage of the decision-making process is important because it can create a sense of ownership and responsibility for the sustainability of natural resources. The sustainability of sea cucumber sasi practice in Werka Village requires the synergy of customary provisions and formal regulatory support. Support from central and regional governments in forming regulations and economic incentives has the potential to create a management system based on local wisdom as part of a broader conservation strategy. The implementation of sasi customs in Werka Village can serve as an input for the government as a reference for recommendations on in-situ resource conservation management, achieved through collaboration between local communities and government institutions in a sustainable manner. The results of the study show that the practice of sea cucumber sasi practice in Werka Village depends on the integration of economic, social, and institutional factors, combined with active participation from the community.

The authors would like to express their utmost gratitude to the CFI Indonesia GEF-6 Grant, provided by the World Wildlife Fund (WWF) Global Environmental Facility (GEF) Agency and the Ministry of Maritime Affairs and Fisheries of the Republic of Indonesia, for their full support and funding.

References

1. N. P. R. Persada, F. M. Mangunjaya, I. S. Tobing, Ilm. & Bud. J. **41**, 59 (2018)
2. Law of the Republic of Indonesia No. 32 of 2009 on Environmental Protection and Management
3. C. Fabricius, The fundamentals of community-based natural resource management. In *Rights Resources and Rural Development* (Routledge, London, 2013)
4. N. M. Dawson, B. Coolsaet, E. J. Sterling, R. Loveridge, N. D. Gross-Camp, S. Wongbusarakum, ... , F. J. Rosado-May, Eco & Soc. J. **26**, 3 (2021).
5. J. Jasmadi, I. Wirawati, I. T. Suryaningtyas, S. Permadi, Ilm. Per. & Kel. J. **12**, 1 (2018)
6. R. E. Johannes, An. Re. of Eco. & Syst. J. **33**, 1 (2002)
7. K. Nugroho, F. Carden, H. Antlov, *Local knowledge matters* (Policy Press, Bristol, 2018)
8. A. Kusumadinata, Sos. Hum. J. **6**, 1 (2015)
9. E. McLeod, R. Salm, A. Green, J. Almany, Designing marine protected area networks to address the impacts of climate change. Front. in Eco. and the Env. J. **7**, 7 (2009)
10. E. Mantjoro, Oce. & Coa. Man. J. **32**, 1 (1996)
11. S. Touwe, Mar. Stu. & Nat. Int. J. **4**, 2 (2020)
12. H. Latuconsina, Tri. J. E **5**, 1 (2009)
13. R. Harini, V. Natarajan, C. K. Sunil, Fo. Sci. **89, 7** (2024)
14. J. Schleicher, C. A. Peres, N. Leader-Williams, Cons. Lett. J. **12**, 5 (2019)
15. Q. Han, J. K. Keesing, D. Liu, Rev. in Fish. Sci. & Aqu. J. **24**, 4 (2016)
16. T. G. Hallatu, I. D. Palittin, M. C. B. Umanailo, *Culture and religion in sasi tradition*. In Proceedings of the International Conference on Industrial Engineering and Operations Management, 26-28 November 2019, Riyadh, Saudi Arabia (2019)
17. S. W. Purcell, C. Conand, S. Uthicke, M. Byrne, Oce. & Mar. Bio. J. **54** (2016)
18. Y. Yasir, Y. Firzal, C. Yesicha, A. Sulistyani, Lnd. Eco. J. **15**, 2 (2022)
19. F. Berkes, N. J. Turner, Hum. Eco. J. **34**, 4 (2006)
20. N. I. Sinthumule, N. I. Env. Sci. J. **11** (2023)
21. Z. Molnár, A. Fernández-Llamazares, C. Schunko, I. Teixidor-Toneu, I. Jarić, I. Díaz-Reviriego, ... , R. Hill, Bio. Cons. J. **285**, 110190 (2023)
22. J. Mohd Salim, S. N. Anuar, K. Omar, T. R. Tengku Mohamad, N. A. Sanusi, Sust. J. **15**, 1 (2023)
23. P. Werbner, R. Werbner, *African customary justice: living law, legal pluralism, and public ethics* (Routledge, London, 2021)

Snapper and grouper fisheries management in Aisandami using the EAFM approach

Ridwan Sala[1*], *Duaitd* Kolibongso[1], *Yuanike* Kaber[1], *Dougklas* Wattimury[1]

[1]Department of Marine Science, Faculty of Fishery and Marine Science, Papua University. Jalan Gunung Salju, Amban, Manokwari, Papua Barat, Indonesia

Abstract. The coastal waters of Aisandami are rich in diverse coral fish resources, including snappers and groupers. A study is needed to evaluate how fisheries are managed in water. The present study was conducted to examine the implementation of the ecosystem approach to fisheries management (EAFM) for snapper and grouper fisheries by the community in Aisandami. The data used consisted of primary and secondary data obtained from various scientific publications. EAFM analysis focuses on six domains: fish resources, habitat and aquatic ecosystems, fishing technology, economy, social, and institutional. The results of the study show that in aggregate, the implementation of EAFM in the management of snapper and grouper fisheries is carried out well but is not yet optimal. Several components from each domain of the fisheries system still need to be improved to achieve the optimal EAFM-based management of fisheries.

1 Introduction

Aisandami Village is located in Cenderawasih Bay National Park (TNTC) and is one of the target locations of the Global Environment Facility (GEF) 6 program. The coastal waters of Aisandami Village are rich in diverse resources due to the presence of coral reefs, seagrass, and mangrove ecosystems that support the life of marine biota, including economically important fish species such as snappers and groupers. The fish group was the fishing target of fishermen in Aisandami Village. Consequently, the coastal water area has become an important place for the Aisandami Village Community to carry out economic activities, such as fishing. Aisandami Village is very interesting to research because of the community-based management of coastal water resources that has been implemented, especially in the last three years. The community applies local wisdom in the form of "sasi" ("kadup") as a tool for managing coastal and marine resources.

The fishery resources targeted by the fishing community in Aisandami Village are coral fish, including snappers and groupers. Grouper and snapper fish resources were included in economically important fish groups. Several studies have reported the intensive use of these resources by fishermen in TNTC [1,2].

The ecosystem approach is the model chosen by the Indonesian Government for implementation in the fisheries management area of the Republic of Indonesia (WPP-NRI). This model aims to preserve natural resources in the sea, habitats, and ecosystems, as well as

* Corresponding author: r.sala@unipa.ac.id

in social, economic, and institutional aspects. An ecosystem approach has been applied to manage commodities such as snappers, groupers, shrimp, and mangrove crabs. To date, research on fisheries resource management in TNTC is still very limited. One study that focused on assessing the status of fisheries management in Menarbu, which also implemented sasi, was conducted by Putri *et al.* [3]. This study reports that sasi has a positive impact on fishery management.

Information about fishing activities by fishermen in Aisandami Village is still very limited, including the level of utilization of fish resources and socioeconomic aspects. Therefore, an effort to assess the sustainability status of capture fisheries in the coastal area of the village is very important to ensure the sustainability of fish resources and social and economic benefits to the community in the village. The aim of this study was to examine Ecosystem Approach to Fisheries Management (EAFM) implementation in the management of snappers and groupers in the coastal waters of Aisandami Village.

2 Methods

2.1 Time and Place of Study

This study was conducted between August and September 2024. Primary data were collected in September 2024 in the Aisandami Village area and the waters where the Aisandami Community carried out grouper and snapper fishing activities (see Figure 1).

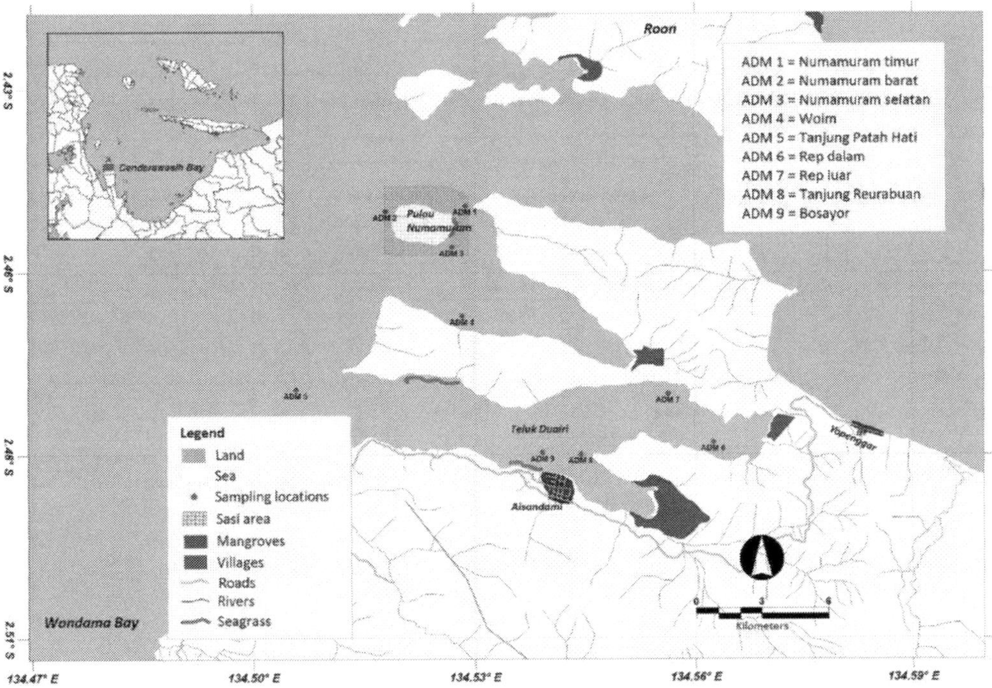

Fig 1. Map of the research area

2.2 Data Collection Methods

The data used in this study consisted of secondary and primary data. Secondary data were obtained from various sources, research report documents, and relevant scientific journals. Primary data were collected directly from the research location.

Primary data collection at the research location consists of:

a. Interviews with fishing communities to collect socio-economic, institutional, and fishing activity data. The interviews were conducted using a questionnaire administered to 41 respondents. In addition, interviews and discussions were conducted with the village head and coordinator of small and medium enterprises (UMKM) to obtain a general picture of the village profile, economic business activities, and institutions.
b. Measurement of physical chemical parameters
   Physical and chemical water data collection, namely, temperature, salinity, pH, and DO, was carried out directly at the coral data collection location. Chlorophyll data *a* were obtained using AQUA MODIS image data for the period of data collection.
c. Observation of coral reef ecosystems
   Observations of coral biota as the main component of coral reef ecosystems and various other benthic organisms were carried out using the point intercept transect (PIT) method [4]. This method was used to estimate live or dead coral cover. Transect positions were located at nine stations, determined using the Global Positioning System (GPS). At the transect observation point, a line (roll meter) was stretched 50 m long, which was placed parallel to the coastline following the depth contour. Observations were made at water depths of 5–9 m. The method for observing basic habitat coverage refers to English et al. [5], based on the growth form (lifeform).
   Identification of coral types was carried out by referring to the identification guidebook by Veron [6] and Suharsono [7] and using the online media WoRMS (World Register of Marine Species) as a source of data classification for each genus found. Several types of hard coral were identified directly at the data collection location using the Coral Finder by Kelley [8].
d. Observations of the distribution and condition of seagrass and mangroves were carried out at approximately nine stations where coral data were collected. These data support the available data based on previous research results, namely, seagrass data according to Tebay *et al.* [9] and seagrass data according to Dharmawan and Widyastuti [10].
e. Observation of fishermen's activities, consisting of fishing technology used (boats and fishing gear), types of catch, and measurements of groupers and snappers caught by fishermen. Data on coral fish catches were collected directly from fishermen who performed fishing activities. The collected data included species composition, total length (TL), and weight. The identification of fish types was based on Reef Fishes in East Indies [11] and Snapper of the World and Grouper of the World [12].

2.3 Data Analysis Methods

The data and information collected were used as input for analysis of the EAFM of snapper and grouper fisheries management in Aisandami Village. The indicators used in the EAFM analysis refer to Adrianto *et al.* [13] and consist of six domains: fish resources, habitat and aquatic ecosystems, fishing technology, economics, social, and institutional. Adrianto et al. [14] described the EAFM indicator assessment and data analysis methods refer to Adrianto *et al.* [14].

3 Results and Discussion

3.1 Brief of fisheries in Aisandami Village

Most of the residents of Aisandami Village earn their living as fishermen catching sea fish, especially coral and pelagic fish. Fishing activities spread across the waters of Duairi Bay in front of the village to Namamuran Island, which is located outside Duairi Bay.

The fishing technology used by Aisandami Village fishermen is still relatively traditional, namely, using handlines, gillnets, and lift net fishing gear. Gillnets and lift nets are used to catch small pelagic fish, whereas handlines are used to catch demersal fish and coral fish, including snappers and groupers. Handline fishing activities use fishing fleets in the form of paddle boats and longboats powered by a 15 PK outboard motor.

3.2 Grouper and Snapper Fisheries Management Status

3.2.1 Composite Index of Grouper and Snapper Fisheries Management Status

The results of the analysis of the management status of the grouper and snapper fisheries in Aisandami waters show that, in general (aggregate), the fisheries are managed "well" (composite score 71) (see Table 1). The aggregate value was the contribution of the six domains used in the evaluation. The social domain received a score of 87 or was categorized as 'very good' management. The domains of fish resources, habitats and ecosystems, fishing techniques, and institutions received the "good" management category. The management status classified as "moderate" is for the economic domain. Compared with the reference index for optimum (very good) fisheries management, only the social domain has optimally implemented EAFM in managing grouper and snapper fisheries (Figure 2). The index value of each domain mentioned above was determined by the values of the attributes used in each domain, which are described in detail in the next section.

Table 1. Composite (aggregate) index values and indices from each management domain for grouper and snapper fisheries in Aisandami waters.

Domain	**Composite value**	**Description**
Fish Resources	63	"Good" in implementing EAFM
Habitat & Ecosystem	73	"Good" in implementing EAFM
Fishing Technology	68	"Good" in implementing EAFM
Social	87	"Very Good" in implementatioing EAFM
Economic	60	"Moderate" in implementing EAFM
Institution	68	"Good" in implementing EAFM
Aggregate	71	"Good" in implementing EAFM

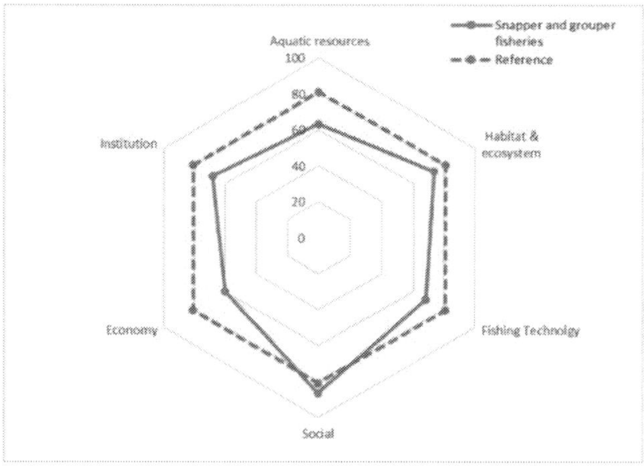

Fig 2. Kite diagram shows the position of the grouper and snapper fisheries management index in Aisadami compared to the reference for sustainably managed fisheries (> index value 80)

3.2.2 Fish Resource Domain

The results of the evaluation of the management status of the fish resource domain yielded a score of 72. This means that the fish resources are managed well. However, it still requires improvement to achieve an optimum (very good) management status. The status of fish resource management was determined by the six attributes used in the assessment. The results of the analysis of the six attributes used to evaluate the sustainability of fish resource management are shown in (Figure 3).

The excellent application of management using the EAFM approach in the aspect of fish resources is demonstrated by three attributes, namely the "range of collapse" of fish resources and the species composition of the catch. This explains why there is no real change in the distribution of the target fish, or in other words, the fishermen's fishing area has not changed (fixed). The composition of the target fish species in the form of economically important fish (i.e., snapper and grouper) was still high (or dominant) (Figure 4).

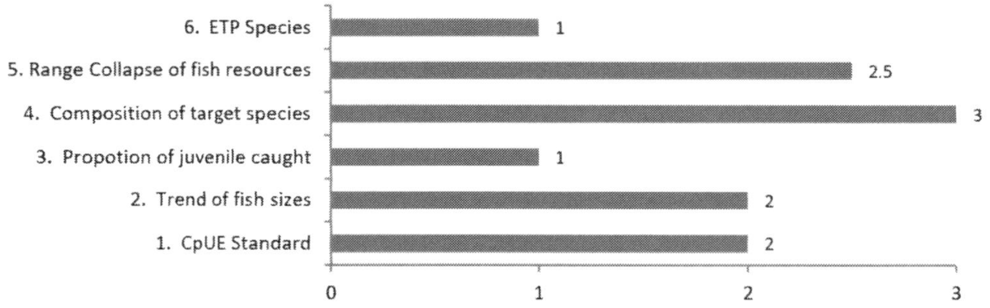

Fig 3. Results of evaluation of the Fish Resources domain in the management of grouper and snapper fisheries in Aisandami waters

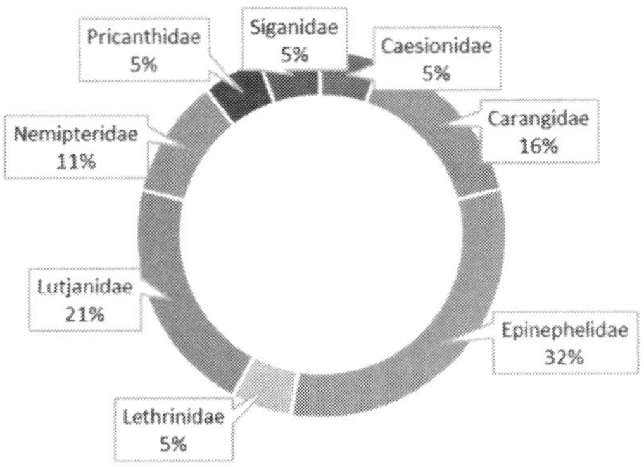

Fig 4. Composition of coral fish target caught in Aisandami waters.

However, several attributes require attention, namely endangered, threatened, and protected (ETP species), proportion of juvenile fish, fish size trends, and standard CPUE. In terms of ETP species, fishermen still catch napoleon fish, although it is very rare. According to the Decree of the Minister of Maritime Affairs and Fisheries of the Republic of Indonesia Number 37/KEPMEN-KP/2013 concerning Determination of the Protection Status of Napoleon Fish (*Cheilinus undulatus*), napoleon fish weighing 100–1000 g and those weighing more than 3000 g must be protected (not allowed to be caught). Regulations such as this have not been implemented well because there is no control over the size of the fish caught by the community in Aisandami.

However, the proportion of juvenile fish caught by fishermen requires serious attention. The proportion of juvenile fish remained very high (> 60%). For example *Lutjanus malabaricus* (Figure 5). The length of first maturity of *L. malabaricus* was 57.6 cm [15], but the size range of the species caught in Aisandami was less than 50 cm (Figure 6). So it can be said that most species *L. malabaricus* caught was young fish (juvenile). The proportion of juveniles was significantly higher than that found in Menarbu, which is also located in TNTC [3]. This may be due to the longer implemented in Menarbu than in Aisandami.

Fig 5. *Lutjanus malabaricus*

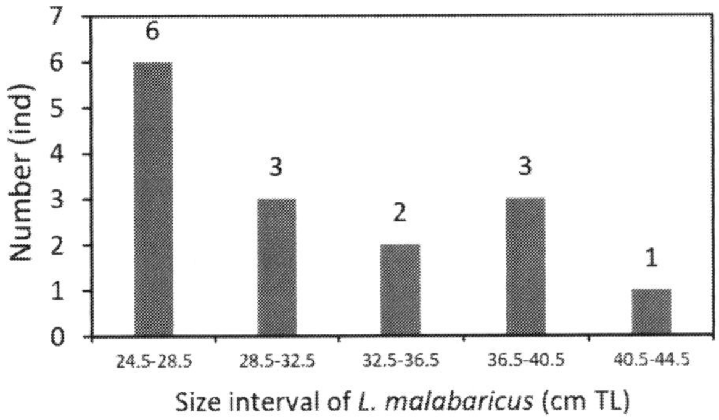

Fig 6. Length size distribution of *Lutjanus malabaricus* caught in Aisandami waters

Based on the results of interviews with fishermen, the size of the fish caught was relatively constant (it did not show a trend in size change). There are no data record *time series* regarding the individual size of the fish caught. To achieve good management of fish resources, the fish size must be monitored regularly.

The standard CPUE is an attribute related to fleet characteristics (fishing effort) associated with the number of catches obtained. The standard CPUE can be an indicator of the abundance of fish in water. To date, there are no records of catch data per unit of fishing effort available for snapper, grouper, and other fisheries in Aisandami. It is necessary to receive attention from fishery managers and authorities to record the CPUE of the dominant fishing gear used to catch groupers and snappers.

Data on fish resources were obtained by interviewing fishermen and length measurements were taken during the research period. It is recommended that a structured and regular data recording protocol be developed, especially for recording data on species composition and catch size and CPUE This can be in form of fishing logbook. This can provide an overview of the species composition, fish size trends, and CPUE of fish caught throughout the year.

3.2.3 Ecosystem and Habitat Domain

The score value for the ecosystem and habitat domain is 73, which indicates that EAFM-based management in the ecosystem and habitat aspects is carried out well. To achieve optimum management, several contributing attributes (Figure 7) must be considered.

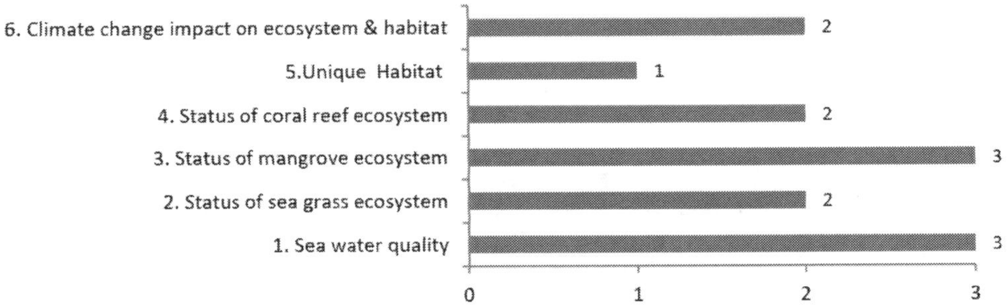

Fig 7. Evaluation results of the Ecosystem and Habitat domains in the management of grouper and snapper fisheries in Aisandami waters

Mangrove status and water quality are two attributes that have a large positive impact on the ecosystem and habitat domain value scores. Based on secondary data [10] and field observations, the density of mangroves scattered on the coast around Aisandami is approximately 3000 individuals/ha, with a cover of approximately 90%. Mangrove species that grow along the coastline of Aisandami include *Rhizophora stylosa*, *Rhizophora mucronata*, *Sonneratia alba* and *Xylocarpus mekongensis*. The mangroves in Aisandami are in very good condition. This is very important in supporting the life of fish resources, including snapper and grouper, considering the ecological functions of mangroves, including as a place for spawning, rearing, and foraging.

The quality of seawater in Aisandami can be said to be in good condition based on several factors, namely there are no sources of pollution around the village and surrounding area, and based on satellite image data processing, the chlorophyll value was 24–136 µg/l (classified as very high). This means that the conditions of water quality and productivity support the life of fish resources.

Three attributes, namely seagrass status, coral reef status, and climate change, are not yet optimal in contributing to the value of the EAFM-based management index. The condition of seagrass in Aisandami, according to Tebay *et al.* [9], described a density of 220–250 individuals/m with a cover of 27–32% and species *Enhalus acoroides* and *Thalassia hemrichii*. Field observations revealed that apart from the two species mentioned above, three other species were also found, namely, *Halophila ovalis*, *Cymodocea serulata*, And *Halodule uninervis*. The seagrass cover mentioned above is in the "medium" category. Low seagrass cover was found in all areas of the Aisandami waters.

The coral reef status is described by live coral cover, where, on average, from observations at nine observation stations, the percentage of coral cover is 37.6% or categorized as medium cover. The low coral cover might be explained by the influence of destructive fishing activities (that damaged coral) that occurred in the period before "sasi" was designated; currently, coral conditions are starting to recover.

Based on observations and interviews with the community, there was no impact of climate change on water conditions and habitats. For example, coral bleaching was not observed (*coral bleaching*). However, to date, there have been no specific studies on the impact of climate change on water and habitat in the Aisandami waters.

To date, there have been no studies on the existence of unique/special habitats in the waters around Aisandami. These unique/special habitats include SPAGs (Spawning Aggregations) areas, nursery grounds and feeding grounds for snapper and grouper. Therefore, to achieve optimum management of snapper and grouper fisheries, studies of unique habitat locations must be conducted.

3.2.4 Fishing Technology Domain

Based on data analysis, it was found that the fishing technology domain was well managed based on the EAFM (score = 68). To achieve optimum management of fishing technology, several attributes (see Figure 8) need to be considered and improved.

The attributes of fishing capacity and effort describe the characteristics of the fishing fleet associated with the maximum number of catches produced by the existing fishing fleet. In this study, the carrying capacity of a fleet/boat was used as a reference for the maximum capacity. Based on interviews with fishermen and field observations, fishermen in Aisandami rarely obtained results that exceeded the maximum carrying capacity of the boats used; in other words, the fishery capacity ratio was more than one. Conditions such as these need to be maintained to ensure that excessive fishing does not occur.

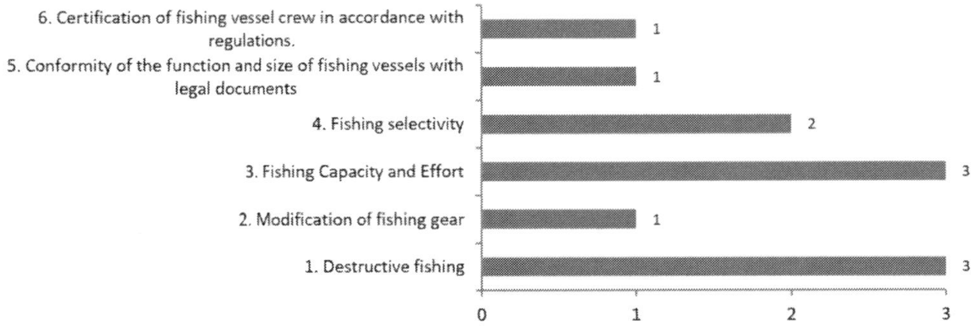

Fig 8. Results of evaluation of the Fishing Technology domain in the management of grouper and snapper fisheries in Aisandami waters

Since the implementation of the church's "sasi," destructive fishing has no longer been carried out by the people of Aisandami Village. Similarly, there have been no reports of destructive fishing carried out by fishermen from outside Aisandami. The use of gill nets is still carried out by fishermen but is aimed at catching small pelagic fish (as handline bait for catching coral fish).

In connection with the selectivity of fishing gear, fishermen catching snappers and groupers use hand lines with hook sizes that vary from 6 to 12. With hook sizes that vary and are relatively small, there is still a chance of catching illegal fish (fish with length size less than length at first maturity or Lm). Fishermen do not yet have a reference regarding the appropriate hook size, so they can avoid catching juvenile fish or those with a size less than Lm. Mudjirahayu *et al.* [1] recommended using hook sizes of six and seven for grouper fishing to avoid catching fish smaller than Lm. Studies on the selectivity of fishing gear for snappers and groupers need to be conducted to determine the optimum hook size according to the target species. In addition, outreach is needed to raise awareness among fishing communities to release fish whose size is not yet suitable for catching.

Modifications of fishing gear and auxiliary equipment are associated with the visible impact of gear on catch results. Based on individual measurements of caught fish and interviews with fishermen, there is still a high composition of fish sizes that are not yet suitable for catching. For example, more than 50% of snappers *Lutjanus malabaricus* caught were smaller than Lm.

Aisandami fishermen do not have vessel documents regarding the suitability of the functions and sizes of fishing vessels. Fishermen use paddle boats (kole-kole) and outboard motorboats (longboats). In the future, it will be necessary to have vessels measured by the Maritime Affairs and Fisheries Service complete the administrative documents required by the government.

Until the time this study was conducted, no fishermen had a fishery boat crew proficiency certification. They do not consider crew proficiency certification important. This is because, apart from fishermen, fishing has become a daily profession; it is also because they only use small fishing fleets that do not require special skills.

3.2.5 Social Domain

The results of the evaluation of the management status of the fish resource domain yielded a score of 87. This means that the social aspects are managed very well. The management status of fish resources is determined by the three attributes used in the assessment. The results of the analysis of the three attributes used to evaluate the social aspects are shown in Figure 9.

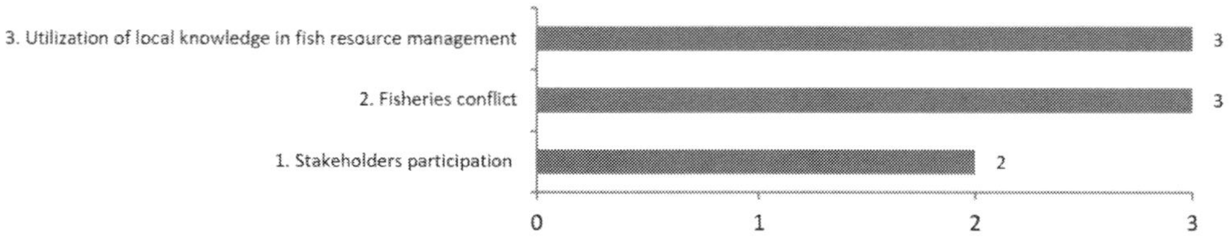

Fig 9. Results of evaluation of the Social domain on management of grouper and snapper fisheries in Aisandami waters

Stakeholder participation: Results of interviews with village and community leaders. Stakeholders often accompany the community (BBTNC, World Wildlife Fund , GEF6). The focus of assistance from stakeholders is on community empowerment and raising awareness of the importance of protecting the environment and natural resources.

In interviews with community figures, conflicts over the use of fish resources have never occurred, neither between fishermen in Aisandami Village nor with fishermen from other villages. This is because even though fishermen are from other villages, historically these fishermen are still related. For example, fishermen from Ron Island have brought no conflicts of fish near or around the Aisandami area.

The use of local knowledge in managing fish resources is in the form of church "Sasi" (Kadup). Sasi is a tool for managing natural resources by designating water areas as protected areas. The sasi area of Aisandami Village is in the marine protection zone (based on TNTC area zoning) on Namamuran Island and around Aisandami village. In areas where sasi is applied, people are prohibited from taking all the natural resources within them. Sasi provides benefits in protecting the richness of marine biodiversity so that its abundance increases, including the abundance of economically important fish that can be utilized by fishermen. The closing and opening ceremony of the sasi is a traditional event that attracts local and foreign tourists, making it an economic source for the Aisandami community.

3.2.6 Domain Economy

The results of the evaluation of the management status of the fish resource domain yielded a score of 60. This means that from an economic perspective, management is still not good (moderate). The three attributes used in the EAFM analysis for economic aspects must be improved to achieve an optimum (very good) management status. The results of the analysis of the three attributes used to evaluate the economic aspects are shown in Figure 10.

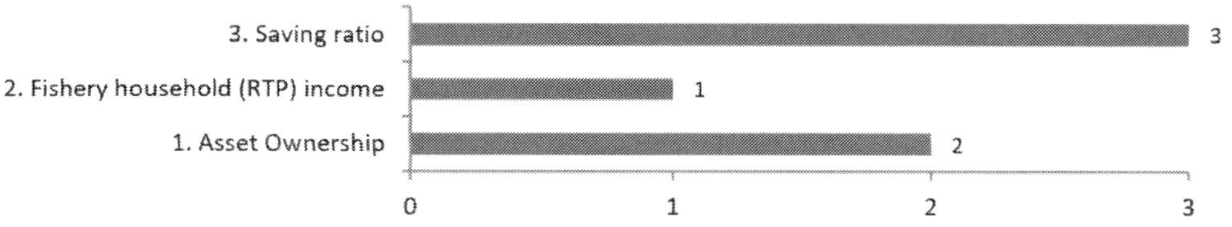

Fig 10. Results of evaluation of the Economic domain on management of grouper and snapper fisheries in Aisandami waters

The productive assets owned by fishermen are fishing equipment, boats, and engines. Based on the results of interviews with fishermen, ownership of their fishing units has not changed over the last few years. This explains why economic activities through fishing have not progressed.

Fisheries household (RTP) income is still below the Papua Barat minimum province wage (UMP). Papua Barat's UMP in 2024 is the IDR. 3,393,000, while the average income of fishermen in Aisandami was IDR. 854,000.

The savings ratio is the amount of a fisherman's net income compared with his income. Results of interviews with fishermen. The average income is 854,000 and the average expenditure is 750,000. Savings ratio = (income expenses)/income × 100% = 12.1%. This savings ratio value is higher (or has a positive value) than the bank loan interest rate of 9% (September 2024). This condition shows that the welfare of fishermen is relatively good [13].

Even though the savings ratio is still higher than the corporate credit interest, this is more because people always try to limit their spending, adjusting to income that is still low (less than UMP). Therefore, efforts to increase fishermen's income still need to be made, such as providing a marketing network for fishery products.

3.2.7 Institutional Domain

Based on the data analysis, it was found that the institutional domain was well managed based on EAFM (score = 68). To achieve optimum management from an institutional perspective, several attributes (see Figure 11) must be considered and improved.

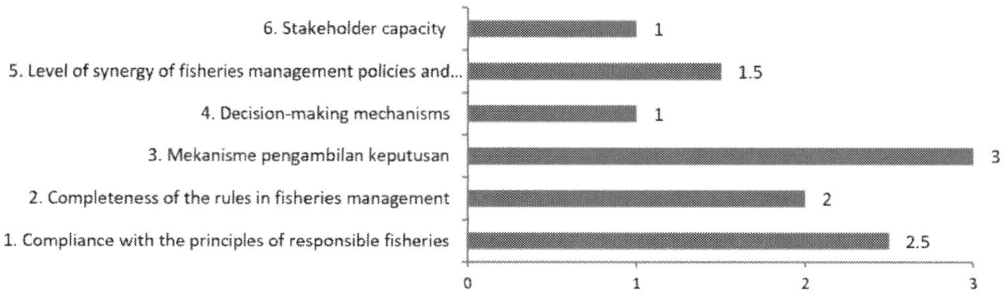

Fig 11. Results of institutional domain evaluation on grouper and snapper fisheries management in Aisandami waters

Compliance with the principles of responsible fisheries in fisheries management that have been established both formally and non-formally: interviews with fishermen and community leaders. Compliance with the implementation of sasi is very high, and there are no legal violations of the rules prohibiting the use of destructive fishing gear in grouper or snapper fishing.

Apart from being agreed upon through the implementation of sasi, the completeness of the rules of the game in fisheries management is also regulated through Village Regulations (PERKAM) Number 4 of 2019 concerning natural resource management. The application of sasi and PERKAM includes regulations on habitat domains, fish resources, fishing technology, and institutions. The implemented church sasi makes fishermen obedient to rules. However, it is necessary to provide facilities for supervision and an institutional structure for supervisors to monitor fishing violations by fishermen from outside of Aisandami.

In terms of decision making, even though written standard operational procedures (SOPs) regarding implementation mechanisms are not made, the community can agree on various decisions made in the village. For example, in relation to the management of natural resources, including fisheries resources, it was agreed to determine the "sasi" area and Village Regulations regarding the management of natural resources.

Aisandami Village is located in the TNTC area and in the Fisheries Management Area (WPP) 717. Therefore, fisheries management follows the Fisheries Management Plan (RPP)

717, which is stipulated by the Decree of the Minister of Maritime Affairs and Fisheries of the Republic of Indonesia Number 84/Kepmen–Kp/2016 Concerning Fisheries Management Plans for the Republic of Indonesia State Fisheries Management Area 717.

On a village scale, fisheries management around Aisandami Village is also regulated by PERKAM Number 4 of 2019 concerning Natural Resource Management. However, neither RPP WPP 717 nor PERKAM have been fully implemented. For example, the ban on fishing using lift nets in the area around Aisandami is regulated through PERKAM; however, in reality, the community still allows left net boat fishing in Aisandami water.

The level of synergy between fishery management policies and institutions had a low score. One example of synergy in issuing fishing permits between government institutions that have the authority to issue fishing permits (SIPI) (namely, the Maritime Affairs and Fisheries Service) and BBTNC as the authority for managing the TNTC area has not gone well. Based on the results of interviews with the Head of the BBTNC Technical Division, coordination between the West Papua Province Maritime and Fisheries Service and BBTNC still needs to be improved, so that fishing activities do not have a negative impact on coastal ecosystems and fisheries resources within the TNTC area.

Increasing the capacity of stakeholders at the Aisandami Village level is carried out through assistance activities from several government institutions (DKP, BBTNC, and Village Empowerment Service) as well as government partner institutions (GEF and locally managed marine area/LMMA). One form of capacity building is assistance in drafting village regulations. In addition, assistance is provided to increase the capacity of village economic institutions to convert local resources into economically valuable products.

4 Conclusions

The management of snapper and grouper fisheries using an eco-system approach (EAFM) carried out by the community in Aisandami Village is categorozed as "good" but not yet optimal. Therefore, efforts are needed to improve management status through various interventions by relevant government institutions and government partner institutions.

This research was funded by a CFI Indonesia GEF 6 Grant through the World Wildlife Fund (WWF) Global Environmental Facilities (GEF) Agency and the Ministry of Maritime Affairs and Fisheries of the Republic of Indonesia.

References

1. Mudjirahayu, R. Bawole, U. N. W. J. Rembet, A. S. Ananta, F. Runtuboi and R. Sala. EJAR **43**, 8 (2017). https://www.sciencedirect.com/science/article/pii/S1687428517300456
2. R. Bawole, R. Sala, F. Runtuboi, Mudjirahayu, Irwanto, G. Gaman and E. M. Randa. *Capture fisheries in Teluk Wondama Regency. Potential and Management Strategy*. West Papua Province Regional Research and Development Center, Manokwari (2020). https://balitbangda.papuabaratprov.go.id/konten_penelitian/detail/29
3. N. A. Putri, N. Widiastuti, F. F. C. Simatauw, F. Manohas, J. A. Jentewo, R. Sala, T. F. Pattiasina and M. Wayoi. *Domain Habitat, Fish Resources, and Catching in Fisheries Management in Menarbu Village, Teluk Wondama Regency*. Proceedings of the XI Bogor National Fish Seminar (2022).
4. G.N. Ahmadia, J.R. Wilson and A.L. Green. *Coral Reef Monitoring Protocol for Assessing Marine Protected Areas in the Coral Triangle*. Coral Triangle Support Partnership (2013)

5. S. English, C. Wilkinson & V. Baker (Eds). *Survey manual for tropical marine resources. 2nd Edition. Australian Institute of Marine Science*. Townsville. 408 pp. (1997)
6. J.E. Veron, *Corals of the World Volume 1*. Australian Institute of Marine Science and CRR Qld. Australia 3 (2000)
7. Suharsono, *Types of Coral in Indonesia*. center for oceanographic research and development-LIPI. Jakarta. (2008)
   https://opac.perpusnas.go.id/DetailOpac.aspx?id=86978
8. R. Kelly. *Coral Identification Manual*. Online pada https://indooceanproject.org/ wp-content/uploads/2020/08/Coral-Identification-Manual.pdf. Accessed August 1, 2024. (2020)
9. S. Tebay, P. Boli and J. Ainusi. JSAI **4**, 17 (2020). https://www.ejournalfpikunipa.ac.id/index.php/JSAI/article/view/80
10. I. Dharmawan, W. E. and A. Widyastuti. *Mar. Res. Indonesia* **42**, 10 (2017). https://www.researchgate.net/publication/320083185_PRISTINE_MANGROVE_COMMUNITY_IN_WONDAMA_GULF_WEST_PAPUA_INDONESIA
11. G.R. Allen and M.V.Erdmann. *Reef fishes of the East Indies. Volumes I–III. Tropical Reef Research, Perth, Australia*. (2012)
12. G.R. Allen. *FAO Species catalogue*: Vol. 6 Snappers of The World. Food and Agriculture Organization of the United Nations. Rome (1985)
13. L. Adrianto, A. Habibi, A. Fahrudin, A. Azizy, H. A. Susanto, I. Musthofa, M. M. Kamal, S. H. Wisudo, P. Raharjo, Z. Nasution and Yonvitner. *Indicators for Fisheries Management Using an Ecosystem Approach*. National Working Group on Ecosystem Approach to Fisheries Management, Directorate of Fish Resources, Ministry of Maritime Affairs and Fisheries. Jakarta (2012)
14. L. Adrianto, A. Habibi, A. Fahrudin, A. Azizy, H. A. Susanto, I. Musthfa, M. M. Kamal, S. H. Wisudo, Y. Wardiatno, P. Raharjo and Z. Nasution. *Assessment of Indicators for Management Fisheries with an Ecosystem Approach (Ecosystem Approach to Fisheries Management)*. National Working Group II EAFM, Direktorat Sumberdaya Ikan Kementerian Kelautan dan Perikanan Republik Indonesia (2013). https://www.academia.edu/8079741/Modul_Penilaian_Pendekatan_Ekosistem_dalam_Pengelolaan_Perikanan_Penilaian_Indikator_untuk_Pengelolaan_Perikanan_Berpendekatan_Ekosistem_Ecosystem_Approach_to_Fisheries_Management (2013)
15. Fishbase. *Lutjanus malabaricus* (Bloch & Schneider, 1801). Online pada https://fishbase.mnhn.fr/summary/lutjanus-malabaricus. Accessed September 1, (2024).

Ichthyoplankton composition and distribution in Kilwaru Waters, East Seram Regency, Maluku Province

Karsono Wagiyo[2*], *Masayu Rahmia Anwar Putri*[1], *Muhammad Taufik*[2], *Reny Puspasari*[1], *Puput Fitri Rachmawati*[1], *Andrias Samusamu*[1], *Prihatiningsih*[2], *Mujiyanto*[1], *Tirtadanu*[2], *Ofri Johan*[1]

[1]Research Center for Conservation of Marine Resources and Inland Waters, National Research and Innovation Agency
[2]Research Center for Fishery, National Research and Innovation Agency

Abstract. Ichthyoplankton play an important role in marine ecosystem dynamics and their relationship to conservation. This study aims to investigate the abundance and composition of ichthyoplankton in Kilwaru waters. Sampling was conducted in the morning at seven stations by towing at a boat speed of approximately two knots. Ichthyoplankton were collected using a bongo net with a mesh size of 500 μm and were carried out simultaneously with the measurement of hydrological parameters. The results found A total of 620 fish larvae from 24 families with abundances ranging from 45 to 2210 ind/10^3m^3 with an average of 564 ind/10^3m^3. The highest abundance was observed at Station 2, where the lowest was at Station 6. Mullidae and Carangidae had the widest distribution, with 100% habitat coverage. The diversity index was 0.98, the dominance index was 0.15, and the evenness index was 0.21. The highest diversity was observed at station 2, where as the highest evenness was observed at station 6. Approximately 81% of the ichthyoplankton were found in the preflexion stage, followed by postflexion 13% and flexion 6.1%. The mean values for hydrological parameters such as temperature were 28.9°C, pH 8.1, dissolved oxygen 5.8 mg/l, and salinity 30.3 ppt. The findings of this study indicate that the Kilwaru waters are spawning grounds.

1 Introduction

Kilwaru waters are traditionally a "Sasi" area located in Seram Laut Island, East Seram. Sasi is the local wisdom of Moluccan society to maintain the sustainability of livelihood through resource conservation [1]. As a sasi area, Kilwaru waters are expected to be a marine conservation area that functions as a source of "spill-over" aquatic resources for the surrounding area.

The spillover function of an area can be recognized by the presence of spawning grounds. The spawning area is a source of resources that can be distributed to a wider area [2]. The

* Corresponding author: karsonowagiyo@gmail.com

identification of spawning grounds can be performed in various ways. One indicator of the certainty of the spawning ground area was the presence of ichthyoplankton in the early phase. Ichthyoplankton in the spawning ground area with strong currents and internal migration power will be distributed in suitable areas to settle and continue their life cycle to grow and reproduce as a stock of resources.

Studies on the role of ichthyoplankton sources and spillovers in maintaining stock resources have been conducted in various countries for various fish commodities (pelagic, demersal, and coral). Research on ichthyoplankton transport of the pelagic fish *Scomber australiasicus* from spawning grounds in the South China Sea may serve as a source of recruitment stock in the area itself and may spread widely as a Pacific Ocean recruitment stock [3]. Ichthyoplankton transport studies of the demersal commodity Sciaenidae (*Totoaba macdonaldi*) in California revealed that ichthyoplankton from spawning areas in delta areas can be transported through the coastal, frontal and mixed zones offshore [4]. [2] stated that reef fish ichthyoplankton can disperse from tens of kilometers, such as parrotfishes and wrasses, to hundreds of kilometers, such as emperor, snapper, and grouper, which have an impact on the stock of spawning waters and habitat connectivity.

Research on the existence of resources in Kilwaru will prove the effectiveness of the sasi area as a model of marine conservation areas that can be extended to other areas in Indonesia and worldwide. The Kilwaru area plays an important role as a source of aquatic resources, as revealed by the presence of ichthyoplankton. This research will reveal the presence of ichthyoplankton in relation to the composition, abundance, diversity, distribution, and hydrological conditions of its habitat.

Based on the results of this research, it is expected that the establishment and development of marine protected areas will be based on the early life history of fish. Research priorities are carried out in sasi areas or in other forms of local wisdom that exist in communities in various regions.

2 Materials and Methods

2.1 Samples collection

The study was conducted in the waters of Kilwaru, a sasi area (Figure 1). The sampling time was 24–May 31, 2023. The sampling locations were determined by systematic cluster random sampling (the research area was divided into several clusters based on biota diversity and hydrological conditions), and the number and sampling points were determined. The coordinates and wind directions of the larval sampling stations in Kilwaru are listed in Table 1.

Table 1. Coordinates and wind direction of the larval sampling stations in Kilwaru

No. Station	Coordinates	Wind direction
1	03°51'42.49" S/130°56'33.90"E	Winward
2	03°52'39.84" S/130°56'10.85"E	Leeward
3	03°52'48.34" S/130°56'56.76"E	Leeward
4	03°52'37.03" S/130°58'08.19"E	Leeward
5	03°51'59.07" S/130°57'06.33"E	Winward
6	03°51'13.38" S/130°57'16.31"E	Winward
7	03°50'56.55" S/130°55'48.75"E	Winward

Fish larvae were collected using a larvae net with a mesh size of 500 µm, length of 3 m, and diameter of 60 cm. Sampling was carried out in the morning from 06-09 am at a depth of 2-3 m, towing the net horizontally using a boat at a speed of 1-3 knots for 10-15 minutes. Hydrological parameter measurements were conducted insitu at each sampling station. The parameter acquisition included temperature, pH, conductivity, turbidity, dissolved oxygen (DO), and salinity. We used the Water Quality Checker brand Horiba Type U50 to measure the hydrological parameters.

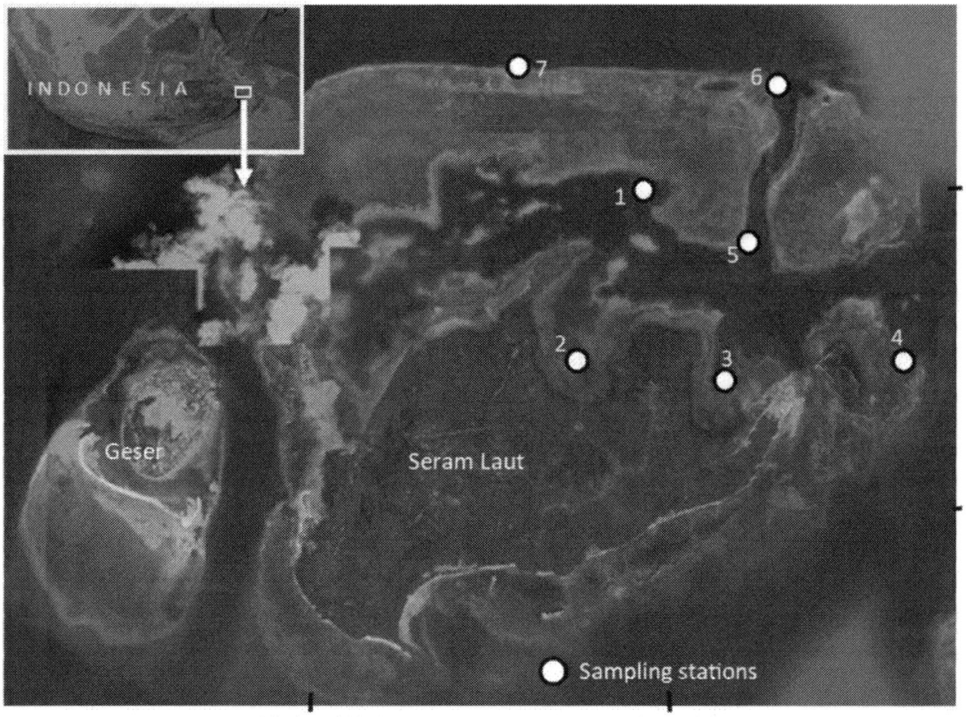

Fig 1. Ichthyoplankton sampling stations in Kilwaru Waters, Seram Laut Island.

2.2 Ichthyoplankton identification and Analysis

2.2.1 Identification

Ichthyoplankton samples were preserved 10 % formalin in 1000 ml bottles for identification purposes. Prior to identification, the ichthyoplankton were separated from the other materials using a stereomicroscope. The sorted ichthyoplankton were counted and placed in 50 ml ampoule bottles containing 70% alcohol. Identification was carried out at the LATERIO-BRIN Ancol laboratory using a Leica 205C stereo microscope and Nikon SMZ1500 at 40 x magnification. Morphometric measurements were performed using a Zeiss Stereo Discovery 8 digital microscope. Determination of ichthyoplankton families using the main guidebooks from; [5], [6], [7], [8], [9], [10] and [11].

2.2.2 Data analysis

Abundance of Ichthyoplankton using a modification of the SEAFDEC equation [11]

$$A = (n \times (1/B \times D) \times 1000) \quad (1)$$

A = Abundance (1000 m^3); n = number of individuals counted in the sample.
B = Bongo net mouth area (m^2), D = Towing distance (m)

Habitat consistency using a modified equation from [12]

$$CH=FS/TS \times 100 \quad \ldots\ldots (2)$$

CH = Consistency of larval family presence (%), FS = Number of larval family samples located, TS = Total number of samples observed

The diversity index uses the following equation

$$H' = -\sum_{i=1}^{n} \left(\frac{ni}{N}\right) X \text{ Ln } \left(\frac{ni}{N}\right) \ldots\ldots 3)$$

H' = Diversity index, ni = number of individuals to i, and N = Total number of individuals.

The dominance index uses the following equation

$$C = \sum_{i=1}^{n} \left[\frac{ni}{N}\right]^2 \ldots\ldots (4)$$

C = Index of dominance, ni = number of individuals to I, N = Total number of individuals.

The evenness index uses the following equation

$$E = \frac{H'}{\text{Ln } S} \ldots\ldots (5)$$

E = Evenness uniformity index, S = Number of species, H' = Diversity index.

3 Results

3.1 Location Description

Kilwaru waters are located in the northern part of the Seram Laut Island. Administratively, it is part of the East Seram Regency. In fishery management, it is part of FMA 715, which borders FMA 714. Traditionally, Kilwaru waters have been an area of sasi aquatic resources. Sasi is the management to conserve natural resources and the sustainability of traditional livelihoods regulated by customary law.

Kilwaru waters are affected by two major monsoons: SSE and NNE. The NNE moonsoon occurs between December and May and the SSE moonsoons from June to November. Water masses were saturated by oceanic water masses from South Pacific Indonesia Through Flow. Tides in the Kilwaru waters are semi-diurnal and have a high range. Ecosystems in the Kilwaru Waters are diverse in the form of mangrove forests, seagrass beds, coral reefs, and desert islands.

Sampling sites and stations were divided into several clusters. The first cluster was leeward waters in a bay like area located semi-enclosed with mixed water masses (stations two and three). The second cluster comprises the backward waters of the islands (station one). The third cluster comprises waters in the reef slope area with oceanic water masses (station seven). The fourth cluster was a reef slope area with mixed water masses (station four). The fifth cluster was in the current channel area (stations five and six).

3.2 Composition

The identification results of ichthyoplankton in the Kilwaru waters revealed 24 family groups. Five dominant ichthyoplankton family groups covered 75%, including Clupeidae 26%), Mullidae 21%), Pomacentridae 16%), Carangidae 7%), and Labridae 5% (Figure 2a). Nineteen other groups accounted for 25%.Seven families with contributions below 5% were Scaridae, Engraulidae, Lethrinidae, Gobiidae, Blenniidae, Sillaginidae, and Terapontidae. 12

family groups that contributed less than 1% were Holocentridae, Bythitidae, Gerridae, Isonidae, Priacanthidae, Scorphainidae, Apogonidae, Balistidae, Bothidae, Drepaneidae, Lutjanidae, and Schindleridae.

The composition of ichthyoplankton in Kilwaru, based on habitat, was divided into reef, demersal, and pelagic. The contributions of each habitat group were 58%, 25% and 17% components, respectively (Figure 2b). The reef habitat group included Apogonidae, Balistidae, Blenniidae, Bythitidae, Drepaneidae, Gobiidae, Holocentridae Labridae, Lethrinidae, Pomacentridae, Priacanthidae, Scaridae, Schindleriidae, and Scorphaenidae. Demersal habitat groups included Bothidae, Gerridae, Lutjanidae, Mullidae, Sillaginidae, and Terapontidae. Pelagic habitat groups included Carangidae, Clupeidae, Engraulidae, and Isonidae.

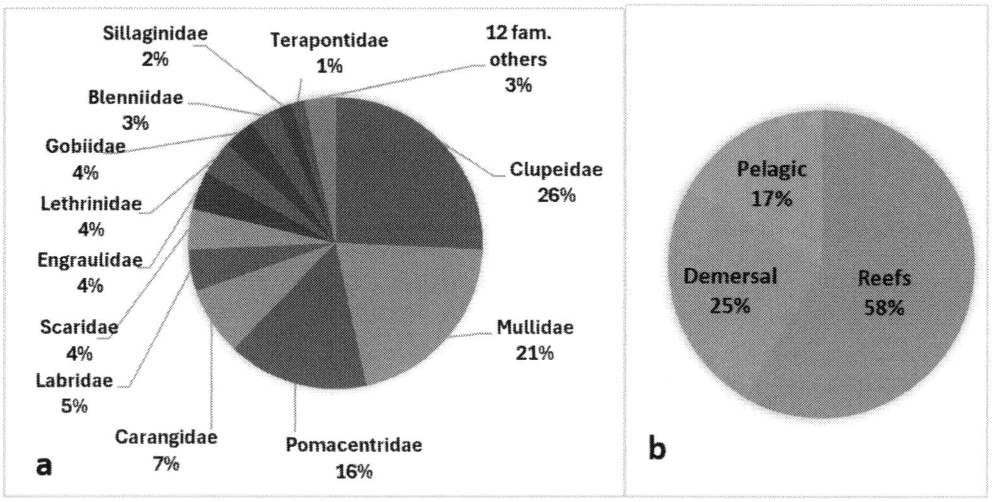

Fig 2. a) Familia composition and b) Habitat composition of ichthyoplankton in Kilwaru waters.

3.3 Abundance and Distribution

The spatial distribution of the abundance of ichthyoplankton in the Kilwaru waters is listed in Table 2. Leeward stations (stations 2,3 and 4) tended to be more abundant than winward stations (1,5,6 and 7). In all stations, the average abundance of ichthyoplankton was 564 ind/10^3m^3, highest at station 2 (2210 ind/10^3m^3), and lowest at station 6 (45 ind/10^3m^3). The most abundant ichthyoplankton in the Kilwaru waters was Clupeidae, which had an inter-station range of 0–656 Ind/10^3m^3, with an average of 131 ind/10^3m^3.

Families that have the widest spatial distribution were Carangidae and Mullidae, found at all stations, with 100% distribution coverage. The ichthyoplankton families with a distribution of > 50% were Clupeidae, Gobiidae, Labridae, Lethrinidae, Pomacentridae, and Scaridae. families with the narrowest distribution with a coverage of 14.29 % are Apogonidae, Balistidae, Bothidae, Drepaneidae, Gerridae, Lutjanidae, Schindleriidae, and Scorphaenidae.

Table 2. Abundance distribution and habitat coverage of Ichthyoplankton in Kilwaru waters

No	Famili	Station (Ind/10^3m^3)							Average	Coverage
		1	2	3	4	5	6	7		%
1	Apogonidae				6				1	14,29
2	Balistidae				6				1	14,29
3	Blenniidae		108						15	14,29

No	Famili	Station (Ind/10^3m^3)							Average	Coverage
		1	**2**	**3**	**4**	**5**	**6**	**7**		**%**
4	Bothidae			6					1	14,29
5	Bythitidae					13			2	14,29
6	Carangidae	38	57	45	51	25	13	38	38	100,00
7	Clupeidae	6	656	159	89	6			131	71,43
8	Drepaneidae	6							1	14,29
9	Engraulidae	6	102		45				22	42,86
10	Gerridae				13				2	14,29
11	Gobiidae	19	45	38	19			6	18	71,43
12	Holocentridae			6		13			3	28,57
13	Isonidae		6				6		2	28,57
14	Labridae		45	25	57	32		7	24	71,43
15	Lethrinidae	32	19	57	6	19			19	71,43
16	Lutjanidae			6					1	14,29
17	Mullidae	6	439	210	19	19	13	38	106	100,00
18	Pomacentridae	76	350	51		19	6	51	79	85,71
19	Priacanthidae			6		6			2	28,57
20	Scaridae	96	19	13	19	13			23	71,43
21	Schindleriidae		6						1	14,29
22	Scorphainidae		13						2	14,29
23	Sillaginidae		57	6					9	28,57
24	Terapontidae		13	19				19	7	42,86
25	Un identified	13	274	38	32		6	19	64	
	Abundance	**299**	**2210**	**688**	**363**	**166**	**45**	**178**	**564**	

3.4 Diversity distribution

The results of the diversity distribution analysis are shown in Figure 3. The distribution of the highest number of ichthyoplankton families was 16 families at station two and the lowest with four families at station 6, averaging 10 families per station. The distribution of the H' index value (diversity) was highest at 0.95 at station 5 and lowest at 0.58 at station 6. The distribution of the C index value (dominance) was highest at 0.28 at station 6 and lowest at 0.12 at station 5. The distribution of the E index value (uniformity) was highest at 0.29 at station 6 and lowest at 0.21 at station 2. Overall, not differentiating the distribution per station, the ichthyoplankton population in Kilwaru waters had a diversity index of 0.98, dominance of 0.15 and uniformity of 0.21.

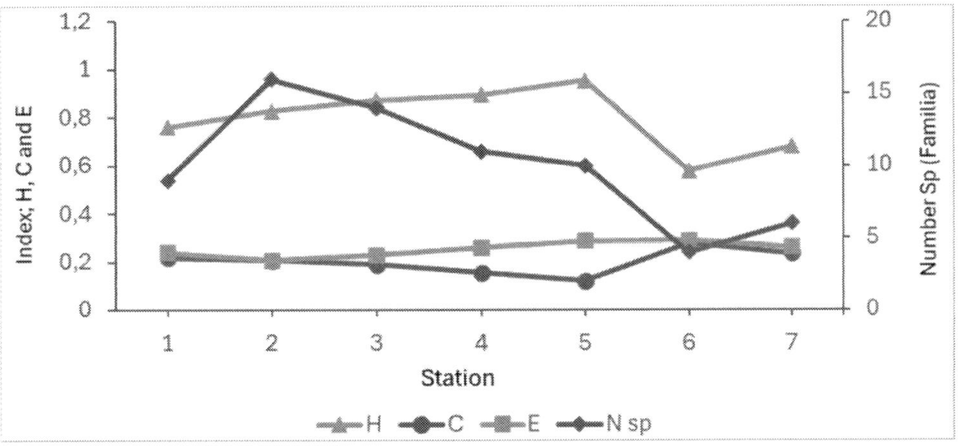

Fig 3. Diversity Distribution of ichthyoplankton in Kilwaru waters

3.5 Stage distribution and size structure

Ichthyoplankton stages can be divided into preflexion, flexion, and postflexion. The distribution of the ichthyoplankton stage composition is shown in Table 3. The largest preflexion stage distribution 96.54% at station two and the smallest 65.8% at station one. The distribution of the largest flexion 15.79% at station 4 and the smallest zero at stations 2, 3, 5, and 6. The largest postflexion distribution was 31.91% at station one and the smallest was 3.46% at station 2.

Table 3. Stage distribution of ichthyoplankton in Kilwaru Waters

Stage	Station (%)						
	1	2	3	4	5	6	7
Preflexion	65,8	96,54	96,26	77,19	92,59	85,71	85,71
Flexion	2,29			15,79			10,71
Postflexion	31,91	3,46	3,74	7,02	7,41	14,29	3,58

The size structure of ichthyoplankton has a mean body length of 2.728 mm; the smallest size of 1.282 mm is Gerreidae, and the largest 5.967 mm is Scaridae. The average body width was 0.622 mm, the smallest was 0.219 mm and the largest was that of Drepaneidae (1.358 mm). The ichthyoplankton found in Kilwaru waters had an insignificant relationship between body width and body length. In total based on The stage composition of Ichthyoplankton in Kilwaru waters consists of 81 % preflexion, 13 % flexion, and 6 % postflexion and (Figure 4).

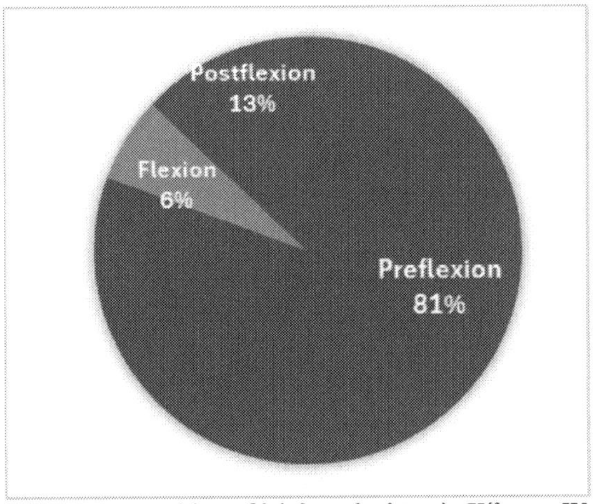

Fig 4. Stage composition of ichthyoplankton in Kilwaru Waters

3.6 Hydrological condition parameters

The results of the in-situ measurements of hydrological parameters in the Kilwaru waters are listed in Table 4. Mean temperature 28.91°C, the smallest 28.70°C at station 5, and the largest at station 7. Mean pH 8.12, smallest 8.08 at station 2 and largest 8.18 at station 7. Mean conductivity 46.69 ms/cm, smallest 45.50 ms/cm at station 1 and largest 47.50 ms/cm at station 7. Mean turbidity 0.20 NTU, smallest 0 NTU at five stations and largest 0.90 NTU at station 6, respectively. Mean dissolved oxygen 5.84 mg/l, the smallest was 5.42 at station 6 and the largest 6.90 at station 1. Mean TDS 28.46 g/l, smallest 27.80 g/l at station 1 and largest 28.90 g/l at station 7. Mean salinity 30.31 ppt, the smallest 29.49 ppt at station 1 and the largest 30.80 ppt at station 7.

Table 4 Measurement results of hydrological parameters in Kilwaru waters

Sta.	Temp. (°C)	pH (scala)	Conductivity (ms/cm)	Turbidity (NTU)	DO (mg/l)	TDS (g/l)	Salinity (ppt)
1	28,82	8,11	45,5	0,00	6,90	27,8	29,49
2	28,71	8,08	46,6	0,50	5,53	28,4	30,27
3	28,84	8,09	46,8	0,00	5,64	28,5	30,38
4	28,96	8,12	46,7	0,00	5,87	28,5	30,37
5	28,70	8,14	46,9	0,00	5,71	28,6	30,46
6	29,12	8,14	46,8	0,90	5,42	28,5	30,40
7	29,25	8,18	47,5	0,00	5,78	28,9	30,80
Average	28,91	8,12	46,69	0,20	5,84	28,46	30,31
Min	28,70	8,08	46,50	0,00	5,42	27,80	29,49
Max.	29,25	8,18	47,50	0,90	6,90	28,90	30,80

3.7 Discussion

In this study, it was found that the dominant fish per family were Clupeidae, Mullidae, and Pomacentridae. Clupeidae which is generally a pelagic fish, can be dominant in the Kilwaru sasi area for several reasons: 1) The family Clupeidae has spawning areas, larval phases, and recruitment in coastal areas [13], and 2) spawning aggregations of Clupeidae seek areas with abundant food availability, such as coral reef ecosystems [14]; and 3). Ichthyoplankton

Clupeidae, due to its spawning aggregation in pelagic areas, are easily carried away by currents compared to demersal and coral reef fish and collected in a quiet area. In contrast, reef fish families include Apogonidae, Balistidae, and Gerridae. Balistidae have a low contribution per station, even though the sampling location in the reef area is possible: 1) reef fish generally have a territorial home range microhabitat [14]. 2) Habitat shifting and lower fecundity [15, 16] than pelagic fishes. The discovery of Clupeidae being dominant in the neritic area was also found in a previous study [17]. The contribution of reef fish to aggregate is greater than that of pelagic and demersal fish because kilwaru waters are an area of coral reef ecosystems as a habitat, so the number of reef fish families is greater than that of pelagic and demersal fish families. The number of families contributing to this study is small compared to 24 families in the Arafura Sea [18] and 42 families in the coral reef, Gorgona Island, Eastern Tropical Pacific [19] because sampling was done only once in a narrow area and shallow water. Sampling did not represent the spawning season of various fish species in this case at the end of the warm season, while fish generally spawn due to the influence of warm waters in the environment [15].

The average abundance of ichthyoplankton in Kilwaru waters was 564 Ind/10^3m^3, which is higher than that in the Arafura Sea (204 individuals/10^3m^3 [18] and the South Indian Ocean of Java (174 ind/10^3m^3 [20]. The abundance of ichthyoplankton might be higher if the study was conducted in warmer water conditions, considering that at the time of the study it was not the spawning season (the end of the warm season towards a cooler season). The spawning season and period are affected by warming [15]. Similar circumstances have been reported in various studies; goatfish (Mullidae) in the Red Sea spawn when temperatures rise during spring [21]. *Thunnus albacares* spawns when the water warms up in October-December [22], *Parupeneus barberinus* spawns in Mariana Island during the spring-summer season [23], and three Clupeiform species in China spawn in the spring season [24]. The high abundance of Ichthyoplankton in Kilwaru waters is thought to be a source of spill over larvae for surrounding habitat connectivity. This suggests that the Kilwaru sasi area functions similarly to a marine protective area that has a spill-over function [25].

The distribution of high abundance occurs at stations two and three this occurs located in the bay area part of the leeward area and has a weak current in accordance with [26], who stated that fish spawn in areas of weak current. The low abundance of ichthyoplankton at station six, because this area is an inlet current into the channel, high currents in the Chanel prevent ichthyoplankton from clumping [27].

At the time of the study, there were seven families of ichthyoplankton that had a consistent distribution by covering habitat coverage above 50%, including Carangidae, Mullidae, and Pomacentridae. According to [12], the distribution value greater than 50% of resident fish. A high distribution of Carangidae was also found in other areas of the WPP 715 [17]. The presence of ichthyoplankton from resident fishes indicates that Kilwaru waters are a resource for the surrounding waters.

This study showed a low diversity index, which is possible because each species will choose the location and time of spawning according to their life history, so that at the same place and time, only a few species are found to spawn. Although the diversity is low, the population is still good because the dominance and uniformity index values are low.

The stage composition of ichthyoplankton showed a high preflexion contribution (81%). The high number of ichthyoplankton indicates that the ichthyoplankton found in the Kilwaru sasi water area have just hatched from eggs spawned in the local area. This situation emphasizes that the Kilwaru waters have a resource function that is distributed (spill-over) to the surrounding waters. As stated by [28], the presence of currents is very influential on spillover; the currents in Kilwaru waters are very supportive in addition to large tidal currents and are also part of the Indonesian Through-Limafatola Path [29].

The hydrological parameters in the Kilwaru waters showed normal values, indicating very good agreement with the standards of the Ministry of Environment for the life of marine biota for the parameters of turbidity, dissolved oxygen, and conductivity. Salinity shows the value of a period of mixed water because the sampling location was close to the beach. This condition is still tolerated by ichthyoplankton because they have a greater tolerance for salinity than their adult life phase. This good condition allows ichthyoplankton to survive for a long time before experiencing retention and settlement, thus ensuring spill-over to the surrounding habitat [30].

4. Conclusion

Based on the state of ichthyoplankton, the results of this study can be concluded to be related to the waters of the sasi area in Kilwaru. Ichthyoplankton in Kilwaru Waters are mostly coral reef fishes. Ichthyoplankton with high abundance and wide habitat coverage, which are resident fish in Kilwaru waters, include Carangidae, Mullidae, Pomacentridae, and Clupeidae. The ichthyoplankton population in Kilwaru has poor diversity, but good dominance and uniformity. Ichthyoplankton in Kilwaru Waters are mostly in the preflexion stadia, indicating that Kilwaru waters are spawning areas for various types of fish. The hydrological conditions of the Kilwaru waters remain good. The conclusion of this research is that Kilwaru sasi water functions as a source of resources that can be spill-over to its neighboring areas. Sasi waters in Kilwaru can be developed as a Marine Protected Areaor OECM with strict traditional legal arrangements, so that natural resources and livelihoods remain sustainable.

This research was funded by a CFI Indonesia GEF 6 Grant through the World Wildlife Fund (WWF) Global Environmental Facilities (GEF) Agency and the Ministry of Maritime Affairs and Fisheries of the Republic of Indonesia.

References

1. Marjanto, D.K, Vol. 16, No. 3, September 2015: 311 – 32. (2015).
2. Green, A.L., Maypa, A.P., Almany, G.R., Rhodes, K.L., Weeks, R., Abesamis, R.A., Gleason, M.G., Mumby, P.J. & White, A.T. Biol. Rev. 90, pp. 1215–1247. https://doi.org/10.1111/brv.12155. (2015).
3. Huang, S., Deng, Z., Tang, G., Li, H. & Yu, T. Journal of Marine Systems 217, 103515 https://doi.org/10.1016/jmarsys.2021.103515. (2021).
4. Garces-Rodriguez, Y., Sanchez-Velasco, L., Diaz-Viloria, N., Jimenez-Rosenberg, S. P. A., Godinez, V., Montes-Arechiga, J. & Munguia-Vega, *J. Plankton*. (2018).
5. Delsman HC. Fish Eggs and Larvae from the Java Sea. Treubia VOL.XIV,LIVR. 1. (1933).
6. Unesco. Ichthyoplankton. Unesco technical papers in marine science 20. Report of the CICAR Ichthyoplankton Workshop. Mexico City, 17-26 July 1974. (1975).
7. Leis, J.M & Rennis, D.S. (Sydney: New South Wales University Press and Honolulu: University of Hawaii Press) p 269. (1983).
8. Leis, J.M & Trnski, T. (Sydney: New South Wales University Press) p 371. (1989)
9. Leis, J.M & Carson-Ewart, B.M. (Brill Sydney) p 850. (2000).
10. Westhaus-Ekau P, 16-21 September 2002 (Purwokerto: Jenderal Soedirman University, Faculty of Biology). (2002)

11. SEAFDEC. Larval Fish. Southeast Asian Fisheries Development Center in Collaboration with the UNEF/GEF South China Sea Project. (2007).
12. Lirdwitayaprasit, P., Nuangsang, C., Puewkhao, P., Rahman,M.D., Htay-Oo, U.A., & Sien, U.A.W. The Ecosystem-Based Fishery Management in the Bay of Bengal, p. 93-124. (2009).
13. Carpenter, K.E., & Niem, V.H. The Living Marine Resources of the Western Central Pacific.Vol. 4. Bony Fish Part 2 (Mugillidae to Carangidae). FAO Species Identification Guide for Fishery Purposes. Rome. p.2790. (1999).
14. Gladstone, W. *Marine Ecology Prpgress Series. Vol. 351: 235–247.* doi: 10.3354/meps07108 Res. 40(5): 606–618. https://doi.org/10.1093/plankt/fby033. (2007).
15. Armstrong, M.J. & Witthames, P.R. Fisheries Research 117–118, 35–47. https://doi.org/10.1016/j.fishres.2010.12.028. (2012)
16. Hammou, J.H., Cinner, J.E., Barneche, D.R., Caldwell, I.R., Mouillot, D., Robinson, J.P.W., Schiettekatte, N.M.D., Siqueira, A.C., Taylor, B.M. & Graham, N.A.J. Nature Communications 15:6105 https://doi.org/10.1038/s41467-024-50367-0. (2024).
17. Wagiyo K, Priatna A, Herlisman. Bawal 11(1):1–17. (2019).
18. Wagiyo, K. Iktiologi Indonesia, 7(2), 75-82. https://doi.org/10.32491/jiiv7i2.214. (2007).
19. Ramírez-Martínez, G.A., Giraldo, A., Rivera-Gómez, M., Aceves-Medina, G. Regional Studies in Marine Science 52, 102378. https://doi.org/10.1016/jrsma.2022.102378. (2022).
20. Taufik, M.Restiangsih, Y.H., Ma'mun, A., T Hidayat, T., Wagiyo, K., Panggabean, A S., , Nurulludin &, Prihatiningsih. (2024). IOP Conf. Series: Earth and Environmental Science 1350 012018. https://doi.org/10.1088/1755-1315/1350/1/012018. (2024).
21. El-Regal, M.A.A. Egyptian *Journal of Aquatic Biology & Fisheries. Vol. 22(4): 233- 247.* (2018).
22. Wagiyo, K., Suman, A. & Patria, M.P. BAWAL, Widya Riset Perikanan Tangkap. Vol. 7 , No.3. (2015).
23. Reed, E.M. & Taylor, B.M. Marine and Freshwater Research. https://doi.org/10.1071/MF20049. (2020).
24. Xu, Q., Jiang, Y., Fang, L. & Liu, M. Marine and Coastal Fisheries: Dynamics, Management, and Ecosystem Science 13:679–692. https://doi.org/10.1002/mcf2.1018. (2021).
25. Lenihan, H.S., Gallagher, J.P., Peters, J.R., Stier, A.C., Hofmeister, J.K.K. & Reed, D.C. Scientific Reports,11:2663. (2021) https://doi.org/10.1038/s41598-021-82371-5
26. Colin, P.L. Journal of Fish Biology 76, 987–1007. https://doi.org/10.1111/j1095-8649.2010.02553.x. (2010).
27. Lambert, T. C., and D.M. Ware. Can. J. Fish. Aquat. Sci. 41 : 1565-1569. (1984).
28. Tian, R. C., Chen, C., Stokesbury, K. D. E., Rothschild, B. J., Cowles, G. W., Xu, Q., Hu, S., Harris, B. P., and Marino II, M. C. ICES Journal of Marine Science, 66: 2155–2164. (2009).
29. Gordon, A.L. 2005. Oceanography. 18, (4).14-27
30. Kume, G., Nakaya, K., Takeda, T., Ichinomiya, M., Komorita, T., Habano, A., Makino, F., Kodama, M. & Kobari, T. Plankton Benthos Res 18(3): 148–159. https://doi.org/10.3800/pbr.18.148. (2023).

Utilization of coastal & marine resources by the local community of Kei Kecil conservation area

Hakim Miftakhul Huda[1*], *Radityo* Pramoda[1], *Siti* Hajar Suryawati[2], *Masayu* Rahmia Anwar Putri[3], *Reny* Puspasari[3], *Arip* Rahman[3], *Raymon* Rahmanov Zedta[3], *Andrias* Steward Samusamu[3]

[1]Research Center for Behavioral and Circular Economics, National Research and Innovation Agency (Jakarta-Indonesia)
[2]Research Centre for Economics of Industry, Services, and Trade, National Research and Innovation Agency (Jakarta-Indonesia)
[3]Research Center for Conservation of Marine and Inland Water Resources, National Research and Innovation Agency (Bogor-Indonesia)

Abstract. Kei Kecil Island, a vital part of Indonesia's marine conservation areas, is rich in coastal and marine resources essential for local communities' economic sustainability. However, increased resource utilization due to population growth and expanding tourism and fisheries sectors poses challenges to environmental sustainability and the potential overexploitation of marine resources. This research analyzes the use of coastal and marine resources in Kei Kecil Island's conservation areas by coastal communities through a mixed-method approach, including field surveys, stakeholder interviews, and secondary data analysis. The study reveals that fishing and seaweed cultivation are the main sources of income, with significant variations in investment costs and business performance across different sub-districts. Higher investments generally lead to better financial performance, but challenges such as overfishing, destructive fishing practices, limited education, and governance issues threaten resource sustainability. The research suggests a multi-faceted approach involving collaborative management frameworks, investment in education and training, and strengthening regulatory frameworks to ensure the long-term health of the ecosystem and the welfare of local communities.

1 Introduction

Kei Kecil Island, a region with abundant coastal and marine resources, is an essential part of Indonesia's marine conservation areas[1]. Resources such as fisheries, coral reefs, and mangrove ecosystems offer substantial benefits, not only for the ecosystem but also for the economic sustainability of local communities. In recent decades, resource utilization has intensified owing to population growth and the expansion of the tourism and fisheries sectors. However, this increased activity brings serious challenges related to environmental sustainability and potential overexploitation of marine resources.

* Corresponding author: hamihud@gmail.com

The coastal conservation area of Kei Kecil Island and its surrounding waters have significant marine resource potential. However, if these resources are exploited unsustainably, it could jeopardize the health of the ecosystem and welfare of the local community. Effective management and sustainable practices [2]are crucial to ensure that both the marine environment and the livelihoods of people who depend on these resources can thrive together.

Economic pressure on communities in marine conservation areas can greatly affect the success of conservation[3]. When financial hardship is high, residents may turn to unsustainable resource use, such as overfishing or coral harvesting, to meet their immediate needs. This creates a conflict between livelihood demands and conservation goals, thereby reducing community support for protective measures. To address this, conservation efforts should integrate sustainable income options, such as sustainable fishing practices, to help communities gain economic stability while preserving marine resources. By aligning conservation goals with local economic interests, these programmes can foster both ecological preservation and community resilience.

This research aims to analyze the use of coastal and marine resources in small Kei Island conservation areas by coastal communities.

2 Research Method

The research method employed in this study on coastal and marine resource utilization on Kei Kecil Island combines both qualitative and quantitative approaches [4] to obtain a comprehensive understanding of the subject. The methodology included field surveys, stakeholder interviews, and secondary data analysis. Each of these components is designed to gather data on resource usage patterns, stakeholder perspectives, and the impact of these activities on both the environment and local community's well-being. The survey was carried out in August 2024 in four sub-districts that are included in the Kei Kecil conservation area: Kei Kecil sub-district, Manyeuw sub-district, Hoat Sorbay sub-district, Kei Kecil Barat sub-district, and Maluku Tenggara Regency, Maluku Province. Respondents were determined through purposive sampling, with 88 fishing households and 42 seaweed-cultivating households.

Field surveys were conducted to gather direct observations of the types of coastal and marine resources utilized and the scale of their usage by local communities. These surveys focused on areas with high levels of activity, including fishing, tourism, and marine aquaculture. Field data were recorded to assess the environmental conditions and potential impact of each type of activity on the local ecosystem.

In-depth interviews were conducted with a range of stakeholders, including local community members, fishermen, business owners, and representatives from local government agencies. These interviews provided insights into the motivations behind resource use, community dependency on coastal and marine resources, and challenges faced in managing these resources sustainably. Stakeholder perspectives on the importance of conservation efforts and their involvement in such initiatives are also captured.

To supplement the field observations and interview data, secondary data were obtained from local government reports, conservation agency publications, and relevant academic studies. These data helped contextualize findings within a broader scope, providing a historical background on resource utilization and existing conservation measures on Kei Kecil Island. Secondary data analysis also aided in identifying trends, comparing them with similar regions, and assessing the regulatory framework governing coastal and marine resource use.

Data analysis was conducted using descriptive and financial methods. Descriptive analysis provides an overview of the characteristics and trends within the data, helping identify patterns, challenges, and opportunities in the community's seaweed cultivation and fishing activities. Financial analysis evaluates the economic viability of seaweed cultivation and fishing businesses by assessing profitability, income distribution, costs, and potential returns. This combined approach offers insights into both the social and economic aspects of the community's activities, informing strategies for sustainable income improvement and resource management.

3 Result and Discussion

3.1 Sociodemographic Kei Islands Concervation Area

The Kei Islands Conservation Area, located in Indonesia's Maluku region, is characterized by a rich socio-demographic landscape that is deeply influenced by its coastal environment and traditional maritime culture. The communities are predominantly rural, with livelihoods closely tied to natural resources, particularly fisheries, marine aquaculture, small-scale agriculture, and tourism. As the mainstay of local income, fishing has been practiced for generations, with knowledge and skills passed down through families, forming the cultural backbone of the community. This reliance on traditional practices highlights the close relationship between people and their surrounding environment, where resources such as coral reefs, mangroves, and open waters provide both sustenance and economic security.

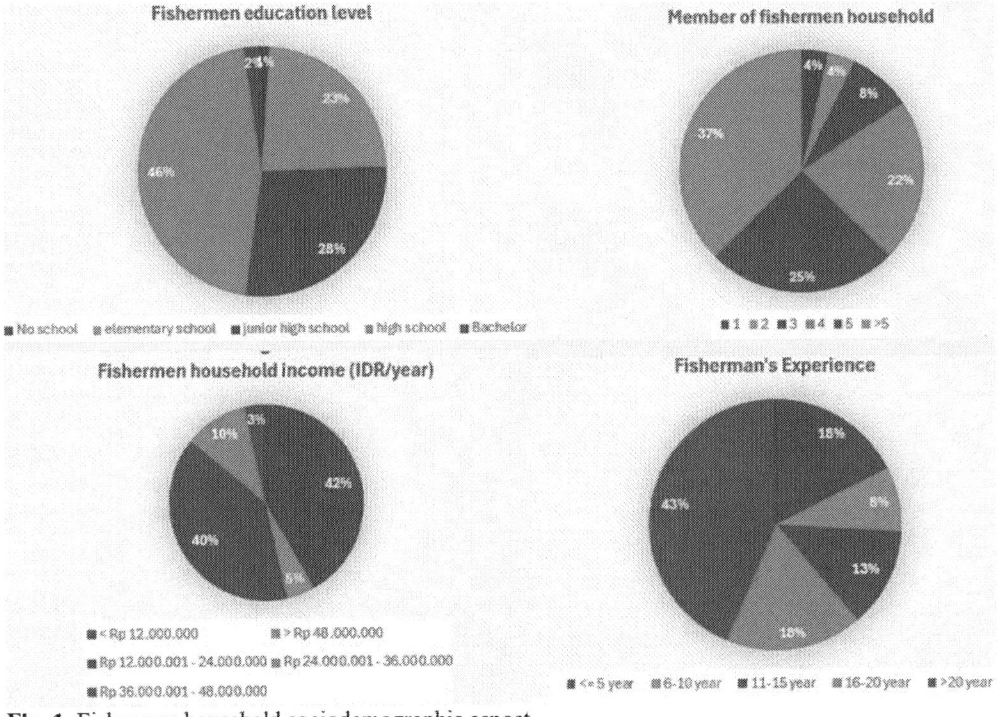

Fig. 1. Fishermen household sociodemographic aspect.

Figure 1 provides an overview of the sociodemographic aspects of fishermen households. Most fishermen in this area have a secondary education (junior and senior high school), with only a few reaching higher education. This suggests that although formal education is accessible to most, there are challenges in advancing to higher levels. Higher education is expected to provide additional skills for managing marine resources and exploring other job opportunities in related sectors. Fishermen's households are generally large, with most having more than five members, indicating possible economic challenges, particularly in terms of welfare. Larger households tend to have more dependents, which can affect their ability to improve welfare. However, more household members can benefit if they contribute to productive work. Overall, the welfare of fishing households may rely on effective resource management and economic distribution within large households. Fishing experience is strongly correlated with productivity and income. Fishermen with more than 15 years of experience tend to have higher productivity and income because of their knowledge of fishing strategies, resource management, and market networks. By contrast, those with less than five years of experience may face productivity and income challenges. Thus, the longer the fishing experience, the greater the likelihood of increasing catch and income. Household income data show that most fishermen have relatively low incomes, indicating income inequality. Most fall into the low-income category, with only a few reaching higher income levels, likely because of limited access to resources or broader markets.

In addition to fishing, most of the communities in the conservation area of Kei Kecil Island engage in seaweed cultivation to meet their daily needs. Seaweed cultivation has been developing in this area for more than 20 years. The growth of seaweed has fluctuated, mainly due to variations in productivity and the dynamics of seaweed selling prices. The following image shows the characteristics of the seaweed cultivator respondents in the conservation area of Kei Kecil Island:

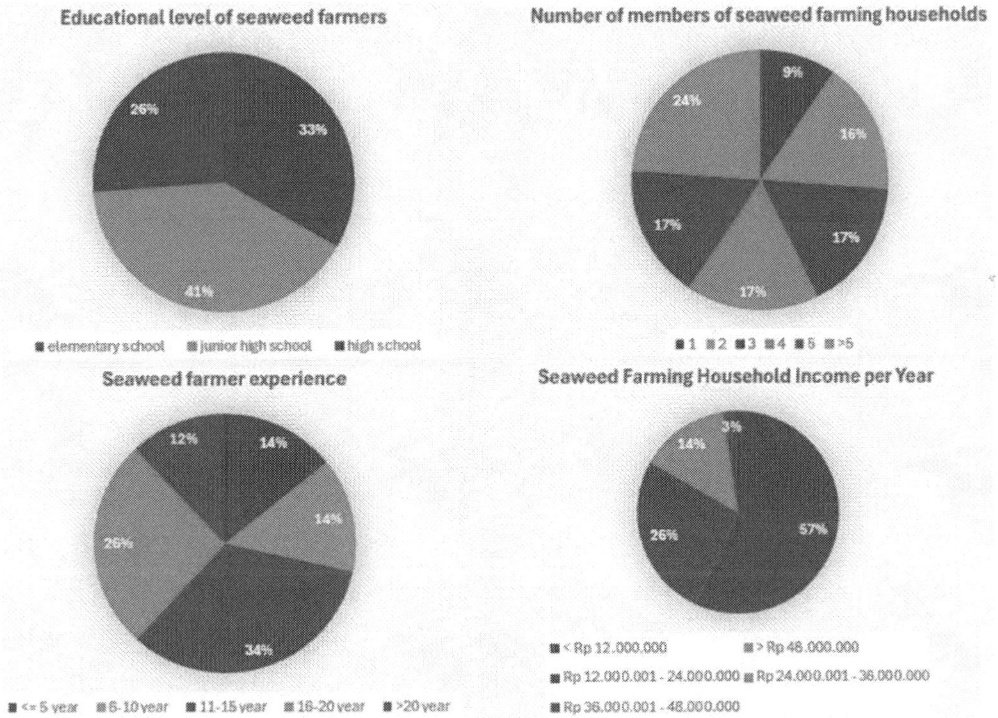

Fig. 2. Seaweed farmer household's sociodemographic aspect.

Figure 2 illustrates various sociodemographic aspects of seaweed farming households. Seaweed cultivators and fishermen have similar educational profiles, with most having only basic or secondary education. Lower education, especially among seaweed cultivators and fishermen, can hinder the adoption of new technologies and effective business management. However, seaweed cultivators with secondary education have the potential to improve productivity and business management, especially through training and access to information resources. Most seaweed cultivator households are smaller than fishermen's households.

While smaller households have lighter expenses, they may face productivity challenges due to limited labor. Conversely, fishermen with larger households face higher economic burdens but can also benefit from additional labor for economic activities. Seaweed cultivators generally have significant experience, with most having 11 to 20 years of experience in the field. Experienced cultivators tend to have better skills and knowledge, which leads to higher productivity and income. Those with less than ten years of experience are still in the learning phase, which may limit their yield. Cultivators with over 20 years of experience need to adapt to technological and market changes to maintain a stable income. Training programs for new cultivators are essential for boosting skills and productivity. Most seaweed cultivator households earn less than IDR 12,000,000 per year, limiting their ability to meet their basic needs. Only a few earns above IDR 24,000,000. These low-income groups force many to allocate most earnings to basic needs, hindering investment or business development. Increasing income through training, market access, and cultivation technology is essential to improve welfare.

3.2 Most utilizing marine and fisheries resources in Kei Islands Concervation Area

Marine and fisheries resources are the cornerstone of economic activity and sustenance in the Kei Islands Conservation Area, with these resources deeply embedded in the daily lives, traditions, and economic structures of local communities. The islands boast a highly diverse marine ecosystem home to a multitude of fish species, vibrant coral reefs, and expansive mangrove forests, each playing a critical role in supporting various livelihoods. Artisanal and small-scale fishing, which draws on generation-old techniques and knowledge, is the most widespread activity and a primary source of food security and household income for many of the population.

Table 1. Fishing business investment costs.

Fishing gears	Investment value (IDR)
Hoat Sorbay Sub district	
traps	6,660,000
gillnet	25,296,000
handline	42,623,571
mini purseseine	96,882,857
mini longline	38,543,333
Kei Kecil Sub district	
gillnet	53,750,625
spearguns	33,400,000
handline	10,725,833
Kei Kecil Barat Sub district	
gillnet	22,720,714
handline	43,060,000
mini longline	18,425,000
Manyeuw Sub district	
liftnet	211,148,000
gillnet	23,756,818
handline	38,833,333

Table 1 presents the investment costs of various fishing gears across different sub-districts. The investment values in fishing gear across different sub-districts on Kei Island

vary significantly, reflecting the diversity in fishing methods and financial resources available. The largest investment cost component in a fishing business is the fishing fleet, its machines, and fishing gear. In Manyeuw Sub-district, liftnet shows the highest investment at IDR 211 million, suggesting that it is a costly, possibly large-scale operation. Mini purseseines in Hoat Sorbay (IDR 96.9 million) and gillnets in Kei Kecil (IDR 53.8 million) also have high investment values, indicating substantial capital input in these areas. In contrast, lower investments, such as handlines in Kei Kecil (IDR 10.7 million) and traps in Hoat Sorbay (IDR 6.7 million), are more accessible, likely used by smaller-scale or traditional fishers. This distribution highlights the varying levels of fishing gear investment across regions with potential differences in fishing capacity and productivity. The following table shows the business operational costs and business performance of each fishing gear for each fishing effort.

Table 2. Fishing business performance per fishing trip (by vessel).

Fishing gears	**Total cost (IDR/trip)**	**Catch/trip (Kgs)**	**Total revenue (IDR/trip)**	**Profit (IDR/trip)**
Hoat Sorbay Sub district				
traps	134,861	6	700,000	565,139
gillnet	227,107	55	461,083	233,977
handline	314,109	17	730,714	416,605
mini purseseine	529,008	276	1,615,000	1,085,992
mini longline	338,833	25	1,276,667	937,833
Kei Kecil Sub district				
gillnet	336,690	27	527,250	190,560
spearguns	393,000	26	550,000	157,000
handline	177,152	9	606,667	429,514
Kei Kecil Barat Sub district				
gillnet	129,238	13	303,571	174,334
handline	399,309	35	635,000	235,691
mini longline	229,355	10	400,000	170,645
Manyeuw Sub district				
liftnet	406,639	74	813,125	406,486
gillnet	258,318	19	490,909	232,591
handline	327,444	18	813,889	486,445

Table 2 presents the fishing business performance across different sub-districts, highlighting total costs, catch per trip, revenue, and profit per trip for various fishing gears. Mini purseseine and mini longline operations in Hoat Sorbay generated the highest profits per trip (IDR 1,085,992 and IDR 937,833, respectively), which is attributed to their high catch volumes (276 and 25 kg). Conversely, gillnets and spearguns in Kei Kecil yield lower profits (IDR 190,560 and IDR 157,000, respectively), partly because of higher costs relative to their catch size. Smaller-scale options, such as traps in Hoat Sorbay and handlines across several regions, provide moderate profits, showing that traditional, lower-cost methods still bring reasonable returns. Overall, higher-cost gear often results in higher profits, but the balance between cost, catch size, and revenue varies significantly by method and region. In the next discussion, we examine the business costs and performance of the fishing business during a year of activity.

Table 3. Fishing business performance in a year.

Fishing gears	Effort (Trip/year)	Total cost (IDR/year)	Total revenue (IDR/year)	Profit (IDR/year)	R/C ratio
Hoat Sorbay Sub district					
traps	288	38,840,000	201,600,000	162,760,000	5.19
gillnet	140	36,665,200	72,043,833	35,378,633	2.78
handline	187	66,217,143	155,658,571	89,441,429	2.58
mini purseseine	152	80,490,000	240,228,571	159,738,571	3.20
mini longline	160	54,213,333	204,266,667	150,053,333	3.86
Kei Kecil Sub district					
gillnet	152	50,775,500	80,217,000	29,441,500	1.58
spearguns	108	42,444,000	59,400,000	16,956,000	1.40
handline	173	27,558,500	93,326,667	65,768,167	3.89
Kei Kecil Barat Sub district					
gillnet	140	19,238,571	44,612,857	25,374,286	3.16
handline	94	34,749,000	55,265,000	20,516,000	2.30
mini longline	128	32,646,667	56,750,000	24,103,333	2.19
Manyeuw Sub district					
liftnet	141	54,445,000	121,382,500	66,937,500	2.08
gillnet	156	43,281,273	78,080,909	34,799,636	2.28
handline	121	45,237,111	118,702,222	73,465,111	2.92

Table 3 presents the annual fishing business performance across different sub-districts, highlighting the effort (trips per year), total cost, total revenue, profit, and the Revenue-Cost (R/C) ratio, which indicates the profitability of different fishing gears. The Hoat Sorbay Sub-district shows high profitability, especially with traps (R/C ratio 5.19) and mini purseseines (R/C ratio 3.20), indicating significant returns on investment. The Kei Kecil Sub-district has lower R/C ratios, with spearguns generating the lowest (1.40), suggesting limited profitability. The Kei Kecil Barat Sub-district has moderate R/C ratios, with gillnets performing relatively well (3.16). The Manyeuw Sub-district displays an average performance, with handlines yielding the highest profit and R/C ratio (2.92). Overall, traps in Hoat Sorbay and handlines in Kei Kecil offer the highest returns relative to costs, whereas spearguns in Kei Kecil yield the lowest profitability.

The performance of fishing operations in the Kei region is influenced by factors such as wave conditions and availability of resources or assets within coastal communities. In June, July, and August, the primary fishing grounds across the waters of the Kei Islands show strong potential[5]. Rainfall and wind speed [6]can limit fishing activities, thereby reducing both the frequency and safety of trips. Additionally, access to fishing assetssuch as boats, nets, and equipmentvaries among households, impacting their ability to fish effectively and consistently. Communities with better resources can operate more resiliently, whereas those with limited assets may struggle to sustain fishing efforts and income during challenging conditions. Consequently, these environmental and resource-related factors play crucial roles in determining the productivity and economic stability of local fishing efforts.

In addition to fishing, seaweed farming has become an important source of livelihood for coastal communities in the Kei Kecil Island conservation area. This activity has grown rapidly due to the environmental conditions that support seaweed cultivation, such as clean, nutrient-rich waters. Seaweed farming provides economic benefits to the community, helping increase income and diversify livelihoods beyond fishing. In this area, seaweed farming is also considered environmentally friendly, as it does not harm the ecosystem and supports marine conservation efforts, contributing to the balance of the coastal ecosystem.

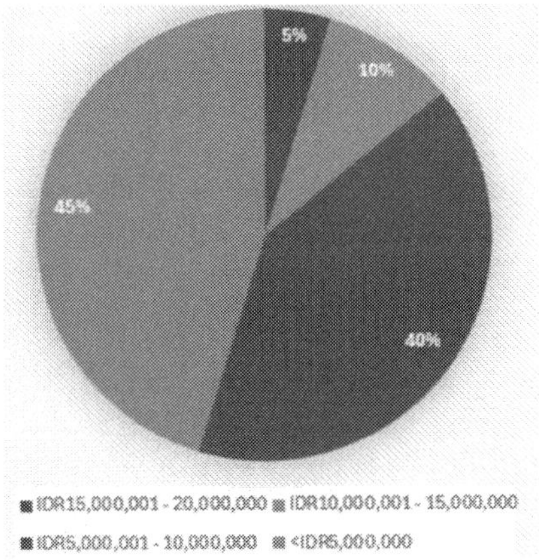

Fig. 3. Distribution of seaweed cultivators based on the size of business investment.

Figure 3 illustrates the distribution of seaweed cultivators according to their business investment size. The data show that most seaweed cultivation businesses on Kei Island operate with lower investments, with 45% of farmers investing under IDR 5 million and 40% investing between IDR 5-10 million. Only a small proportion of farmers (15% in total) invest in IDR 10 million. This suggests that seaweed cultivation on Kei Island is largely a low-capital venture, likely because of limited access to financial resources. However, as previous analyses show, higher investments tend to yield better returns and greater efficiency, indicating that there may be potential for growth if farmers gain access to more capital. Supporting these farmers with resources or financial assistance could help them scale up and increase their profitability.

Table 4. Seaweed Cultivation Business Performance in a Year.

No.	Investment value (IDR)	Variable cost (IDR per year)	Fixed cost (IDR/year)	Total cost (IDR/year)	Total revenue (IDR/year)	Profit (IDR/year)	R/C ratio
1	15,000,001 - 20,000,000	7,760,000	2,630,000	10,390,000	28,466,667	18,076,667	2.73
2	10,000,001 - 15,000,000	8,300,000	3,450,000	11,750,000	34,666,667	22,916,667	2.87
3	5,000,001 - 10,000,000	5,688,529	2,752,941	8,441,471	19,520,588	11,079,118	2.20
4	<5,000,000	3,610,526	1,792,105	5,402,632	14,238,596	8,835,965	2.65

Table 4 presents the annual business performance of seaweed cultivation based on different levels of investment. Analysis of seaweed cultivation businesses by investment level reveals that higher investments generally lead to better financial performance. Businesses with investments in the IDR 10-15 million range achieve the highest revenue (IDR 34.7 million), profit (IDR 22.9 million), and cost efficiency (R/C ratio of 2.87), making this the most profitable and efficient bracket. Even at lower investment levels, all groups showed profitability with R/C ratios above 2.0, indicating that seaweed cultivation is financially viable across different scales. For the best returns, farmers may consider investing at least IDR 10 million, whereas those with limited capital can still achieve decent profits with smaller investments.

Seaweed cultivation is attractive to coastal communities because of its low capital and operational costs, simple cultivation techniques, relatively low labor requirements, and short production cycles, which provide a steady income[7], [8], [9]. However, currently, the price of seaweed is relatively low, and it cannot grow optimally. This prevents seaweed farmers from optimizing their cultivation areas. Seaweed price levels are determined by aggregation centers and main transportation hubs in remote areas. [10]

3.3 Opportunity and challengin of marine and fisheries utilizing

The utilization of marine and fishery resources in the Kei Islands Conservation Area embodies a complex blend of opportunities and challenges that impact both the local economy and the environment. The region's rich marine biodiversity, including various fish species, coral reefs, and mangrove ecosystems, provides a wealth of economic opportunities for the local community. Fishing and seaweed farming are critical sources of income, providing stability for households and supporting food security across the islands.

However, such opportunities pose substantial challenges. Overfishing, driven by both local demand and commercial interests, strains fish populations and threatens the long-term sustainability of fishing [11], [12]. Additionally, unsustainable fishing practices, such as the use of destructive methods like blast fishing and cyanide fishing, [13]cause significant harm to coral reefs and other vital habitats. Although illegal, these destructive methods can be difficult to eradicate because of insufficient enforcement and economic pressure on fishers to maximize short-term catches.

Another critical challenge is limited access to education and alternative employment opportunities, which leaves many residents reliant on extractive activities. For many locals, fishing has been the primary livelihood for generations, and transitioning to other forms of income, such as ecotourism or sustainable aquaculture, requires new skills, knowledge, and often upfront investment. Furthermore, governance issues such as limited funding for conservation programs, insufficient monitoring capabilities, and a lack of integrated resource management policies hamper efforts to implement sustainable practices.

To fully realize the opportunities and address the challenges of marine and fishery resource utilization in the Kei Islands, a multi-faceted approach is necessary. This involves building collaborative management frameworks that actively engage local communities, government agencies, and NGOs to create shared responsibility for resource sustainability. Investing in environmental education, vocational training, and capacity-building programs will empower residents to adopt sustainable practices and to explore alternative livelihoods. Strengthening regulatory frameworks with stricter enforcement and transparent management is also essential to curb illegal and destructive fishing practices. With a holistic strategy, the Kei Islands can harness the economic potential of their marine resources while preserving the ecological integrity of the conservation area, ensuring a sustainable future for both the community and environment.

4 Conclusion

The Kei Kecil Island conservation area has significant coastal and marine resources and is an important marine conservation area. Fishing and seaweed cultivation play an important role as sources of livelihood for coastal communities in the Kei Island conservation area. This effort provides sufficient results to meet the needs of coastal communities.

The research activities have undergone the GEF6 Screening Tools document, which is an effort to screen the implementation of the activities. The authors would like to express their utmost gratitude to the CFI Indonesia GEF-6 Grant, provided through the World Wildlife Fund (WWF) Global

Environmental Facility (GEF) Agency and the Ministry of Maritime Affairs and Fisheries of the Republic of Indonesia, for their full support and funding.We would like to acknowledge the GEF-6 for support towards this community service in funds, facilitation and field cooperation.
The authors declare that they have no conflicts of interest. The research was conducted independently, and the authors have not received any financial or non-financial benefits that could influence the results or interpretation of this study. The views expressed in this manuscript are solely those of the authors and do not necessarily reflect the views of any organization or institution associated with this research.

References

[1] W. A. Teniwut, S. K. Hamid, R. M. K. Teniwut, M. Renhoran, and C. D. Pratama, "Do coastal communities in small islands value marine resources through marine protected areas?: Evidence from Kei Islands Indonesia with choice modelling," *Mar Policy*, vol. 157, p. 105838, (2023).

[2] M. Yulisti, A. S. Hidayat, C. M. Firdausy, U. Mu'awanah, N. Kurniasari, and E. Nurjati, "Effects of eco-friendly fishing gears on fishermen's welfare and sustainable fisheries: Lessons learned from Indonesia.," *Mar Pollut Bull*, vol. 198, p. 115888, (Jan. 2024), doi: 10.1016/j.marpolbul.(2023).115888.

[3] N. J. Bennett *et al.*, "Advancing social equity in and through marine conservation," *Front Mar Sci*, vol. 8, p. 711538, (2021).

[4] D. Strijker, G. Bosworth, and G. Bouter, "Research methods in rural studies: Qualitative, quantitative and mixed methods," *J Rural Stud*, vol. 78, pp. 262–270, (2020).

[5] R. W. Saragih, D. M. A. Mannu, and M. Ryan, "Forecast fishing areas in the waters of the Kei Islands based on satellite data," in *IOP Conference Series: Earth and Environmental Science*, IOP Publishing, (2020), p. 12005.

[6] S. M. Picaulima, E. S. Wiyono, M. S. Baskoro, and M. Riyanto, "Analysis of factors affecting small-scale fishing trips in kei islands, indonesia," *AACL Bioflux*, vol. 14, no. 6, pp. 3614–3622, (2021), [Online]. Available: https://www.scopus.com/inward/record.uri?eid=2-s2.0-85121857356&partnerID=40&md5=679ecccdf3e528a91a3f7135f249b537

[7] M. A. Rimmer *et al.*, "Seaweed aquaculture in Indonesia contributes to social and economic aspects of livelihoods and community wellbeing," *Sustainability*, vol. 13, no. 19, p. 10946, (2021).

[8] S. Spillias *et al.*, "The empirical evidence for the social-ecological impacts of seaweed farming," *PLOS Sustainability and Transformation*, vol. 2, no. 2, p. e0000042, (2023).

[9] S. Larson *et al.*, "Women's well-being and household benefits from seaweed farming in Indonesia," *Aquaculture*, vol. 530, p. 735711, (2021).

[10] A. Langford *et al.*, "Price analysis of the Indonesian carrageenan seaweed industry," *Aquaculture*, vol. 550, p. 737828, (2022).

[11] N. Stacey *et al.*, "Developing sustainable small-scale fisheries livelihoods in Indonesia: Trends, enabling and constraining factors, and future opportunities," *Mar Policy*, vol. 132, p. 104654, (2021).

[12] M. Knudsen, "Poverty and Beyond: Small-Scale Fishing in Overexploited Marine Environments," *Hum Ecol*, vol. 44Knudsen, pp. 341–352, (2016), doi: 10.1007/s10745-016-9824-y.

[13] S. N. Tranter *et al.*, "The inclusion of fisheries and tourism in marine protected areas to support conservation in Indonesia," *Mar Policy*, vol. 146, (2022), doi: 10.1016/j.marpol.2022.105301.

Spawning season estimation for bullet tuna (*Auxis rochei* Risso, 1810) in southern waters of Bali: a postovulatory follicles - based approach

Indrastiwi Pramulati[1*], *Hety* Hartaty[2], *Gussasta* Levi Arnenda[3]

[1]Biology Postgraduate Program, Faculty of Mathematics and Natural Sciences, Udayana University, Denpasar, 80234 Bali - Indonesia
[2]Research Center for Fishery, National Research and Innovation Agency, Indonesia
[3]Research Center for Conservation of Marine and Inland Water Resources, National Research and Innovation Agency, Indonesia

Abstract. Bullet tuna is a crucial neritic tuna species in Indonesia that plays a significant role in small-scale and commercial fisheries. Regulation of seasonal fishing is an example of fishery resource management to prevent negative impacts on fishery resources due to fishing pressure. Postovulatory follicles (POFs) are a potential characteristic for determining spawning season in adult fish that have already spawned. The objective of this study was to determine the spawning season of bullet tuna (*Auxis rochei*) in the presence of POFs. This study employed purposive stratified sampling techniques, which were obtained from gill net fishing boats operating in the Kedonganan Fish Landing Base, Bali. Sampling was conducted from June to November 2021, and 151 ovaries were collected including the morphometric such as fork length (cm FL). Fresh ovary samples were fixed for histological analysis using the paraffin method and Harris-Hematoxylin and Eosin staining. The spawning season was determined by the presence of POFs at the spawning gonad development stage in the length range of 22-32 cm FL, indicating that peak spawning occurred in August. These findings can be used as a reference for formulating the management of bullet tuna in Bali.

1 Introduction

Neritic tuna is an important fishery commodity because it ranks as the second-largest catch of small pelagic fish in Indonesia [1]. One group of neritic tuna is bullet tuna (*Auxis rochei*), which is listed by the International Union for Conservation of Nature (IUCN) as Least Concern (LC). It is possible that species with an LC status still face the risk of extinction and will become a conservation concern if there is continuous fishing pressure. Reproduction data, including information on spawning season, size at maturity, and fecundity, are crucial for developing effective management strategies for bullet tuna. Understanding these reproductive patterns will help in setting appropriate fishing regulations, such as size limits,

* Corresponding author: indrastiwipramulati@gmail.com

seasonal closures, and marine protected areas, to ensure sustainable populations and prevent overexploitation.

Bullet tuna exhibit an asynchronous spawning pattern, where the ovaries contain oocytes in all stages without a dominant oocyte. Ovaries have a mix of oocyte phases, allowing spawning to occur throughout the year [2-4]. Determining spawning patterns can be assessed macroscopically and microscopically by identifying the level of gonadal development and differentiation. Microscopic observation involves histological analysis of the gonadal tissue. Microscopic gonadal development is determined based on the composition of oocyte phases through the Most Advanced Group of Oocytes (MAGO) and classification of gonadal developmental stages, including postovulatory follicle (POF) availability [5]. The postovulatory follicle (POF) is a key indicator of gonadal development in fish. POF is a follicle in the ovary where the egg is released during ovulation [5]. POF is used to estimate spawning frequency by calculating the number of eggs spawned per breeding season [6]. The POF also indicates a peak in the annual spawning season [7]. This study enhances our understanding of reproductive biology, which can be linked to fishery resource management.

Spawning season estimation for bullet tuna, when observed microscopically, is considered accurate and detailed in determining the peak spawning of bullet tuna, allowing for identification of the size of the bullet tuna that is considered suitable for capture. The aim is to avoid overfishing [8]. Several studies have mentioned that the capture of bullet tuna in the waters of Nusa Penida Island, Bali, has reached overfishing status [9]. Regulations on fishing months based on peak spawning seasons are an example of fishery resource management, which can be established through government regulations to prevent negative impacts on fishery resources.

The aim of this study was to describe the histological characteristics of POF in bullet tuna, and to use this observation to calculate the frequency of POF, analyzed in relation to estimating the peak spawning season based on the presence of POF.

2 Materials and Methods

1.1 Data Collection and Histological Preparation

This study used a purposive stratified sampling technique [10] targeting 20 adult female fish per month. The samples were taken from gill net fishing boats (neritic gill nets) operating at the Kedonganan Fish Landing Base, Bali. Data collection and sampling of bullet tuna was conducted from June to November 2021, and 151 samples were collected. The morphometric aspects measured included forked length (cm FL), and the preparation of the female gonad samples in the laboratory followed the histological technique, which used the paraffin method with Harris-hematoxylin and eosin staining [11].

Samples were taken from the middle section of both the right and left lobes of the ovary, with a thickness of 2-3 mm, and fixed in 10% formalin. The histological preparation process included: (1) dehydration using graded alcohol (70%, 90%, and 96%), (2) washing with xylene, (3) infiltration with liquid paraffin, (4) embedding the tissue in liquid paraffin, (5) sectioning the tissue with a microtome to obtain intact sections 4-6 μm thick (as preparations), (6) staining with hematoxylin and eosin (HE), and (7) covering with cover glass and entellan.

1.2 Analysis of POF

The presence of POF was determined through the microscopic classification of gonadal development, which also described POF [4]. A POF is a follicle in the ovary that originates from hydrated oocytes that have undergone ovulation. POF is typically found in mature fish during spawning gonadal development [12]. POF analysis was conducted after identifying the level of gonadal development, where gonads containing advanced yolked oocytes, migratory nuclei, and hydrated oocytes required further observation to locate the POF. This approach aims to reduce the analysis time per histological preparation, as gonads identified with POF usually indicate active spawning conditions.

3 Results

The total number of female samples collected from June to November 2021 was 151, with lengths ranging from 15,7 to 33,5 cm FL. According to a previous study [13], histological preparations of 151 gonad samples of bullet tuna revealed eight stages of gonadal development: immature (26%) and developing (3%), while mature fish included spawning capable (10,5%), spawning (51,8%), regressing-potentially reproductive (2%), and regressed (5,1%). Postovulatory follicles (POF) were found in fish with active spawning status. Microscopically, signs indicating spawning were emphasized by the presence of oocytes, such as migratory nuclei and hydrated yolk, as well as the presence of POF (Figure 1).

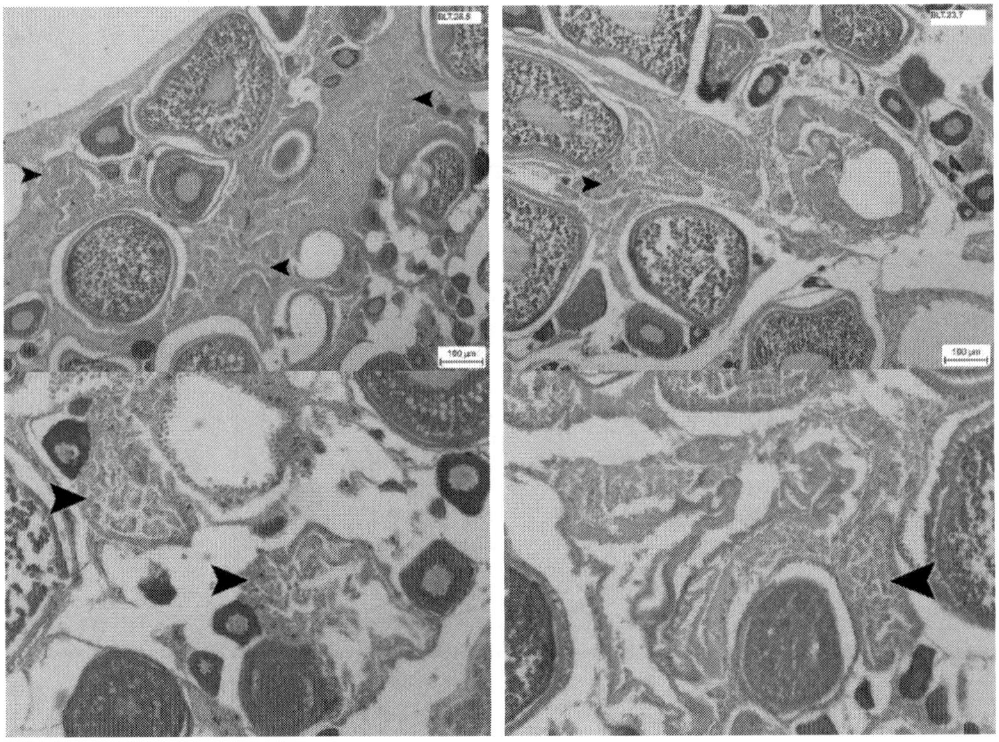

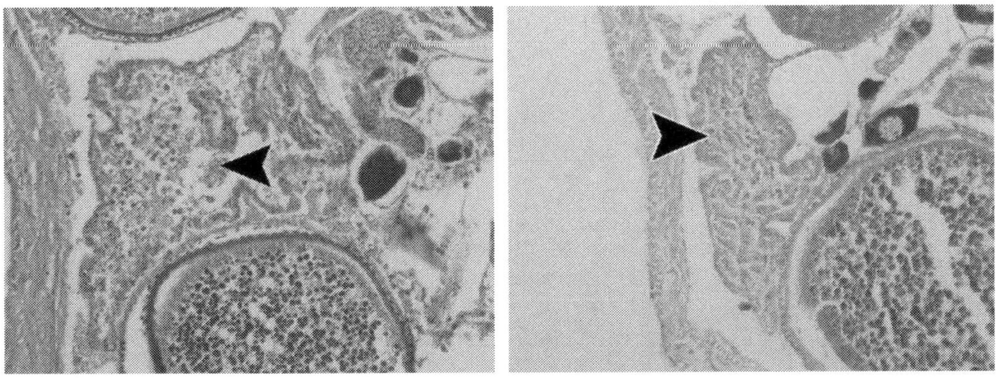

Fig. 1. The ovaries histological structure of the bullet tuna (*A. rochei*) showing the presence of POF (black arrow).

The results indicate that 22 female fish gonad samples showed the presence of postovulatory follicles (POF). The highest proportion of gonad samples identified with POF occurred in July and August, the lowest proportion with POF occurred in June and September, and with no POF in October and November (Figure 2). POF was found in fish with forked lengths ranging from 22 to 32 cm FL, with the majority ranging from 23 to 24 cm FL.

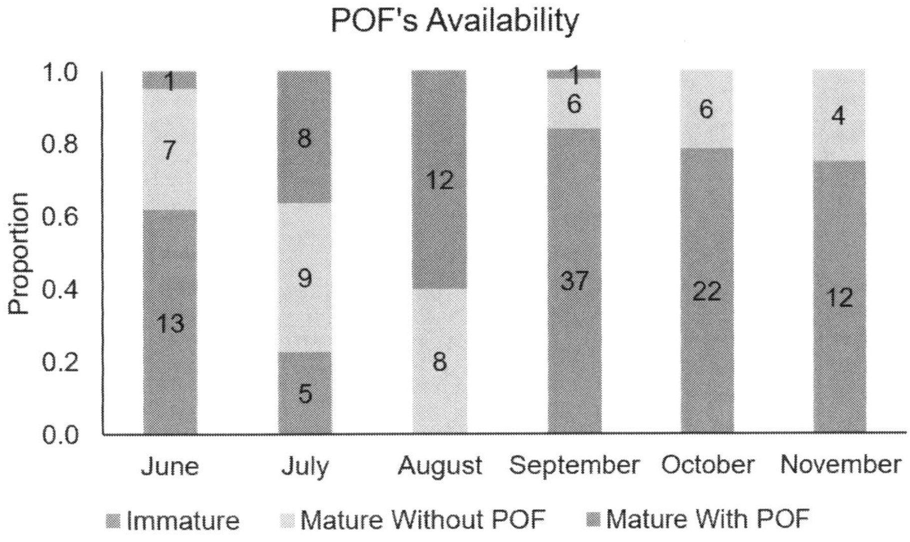

Fig. 2. Proportion of POF presence in bullet tuna (*A. rochei*) during June-November 2021.

4 Discussion

Microscopic observations indicate that bullet tuna spawn throughout the year, as spawning gonad development levels with postovulatory follicles (POF) were found nearly every month, except in October and November. This demonstrates that bullet tuna underwent multiple spawning phases in each season, with a peak spawning period in July and August. This finding is supported by a previous study on spawning seasons based on sex ratios and GSI values [13]. The sex ratio approached the ideal value of nearly 1:1 in August and the GSI peaked in July and August.

Spawning occurs concurrently with ovulation, during which hydrated oocytes are released from the surrounding follicles. Follicles, which are POF, remain unfragmented and do not exit the ovary along with hydrated oocytes. Initially, POF have a wrinkled shape, but over time, these follicles shrink and are reabsorbed [14].

4.1 Linking to Spawning season

The use of POF classification to determine spawning times was introduced by Hunter and Goldberg [15], which indicated that anchovies can spawn approximately 20 times a year. Determining spawning season can support the implementation of sustainable management strategies. The need for sustainable management was based on a previous study that showed that

The length at first capture was lower than the length at first maturity, indicating that young fish (immature and never spawned) were dominant in the catch and would be an overfishing status [16]. On the other hand, fishing pressure on bullet tuna is indicated by an exploitation rate above 0.5, which means overexploitation (overfishing) [9, 17]. Therefore, responsible and sustainable fishery management is essential for ensuring that fishery resources remain viable.

Responsible and sustainable fisheries management is based on research findings, particularly regarding the spawning seasons. Restrictions on fishing days or months and areas during spawning seasons are also necessary, as exemplified by the policy outlined in Indonesian PERMEN-KP Republik Indonesia No. 26, which limits the capture of yellowfin tuna (*Thunnus albacares*) in spawning and breeding areas in WPP RI 714 from October to December. Customary laws, local wisdom, and community involvement also support the enforcement of seasonal fishing restrictions.

This consideration is reflected in the Undang-Undang Republik Indonesia (UU RI) No. 31 Pasal 6 ayat 2 (2004) regarding fishery management. An example is Nyepi Segara on Nusa Penida Island, Bali, where marine activities, such as fishing and transportation to and from Nusa Penida, are halted for one day, typically observed in September or October [18].

Another example is the traditional "kenduri adat laut" or " hari pantang laut" in Pidie Jaya Regency, Aceh, which is held at least every three years, depending on the agreement and willingness of local fishermen. Fishing was halted for three days, from sunrise on the day of the feast until sunset on the third day. In addition to the "kenduri adat laut," there are other days of abstinence, including Fridays, from sunset on Thursday until sunset on Friday, as well as during the celebrations of Idul Fitri and Idul Adha, on Independence Day, August 17, and on December 26 to commemorate the tsunami that occurred on December 26, 2004 [19].

In addition, similar spawning seasons have been observed in studies on bullet tuna tuna in the waters of Mejane in August [20], Tunisia from June to August [21], Indian waters from July to August [22], Mediterranean waters of Turkey from July to August [23], and Ligurian waters starting in July [24].

5 Conclusion

The spawning season for bullet tuna (*Auxis rochei*) in the presence of POF occurred in July and August, with lengths ranging from 22 to 32 cm FL. Spawning occurs concurrently with ovulation, during which hydrated oocytes are released from the surrounding follicles. Follicles, which are POF, remain unfragmented and do not exit the ovary along with hydrated oocytes. Determining spawning season can support the implementation of sustainable management strategies. Responsible and sustainable fisheries management is based on research findings, particularly regarding the spawning seasons.

We thank Ririk Kartika Sulistyaningsih, S.Pi, and M. App. Sc.; and Bram Setyadji, S. Kel. M.Sc. who guided observations in the histology laboratory at the Research Institute for Tuna Fisheries, including making gonad histology preparations and analysis of results. Thanks are also addressed to those who guided the authors in this research. The authors are also grateful to Yesika Nanda Pramurdya, S.Pi, M.Si, Naily Maufiroh, S.Pi, and Desy Shintya Irene, S.Kel, who helped in collecting samples and histological preparations.

References

1. DJPT. 2013. Fisheries Statistics Catch in the Sea by WPP RNI 2005-2012. Ministry of Marine Affairs and Fisheries-Direktorat Jenderal Perikanan Tangkap, 11 (2) (2013).
2. J.M. Redding and R. Patino. CRC Mar. Sci. Ser., **12** (1993)
3. H. Murua, G. Kraus, F. Saborido-Rey, A. Thorsen and S. Junquera. J. Northwest Atl. Fish. Sci., **33**: 33-54 (2003)
4. Z.A. Muchlisin. Aceh Int. J. Sci. Technol., **3** (1): 43-52 (2014)
5. J.H. Farley and T. Davis. CSIRO, Mar. Res., **18** (1999)
6. R.J. Wootton. *J. Fish Biol.* 5: 89–96 (1973)
7. J.R. Hunter and B.J. Macewicz. *Sexual maturity, batch fecundity, spawning frequency and temporal pattern In the northern anchovy, Engroulis mordar, during the 1979 spawning season*, in Proceedings of the Calif. Coop Oceanic Fish Invest. Rep. 21:139-149 (1980)
8. D. Noija, S. Martasuganda,, B. Murdiyanto, and A.A. Taurusman. J. Tek. Perik. Kelaut. 5 (1): 65- 74 (2014)
9. E. Yuliana, Nurhasanah, D.E. Yani dan S. Agustina. J. Ilmu Tek. Kelautan Trop., **12**(3): 659-672 (2020)
10. S. Notoatmodjo. *Metodologi Penelitian Kesehatan*. Edisi 1. Renakacipta, Jakarta (2010)
11. Mujimin. Buletin Teknik Litkayasa Akuakultur, 4 (2005)
12. J. H. Farley, A. J. Williams, S. D. Hoyle, C. R. Davies, and S. J. Nicol, PLoS ONE, **8**, 4 (2013)
13. I. Pramulati, A.A.S.A. Sukmaningsih and F.X. Sudaryanto. Bawal, **14**(1): 21-37 (2022)
14. J.R. Hunter and B.J. Macewicz. Fish. Bull., 83: 119-136 (1985)
15. J.R. Hunter and R. Goldberg. Fish. Bull., U.S. 77:64-652 (1980)
16. I. Pramulati, H. Hartaty, A.A.S.A. Sukmaningsih and F.X. Sudaryanto. *Length at First Maturity and Le ngth at First Capture for Bullet Tuna* (*Auxis rochei* Risso, 1810) *in the Southern Waters of Bali*, in Proceedings of the ISFM XII, August, Riau, Indonesia (2023)
17. K. Amri, A. Novalina dan B. Setyadji. Indones. Fish. Res. J., **25** (2): 103-112 (2018)
18. N.K.S. Adnyani, I.N.D. Prasetia dan R.A. Windari. J. Ilmu Sos. Hum., **3**(1): 300-312 (2014)
19. Sulaiman. Prospek Hukum. Yustisi, **2** (3) (2013)
20. M.A. Dahlan, B. Yunus, M.U. Umar and M. Nur. *Musim Pemijahan Ikan Tongkol Lisong* (*Auxis rochei* Risso,1810) *di Perairan Majene Sulawesi Barat,* In Proceeding of the Simposium Nasional Kelautan dan Perikanan VI, 21 Oktober, Makassar, Indonesia (2019)
21. H. Allaya, A. Hattour, G. Hajjej and M. Trabelsi. Cah. Biol. Mar., 54: 287-292 (2013)

22. S. Jasmine, P. Rohit, Abdussamad, S.P. Koya, K.K. Joshi, S. Kemparaju, D. Prakasan, Elayathu and M. Sebastine. Indian J. Fish., **60** (2): 13-20 (2013)
23. A.E. Kahraman, D. Göktürk and F.S. Karakulak. Afr. J. Biotechnol., **10** (15): 3009-3013 (2011)
24. G. Palandri, L. Lanteri, F. Garibaldi and R.L. Orsi. Collect. Vol. Sci. Pap. ICCAT, **64**(7): 2272-2279 (2009)

Community understanding and recommendations for sea turtle conservation amidst life-threatening risks on Buru Island

Ravidas Kirtan[1*], *Ahmad R. Faruqie*[1], *Mochamad Dwianto*[1], *Hasna S. Al Munawar*[1], *Aulia P. Khaerunnisa*[1], *Caesar A.W. Putra*[1], *Gabriella A.S. Gerungan*[1],*Raihana Y. Azzahra*[1], *Ahmad Faisal*[1], *Gifta B. Sukmaayu*[1], *Iffa Purnama*[1], *Taufik Gifari*[1], *Albi M.T. Musa*[1], *Alifvian K. Putra*[1], *Zeta Raihan*[1], *Arinal F. Rahmat*[1], *Fida Z.L. Yasmin*[1], *Ilham Maulana*[1], *Nadya Cakasana*[1], *and Muhammad Iqbal*[1]

[1]Department of Marine Science and Technology, Faculty of Fisheries and Marine Science, IPB University, Indonesia

Abstract. Buru Island's rich coastal ecosystems serve as critical habitats for endangered sea turtles, such as the leatherback, green turtle, hawksbill, and olive ridley, which frequently nest and forage along its coasts. The aim of this study was to assess the community's understanding of sea turtle conservation efforts and provide recommendations to improve conservation initiatives in the face of life-threatening risks. Using a Likert scale to ensure data validity, this study employed semi-structured interviews and a random sampling method, engaging the communities of two key regions: Kecamatan Fena Leisela in Kabupaten Buru and Kecamatan Kepala Madan in Kabupaten Buru Selatan. These findings reveal significant knowledge gaps regarding the ecological service and regulatory importance of sea turtle conservation among local populations. While most residents in Kecamatan Fena Leisela demonstrate awareness and support for conservation efforts, the majority in Kecamatan Kepala Madan possess only basic knowledge but lack active participation in conservation. This is evidenced by ongoing illegal poaching activities for personal consumption and local trade involving both eggs and turtle body parts. Furthermore, turtle habitats in these regions face severe threats from marine debris and the destruction of nesting grounds due to ongoing local sand mining activities.

1 Introduction

Buru Island, which is nestled within the Indonesian archipelago, boasts an incredibly rich and diverse coastal ecosystem. The island's mangrove forests, seagrass meadows, and corals play a vital role in supporting various species, particularly endangered sea turtles. Leatherback (*Dermochelys coriacea*), green turtle (*Chelonia mydas*), hawksbill (*Eretmochelys imbricata*), and olive ridley (*Lepidochelys olivacea*) frequent the shores of the island using its beaches for nesting and the surrounding waters for foraging. These species,

* Corresponding author: kirtan111ravidas@apps.ipb.ac.id

listed as endangered and protected under international conventions such as the CITES and IUCN Red List, are critical not only for the balance of the marine ecosystem, but also for global biodiversity. These turtles are not only essential for maintaining the balance of marine ecosystems but are also deeply embedded in the island's environmental heritage. However, these species face severe threats including habitat destruction, illegal poaching, and pollution. These life-threatening risks have escalated to critical levels, making conservation efforts more urgent than ever [1].

Despite their ecological importance, these turtles face a plethora of threats that jeopardize their survival. Human activities such as illegal poaching, where turtle eggs and body parts are harvested for consumption and local trade, have placed immense pressure on turtle populations [2]. Habitat destruction due to coastal development and local sand mining further exacerbates this issue, particularly in areas vital for nesting [3]. Additionally, the increasing presence of marine debris, primarily plastics, poses a significant danger to sea turtles, often leading to injuries, entanglement, and even death [4]. Conservation efforts on Buru Island have been implemented over the years, but their success largely depends on the active participation of the local communities [5]. The local community can serve as stewards for the environment, protecting sea turtle populations by adopting sustainable fishing practices, preventing habitat destruction, and actively reporting illegal poaching activities [6]. Empowering local people with the knowledge and resources to participate in conservation not only aids in protecting these endangered species but also fosters a sense of collective responsibility for the island's natural heritage [7].

Success in conservation relies heavily on each component, fulfilling its role and having sufficient data. This study aimed to identify the distribution, habitat, and threats faced by sea turtle species on Buru Island. Additionally, we sought to assess the community's understanding, conservation trends, and the roles played by the government and NGOs in sea turtle conservation efforts. This study is essential for bridging the knowledge gap by assessing the community's understanding of sea turtle conservation and offering evidence-based recommendations to enhance ongoing efforts. By understanding the community's perspective and identifying gaps in knowledge, this research will provide a roadmap for strengthening conservation efforts. Practical recommendations target local authorities, law enforcement, and community leaders to improve awareness campaigns and implement more effective conservation strategies. Ultimately, the goal is to foster a community that actively participates in preserving the critical habitats of Buru Island's endangered sea turtles and safeguarding them against the life-threatening risks they face.

2 Study Design and Methods

The surveys were conducted from September 27 to October 10, 2024. Encounters with sea turtles, their activities, and threats were noted and mapped using handheld or mobile phone GPS. The surveys also included communities from six villages with high levels of human-sea turtle interactions located in two districts: Kecamatan Kepala Madan (Pasir Putih, Balpetu, and Biloro villages) and Kecamatan Fena Leisela (Wamlana, Waspait, and Waenibe villages) (Fig. 1). Respondents were chosen using random sampling methods and participated in semi-structured interviews, which provided focused evaluations while allowing flexibility in exploring in-depth community insights. Afterwards, responses were analyzed and classified into eight categories: (1) Knowledge of ETP species, (2) interaction level with sea turtles, (3) knowledge of sea turtle life cycle, (4) knowledge of nesting behavior and season, (5) knowledge of sea turtle ecosystem service, (6) knowledge of threats to sea turtles, (7) society's efforts and trends regarding conservation vs. exploitation, and (8) roles of governments/NGOs in sea turtle conservation.

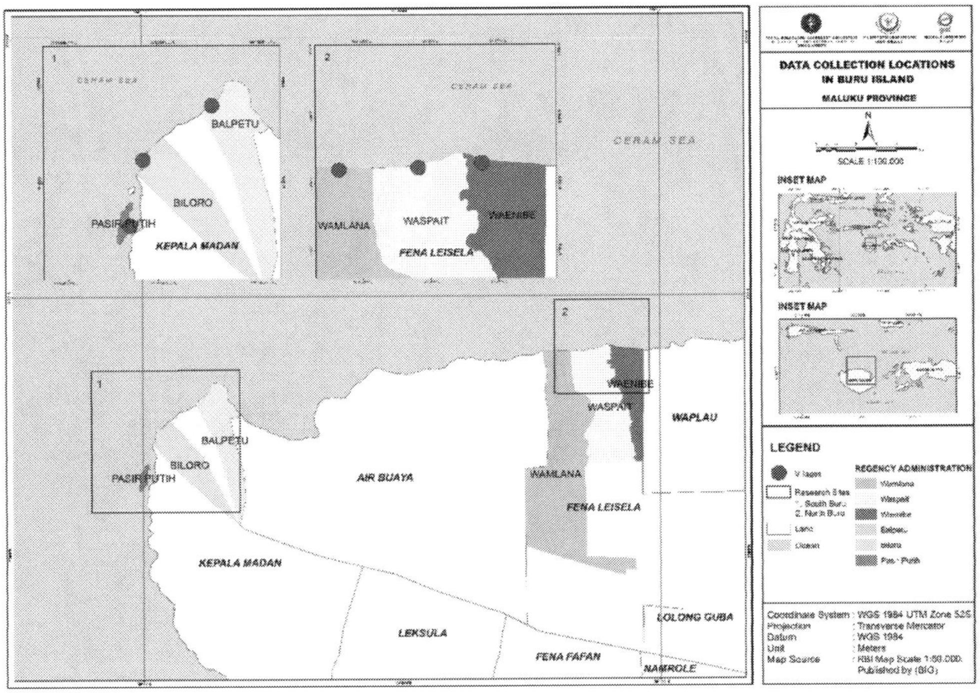

Fig. 1. Survey sites of semi-structured interviews including 6 villages in 2 districts.

Each category was then scored using a Likert scale from 1 to 5 with increasing quality with score one being not at all familiar or very poorly performed and 5 being totally familiar or excellently performed [8] (Table 1). This method helped us generate measurable and comparable data to better analyze the community's engagement with these important species.

Table 1. assessment scale and criteria class used in this research.

Criteria		Response Set
Familiarity	Quality	
Totally Familiar	Excellent	5
Very Familiar	Good	4
Familiar	Fair	3
Slightly Familiar	Poor	2
Not al all Familiar	Very Poor	1

The identified threats are then tabulated into an ordinal matrix. The rows represent villages, and the columns represent different types of threats. The prevalence of threats was classified into three levels: absent (abt), rare (rar), and common (com). This classification is based on direct observations and local information, covering threats that have occurred in the past year.

3 Results and Discussion

3.1 Respondent Demographics

3.1.1 Fena Leisela District

The demographic data of respondents in the Fena Leisela District are illustrated in Fig. 2, which includes several categories: age range, occupation, gender, and educational background.

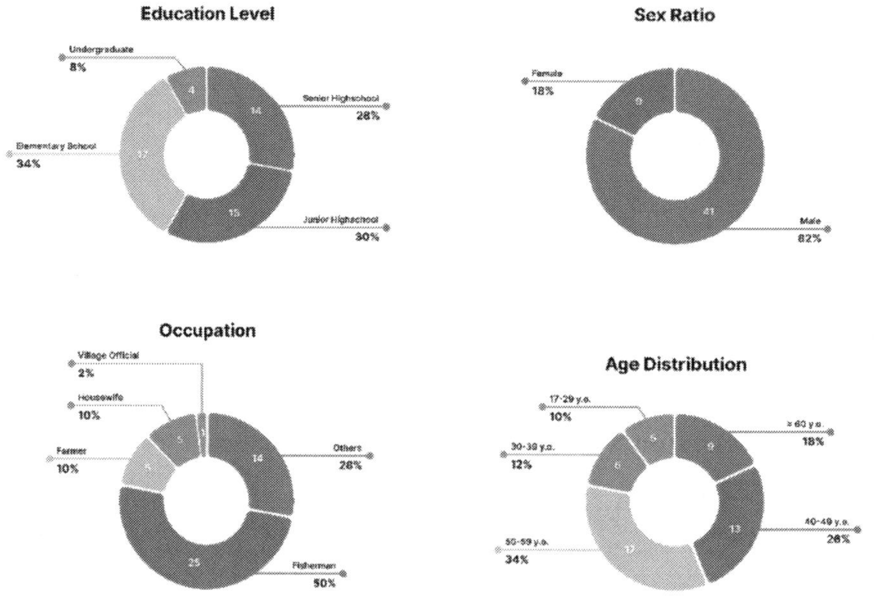

Fig. 2. Demographics of respondents in Fena Leisela District.

A total of 50 respondents participated in the interviews, consisting of 9 women and 41 men. Among these respondents, the majority (17 individuals) were within the 50-59 age range. This was followed by 13 respondents aged 40-49. Additionally, nine respondents were over the age of 60, six were between 30-39 years old, and five were between 17-29 years old.

In terms of employment, residents of Fena Leisela engage in various occupations, including fishermen, farmers, laborers, and traders. The local economy is primarily supported by fishermen (50%) and farmers (10%), attributable to the district's coastal location and fertile agricultural land suitable for diverse crops. The remaining respondents were homemakers and enumerators (five individuals each), with four teachers, two laborers, a village official, a police officer, a merchant, and a pastor.

Educational background data from 50 respondents showed significant variation. Seventeen respondents (34%) had completed primary school (SD) as their highest level of education. This was followed by 15 respondents (30%) who had completed junior high school (SMP). Additionally, 14 respondents (28%) had completed high school (SMA). Only four respondents (8%) obtained a university degree. These data highlight that the majority of respondents in the district had limited educational attainment, with most having only completed primary or junior high school, while the number of university graduates remained notably low.

3.1.2 Kepala Madan District

Fig. 3 illustrates the demographics of respondents in the Kepala Madan District, categorized by age range, gender, educational background, and occupation.

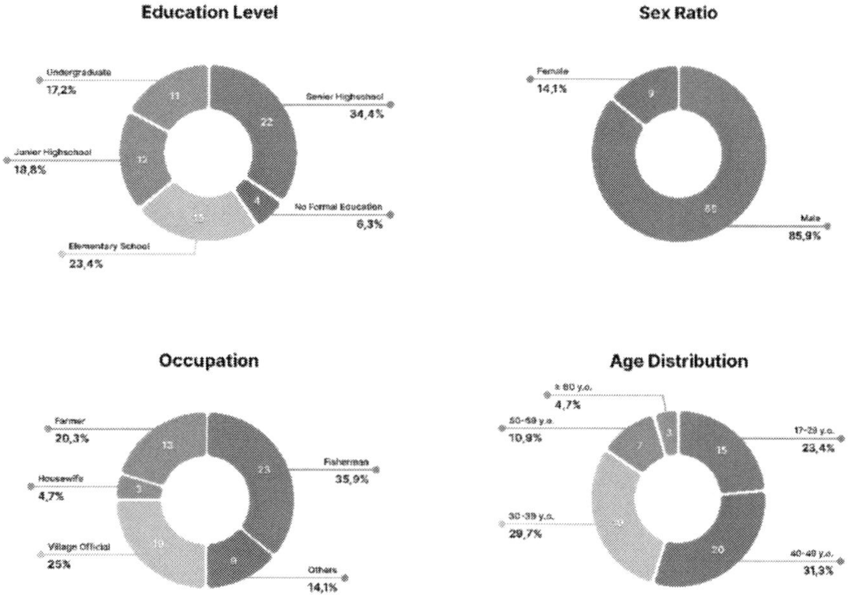

Fig. 3. Demography of respondents in Kepala Madan District.

Of the 64 respondents, 55 were male and 9 were female. The largest age group was 40-49 years, comprising 20 respondents, followed by the 30-39 age group, with 19 respondents. The 17-29 age group included 15 respondents, while the remaining 10 were over 50 years old.

More than one-third of the respondents in Kepala Madan District had completed high school (SMA/SMK), with an additional 17% having pursued higher education (associate 'sor bachelor's degrees). Notably, 13 out of 22 respondents with a high school-level education were from Pasir Putih Village, where the presence of a vocational high school contributed to this statistic. Pasir Putih is an island separated from mainland Buru Island. Those who completed high school or higher education often worked as village officials, military personnel (TNI), or police (POLRI), with most being under 40 years of age. Conversely, two-thirds of the respondents over 40 years old had an educational level equivalent to junior high school or lower, with 10% having never attended or completed formal education. This trend indicates a significant increase in educational attainment in Kepala Madan in recent decades.

The main occupations in the Kepala Madan District include fishermen (36%) and farmers (20%). Those who owned boats or private farmlands generally had more stable incomes. Those who do not own such assets often engage in temporary jobs to sustain their daily needs, work as non-permanent crew members (ABK), or assist farmers during harvests. Employment in Kepala Madan is dynamic and influenced by available opportunities and environmental conditions. For instance, fishermen who could no longer continue their work due to physical or economic reasons would often transition to farming, while some older residents became fishermen later in life. Other occupations in the district include traders, village officials, military and police personnel, teachers, boat builders, boat engine mechanics, laborers, and NGO employees. This dynamic work environment fosters open interactions among community members from different backgrounds, contributing to a strong sense of unity.

3.2 Sea Turtles Findings

The distribution map below (Fig. 4) illustrates the distribution of sea turtles across Buru Island, specifically highlighting two districts, Kepala Madan (Balpetu, Biloro, and Pasir Putih) and Fena Leisela (Wamlana, Waspait, and Waenibe).

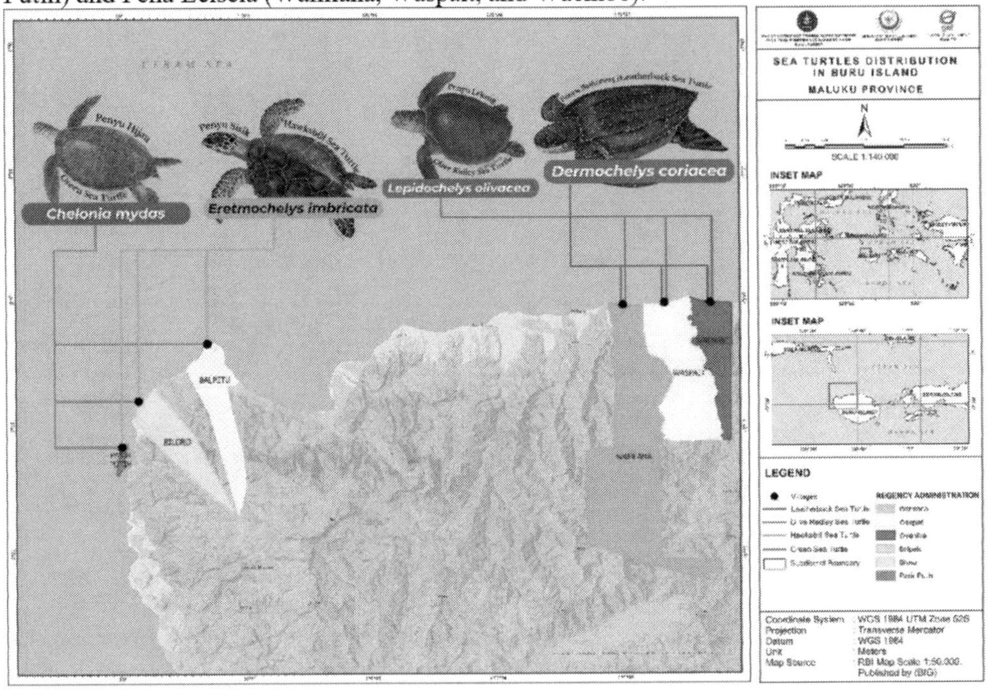

Fig. 4. Sea turtles distribution based on encounter and local information.

In Kepala Madan, species such as the green turtle (*Chelonia mydas*) and the hawksbill sea turtle (*Eretmochelys imbricata*) are observed to utilize the coastal zones both for nesting and foraging in the surrounding shallow waters, indicating a dual role of this area as a critical habitat. This dual usage underscores the importance of coastal and marine ecosystems in supporting these endangered species in Kepala Madan.

Conversely, leatherback turtles (*Dermochelys coriacea*) and olive-ridley turtles (*Lepidochelys olivacea*) have been documented landing on beaches exclusively for nesting purposes. The specific areas where these turtles forage remain undetermined, leaving an important gap in understanding their habitat utilization in this region. The map thus underscores distinct ecological functions in each district, with Kepala Madan providing comprehensive support for both nesting and foraging, whereas Fena Leisela primarily supports nesting activities. This distinction may inform future conservation and research efforts to better protect the vital sea turtle populations and their habitats.

3.3 Prevailing Local Threats

Threats to sea turtles are among the primary concerns of conservation efforts. They arise from direct and indirect causes as well as natural and anthropogenic factors (Table 2). Natural predation by adult nesting turtles has occurred in Fena Leisela. Nesting adults are preyed upon by crocodiles, as they are slow and tired after many hours of nesting. Some turtles ended up being wounded, crippled, and even killed during nesting (Fig. 5b). Turtle eggs are usually preyed upon by monitor lizards (Fig. 5a), sand crabs, stray cats and dogs, and ants.

Fig. 5. Threats to sea turtle based on observation and local information: (a)Predation of eggs, (b) predation of nesting turtle, (c) sand mining, (d) marine debris, (e) egg poaching, (f) turtle hunting, (g) beach obstruction by logging.

The most prevalent threats stem from indirect causes, such as sand mining (Fig. 5c) and plastic debris (Fig. 5d), which were observed at all surveyed sites. Additionally, the conversion of nesting habitats for logging activities obstructs sea turtles from nesting on the beaches of Fena Leisela, particularly in Wamlana Village (Fig. 5g).

Table 2. Matrix of threats to sea turtles for the past year.

Threat / Village	Predation of Eggs	Predation of Turtles	Sand Mining	Marine Debris	Egg Poaching	Turtle Hunting	Entangle-ment	Beach Obstruction
Wamlana	com*	rar**	com	com	rar	abt***	abt	com
Waspait	com	rar	com	com	rar	abt	abt	rar
Waenibe	com	rar	rar	com	rar	abt	abt	rar
Balpetu	com	abt	rar	rar	com	com	abt	abt
Biloro	com	abt	com	rar	com	rar	abt	rar
Pasir Putih	com	abt	com	rar	com	com	rar	abt

*common
** rare
*** absent

The poaching of eggs (Fig. 5e), meat and body parts were common in both districts, although it was largely diminished in Fena Leisela. In contrast, such activities remain significant in Kepala Madan, where eggs are collected for personal consumption or traded in local villages. Although turtle meat is not a common delicacy, many respondents reported consuming it. Perhaps the most concerning practice is the collection of turtle carapace scales (Fig. 5f), which are used as materials for accessories, such as bracelets and rings. Even when turtles are released alive afterwards, they often sustain permanent injuries that severely diminish their chances of survival in the wild. Bycatch by fishermen rarely occurs, as most of the fishing methods used are handlines (<1GT) and medium-scale purse seines (<30 GT). Even when turtles were captured in the purse seine, fatalities almost never occurred, as there were fishermen inside the water to monitor any non-target catch that could potentially tear the net.

3.4 Community understanding

The local community's knowledge and understanding of Endangered, Threatened, and Protected (ETP) species in the Fena Leisela and Kepala Madan sub-districts show significant differences. The community in Fena Leisela demonstrated a higher level of understanding, with an average score of 2.92, compared to Kepala Madan, which scored 2.58 (Fig. 6).

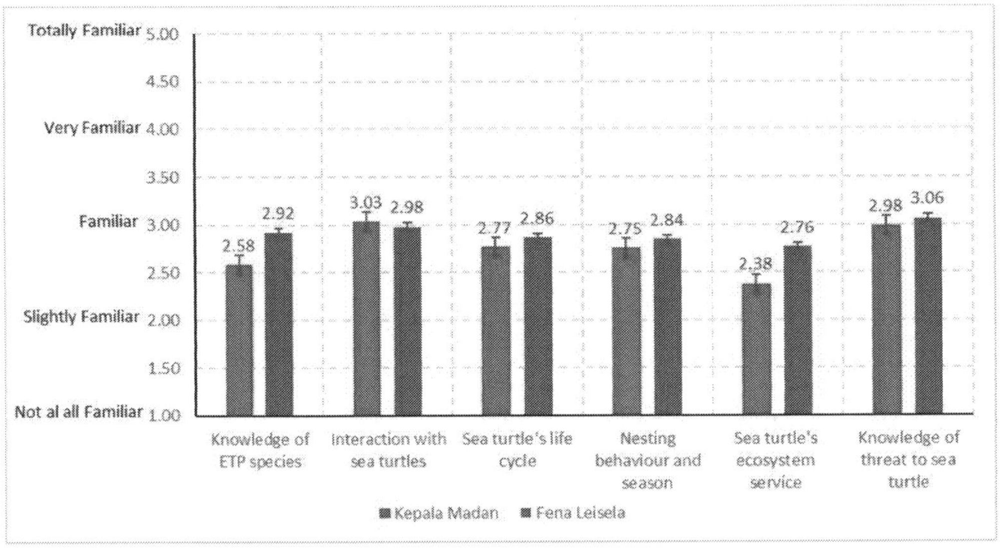

Fig. 6. Comparison of community understanding on various sea turtle's aspect of life.

This disparity is likely due to Fena Leisela's better access to conservation information and more frequent participation of its residents in educational and environmental conservation activities. As an endangered species, sea turtles in Indonesia are protected by Law No. 5 of 1990 on the Conservation of Biological Resources and Ecosystems, and Law No. 31 of 2014 on Fisheries. Internationally, sea turtles are listed in Appendix 1 of the CITES (Convention on International Trade in Endangered Species (CITES) (International Union for Conservation of Nature (IUCN) Red List (2007); they are classified as critically endangered and cannot be traded in any form.

The local communities' knowledge of sea turtles in both sub-districts showed similar trends, but the community in Kepala Madan had more frequent interactions with sea turtles, with an average score of 3.03. This is because the coastline of Kepala Madan serves as a feeding ground and nesting habitat for green turtles (Chelonia mydas) and hawksbill turtles (Eretmochelys imbrica). The people of Kepala Madan primarily use small fiber boats for transportation because of limited land access. They generally encounter sea turtles when heading into the sea for fishing or work. These encounters usually involve turtles in their juvenile to adult stages and occur more than five times per minute. Some fishermen have reported occasional sightings of leatherback turtles in the waters of Kepala Madan, although these encounters are rare and typically occur 20-30 miles offshore.

In contrast, the community in Fena Leisela interacted less frequently with sea turtles, resulting in a lower interaction score (2.98). This is because communities tend to rely on land-based transportation. Additionally, the coastline of Fena Leisela borders deeper waters with a substrate of volcanic rock and gravel, which does not support the growth of seagrass and coral reefs, key habitats for sea turtles to feed. Encounters with sea turtles are typically limited to those that come ashore to nests. Local residents often build homes on beaches, leading to frequent reports of turtle nesting. Species that nest on the beaches of Fena Leisela include leatherback turtles (*Dermochelys coriacea*), olive ridley turtles (*Lepidochelys olivacea*), and green turtles (*Chelonia mydas*). Both communities were able to distinguish between green turtles, hawksbill turtles, and leatherback turtles based on the shape of their carapace and head. In the local language, green turtles are called *teteruga ikan*, whereas hawksbill and leatherback turtles are referred to as *teteruga sisir* and *teteruga salawaku*, respectively.

Regarding the life cycle of sea turtles, both communities show similar levels of understanding, with average scores of 2.77 and 2.84, respectively. Most people know that sea turtles come ashore to lay eggs, and some are aware of their migratory behavior, particularly in Fena Leisela. This was confirmed by fishermen in both subdistricts who observed turtle mating while at sea. Some community members also shared stories about the hatching process from eggs to hatchlings, and witnessed predators preying on hatchlings as they made their way to the sea. Both communities demonstrated a good understanding of turtle nesting behavior and seasons, with average scores of 2.75 and 2.82. Many locals are familiar with general nesting locations, although those directly involved in egg collection, such as enumerators, have a more precise understanding of the exact nesting sites and characteristics of turtle nests.

According to local reports and data from the Pokmaswas (Community-Based Conservation Group) in 2021, the peak nesting season occurs from March to April for Olive Ridley turtles and from June to July for leatherback turtles. However, in Kepala Madan, uncertainty and inconsistency exist regarding the peak nesting season. Most community members are unsure of the timing of nesting, with some stating that it occurs uniformly throughout the year. Others suggested specific months, but their answers varied. Only enumerators and those involved in egg collection can describe the nesting process in detail, including the stages of emergence, nest digging, egg-laying, nest covering, and return of turtles to the sea. The general public also struggles to distinguish between the tracks and nests of different turtle species, or to identify specific nesting behaviors. Only two out of 64 respondents in Kepala Madan described the differences in the nesting processes and characteristics of tracks and nests for each species. Generally, people report that turtles tend to nest during the full moon and when the tide is high. This is predicted to minimize the risk of predation and provide better visibility and access to the nesting sites [9].

Sea turtles play an essential role in providing ecosystem services and offer benefits to communities that rely on marine and coastal ecosystems [10]. There are four primary categories of ecosystem services: provisioning, regulating, cultural, and supporting. Based on the final graph regarding sea turtle ecosystem services, the community in Fena Leisela has a higher level of knowledge than the community in Kepala Madan, with average scores of 2.38 and 2.73, respectively. People in both districts are generally aware that sea turtles provide food sources, such as eggs, meat, and materials for jewelry. However, people in Kepala Madan are mostly familiar with the turtle's diet, including algae, seagrass, sponges, and jellyfish, and most are unaware of the regulatory and supporting services that sea turtles provide to the ecosystem. In contrast, more people in Fena Leisela can explain the important role that sea turtles play in the ecosystem. This difference is likely due to more intensive educational and outreach activities focused on turtle conservation in Fena Leisela than in Kepala Madan. However, both communities are unaware of the cultural services provided by sea turtles, as there are no local traditions or rituals involving turtles, and sea turtles on Buru Island are not widely recognized as tourist attractions. Only a few external researchers, particularly those working with WWF, are aware of the presence of turtle species on Buru Island.

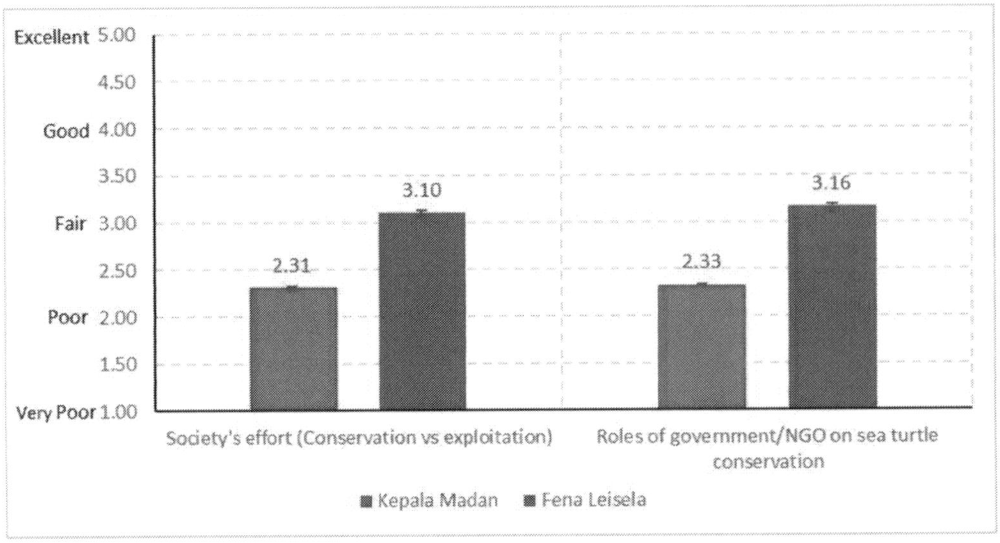

Fig. 7. Comparison of society's effort and trends toward conservation, roles of organization on sea turtle conservation.

Fig.7 presents a detailed comparison of the two major aspects of sea turtle conservation between the Kepala Madan and Fena Leisela sub-districts. In the *Society's Effort* aspect, which compares awareness of conservation efforts versus tendencies toward exploitation, Fena Leisela scored an average of 3.10, indicating a relatively good level of understanding. By contrast, Kepala Madan scored only 2.31, indicating poor awareness. This difference suggests that the community in Fena Leisela has a stronger understanding of and motivation to engage in sea turtle conservation efforts. They appeared to value turtle protection more highly, likely because of their involvement in frequent educational programs and environmental campaigns that emphasize the importance of conserving endangered species. Educating the community on the importance of turtle conservation can prevent the exploitation of turtles for commercial purposes, such as the sale of eggs, meat, or shells, and help protect turtle habitats from degradation [11].

In contrast, the lower score in Kepala Madan suggests that awareness of and action on conservation are still suboptimal and may lean more toward exploitation. This could be due to limited access to information or lack of active conservation programs involving the community. Furthermore, differences in educational support and involvement of governmental or NGO initiatives may also contribute to the contrasting perceptions and behaviors between the two subdistricts. The data indicate that the people of Fena Leisela are not only more active in supporting sea turtle conservation efforts but also have a stronger awareness of the threats facing these species compared to the community in Kepala Madan, which appears to be more inclined toward exploitation.

Regarding the *Roles of Government/NGO in Sea Turtle Conservation*, Fena Leisela scored an average of 3.16, while Kepala Madan scored only 2.33. This indicates that the community in Fena Leisela perceives the role of the government and NGOs in turtle conservation to be more effective and impactful than those in Kepala Madan. This difference is likely due to the better access to conservation programs facilitated by the government and NGOs in Fena Leisela, including educational campaigns, training on the importance of sea turtle ecosystems, and legal enforcement to protect turtle habitats. Sustained conservation programs tend to increase community involvement and understanding of the importance of turtle protection, emphasizing the effectiveness of collaboration between the government, local communities, and NGOs in achieving successful conservation outcomes [12]. By contrast, the lack of collaborative efforts in Kepala Madan may cause the community to feel that the role of the government and NGOs is less visible or effective.

We would like to extend our heartfelt gratitude to CFI Indonesia, the GEF 6 Grant, WWF, the GEF Agency, and the Ministry of Marine Affairs and Fisheries for providing us with an invaluable opportunity to collaborate in this research. This work is not only the result of our HIMITEKA-IX expedition team effort, but is made possible by the generous support and collaboration of these esteemed organizations, reflecting their commitment to advancing marine conservation. Thank you for empowering us and believing in our missions.

References

1. D. Willer, K. Smith, D. Aldridge, Front. Ecol. Evol. 7, 297 (2019).
2. J. Joseph, H. Nishizawa, J. Alin, R. Othman, G. Jolis, I. Isnain, J. Nais, Glob. Ecol. Conserv. 19, e00586 (2019).
3. N. Seitz, D. vanEngelsdorp, S. Leonhardt, Glob. Ecol. Conserv. 19, e00642 (2019).
4. S. Nelms, E. Duncan, A. Broderick, T. Galloway, M. Godfrey, M. Hamann, P. Lindeque, B. Godley, Ices J. Mar. Sci. 73, 165-181 (2016).
5. S. Rasoolimanesh, M. Jaafar, Tourism Develop. Conserv. Program, 62293 (2016).
6. A. Mancini, J. Senko, R. Borquez-Reyes, J. Poo, J. Seminoff, V. Koch, Hum. Ecol. 39, 743-756 (2011).
7. N. Dawson, B. Coolsaet, E. Sterling, R. Loveridge, N. Gross-Camp, S. Wongbusarakum, K. Sangha, L. Scherl, H. Phan, N. Zafra-Calvo, W. Lavey, P. Byakagaba, C. Idrobo, A. Chenet, N. Bennett, S. Mansourian, F. Rosado-May, Ecol. Soc. 26, 319 (2021).
8. B. Tanujaya, R. Prahmana, J. Mumu, FWU J. Soc. Sci. 16, 89-101 (2023).
9. T. Escalona, N. Valenzuela, D.C. Adams, Diversity 11, 78 (2019).
10. E. Patel, M. Kotera, A. Phillott, Indian Ocean Turtle Newsl. 36, (2022).
11. T.Z.A.E. Harnino, I.N.Y. Parawangsa, L.A. Sari, S. Arsad, J. Mar. Coast. Sci. 10, 242-255 (2021).
12. E. Miller, G. Meindl, Binghamton Univ. Undergrad. J. 5, 8 (2019).

A snapshot on the risk indicators of fisher's income and the fish size: mini purse seine fishery in Bitung waters

Tirtadanu[1*], *Moh* Fauzi[1], *Fayakun* Satria[1], *Duranta* D. Kembaren[1], Prihatiningsih[1], *Erfind* Nurdin[1], *Lilis* Sadiyah, *Yayan* Hernuryadin[2], *Tegoeh* Noegroho, *Umi* Chodrijah

[1]Research Center for Fishery, National Research and Innovation Agency (BRIN), Address Jalan Raya Bogor km. 47, Nanggewer Mekar, Cibinong 16915, Indonesia
[2]Directorate of Fish Resource Management, Directorate General of Capture Fisheries, Ministry of Marine Affairs and Fisheries of the Republic of Indonesia. Mina Bahari II Building 14th floor, Jalan Medan Merdeka Timur No.16, Jakarta Pusat, Indonesia.

Abstract. A high reliance on fishermen in the MPS fishery could threaten their economic sustainability. Therefore, a snapshot of the fisher's income can be useful as an indicator to understand the feasibility of the fishery. This study aimed to investigate the risk indicators of fishers' income and fish size in the MPS fishery in the southern waters of Bitung. This study was conducted in October 2024 using 30 samples of MPS fishing trips and 12 units of 10 to 28 GT MPS vessels. This study applied a risk indicator, ranging from 0 to 3, based on an ecosystem-based fisheries approach. The results showed that The dominant target species of MPS were mackerel scad (*Decapterus* spp.), Indian mackerel (*Rastrelliger kanagurta*), and bigeye scad (*Selar crumenopthalmus*), with catch per unit effort (CPUE) of 146 kg vessel^{-1} trip^{-1}, 115 kg vessel^{-1} trip^{-1} and 31 kg vessel^{-1} trip^{-1}, respectively. The monthly income of fishermen is still less than the regional minimum wage, with a risk indicator of 2.67 (red zone). The risk indicator for fish size was 1.40, which was in the yellow zone. The risk indicators of fishermen's income, which are currently in the red zone, should be reduced to enhance their economic sustainability.

1 Introduction

The Mini-Purse Seine (MPS) fishery has predominantly contributed to the landing of small pelagic fish in Bitung. The total catch of the MPS fishery has reached 37,379 tons, accounting for 69% of the fish production in Bitung [1]. The production of pelagic fisheries has been strengthened by significant market demand and the growth of the fish processing industry. This fishery business has developed into an export scope, resulting in intensive exploitation of the small pelagic population [2]. Moreover, these factors have resulted in significant economic reliance on the MPS fishery. The dynamics of the MPS fishery have significantly

* Corresponding author: tirtadanu91@gmail.com

impacted the livelihoods of fishers and could threaten their economic sustainability. Therefore, a risk assessment of the fishers' income is essential to develop a management strategy to enhance the economic aspects of local fishers in Bitung.

The MPS fishery in Bitung is characterized by 10 to 28 GT fishing vessels on a one-day fishing trip, with a mesh size from 1 to 2 inches of cod ends, and there are about 54 units of MPS vessels in Bitung, Some of the previous studies focused on investigating the target species, including the fishing season of mackerel scad (*Decapterus* spp.), which was found from May to July, the spawning season of bigeye scad (*Selar crumenopthalmus*), which was found from October to November, and a morphometric study on the bigeye scad [3-5]. A study concerning the economic indicators of the MPS fishery in Bitung has not yet been reported. Moreover, the exploitation status of mackerel scad was reported to have been overfished in 2017 [6]. Consequently, continued fishing efforts under current conditions may threaten economic sustainability.

This study investigated risk indicators for fishers' income and fish size in terms of MPS fisheries in the southern waters of Bitung by applying an ecosystem-based assessment method [7]. The risk indicator for fishers' income was assessed quantitatively by comparing the current condition with the regional minimum wage. Quantitative risk indicators for fish size were derived by comparing length-frequency data to the length at first maturity (Lm_{50}). This method has not yet been used in previous studies. Hence, it is expected to represent a novel approach to ecosystem-based assessment methodology in Indonesia.

2 Materials and methods

2.1 Data collection

This study was conducted in October 2024 by collecting catch rate and catch composition data from 30 landings of the Mini-Purse Seine (MPS) fishery and 12 units of MPS vessels in Bitung. This study focused on the MPS fishery, which operates in the eastern and southern waters of Bitung, part of the Fisheries Management Area (FMA) 715 (Figure 1). The MPS fishery sample consisted of 12 unit of 10 to 28 GT fishing vessels, operated by 15 crews in each vessel, a one-day fishing trip, and a mesh size of 1–2 inches of cod ends. Information on fish prices, total cost per trip, and the number of trips per month was collected through interviews with fishers. Fish size was measured directly from bigeye scad (*Selar crumenopthalmus*) samples captured using MPS. The fishing ground of MPS fishery was in the eastren and southern part of Bitung, which is part of Maluku Sea and part of Fisheries Management Area 715 (Figure 1). The length at first maturity (Lm_{50}) of *S. crumenopthalmus* was obtained from previous studies [6,8]

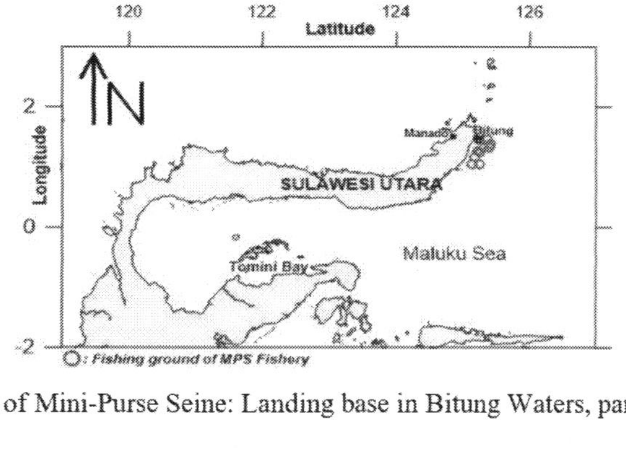

Fig. 1. Fishing ground of Mini-Purse Seine: Landing base in Bitung Waters, part of fisheries management area 715.

2.2 Data analysis

The catch composition of the MPS fishery was estimated by the proportion of the catch in the weight of each fish species within the total catch per landing. Each vessel's catch weight per landing was applied as the catch per unit effort, with units of kg vessel^{-1} trip^{-1}. Total revenue was determined by calculating the catch per unit effort (CPUE) (kg vessel^{-1} trip^{-1}) for each species and multiplying that amount by the price of fish per kg. The cost of *S. crumenopthalmus*, *R. kanagurta*, and *Decapterus* spp. was 15,000 IDR/kg. The profit per trip was calculated by subtracting the total cost from total revenue. Monthly profit was calculated by multiplying the profit generated from each trip by the total number of trips conducted within a month. Ten percent of the monthly profit was allocated to boat maintenance. The monthly income of fishers was calculated based on 90 percent of the monthly profit divided by the total number of crew members, which was 15 crews.

The risk indicator was followed by the equation [7] :

$$RS_i = \left(\frac{I_{target} - I_i}{I_{taget} - I_{limit}}\right) + 1$$

Where RS_i is the risk score for indicator i, I_{limit} and I_{target} are the limits, and is the target of indicator i. The risk indicator for fishers' income (RS_{fi}) was estimated using the Bitung regional minimum wage (3.5 million IDR) as the limit indicator and West's Java minimum regional wage as the largest minimum regional wage in Indonesia (5.3 million IDR) as the target indicator. The risk score (RS) for fish size (RS_{fs}) was estimated using 50% small fish in catch as the limit indicator and 25% small fish in catch as the target indicator. Small-sized fish were categorized as those smaller than the Lm_{50} value.

3 Results and discussion

Based on 30 fishing trips with a cumulative catch of 9,545 kg, the catch composition of the MPS fishery consisted of scad (*Decapterus* spp.) 46.0%, Indian mackerel (*Rastrelliger kanagurta*) 36.14%, bigeye scad (*Selar crumenopthalmus*) 9.85%, squid (*Loligo* spp.) 0.9%, and rainbow runner (*Elagatis bipinnulata*) 0.5% (Figure 2). The mean catch per unit effort (CPUE) of the MPS was 146 kg vessel^{-1} trip^{-1} for *Decapterus* spp., 115 kg vessel^{-1} trip^{-1} for *R. kanagurta*, 31 kg vessel^{-1} trip^{-1} for *S. crumenopthalmus*, 3 kg vessel^{-1} trip^{-1} for Loligo spp., and 2 kg vessel^{-1} trip^{-1} for *E. bipinnulata*. A previous study identified mackerel scad (*Decapterus macarellus*) and bigeye scad (*Selar crumenopthalmus*) as the predominant species captured by MPS in Tomini Bay, indicating a similarity in the MPS species composition within the waters of FMA 715 [9].

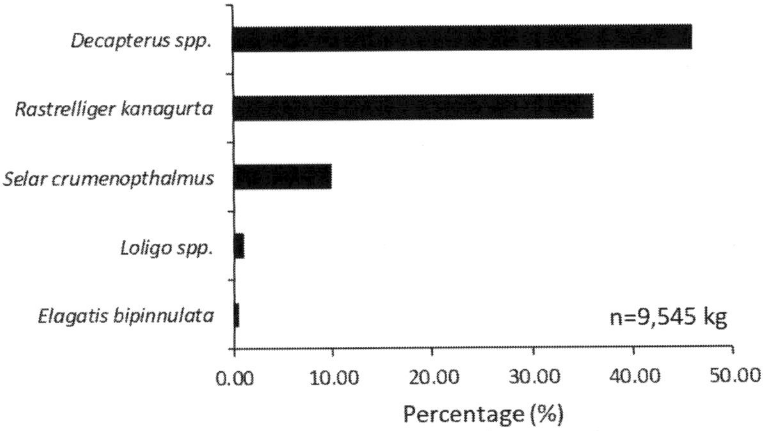

Fig. 2. Catch composition of Mini-Purse Seine (MPS) fishery in the southern waters of Bitung,

The total catch of MPS in the southern waters of Bitung varied between 50 and 1,200 kg vessel^{-1} trip^{-1}, with an average of 313.5 ± 260.9 (Mean ± standard deviation) (Figure 3). The mean CPUE or MPS in the Java Sea was recorded at 2.84 tonnes per trip, driven by broadened fishing durations ranging from 5 to 10 days per trip [10]. The total revenue per trip for MPS in Bitung was 4,702,500 IDR. Total revenue was estimated from the total catch of *Decapterus* spp., *S. crumenopthalmus*, and *R. kanagurta*, with each price at approximately 15,000 IDR per kg. According to interviews with fishermen, the mean total cost of each trip was 2.8 million IDR. Among the 30 fishing trips, approximately 30% encountered losses, where the total costs surpassed the total revenue (Figure 4).

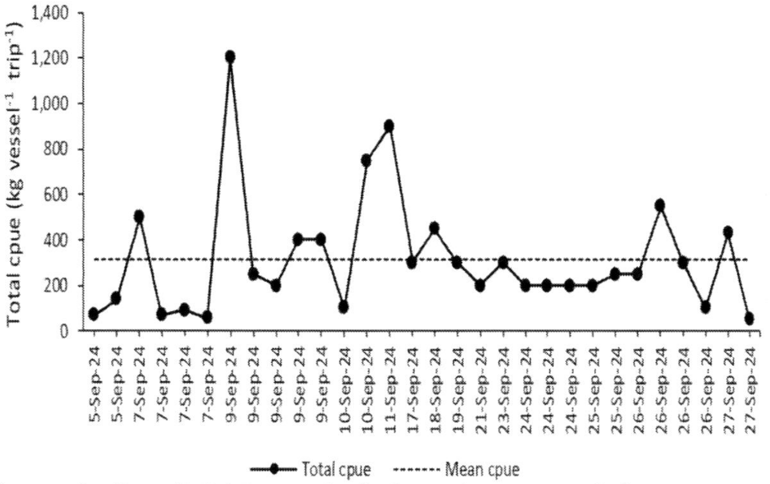

Fig. 3. Catch per unit effort of Mini-Purse Seine in the southern waters of Bitung.

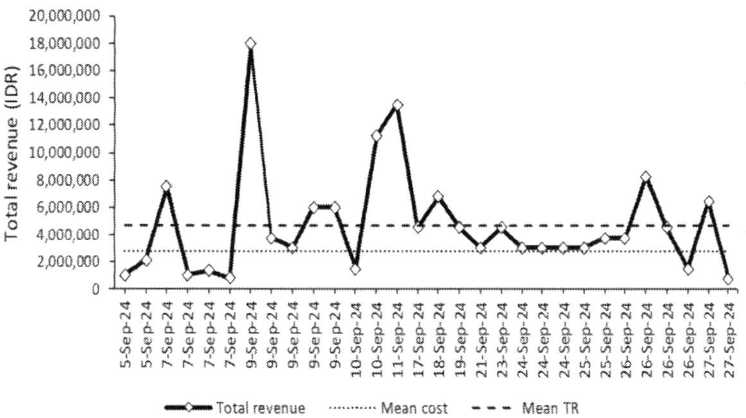

Fig. 4. Total revenue and mean cost for Mini-Purse Seine fishery in the southern waters of Bitung.

The average total revenue was 4,702,500 IDR, whereas the total cost per trip was 2,800,000 IDR, resulting in a profit of 1,902,500 IDR for the MPS fishery. In a month, fishing trips were performed 20 times, resulting in a monthly profit of 38,050,000 IDR per vessel. The current monthly profit of the MPS fishery in Bitung is 37% lower than that of the MPS fishery in Kupang Bay [11]. By applying a deduction of 10% monthly from the total profit, the income for each crew member amounted to 2,283,000 IDR per month. The Bitung regional minimum wage is set at 3.5 million IDR, which is applied as the limit indicator, while 5.3 million IDR is applied as the target indicator of fishers' income. The current risk score (RS) for fishers' income was 2.67, indicating that fishers' income was in a high-risk condition (red zone). A high-risk indicator for fishers' income was also found in the lobster fishery in Gunungkidul waters, caused by a decline in lobster stocks [12]. The risk indicator can be controlled to the green zone if the CPUE rises from 313.5 kg vessel^{-1} trip^{-1} to 478.2 kg vessel^{-1} trip^{-1}, representing a 52% growth in the catch rate from existing conditions. Income from alternative jobs may serve as valuable support for the economic stability of fishermen in Bitung.

The size of bigeye scad (*Selar crumenopthalmus*) in Bitung waters ranged from 12 to 26 cm FL, with a mean size of 21.62 ± 3.46 (mean ± standard deviation) (Figure 5). The length at first maturity (Lm_{50}) was 18 cm, according to previous findings of [6,8]. By comparing the length-frequency data to their length at first maturity (Lm_{50}), approximately 35% of the catch was categorized as small fish or below the Lm_{50} value. Considering stock sustainability, it is advisable to allow reproduction of the fish population at least once during their life [13]. The limit indicator was set at a small catch size of 50 catch, whereas the target indicator was set at 25% of the catch. The current risk indicator for fish size was 1.40, indicating a moderate risk score (yellow zone) (Figure 6).

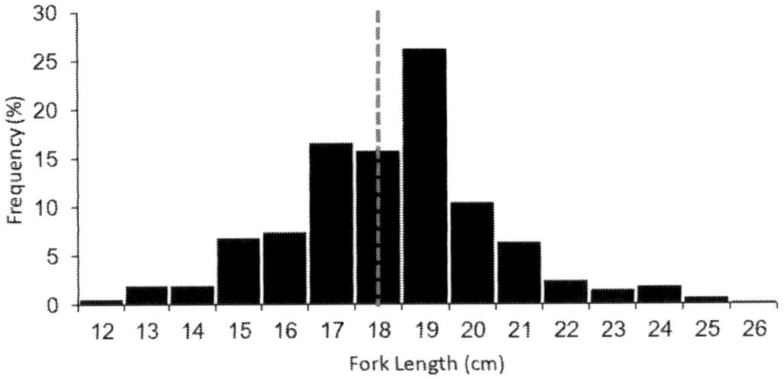

Fig. 5. Length frequency and length at first maturity (Lm_{50}) of bigeye scad (*Selar crumenopthalmus*) in Bitung Waters.

Considering the risk indicators in the case of MPS fisheries in Bitung, the income of fishers has fallen into a state of concern, highlighting the necessity for government intervention to ensure economic stability. The primary factor that may contribute to this situation is the relatively low catch rate of MPS fisheries. The decline in fish biomass due to overfishing can increase the risk to fishermen's income. Additionally, a previous study identified an overfishing status for the bigeye scad (*S. crumenopthalmus*) fishery in Bitung based on data from 2017 [6]. The risk score for fish size was moderate, indicating that intensive fishing of the small pelagic fishery primarily caused the low catch rate. The fishing quota for small pelagic areas should be limited and complemented by alternative job initiatives to ensure the economic stability of the local community.

Fig. 6. Risk score (RS) for fisher's income (RS_{fi}) and fish size (RS_{fs}): Mini Purse Seine Fishery in the southern waters of Bitung.

Conclusion

The three predominant species captured by the Mini-Purse Seine (MPS) in the southern waters of Bitung were scad (*Decapterus* spp.), Indian mackerel (*Rastrelliger kanagurta*), and bigeye scad (*Selar crumenopthalmus*), with catch per unit effort found at 146 kg vessel^{-1} trip^{-1}, 115 kg vessel^{-1} trip^{-1}, and 31 kg vessel^{-1} trip^{-1}, respectively. The monthly fisher income of the MPS fishery remains below the regional minimum wage of Bitung, with the risk score in the concerned state (red zone). The current risk indicator for fish size is moderate (yellow zone). The strong reliance of fishers on the MPS fishery, linked to the high-risk score of their income, poses a significant threat to their economic sustainability. Therefore, it is essential to explore alternative strategies to enhance the economic sustainability of MPS fishermen in Bitung.

Acknowledgment

The authors acknowledge the support provided by the Global Environment Facility (GEF-6) Coastal Fisheries Initiative (CFI) Indonesia, and the Ministry of Marine Affairs and Fisheries Republic of Indonesia for funding this research.

References

1. MMAF Ministry of Marine Affairs and Fisheries (MMAF) of the Republic of Indonesia (2023)
2. F.D. Hariyoto SEMNAS KP Politeknik Kelautan dan Perikanan Bitung 308-316 (2023)
3. R. Saranga, H. Santoso, N. Tumanduk, N, H. Ondang, J. Manohas J Jurnal Bluefin Fisheries **1** 1-23 (2019)
4. A. Zamroni, H. Widiyastuti, A. Kuswoyo BAWAL **11** 113-126 (2019)
5. M.Z. Arifin, M.K. Mujiyono, Karyanto, L. Katili SEMNAS KP Politeknik Kelautan dan Perikanan Bitung, 34-38 (2022)
6. R. Saranga, Asia, J. Manengkey M.Z. Arifin Bulltetin Matric **15** 11-22 (2018)
7. C. I. Zhang, S. Kim, D. Gunderson, R. Marasco, J.B. Lee, H.W. Park, J.H. Lee Fisheries Research **100** 26–41 (2009)
8. B.G. Hutubessy Ind. Fish. Res. J. **27** 99-108 (2021)
9. Sunarto, I.J. Paransa, A. Luasunaung Jurnal Ilmu dan Teknologi Perikanan Tangkap **2** 231-235 (2017)
10. H. Widiyastuti, A.R.P. Pane, S. Mardlijah, M. Fauzi, A. Zamroni IOP Conf. Series: Earth and Environmental Science **1260** 1-8 (2023)
11. W.Y.L. Saunoah, C.A. Paulus, A.L. Jurnal Bahari Papadak **3** 76-84 (2022)
12. Tirtadanu, D.D. Kembaren, A.R.P. Pane, K. Wagiyo, S. Mardlijah, Mahiswara, Suadi The Egyptian Journal of Aquatic Research **50** 141-147 (2024)
13. Tirtadanu, Prihatiningsih, H. N. Yusuf, A. Zamroni, K. Amri, U. Chodrijah Regional Studies in Marine Science **66** 1-12 (2023)

Phytoplankton communities in different season in the coastal area of Kawasi, the western part of Obi Island, North Maluku, Indonesia

Dea Fauzia Lestari[1*], *Afviya Salsabila*[1], *Niken Octaviandini*[2], *Nadya Cakasana*[1], *Riza Aitiando Pasaribu*[1], *Windy Prayogo*[3], *Retno Dewi Handayani Sujatno*[3], *Prasari Risky Hutami*[3], *Pandu Setyabud*[4], *Farel Ahadyatulakbar*[1]

[1]Department of Marine Science and Technology, Faculty of Fisheries and Marine Sciences, IPB University, Indonesia
[2]Master Student of Wageningen University and Research, The Netherlands
[3]HSE Department - Harita Nickel, Gedung Bank Panin Lantai 5, Jalan Jenderal Sudirman Kavling 1, Senayan, Jakarta Pusat 10270
[4]Nusantara Geosains Institut, Bogor, Indonesia

Abstract. Plankton communities are one of the groups monitored to assess environmental conditions. West Obi Island is known for its fisheries resources and is one of Indonesia's largest Nickel mining areas in North Maluku, Indonesia. However, few studies have discussed phytoplankton structure in this region, and this study aimed to assess the distribution and community structure in the western part of Obi Island, Halmahera Selatan. Vertical sampling was conducted to collect phytoplankton and zooplankton from 13 stations during the dry and wet seasons in 2023. We identified 46 species phytoplankton during dry season and 24 species during wet season. During the wet season, phytoplankton is more abundant at specific locations, such as ports and estuaries, and less abundant in coral reef areas. In contrast to the wet season, there was a significant increase in phytoplankton abundance across all locations during the dry season, which could relate to nitrate. *Chaetoceros* sp., *Coscinodiscus* sp., and *Rhizosolenia* sp. were the predominant genera in both seasons. During the dry season, phytoplankton were predominantly concentrated in the strait area between Obi and Mala-Mala Island. In contrast, the phytoplankton were more concentrated in the harbor and estuary areas in the wet season.

1. Introduction

Plankton, a diverse collection of microscopic organisms that float in aquatic environments, is an essential indicator of ecological balance and environmental health. These organisms are bioindicators of water quality and the ecosystem's trophic status as they respond quickly to environmental changes [1]. Phytoplankton communities act as crucial indicators of environmental shifts, as they mirror the effects of increasing nutrient concentrations and tend to respond more strongly to the interaction of multiple stressors than to any single stressor

* Corresponding author: dea.fauzia@apps.ipb.ac.id

[2,3]. Plankton can act as a sensitive barometer responding to temperature, water quality, nutrients, and other environmental conditions [4,5,6].

Phytoplankton diversity is an important indicator of how efficiently resources are used in ecosystems and is crucial for maintaining the stability of ecosystem functions across varying environmental conditions [7,8]. Phytoplankton play a crucial role in the food chain of aquatic ecosystems as primary producers. They are microscopic organisms capable of photosynthesis, producing oxygen, and serving as an energy source for various trophic levels within the aquatic ecosystem. The dynamic of environmental factors can directly affect the composition and variety of phytoplankton communities and indirectly indicate the collective influence of various habitats [9,10]. Seasons significantly influence phytoplankton dynamics via changes in light, temperature, water stratification, and nutrient availability changes. These patterns directly affect the marine food chain and aquatic ecosystems as a whole. Seasons substantially affect phytoplankton in tropical regions, although seasonal changes in these areas are not as extreme as those in subtropical or polar regions. Some key factors affecting phytoplankton dynamics in tropical regions include rainfall, which increases river flow and land runoff and carries nutrients into coastal waters. This can stimulate phytoplankton growth but may also cause water stratification due to differences in salinity. During the dry season, increased salinity and deeper light penetration can affect phytoplankton composition and abundance. Upwelling can bring nutrients from deeper layers to the surface in some areas, thereby enhancing primary productivity.

West Obi Island is part of the Obi Islands in South Halmahera Regency, North Maluku Province, Indonesia. The island is known for its rich natural resources, including fisheries, tropical rainforests, marine biodiversity, and abundant mineral resources. Obi Island is one of the most productive waters in eastern Indonesia and serves as a meeting point between two large oceans: the Pacific Ocean and the Indian Ocean. The Indonesian Throughflow (ARLINDO) passes through the northern part of Obi via the north of Bisa Island, the western part of Obi, and the southern part through the south of Gamumu Island. This makes the waters around Obi Island highly productive. However, few studies have discussed the phytoplankton structure in this region. This study focuses on the coastal area of Kawasi, which is located in the western part of Obi Island and tries to explore the comparative patterns of plankton diversity in two different seasons and relate them to habitat types, such as estuaries, coral reefs, and coastal activities, such as ports and settlements.

2. Method

Research activities were carried out in the western part of Obi Island, also known as the coastal area of Kawasi, in May and December 2023, representing the dry and wet seasons, respectively. Thirteen sampling locations represent different types of coastal areas, such as estuaries, coral reefs, ports, and settlements (Figure 1).

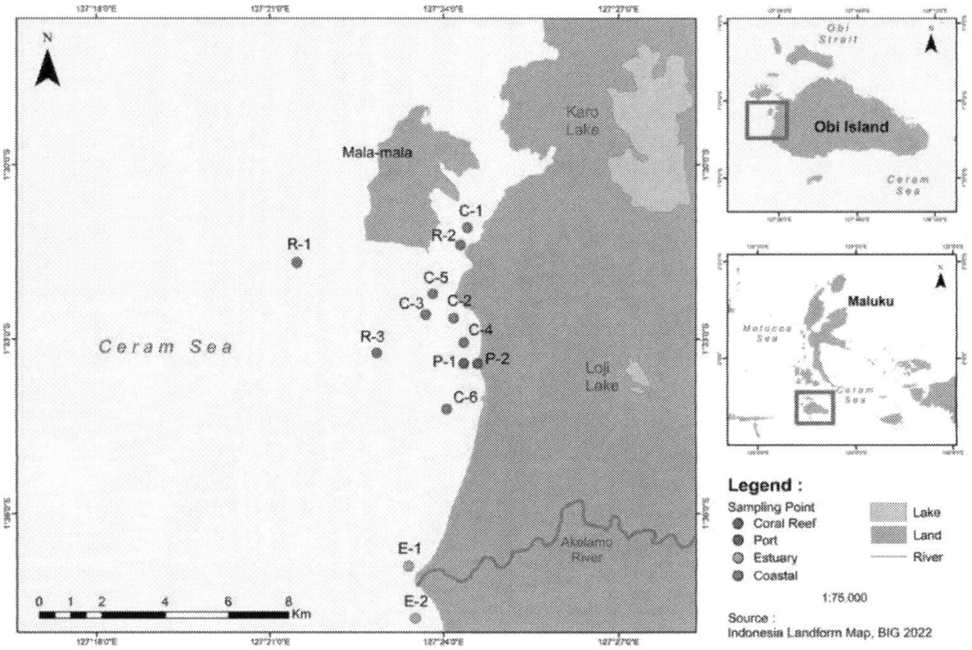

Fig. 1. Study locations of plankton communities in the coastal area of Kawasi, Western Part of Obi Island

As shown in Figure 1, Stations R-1, R-2, and R-3 represent the coral reef areas, P-1 and P-2 represent port areas, E-1 and E-2 represent estuaries, and C-1 to C-6 represent coastal areas of Kawasi. Phytoplankton samples were collected vertically using a plankton net. The net circular opening with a diameter of 30 cm, length of 150 cm, collection volume of 250 ml, and the mesh of 20 µm. Plankton sampling was performed using vertical towing [11]. The vertical towing method was carried out at a water depth of 10 m in calm water conditions (with no strong currents or waves) to ensure that the net remain vertical. The volume of the filtered sample was measured using a cylindrical volume. Water being collected in the field may have a time gap of several days before reaching the laboratory. To preserve the samples and maintain their representation of the natural conditions of the water, preservation is necessary before laboratory processing. Phytoplankton samples were preserved in Lugol. The following are the field treatment activities for each parameter before transport to the laboratory [12].

Plankton abundance refers to the concentration of individuals or cells within a given volume. The population of individual plankton was determined using the following formula [13]:

$$K = \frac{1}{A}\, x\, \frac{B}{C}\, \mathrm{x}\, \frac{V}{v}\, x\, n$$

Note: K = phytoplankton abundance (cell/m^3); n = number of identified phytoplankton; B = total area/container area of Sedgwick-Rafter Counting cell (mm^2); V = volume of filtered water (30 ml); v = concentrate volume of Sedgwick Rafter Counting Cell (ml); A = volume of filtered water sample (50 l); C=observation area (mm^2).

The diversity index was based on the Shannon and Wiener index [13]:

$$H' = \sum_{i=1}^{n} \quad pi\ ln\ ln\ pi$$

Note: H' = Shannon-Wiener diversity index; pi = ni/N; ni = number of individual species-ith; N = total number of individuals.

The Evenness index was determined using the following formula [13]:

$$E = \sum_{i=1}^{S} \frac{H'}{ln\, ln\, S}$$

Note: E = Evenness index; H'= Shannon-Wiener diversity index; ln S= Natural logarithm of species richness; S = total number of species.

The dominance index was determined using the following formula [13]:

$$C = \sum_{i=1}^{S} \left(\frac{ni}{N}\right)^2$$

Note: C = Simpson dominance index; ni = number of individual-ith; N= total number of individuals; S = number of genera

Water samples were collected from all sampling zones and analyzed using standard methods [14]. Phosphorus in water is commonly found as an orthophosphate (PO_4^-). Phosphate is crucial for the growth of organisms and can limit primary productivity in aquatic environments. The APHA 4500-P F. Automated Ascorbic Acid reduction method determines the orthophosphate concentration in water. This method involves the reaction of ammonium molybdate and potassium antimonyl tartrate with orthophosphate in an acidic medium. The resulting antimony-phosphomolybdate complex was then reduced with ascorbic acid to produce a blue color. Photometric measurements were then performed at 650–660 nm or 880 nm wavelengths. Interference from silica, arsenate, NO_2^-, and S^{2-} can be addressed by adding reagents, whereas turbidity and color that may affect photometric readings can be removed through filtration. Nitrate (NO_3^-) measurements followed APHA 4500-NO_3^- (2012). In this method, nitrate in a sample is reduced to nitrite (NO_2^-) using a column containing cadmium (Cd) granules activated with copper sulphate ($CuSO_3$). The resulting NO_2^- is then analyzed colorimetrically after undergoing diazotization with sulfanilamide and reacting with N-(1-naphthyl)-ethylenediamine dihydrochloride to form an azo dye. Interference can be caused by suspended particles, heavy metals, oils, fats, and residual chlorine, which can be mitigated using filtration, ethylenediaminetetraacetic acid (EDTA), organic solvents, or sodium thiosulfate. After reduction, the color was developed using a specific reagent and measured at a wavelength of 543 nm to determine the nitrate concentration in the sample. Ammonia (NH_3-N) measurement uses the APHA 4500-NH_3 F. The phenate method for ammonia testing relies on forming indophenol, a blue compound, through the reaction of ammonia, hypochlorite, and phenol in the presence of sodium nitroprusside as a catalyst. The process involves adding a phenol solution, sodium nitroprusside, and an oxidizing agent to the sample and then allowing the color to develop at room temperature for at least one hour. The resulting color was measured spectrophotometrically at a wavelength of 640 nm. A standard curve was used to determine the ammonia concentration. Interference from magnesium and calcium ions can be eliminated with citrate, whereas turbidity can be removed through distillation or filtration.

3. Result

3.1 Phytoplankton diversity and distribution

The results of identifying phytoplankton species and abundance in the Kawasi coastal area during two different seasons, the dry and wet seasons, showed fluctuations and dynamics in biodiversity and phytoplankton species composition (Table 1 and 2). There were 46 species identified in the dry season, consisting of four families: Bacillariophyceae, Chlorophyceae, Cyanophyceae, and Dinophyceae (Table 1). The diversity index during the dry season ranges from 1,72 to 2,61, while the evenness index ranges from 0,61 to 0,84, and the dominance index ranges from 0,11 to 0,32. The highest diversity was found at the coral reef station R-3, high evenness was observed at the coastal stations C-2 and C-3, while the highest dominance was found at estuary station E-2.

In contrast, fewer phytoplankton species were found during the wet season, in 24 species from the three families Bacillariophyceae, Cyanophyceae, and Dinophyceae. The Bacillariophyceae showed the highest diversity in both seasons, *Chaetoceros* sp., *Coscinidiscus* sp., *Rhizosolenia* sp., and *Trichodesmium* sp. were found to be abundant during the dry season. The diversity index was generally higher during the dry season than during the wet season. The diversity index during the wet season ranges from 1,04 to 2,01, while the evenness index ranges from 0,59 to 0,95, and the dominance index ranges from 0,17 to 0,61. The highest diversity was found at the estuary station (E-1), high evenness was observed at the coral reef stations (R-3), and the highest dominance was recorded at the coastal station (C-2).

Table 1. Phytoplankton diversity and abundance during the dry season in the coastal area of Kawasi, the western part of Obi Island (R = Coral Reef; P = Port; E = Estuary; C = Coastal)

Organism	Stations												
	R-1	R-2	R-3	P-1	P-2	E-1	E-2	C-1	C-2	C-3	C-4	C-5	C-6
BACILLARIOPHYCEAE													
Achnanthes sp.	-	-	++	-	++++	++++	++++	-	-	-	-	-	-
Amphiprora sp.	-	++	++	-	-	++++	++	++++	-	-	-	-	-
Amphora sp.	++	++	-	-	-	++++	-	++	++	-	-	-	-
Asterionella sp.	-	-	++	-	-	-	-	-	-	-	-	-	-
Bacteriastrum sp.	-	++	++	++	-	++	-	-	++++	-	-	-	-
Biddulphia sp.	++	-	-	-	-	-	-	++	++++	-	-	-	++
Chaetoceros sp.	++++	+++++	++++	+++++	+++++	++++	++++	+++++	+++++	++++	++++	+++++	+++++
Corethron sp.	-	-	++	-	-	-	-	-	++	++	-	++	-
Coscinodiscus sp.	++++	++++	++++	++++	++++	++++	++++	++++	++++	++++	++++	++++	++++
Cyclotella sp.	++	++++	++	++++	-	++	-	++++	++	++	++++	-	++
Diploneis sp.	++	++	-	-	-	-	-	-	-	-	-	-	-
Fragilaria sp.	-	-	-	-	-	-	-	++	-	-	-	-	++
Gomphonema sp.	++	-	-	-	-	-	-	-	-	-	-	-	-
Guinardia sp.	++++	+++++	++++	-	+++++	++++	-	++++	++++	++++	-	-	+++++
Hemiaulus sp.	-	++	++++	++++	++++	++	++	++	++++	++++	++++	++++	-
Lauderia sp.	-	-	-	-	++++	-	++	-	-	-	-	++++	-
Leptocylindrus sp.	-	++	++++	-	-	-	++	++++	++++	-	++	++++	++++
Melosira sp.	-	-	-	-	++++	-	++	-	-	-	++++	++++	++++

 https://doi.org/10.1051/bioconf/202516804010

Organism	Stations												
	R-1	R-2	R-3	P-1	P-2	E-1	E-2	C-1	C-2	C-3	C-4	C-5	C-6
Navicula sp.	++	++	++	-	++++	++++	++	++	++	++++	++	++++	++
Nitzschia sp.	++	++	++	++	-	++++	-	-	-	-	++	-	++
Pinnularia sp.	-	++	-	-	-	++++	-	-	-	-	-	++	-
Pleurosigma sp.	++++	++++	-	++++	-	-	++	++++	++++	++	++	++	++++
Rhizosolenia sp.	+++++	+++++	-	+++++	+++++	+++++	+++++	+++++	++++	+++++	++++	+++++	+++++
Skeletonema sp.	+++++	+++++	++++	++++	++++	-	-	+++++	-	-	++++	-	+++++
Thalassiothrix sp.	++	++	++++	-	++++	++	++	-	++	++	++	++++	++
Thalassionema sp.	-	-	++	++++	-	-	-	-	-	-	-	-	-
Eucampia sp.	++++	++++	++	++++	-	++	-	+++++	++++	++++	++++	-	++++
Synedra sp.	-	++	++	-	++	-	++	++	-	-	++	++++	++
Thalassiosira sp.	-	-	++++	-	-	-	-	-	-	-	-	-	-
Triceratium sp.	-	++	-	-	-	-	-	-	-	-	-	++	-
Ditylum sp.	-	-	-	-	-	-	-	-	-	-	++	-	-
Cerataulina sp.	-	-	-	-	-	++++	-	-	-	++++	-	-	-
Bellerochea sp.	-	-	-	-	-	-	-	-	-	++++	-	-	-
Grammatophora sp.	-	-	++	-	-	-	-	-	-	-	-	-	-
Detonula sp.	-	-	-	-	-	-	-	-	-	-	-	-	++++
CHLOROPHYCEAE													
Cosmarium sp.	-	++	-	-	-	-	-	++	-	-	-	-	-
Staurastrum sp.	-	++	-	-	-	-	-	++	-	-	-	-	-
Selenastrum sp.	-	++	-	-	-	++++	-	-	-	-	-	-	-
CYANOPHYCEAE													

Organism	Stations												
	R-1	**R-2**	**R-3**	**P-1**	**P-2**	**E-1**	**E-2**	**C-1**	**C-2**	**C-3**	**C-4**	**C-5**	**C-6**
Microcystis sp.	-	++++	++	-	++++	++++	++	++	++	++++	++++	++++	-
Oscillatoria sp.	++	++	++	++	++	++	++	-	-	++++	-	++	++
Trichodesmium sp.	+++++	+++++	++++	+++++	-	-	-	+++++	+++++	-	+++++	-	+++++
DINOPHYCEAE													
Ceratium sp.	++++	++++	++	++++	++	-	++	++	++	++	++++	++++	++
Dinophysis sp.	-	++	-	-	-	++	-	-	++	++	++	-	++
Ornithocercus sp.	-	-	-	++++	-	-	-	-	-	-	++	-	-
Peridinium sp.	++	++++	++++	++	-	+++++	++	++	++	++	++++	++	++
Ceratocorys sp.	-	++	-	-	-	-	-	++	++	++	++	++	-
Number of Taxa	19	30	25	16	15	21	17	20	19	19	22	19	22
Total abundance (cell/m³)	899.808	1.680.315	399.353	887.676	873.521	741.078	311.394	679.406	450.915	450.915	618.744	726.923	1.048.428
Diversity Index (H')	1,99	2,30	2,61	1,96	2,10	2,47	1,73	2,12	2,49	2,49	2,40	2,29	2,31
Evenness Index (E)	0,68	0,68	0,81	0,71	0,78	0,81	0,61	0,71	0,84	0,84	0,78	0,78	0,75
Dominance Index (C)	0,18	0,14	0,10	0,21	0,17	0,12	0,32	0,19	0,11	0,11	0,12	0,15	0,12

Note: (-) = 0 cell/ m^3; (+) = 1-100 cell/ m^3; (++) = 101-1000 cell/ m^3; (+++) = 1001-10000 cell/ m^3; (++++) = 10001-100000 cell/ m^3; (+++++) = >100000 cell/m^3

https://doi.org/10.1051/bioconf/202516804010

Table 2. Phytoplankton diversity and abundance during the wet season in the coastal area of Kawasi, the western part of Obi Island (R = Coral Reef; P = Port; E = Estuary; C = Coastal)

Organism	Stations												
	R-1	R-2	R-3	P-1	P-2	E-1	E-2	C-1	C-2	C-3	C-4	C-5	C-6
BACILLARIOPHYCEAE													
Achnanthes sp.	-	-	-	-	-	++	-	-	-	-	-	-	-
Campylodiscus sp.	-	-	-	-	-	-	-	-	-	-	++	-	-
Chaetoceros sp.	-	-	-	++++	++	-	-	++++	-	-	++	++	-
Coscinodiscus sp.	++	++	-	++	++	++	-	++	-	++	++++	++++	++
Guinardia sp.	-	-	-	-	-	++	++	-	-	-	-	-	-
Hemiaulus sp.	-	-	-	-	-	-	-	-	-	++	-	-	-
Leptocylindrus sp.	-	-	-	-	-	-	++	-	-	-	-	-	-
Lithodesmium sp.	-	-	-	-	-	++	-	-	-	-	-	-	-
Navicula sp.	-	-	-	-	-	++++	-	-	-	-	-	-	-
Nitzschia sp.	-	-	-	-	-	++	++	-	-	++	-	-	++
Pleurosigma sp.	++	-	-	++	++	++	++	++	-	++	-	++	++
Rhizosolenia sp.	++	-	-	++++	++	++++	++++	++++	++++	-	++++	++	++
Synedra sp.	-	-	++	-	-	++	-	-	-	-	++	-	-
Thalassionema sp.	-	++	-	-	-	-	-	-	-	-	-	-	-
Triceratium sp.	++	-	-	-	-	++	-	-	-	-	++	-	-
CYANOPHYCEAE													
Microcystis sp.	-	-	-	++++	-	++++	-	-	++	++++	-	-	-
Oscillatoria sp.	-	-	-	++	-	-	-	++	-	-	-	-	-
Trichodesmium sp.	-	-	-	-	-	-	-	-	-	++++	-	-	-

Organism	Stations												
	R-1	R-2	R-3	P-1	P-2	E-1	E-2	C-1	C-2	C-3	C-4	C-5	C-6
DINOPHYCEAE													
Ceratium sp.	-	++	-	++	++	-	-	++	-	-	++	++	++
Ceratocorys sp.	-	-	-	-	-	-	-	-	++	-	-	-	-
Dinophysis sp.	++	++	++	++	-	-	-	++++	-	++	++	++	-
Ornithocercus sp.	-	-	-	-	++	-	-	-	-	++	-	-	-
Peridinium sp.	++	-	-	-	++	-	-	-	-	++	-	-	-
Pyrocystis sp.	-	-	++	-	++	-	++	-	-	-	-	-	-
Number of Taxa	6	4	3	8	8	12	6	7	3	9	9	6	5
Total abundance (cell/m³)	11.121	11.121	4.044	143.565	21.231	129.411	33.364	85.937	36.397	47.518	60.661	34.375	18.198
Diversity Index (H')	1,64	1,29	1,04	1,33	1,94	2,01	1,50	1,56	0,65	1,67	1,45	1,60	1,42
Evenness Index (E)	0,92	0,93	0,95	0,64	0,93	0,78	0,84	0,80	0,59	0,76	0,66	0,89	0,88
Dominance Index (C)	0,22	0,31	0,38	0,34	0,17	0,18	0,27	0,24	0,61	0,25	0,35	0,22	0,28

Note: (-) = 0 cell/ m^3; (+) = 1-100 cell/ m^3; (++) = 101-1000 cell/ m^3; (+++) = 1001-10000 cell/ m^3; (++++) = 10001-100000 cell/ m^3; (+++++) = >100000 cell/m^3

Figure 2 shows the distribution of phytoplankton composition at 13 observation stations in the coastal area of Kawasi, western Obi Island, during the dry season. In contrast, Figure 3 presents the distribution during the wet season. Phytoplankton composition varied between stations during each season. Each season showed distinct patterns of phytoplankton distribution across the observation stations.

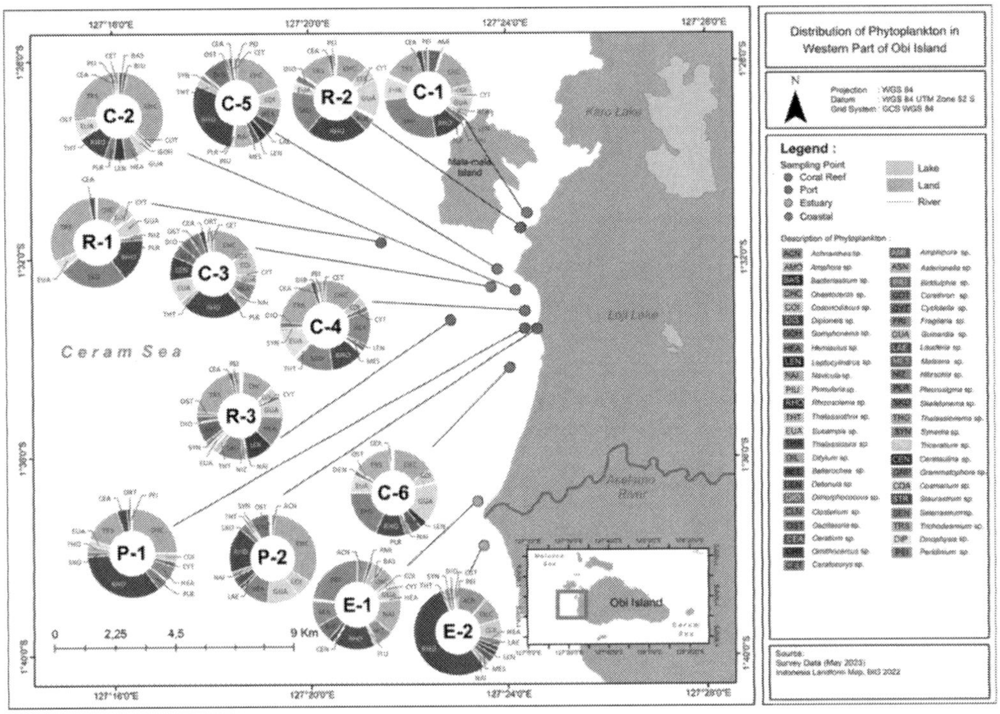

Fig. 2. Distribution of phytoplankton composition during the dry season in the coastal area of Kawasi, the western part of Obi Island (R = Coral Reef; P = Port; E = Estuary; C = Coastal)

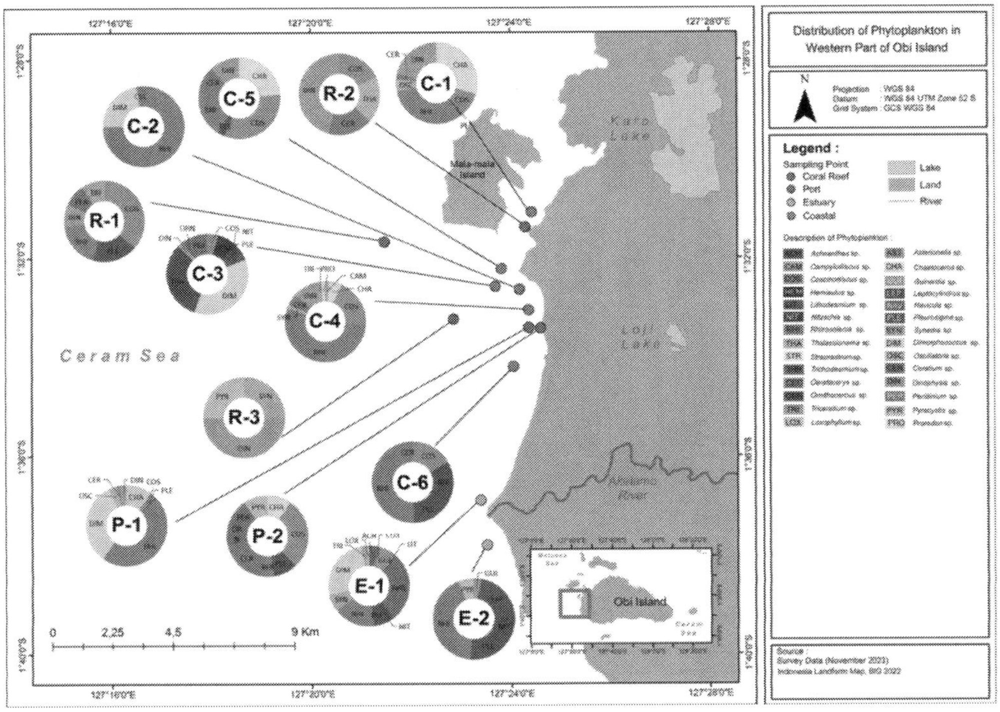

Fig. 3. Distribution of phytoplankton composition during the wet season in the coastal area of Kawasi, the western part of Obi Island

3.2 Class composition and abundance of phytoplankton in different seasons

Bacillariophyceae are a class of phytoplankton commonly found at all stations during the dry season. More than 60% of this class was found during the dry season. Bacillariophyceae are one of the classes within the algae group known as diatoms, which are a very important type of phytoplankton in aquatic ecosystems. In addition to Bacillariophyceae, Cyanophyceae, Chlorophyceae, and Dinophyceae were also found during the dry season (Figure 4a). The composition of phytoplankton classes during the wet season was slightly different from that during the dry season, with Dinophyceae being found in several locations such as R-2, R-3, and C-1 (Figure 4b), and Cyanophyceae was also abundant in P-1 and C-3. Chlorophyceae were not found during the wet season; this class was only present in the dry season at C-1, R-2, and E-1 (Figures 4a and 4b). *Selenastrum* sp. was the most commonly found species in the E-1. This species is commonly found in freshwater environments and serves a vital role in aquatic ecosystems as a primary producer, generating oxygen and forming the foundation of the food chain. Since Station E-1 is located in an estuarine area, its presence is likely affected by the interaction between seawater and river water.

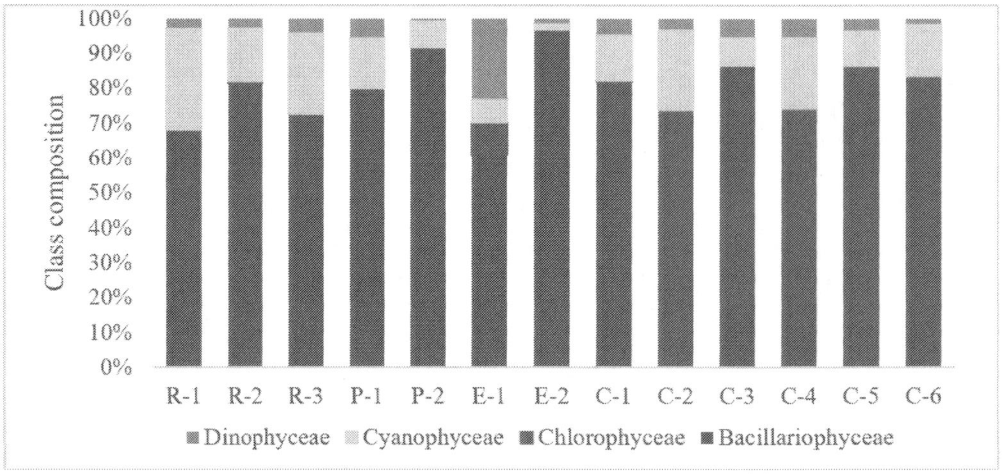

(a)

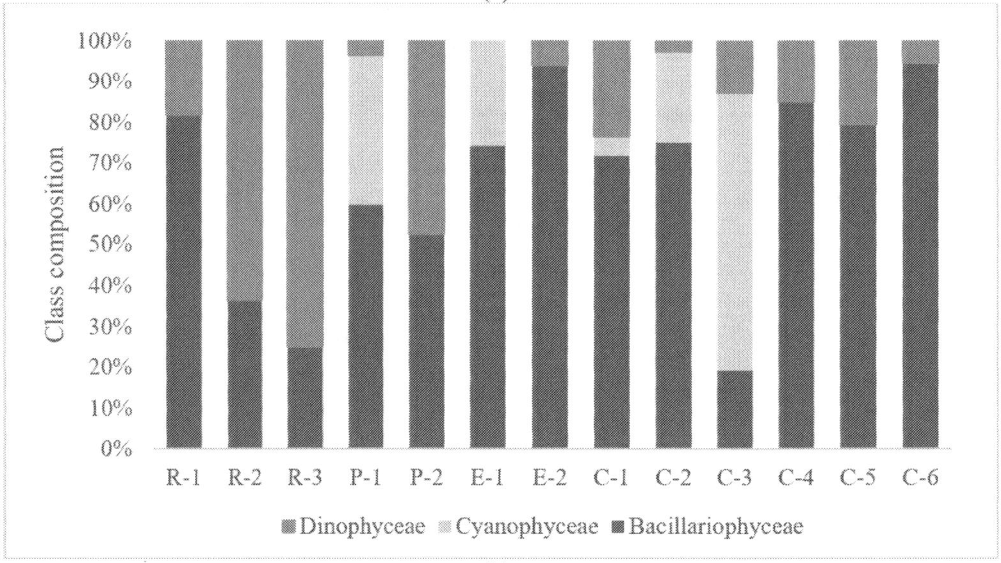

(b)

Fig. 4. The class composition of phytoplankton in the coastal area of Kawasi, the western part of Obi Island: (a) dry season; (b) wet season

Phytoplankton abundance was higher during the dry season than during the wet season. Coral Reef station R-2, located between Obi Island and Mala-Mala Island, had the highest abundance compared to all other stations (Figure 5a). The total abundance range at each station varies between 3,1 × 10^5 – 1,7 × 10^6 cells/m^3, while the phytoplankton abundance range during the wet season is between 4,0 × 10^3 – 1,4 × 10^5 cells/m^3. During the wet season, the highest abundance was found at P-1, a port, and E-1, an estuary area, whereas the lowest abundance was found in R-1, R-2, and R-3, coral reef areas (Figure 5b).

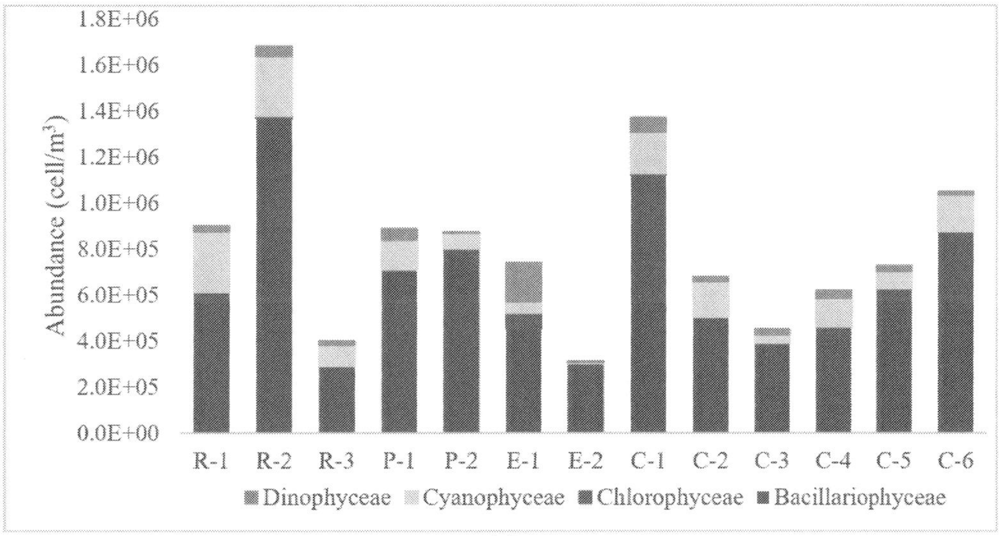

(a)

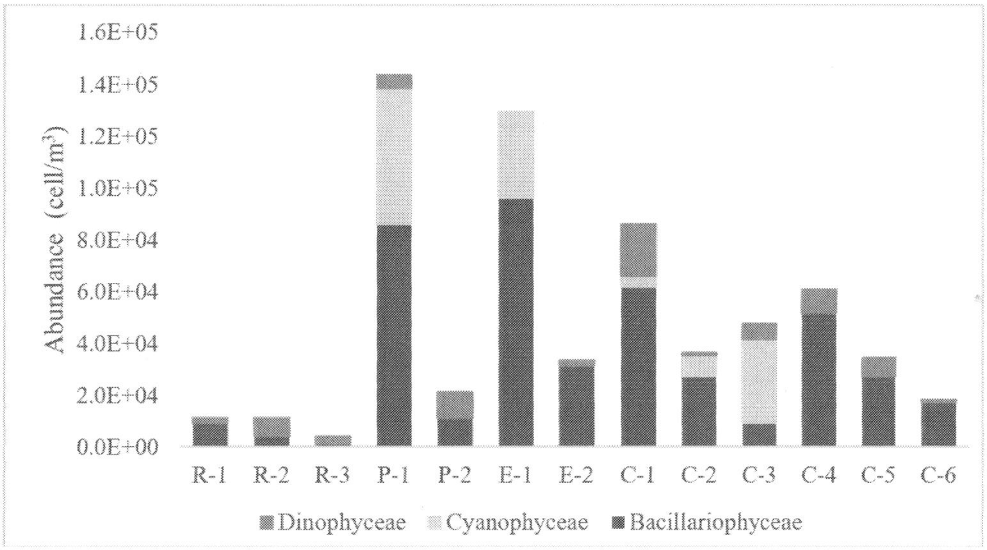

(b)

Figure 5. Class abundance of phytoplankton in the coastal area of Kawasi, western part of Obi Island: (a) dry season; (b) wet season

3.3 Nitrate and phosphate

The results of the nitrate and phosphate measurements fluctuated during each season. During the dry season, nitrate concentrations ranged from 0,005 to 0,102 mg/L; in the wet season, they ranged from 0,05 to 0,37 mg/L. Phosphate concentrations during the dry season ranged from 0,05 to 0,15 mg/L, whereas in the wet season, they ranged from 0,05 to 0,11 mg/L (Table 3). The concentrations of nitrate and phosphate at each station during the different seasons are shown in Table 3.

Table 3. The concentrations of nitrate and phosphate during the dry and wet seasons in the coastal area of Kawasi, the western part of Obi Island

Station	Nitrate (mg/L)		Phosphate (mg/L)	
	Dry Season	Wet Season	Dry Season	Wet Season
R-1	0,005	0,037	0,015	0,005
R-2	0,005	0,007	0,005	0,011
R-3	0,005	0,027	0,015	0,005
P-1	0,021	0,011	0,008	0,005
P-2	0,021	0,011	0,008	0,005
E-1	0,029	0,005	0,005	0,007
E-2	0,102	0,013	0,005	0,008
C-1	0,005	0,007	0,005	0,011
C-2	0,037	0,006	0,006	0,007
C-3	0,005	0,005	0,005	0,008
C-4	0,016	0,006	0,005	0,007
C-5	0,056	0,005	0,008	0,009
C-6	0,046	0,005	0,005	0,006

R = Coral Reef; P = Port; E = Estuary; C = Coastal

During the dry season, the nitrate concentration in the coral reef ecosystem (R-1, R-2, R-3) showed the same value at all stations, which was 0,005 mg/L. During the wet season, there was an increase in the range of 0,007–0,037 mg/L, with R-1 having the highest nitrate concentration. In the port ecosystems (P-1 and P-2), nitrate concentrations during the dry season were measured at 0,021 mg/L and decreased to 0,011 mg/L during the wet season at both stations. The nitrate concentrations varied in the estuary ecosystems (E-1 and E-2). During the dry season, E-1 had a concentration of 0,029 mg/L, whereas E-2 showed the highest value among all stations at 0,102 mg/L. In the wet season, the nitrate concentrations at these two stations decreased to 0,005 mg/L and 0,013 mg/L. In the coastal ecosystem (C-1 to C-6), nitrate concentrations varied more than in the other ecosystems. During the dry season, nitrate concentrations ranged from 0,005 to 0,056 mg/L, with C-5 showing the highest value. During the wet season, the range was lower, between 0,005 and 0,009 mg/L, with the highest concentration recorded at C-1 (Table 3).

The phosphate concentration in the coral reef ecosystem during the dry season ranged between 0,005 and 0,015 mg/L, with R-1 and R-3 having the highest concentrations. In the wet season, phosphate concentrations at all coral reef stations decreased, ranging from 0,005 to 0,011 mg/L. In the port ecosystem (P-1 and P-2), phosphate concentrations were measured at 0,008 mg/L during the dry season, whereas in the wet season, it decreased to 0.005 mg/l at both stations. In the estuary ecosystems (E-1 and E-2), phosphate concentrations ranged from 0,005 to 0,008 mg/L in the dry season, whereas in the wet season, there was a slight increase, ranging from 0,007 to 0,008 mg/L. In the coastal ecosystem, phosphate concentrations during the dry season ranged from 0,005 to 0,009 mg/l, with the highest value at C-2. During the wet season, phosphate concentrations remained within the same range of 0,005–0,009 mg/L, with the highest value recorded at C-3 (Table 3).

Based on statistical analysis, neither nitrate nor phosphate concentrations showed significant seasonal differences. However, the average nitrate concentration during the dry season was higher than that during the wet season (Figure 6).

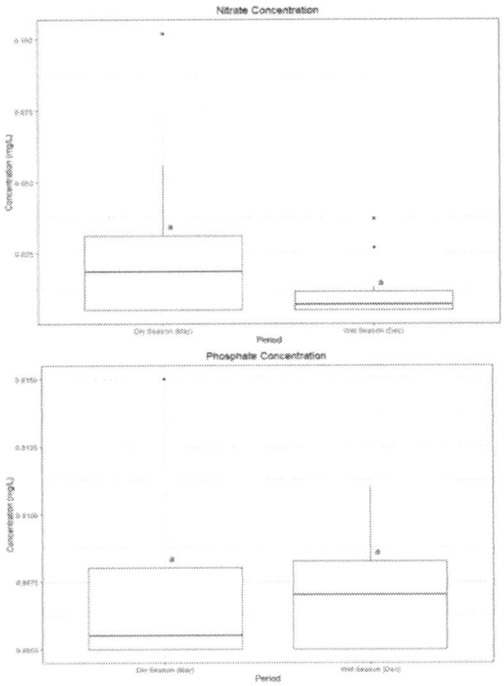

Fig. 6. Nitrate and phosphate concentration during (a) dry and (b) wet seasons in the coastal area of Kawasi, the western part of Obi Island

4. Discussion

The Bacillariophyceae class of diatoms is widely distributed globally and often dominates phytoplankton communities [11,15,16]. Their dominance in marine environments is attributed to their superior adaptability to environmental conditions compared with other phytoplankton classes [17,18]. Additionally, Bacillariophyceae exhibit a high reproductive rate and can double within 18–36 h [18,19,20]. They have a high cell division rate under optimal conditions, allowing them to dominate nutrient-rich water quickly. In this study, the most abundant species during the dry season were *Chaetoceros* sp., *Coscinodiscus* sp, and *Rhizosolenia* sp.

Diatoms can store nitrogen and phosphorus in their vacuoles, allowing them to survive when nutrient availability decreases. They can grow rapidly when nutrient levels increase owing to upwelling from the ocean floor to the surface, often making them the dominant species in phytoplankton blooms. Their hard frustules make it more difficult for zooplankton to consume diatoms than for phytoplankton without rigid cell walls. The process of forming silica cell walls requires less energy than synthesizing carbon-based cell walls, as observed in dinoflagellates, making diatoms more efficient under nutrient-limited conditions. Some diatoms can form dormant spores when environmental conditions are unfavorable and resume growth when environmental conditions improve. Cyanophyceae (Cyanobacteria), commonly known as blue-green algae, thrive in water contaminated with organic matter [21]. Some types of Cyanophyceae, such as *Trichodesmium* sp., reproduce more rapidly during the dry season because of good light penetration in the water and optimal temperatures for cell division [22].

During the dry season, the highest phytoplankton abundance was found at C-1 and R-2, which are located in the strait between Obi Island and Mala-Mala Island, respectively. Strong currents in a narrow strait can cause high turbulence, affecting the distribution of organisms and suspended particles. This factor contributes to the high phytoplankton growth in this area. In addition, the high ship traffic in the strait area could be a source of organic waste from shipping activities.

Generally, the phytoplankton abundance was lower across all observation sites during the wet season. The reduced average nitrate levels may be a contributing factor to this decline. Some studies have suggested that nutrient levels increase along the coast due to nutrient runoff. Obi Island has a relatively low population density compared to the main islands. Human activities are concentrated in Kawasi Village and the Nickel Industry, where waste management systems may already be in place to minimize eutrophication and nutrient runoff into the sea. If human any human activities are conducted irresponsibly in the future, they may result in harmful consequences, including rising CO_2 levels [23], increasing nutrients and runoff from household activities [24], and harmful metal loading to the water [25].

The highest phytoplankton abundance was observed during the wet season in the port area (P-2) and estuary (E-1). In contrast, coral reef sites (R-1, R-2, and R-3) showed lower phytoplankton abundance. Notably, at site R-2, phytoplankton abundance was extremely low, suggesting an oligotrophic condition similar to that of the other two coral reef sites. The highest concentration of nitrate nutrients was found at E-2, which is located at the southernmost point of the study area. However, this did not show the same pattern as the phytoplankton abundance at that particular station. These results indicate that nutrients could influence the seasonal fluctuations of the phytoplankton community in the coastal waters of Kawasi. However, the study has not yet confirmed how nutrients are distributed and whether land-based runoff, currents, or water mass dynamics contribute to seasonal changes in both nutrients and phytoplankton in this region.

5. Conclusion

Bacillariophycea was the most common phytoplankton class at all stations during the dry and wet seasons. *Chaetoceros* sp., *Coscinodiscus* sp., and *Rhizosolenia* sp. were the predominant genera in both seasons. During the dry season, phytoplankton are predominantly located in the strait area between Obi and Mala-Mala Island. In contrast, they are more concentrated in the harbor and estuary areas during the wet season. The highest overall abundance occurred during the dry season, which was linked to the higher average nitrate levels in the water during this time. The waters in the western part of Obi Island have high plankton diversity with no dominance, meaning that there are many species of phytoplankton living in the area and indicates that the aquatic ecosystem has environmental conditions that support the growth and development of various phytoplankton species. The This study provides information on phytoplankton's diversity and community composition in the western part of Obi Island.

References

1. Stanley, G. Preetha, (Springer, Dordrecht, 2016) 411–497.
2. S. Sagert, T. Rieling, A. Eggert, H. Schubert, Hydrobiologia 611, 91–103 (2008).
3. E. Cabecinha, R. Cortes, J.A. Cabral, M.T. Ferreira, M. Lourenco, et al., Ecol. Indic. 9, 240–255 (2009). DOI: 10.1016/j.ecolind.2008.04.006. DOI: 10.1016/j.ecolind.2008.04.006.
4. N. Abrantes, S.C. Antunes, M.J. Pereira, F. Gonçalves, Acta Oecol. 29(1), 54–64 (2006).

5. N.A. Bhat, A. Wanganeo, R. Raina, Ecosyst. Ecol. 5(1) (2015).
6. P.P. Ingale, A.D. Bobdey, N.D. Gorghate, J. Chin. Adv. Mater. Soc. 6(4), 655–665 (2018). DOI: 10.1080/22243682.2018.1537005.
7. R. Ptacnik, A.G. Solimini, T. Andersen, T. Tamminen, P. Brettum, L. Lepistö, et al., Proc. Natl. Acad. Sci. 105(13), 5134–5138 (2008). DOI: 10.1073/pnas.0708328105.
8. Y. Wei, J. Sun, G. Zhang, X. Wang, F. Wang, Environ. Sci. Pollut. Res. 27(18), 23222–23233 (2020). DOI: 10.1007/s11356-020-08927-6.
9. M.K. Thomas, M. Aranguren-Gassis, C.T. Kremer, M.R. Gould, K. Anderson, C.A. Klausmeier, et al., Glob. Change Biol. 23(8), 3269–3280 (2017). DOI: 10.1111/gcb.13641.
10. W. Xiao, X. Liu, A.J. Irwin, E.A. Laws, L. Wang, B. Chen, et al., Water Res. 128, 206–216 (2018). DOI: 10.1016/j.watres.2017.10.051.
11. H. Effendi, M. Kawaroe, D.F. Lestari, T. Permadi, Procedia Environ. Sci. 33, 496–504 (2016). DOI: 10.1016/j.proenv.2016.03.102.
12. N. Cakasana, A. Sunuddin, E.S. Srimariana, D.F. Lestari, et al., Habitus Aquatica 5(1), 1–10 (2024). DOI: 10.29244/HAJ.5.1.1.
13. A.E. Magurran, Measuring Biological Diversity (Wiley-Blackwell, 2004).
14. APHA, (American Public Health Association, 22nd Ed., 2012).
15. N. Audah, L. Japa, M. Yamin, J. Biol. Trop. 21(2), 448–455 (2021). DOI: 10.29303/jbt.v21i2.2699.
16. M.A.M. El Gammal, M. Nageeb, S. Al-Sabeb, Egypt. J. Aquat. Res. 43(4), 275–282 (2017). DOI: 10.1016/j.ejar.2017.10.004.
17. O.H. Arinardi, A.B. Sutomo, S.A. Yusuf, Trimaningsih, E. Asnaryanti, S.H. Riyono, (1997).
18. T. Nakov, J.M. Beaulieu, A.J. Alverson, bioRxiv, 1–36 (2018). DOI: 10.1101/406165.
19. T. Nakov, J.M. Beaulieu, A.J. Alverson, Evolution 73 (2019). DOI: 10.1111/evo.13832.
20. J.H. Kim, P.A. Ajani, S.A. Murray, S. Kang, S. Kim, H.C. Lim, S.T. Teng, P.T. Lim, B.S. Park, Harmful Algae 123, 102392 (2023). DOI: 10.1016/j.hal.2023.102392.
21. S. Khalil, M.H. Mahnashi, M. Hussain, N. Zafar, U. Waqar, F.S. Khan, U. Afzal, G.M. Shah, U.M. Niazi, M. Awais, M. Irfan, Saudi J. Biol. Sci. 28(10), 5728–5737 (2021). DOI: 10.1016/j.sjbs.2021.06.004.
22. L.G. Yap-Dejeto, H.S. Batula, Rev. Biol. Trop. 64(2), 897–911 (2016). PMID: 29451978.
23. E. Worrell, L. Price, N. Martin, C. Hendriks, L.O. Meida, Ann. Rev. Energy Env. 26, 303 (2001).
24. H.J. Qu, C. Kroeze, Reg. Environ. Change. 12,153-167 (2010). DOI: 10.1007/s10113-011-0248-3.
25. S. Cymerman, A.J Kempers, Ecotoxicology and Environ Safety. 59(1), 64-69 (2004). DOI: https://doi.org/10.1016/j.ecoenv.2003.12.002.

The community structure of zooplankton in Bojonegara coastal waters, Banten Bay, Indonesia

Endang Sunarwati Srimariana[1*], *Alexandra* Maria Diandra Rahardjo[1], *Sulistiono* Sulistiono[2], *Tumpak* Sidabutar[3]

[1]Departement of Marine Sciences and Technology, Faculty of Fisheries and Marine Sciences, IPB University, Bogor, 16680 West Java, Indonesia.
[2]Departement of Aquatic Resources Management, Faculty of Fisheries and Marine Sciences, IPB University, Bogor, 16680 West Java, Indonesia.
[3]National Research and Innovation Agency (BRIN), Jakarta, West Java, Indonesia

Abstract. Bojonegara coastal waters, located in an area experiencing significant industrial expansion, are under increasing environmental stress. This study aims to analyze the zooplankton community structure in these waters. The research, conducted from January to October 2020, involved six stations across river, estuary, and sea areas. Data analysis covered various ecological indices, including abundance, diversity, evenness, dominance, principal component analysis, and Bray-Curtis similarity index. Environmental parameters such as temperature, transparency, turbidity, salinity, pH, dissolved oxygen, nitrate, and phosphate were also measured. The study identified zooplankton from 10 different classes, comprising Malacostraca (16 genera), Rotifera (6 genera), Ciliata (4 genera), Rhizopoda (2 genera), and one genus each from Bivalvia, Gastropoda, Polychaeta, Sagittoidea, Sarcodina, and Urochordata. Malacostraca and Ciliata were the dominant groups. Zooplankton abundance, as well as diversity (H'), evenness (E), and dominance (C) indices, varied between 2.8×10^4 and 9.9×10^4 ind/L, 1.061 to 1.646, 0.713 to 0.771, and 0.282 to 0.433, respectively. The zooplankton community structure indicated that Bojonegara waters are experiencing mild environmental stress, with temperature and nitrate levels being key factors affecting zooplankton abundance.

1 Introduction

Indonesia has diverse biological resources, especially in coastal and marine areas; however, using natural resources without considering ecological balance will significantly impact humans and damage existing ecosystems [1]. One of the biological resources that play an essential role in aquatic ecosystems is plankton. A plankton is a microscopic organism that

* Corresponding author: endangsunarwatis@apps.ipb.ac.id

floats in the water column and has fragile swimming abilities. Planktons can be divided into two major groups: phytoplankton and zooplankton [2]. As a primary producer, phytoplankton can utilize sunlight as a source of energy in their life. Zooplankton are the primary consumers of energy sources produced by these primary producers [3].

Bojonegara, one of the districts in Banten Bay, has an area of 4.27 km^2 and a coastal area that is directly adjacent to Banten Bay in the east [4]. The coastal area of Bojonegara is dominated by medium-to-large industrial sites. Industrial activity in this region harms the environment and increases pollution owing to anthropogenic factors. Waste generated from industrial activities is disposed of continuously, following river flow, and ends up in the sea [5]. This waste often carries out nutrients, so that the nutrients that enter the aquatic environment continuously increase. The zooplankton community structure can be used to observe the pressure in Bojonegara waters due to rising nutrients [6].

Zooplankton are the primary consumers that have a strong relationship with phytoplankton in the food web. The presence of zooplankton in water depends on their eating habits and preferences [7]. In addition, the zooplankton community structure is influenced by water temperature because zooplankton are poikilothermic and short-lived animals. Changes in community structure are an excellent indicator for detecting the impact of changes in water [8].

These floating microscopic animals play an essential role in the water, especially in the food chain. These organisms are consumers who play a significant role in connecting the energy transfer from primary producers (phytoplankton) to living organisms at higher trophic levels (fish and shrimp). Zooplankton are mainly preyed upon by larger predatory animals as tertiary producers. This process will occur from producer level IV, level V, etc., which can be described in the food chain [9].

The diversity and abundance of zooplankton indicates the fertility and stability of the water area. According to [10], the presence of zooplankton in a water area is also highly influenced by water conditions. Only certain zooplankton species can live in estuaries and beaches. According to [11], only a few species of marine and freshwater animals can live in estuaries and surrounding beach waters because they are euryhaline. Estuary and coastal zooplankton are generally dominated by the Copepod, Eurytemora, Acartia, Pseudodiaptomus, and Centropage genera [2]. The abundance and biomass or total (weight) of zooplankton are essential biological parameters in water. This is closely related to daily and seasonal fluctuations, where the abundance and distribution of zooplankton can also be used to estimate secondary production, so that the presence of zooplankton can be used as an indicator of water productivity [12].

The distribution of zooplankton, which are abundant in water, is closely related to food or phytoplankton. Changes in zooplankton community composition may vary over the years owing to changes in food and the environment in which they live [13]. Changes in plankton community structure impact aquatic ecosystem stability, where higher plankton diversity indicates a more stable ecosystem. Each plankton community responded differently to environmental influences. Understanding the relationship between water quality and plankton community structure is important for determining environmental pressures, and monitoring and management could become more effective [14].

This study aimed to analyze the zooplankton community structure, which includes species composition and abundance, diversity, evenness, and dominance indices, and their relationship to the physicochemical water parameters in the Bojonegara coastal waters, Serang, Banten. This research is expected to provide information on zooplankton community structure and its relationship to the physicochemical water parameters of Bojonegara coastal waters, Serang, Banten, to be used as a basis for aquatic management in Bojonegara.

2 Materials and Methods

2.1 Time and Location

Sampling was carried out five times in January, February, August, September, and October 2020 at six stations distributed around the Bojonegara coastal area of Banten Bay (**Figure 1**). Sample analysis was performed at the Macrobiology Laboratory, Division of Aquatic Ecology and Conservation, Department of Aquatic Resources Management, Faculty of Fisheries and Marine Sciences, IPB University.

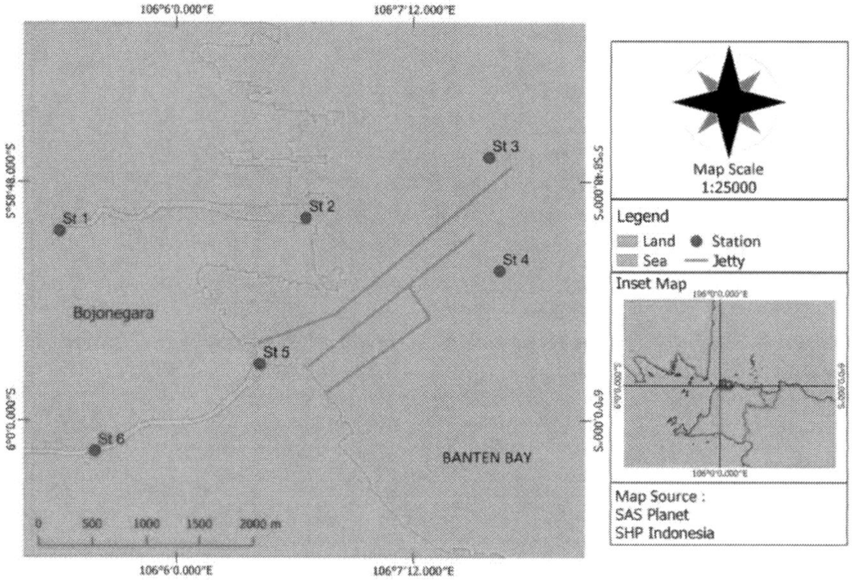

Fig. 1. Sampling locations in Bojonegara coastal waters, Banten Bay of Indonesia

Station 1	:	5°59'02.8" S, 106°05'24.6" E (Wadas River, residential area)
Station 2	:	5°58'58.9" S, 106°06'39.3" E (Wadas estuary, industries area)
Station 3	:	5°58'38.6" S, 106°07'34.1" E (open sea, Banten Bay)
Station 4	:	5°59'14.9" S, 106°07'37.6" E (open sea, Banten Bay)
Station 5	:	5°59'43.1" S, 106°06'25.2" E (Terate estuary, near Steam Electricity Power Station)
Station 6	:	6°00'09.3" S, 106°05'35.3" E (Terate River, residential area)

2.2 Equipment and Materials

The equipment and materials used in this study included GPS (global positioning system), thermometer, Secchi disk, turbidimetry, refractometer, pH indicator, van dorn with a capacity of 1 L, plankton net (mesh size of 25 μm), 100 mL of 500 mL HDPE bottles, cool box, microscope compound, plankton identification book [15,16], 50×20×1 mm3 Sedgewick Rafter Cell (SRC), Ms. Excel, PAST, and XLSTAT software, and samples of water and Lugol as preservatives.

2.3 Data Collecting

2.3.1 Determining Sampling Location

Zooplankton sampling was carried out at six stations: two stations were located in the sea, two stations were found in the estuary, and two stations were located near the river. The location of data collection was determined using the purposive sampling method by considering the conditions of Bojonegara coastal waters. This location was used to compare anthropogenic pressures in residential areas close to rivers, regions affected by the Steam Electricity Power Station near the estuary, and areas not influenced by anthropogenic activities at sea.

2.3.2 In situ Sampling Techniques and Water Parameters Observation

Zooplankton sampling was performed by filtering 50 L of water through a plankton net with a mesh size of 25 µm. The filtered water was then transferred into a 100 mL HDPE bottle, and ten drops of Lugol were added. The HDPE bottles were then stored and protected from sunlight so that the zooplankton samples were not damaged.

Physical parameters measured *in situ* include temperature using a thermometer, brightness using a Secchi disk, and turbidity using turbidimetry. The chemical parameters measured in situ were salinity with a refractometer, and pH was measured using a pH indicator. Chemical parameters, such as nitrate and phosphate, are known by testing water samples in the laboratory. Water samples for nitrate and phosphate tests were taken using a van dorn of as much as 500 mL and then poured into HDPE bottles.

2.3.3 Laboratory Analysis

The zooplankton were identified using a compound microscope (10×10 × magnification). Zooplankton were identified using plankton identification books [16-17]. Zooplankton abundance was calculated using a 50×20×1 mm^3 Sedgewick Rafter Cell (SRC) [17]. A comprehensive census was used as the calculation method. The relationship between the zooplankton community structure and environment can be determined based on principal component analysis (PCA) and the Bray-Curtis similarity index. The community structure data and environmental influences were processed using Microsoft Excel, PAST, and XLSTAT.

2.4 Data Analysis

2.4.1 Community Structure

Zooplankton Abundance

Zooplankton abundance was defined as the number of individual zooplankton found per unit volume of water (ind/L). The zooplankton abundance per liter of water was calculated using the following formula [17]:

$$N = n \times \frac{Vt}{Vsrc} \times \frac{Asrc}{Aa} \times \frac{1}{Vd} \quad (1)$$

Note :

N	=	zooplankton abundance (ind/L)
n	=	number of phytoplankton found (ind)
Vt	=	filtered water volume (mL)
Vsrc	=	water volume in SRC (1 mL)
Asrc	=	SRC area (1000 mm^2)
Aa	=	observed area (mm^2)
Vd	=	filtered water volume (L)

Diversity Index (H')

The diversity index provides an overview of ecosystem stability and pressure, based on the number of zooplankton types. The diversity index can be calculated using the Shannon-Wiener index [18]:

$$H' = -\sum_{i=1}^{n} pi \ln pi \qquad (2)$$

Note :

H'	=	Shannon-Wiener diversity index
n	=	total number of genera
pi	=	ni/N
ni	=	total number of individual genera i
N	=	total number of individuals encountered

According to [19], the diversity index range can be grouped into three categories:

$H' < 1$	:	low diversity
$1 \leq H' \leq 3$	:	moderate diversity
$H' > 3$	:	high diversity

Evenness Index (E)

The evenness index was used to determine the distribution of individuals for each type of zooplankton, and to determine the similarity of distribution at the community level. The evenness index can be expressed by the following equation [18]:

$$E = \frac{H'}{Hmax} \qquad (3)$$

Note :

E	=	evenness index
H'	=	Shannon-Wiener diversity index
Hmax	=	maximum diversity (ln S)
S	=	number of genera

The evenness index ranged from 0 to 1. According to [20], the uniformity index can be grouped into three categories:

$E > 0.6$	:	high evenness
$0.6 \geq E \geq 0.4$	:	moderate evenness
$E < 0.4$	:	low evenness

Dominance Index (C)

The dominance index was used to determine the presence or absence of species dominance in the zooplankton communities. The dominance index was calculated using the Simpson dominance index [18] as follows:

$$C = \sum_{n=1}^{n} \left(\frac{ni}{N}\right)^2 \qquad (4)$$

Note :

C	=	Simpson dominance index
n	=	total number of species
ni	=	total number of individual type-i
N	=	total number of individual

The dominance index ranged from 0 to 1, which [19] categorized into three groups:

$C < 0.5$	:	low dominance
$0.5 \leq C \leq 0.75$	:	moderate dominance
$C > 0.75$	:	high dominance

2.4.2 Principle Component Analysis (PCA)

Principal component analysis (PCA) is an ordinance analysis technique used to observe the influence of physical and chemical factors of the aquatic environment on the zooplankton community structure [20,21]. The data matrix consisted of zooplankton abundance variables as individuals and water physicochemical parameter variables as quantitative variables [1]. Results were visualized using a biplot graph.

2.4.3 Similarity Index

A similarity index can be used to determine the level of similarity between observation stations based on the zooplankton abundance data. The similarity index used was the Bray-Curtis index [22] using the following equation:

$$Ib = 1 - \sum_{i=1}^{s} \frac{|Aij - Bik|}{|Aij + Bik|} \quad (5)$$

Note :

C	=	Simpson dominance index
n	=	total number of species
ni	=	total number of individual type-i
N	=	total number of individual

3 Results and Discussion

3.1 Results

Based on zooplankton sampling that was conducted five times in January, February, August, September, and October 2020 at six observation stations distributed around Bojonegara coastal waters, Banten Bay, found that zooplankton consisted of 10 classes (34 genera), including Malacostraca (16 genera), Rotifera (6 genera), Ciliata (4 genera), Rhizopoda (2 genera), Bivalvia, Gastropods, Polychaeta, Saggitoidea, Sarcodina, and Urochordata (1 genus each). The zooplankton types found are listed in Table 1.

Table 1 Zooplankton classes and genera found in Bojonegara coastal waters, Banten Bay

Class	Genera		
Bivalvia	Larva		
Ciliata	*Colpoda* sp.	*Leprotintinnus* sp.	*Tintinnopsis* sp.
	Favella sp.		
Gastropoda	Larva		
Malacostraca	*Acartia* sp.	*Euterpina* sp.	*Oithona* sp.
	Balanus sp.	*Cyclops* sp.	*Oncaea* sp.
	Calanus sp.	*Microsetella* sp.	*Paracalanus* sp.
	Conchoecia sp.	Nauplius	*Podon* sp.
	Corycaeus sp.	*Neocalanus* sp.	*Temora* sp.
	Eucalanus sp.		
Polychaeta	Larva		
Rhizopoda	*Arcella* sp.	*Difflugia* sp.	
Rotifera	*Asplanchna* sp.	*Lepadella* sp.	*Trichocerca* sp.
	Brachionus sp.	*Notholca* sp.	*Urostyla* sp.
Sagittoidea	*Sagita* sp.		
Sarcodina	*Globigerina* sp.		
Urochordata	*Oikopleura* sp.		

The zooplankton species composition based on sampling time and location is shown in **Figure 2**. Malacostraca and Ciliata are the dominant classes because they are always found at all data sampling times and in areas with a much higher species composition than other zooplankton classes. Malacostraca was most abundant at Station 2, with a species composition of 68.3%, and in August, with a species composition of 84.6%. The highest ciliate composition was found at Station 4, with a species composition of 52.3% and a species composition reaching 59.2% in October.

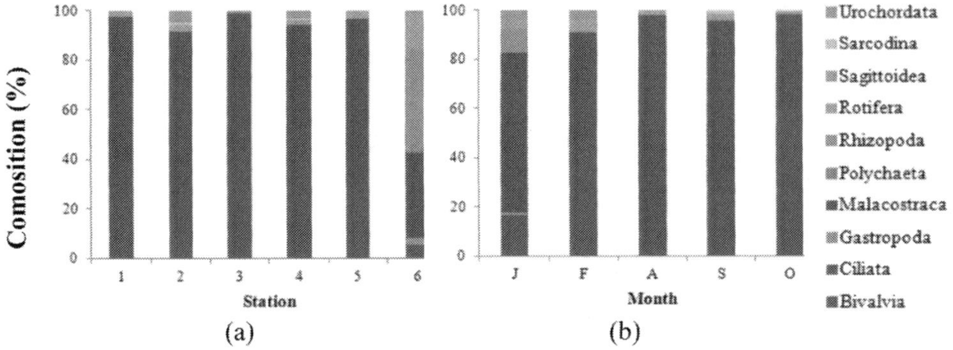

Fig. 2. Zooplankton composition based on sampling (a) location (b) time found in Bojonegar a coastal waters

The average abundance of zooplankton, based on the location and time of sampling, is shown in **Figure 3**. The average abundance of zooplankton ranges from $2.8x10^4$ to $9.9x10^4$ ind/L. The highest abundance was observed at Station 2 (an estuary station), and the lowest abundance was observed at Station 1 (a river station). Based on **Figure 3** below, it can be observed that the average abundance of zooplankton has a fluctuating value, with the highest average abundance at the estuary station and the lowest abundance at the river station. The average abundance of zooplankton based on sampling time showed a decrease in zooplankton abundance from January to February, but this decrease was not drastic. The abundance of zooplankton reached its lowest value in August, increased in September, and then decreased again in October.

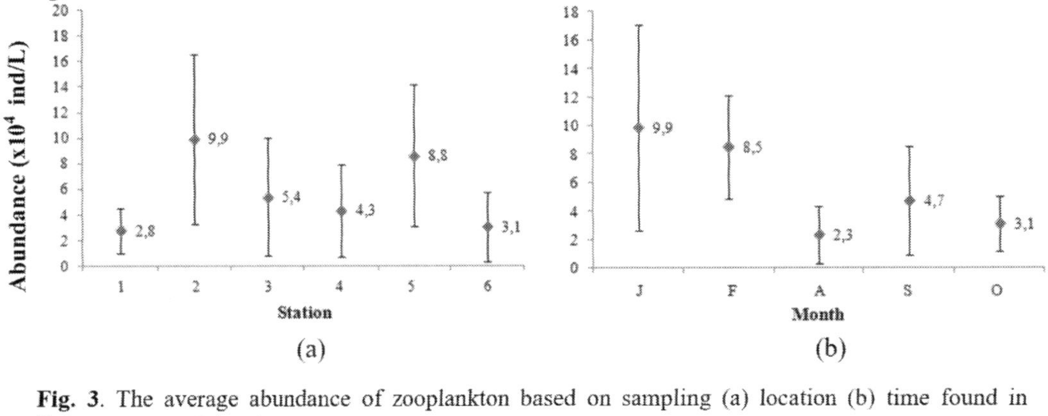

Fig. 3. The average abundance of zooplankton based on sampling (a) location (b) time found in Bojonegara coastal waters

The results of the diversity (H'), evenness (E), and dominance (C) indices are shown in Figure 4. The zooplankton diversity index (H') based on data sampling locations ranges from 1.061-1.646, where the highest diversity index was at Station 2 and the lowest was found at Station 1. The zooplankton diversity index (H') based on sampling time ranged from to 1.147-

1.428, with the highest diversity index in September and the lowest in October. According to [19], the diversity index ranging from 1 to 3 can be categorized as moderate diversity, so it can be concluded that zooplankton in the Bojonegara coastal waters, Banten Bay, is the average variety.

The evenness index (E) of zooplankton based on the data sampling location ranged from 0.713-0.771, where the highest evenness index was at Station 6, and the lowest was at Station 5. The evenness index (E) of zooplankton based on data sampling time ranged from to 0.690-0.815, with the highest evenness index found in August and the lowest in February. According to [20], the evenness index category, which is higher than 0.6, is included in the category of high species uniformity. It can be concluded that zooplankton in the Bojonegara coastal waters of Banten Bay have a high species uniformity.

The dominance index (C) of zooplankton based on data sampling location ranged from 0.282-0.433, where the highest dominance index was at Station 3, and the lowest was at Station 2. The dominance index (C) of zooplankton based on the data sampling time ranged from 0.331-0.445, with the highest dominance index found in October and the lowest in February. The dominance index value was less than 0.5; thus, according to [19], zooplankton in Bojonegara have low dominance or no zooplankton species domination.

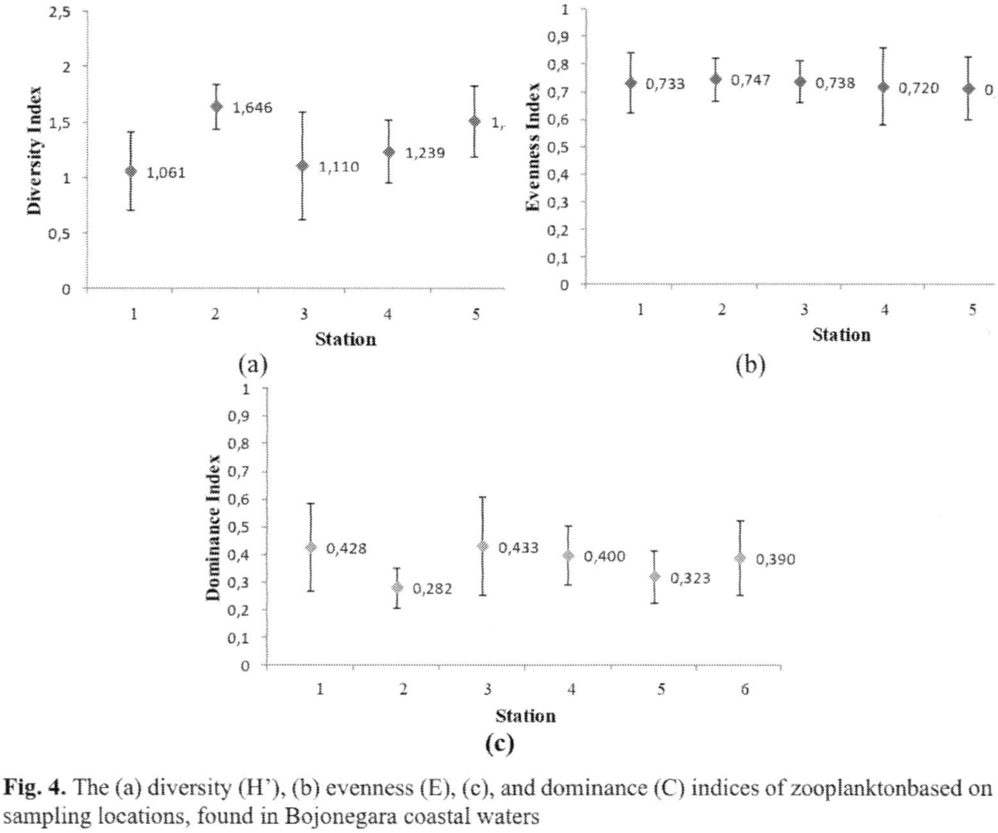

Fig. 4. The (a) diversity (H'), (b) evenness (E), (c), and dominance (C) indices of zooplanktonbased on sampling locations, found in Bojonegara coastal waters

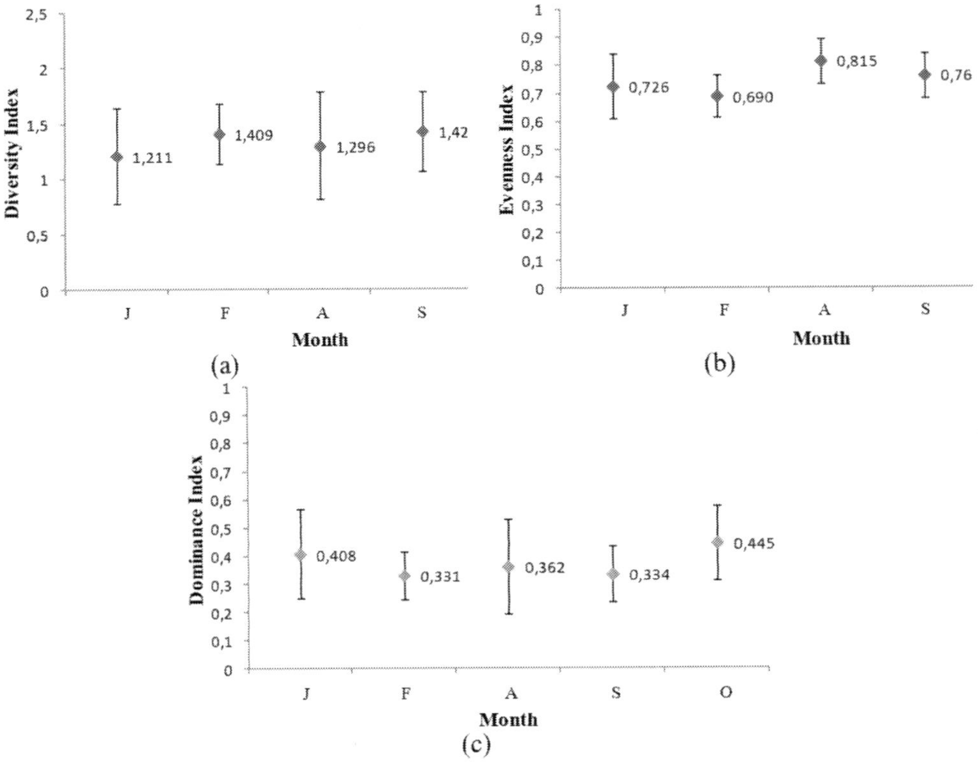

Fig. 5. The (a) diversity (H'), (b) evenness (E), (c), and dominance (C) indices of zooplankton based on sampling time, found in Bojonegara coastal waters

PCA has been used to explain the relationship between plankton abundance and various water parameters [1]. The water parameters observed in this study were temperature, transparency, turbidity, salinity, pH, dissolved oxygen (DO), nitrate, and phosphate. The zooplankton biplot PCA showed that the existing biplot explained 78.54% of the total data, with F1 explaining 56.36% of the data and F2 describing 22.18% of the data. The PCA results showed that zooplankton abundance was strongly influenced by temperature and nitrate parameters. In addition, through PCA, the highest abundance of zooplankton was found at stations 5 and 2.

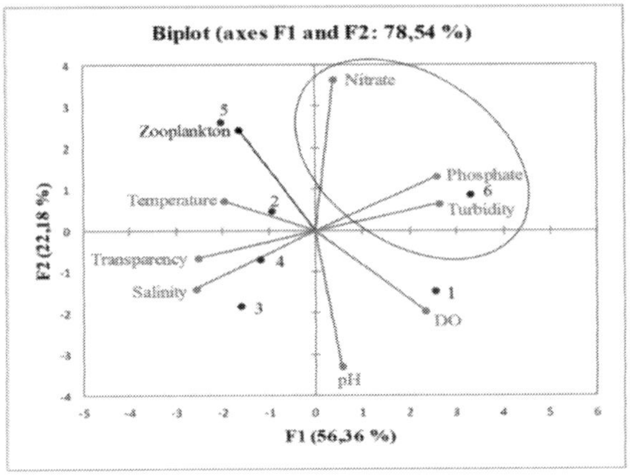

Fig. 6. PCA results from the tendency of water quality parameters with zooplankton abundance in each zona, found in Bojonegara waters

Figure 7 shows the station similarity dendrogram based on the Bray-Curtis similarity index, which was based on zooplankton abundance data from each station. Based on the zooplankton abundance data, it can be observed that Stations 2 and 5 have the highest similarity index, which is 0.90. The similarity index was also high at stations 3 and 4, with a similarity index of 0.85. Station 1 had the highest similarity index at Station 4 (0.77). Station 6 showed no similarity to the other stations.

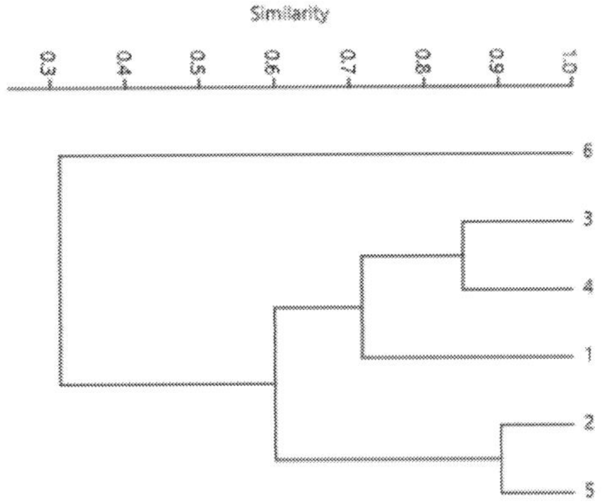

Fig 7. Similarity dendrogram based on zooplankton abundance found in Bojonegara waters

The values of the physical and chemical water parameters based on the data sampling location in the Bojonegara coastal waters of Banten Bay are presented in Table 2. The physical parameters observed during data sampling were temperature, transparency, and turbidity, whereas the chemical parameters observed were salinity, pH, DO, nitrate, and phosphate. Five observation stations have a temperature that tends to be stable, which is 31-32°C, but Station 5 has a much higher temperature of 35°C. The highest transparency was at the sea station and the lowest was at the river station. The highest turbidity was found at river stations and the lowest was observed at sea stations. The transparency value is strongly influenced by the amount of sunlight that can penetrate the water. Sunny weather and clear water cause transparency to increase. The turbidity value was strongly influenced by the content of the dissolved particles in the water.

Salinity in Bojonegara coastal waters increases as they approach the sea. All data sampling stations have salinity values that tend to be high because they are in the salinity range of brackish waters. Differences in salinity strongly affect the characteristics of zooplankton in the water area because each class of zooplankton has a very different adaptability to water conditions. The pH of the waters at the six stations is in the range of 7; therefore, it can be said that the seas tend to be neutral. Dissolved oxygen (DO) in the waters ranges from to 4-6 mg/L, with the lowest dissolved oxygen value at Station 2 and the highest at Station 1. Nitrates and phosphates are essential nutrients required for plankton growth. The highest nitrate level was found in estuarine waters, and the lowest was found at sea stations. The highest level of phosphate was found at river stations, and the lowest was found at sea stations.

Table 2. Value of physical and chemical parameters based on sampling locations in Bojonegara coastal waters, Banten Bay

Parameter	Station					
	1	2	3	4	5	6
Temperature (°C)	32 ± 2.30	31 ± 1.00	32 ± 0.8	32 ± 2.1	35 ± 1.8	31 ± 2.0

Parameter	Station					
	1	2	3	4	5	6
Transparency (m)	0.2 ± 0.13	2.0 ± 1.21	3.0 ± 1.34	2.6 ± 1.07	1.5 ± 0.73	0.6 ± 0.93
Turbidity (NTU)	95.4 ± 42,75	10.1 ± 12,44	3.3 ± 3.24	2.8 ± 1.88	5.5 ± 3.63	139.6 ± 140.70
Salinity (ppm)	17 ± 13.6	29 ± 4.2	30 ± 5.1	29 ± 5.8	26 ± 9.2	4 ± 2.5
pH	8 ± 0.5	7 ± 0.5	7 ± 0.5	7 ± 0.4	7 ± 0.2	7 ± 0.4
DO (mg/L)	6.0 ± 1.07	5.6 ± 1.26	5.0 ± 1.22	5.2 ± 1.44	4.1 ± 0.40	5.7 ± 0.61
Nitrate (mg/L)	0.20 ± 0,094	0.34 ± 0,268	0.16 ± 0,158	0.28 ± 0,194	0.30 ± 0,241	0.29 ± 0.229
Phosphate (mg/L)	0.14 ± 0,070	0.02 ± 0,019	0.01 ± 0,009	0.01 ± 0,012	0.10 ± 0,161	0.18 ± 0.083

3.2 Discussion

Bojonegara coastal waters, Banten Bay, tend to be the same at each station, at approximately 31-32°C. Higher temperatures did not produce a significant difference in zooplankton abundance at Station 5. This could occur because the temperature is not a limiting factor for plankton abundance in the waters [23], so temperature differences will not cause a direct change in the quantity of zooplankton, as the limiting parameters still fulfill the quality standards.

The highest transparency was observed at the sea station, whereas the most elevated turbidity was observed at the river station. Transparency is highly influenced by the zooplankton abundance. This increased transparency value on the water surface and in clear waters, especially during the middle of the day, causes zooplankton to migrate closer to the bottom of the water [24]. This caused zooplankton abundance to tend to be low in waters with high transparency values. However, zooplankton abundance is closely related to phytoplankton abundance during grazing or predation [25]. Phytoplankton require high transparency to perform photosynthesis optimally [23].

Salinity measured at six stations has a relatively high range of, namely 4-30, so it can be categorized as brackish water [26]. The difference in salinity value differentiates between zooplankton species composition, where most of the stations (five stations) are dominated by Malacostraca and Ciliata, whereas the Rhizopoda class dominated one station. The Malacostraca and Ciliata classes are widely distributed at each station because they adapt to different salinities. In contrast, the Rhizopoda class is only found at stations with low salinity values (Station 6) because they can adapt well to freshwater [27].

Data sampling stations in Bojonegara coastal waters, Banten Bay, have neutral pH because the pH value is in the range of 7 to 8. The Indonesian government regulation No. 22 of 2021 states that the pH quality standard for aquatic biota is 7-8.5. Based on this statement, all the data sampling locations have already fulfilled the quality standard so that zooplankton can grow optimally. There was no significant difference in zooplankton at Station 1, even though it had the highest pH among the sampling locations.

Dissolved oxygen or DO conducted at data sampling stations in Bojonegara coastal waters, in general, have fulfilled DO quality standards for aquatic biota according to Indonesian government regulation No. 22 of 2021, which is > 5 mg/L. The DO value fulfilled the quality standard for five sampling stations because it has reached more than five mg/L, except for Station 5. The DO concentration was only 4.1 mg/L. Station 5 is the estuary station where the lowest DO levels located at the estuary station were also found in previous studies conducted by [28] and [29]. [26] stated that the water's dissolved oxygen content of water can originate from oxygen diffusion in the atmosphere and the photosynthetic activity of phytoplankton or other aquatic plants. The primary source of dissolved oxygen in water is photosynthesis. [28] stated that low DO levels in estuaries can occur because of high respiration by aquatic biota and aerobic decomposition by bacteria.

Nitrates and phosphates are essential nutrients for the growth and reproduction of zooplankton, because they play an essential role in cell formation [30]. The quality standard for nitrate levels according to Indonesian government regulation No. 22 of 2021 is more than 0.06 mg/L. Overall, the nitrate levels at each station and data collection location were above the quality standard, so it can be concluded that the current nitrate levels can support the growth of zooplankton. The quality standard for water phosphate levels set by the Indonesian government regulation No. 22 of 2021 is more than 0.015 mg/L. Phosphate levels in the water met the quality standards, except at Stations 3 and 4. However, this does not significantly affect zooplankton abundance because the phosphate levels are slightly below the quality standard, and other factors can still support the growth of zooplankton.

The identification results showed that zooplankton in the Bojonegara coastal waters consisted of 10 classes and 34 genera. The number of categories found was slightly different from [29] the results of a study on the Cengkok coast (Banten Bay). The zooplankton with the highest species composition were Malacostraca, followed by ciliata. These two classes dominated the Bojonegara coastal waters because they were always found at every sampling location and time in high abundance. Both types were also found in the coastal area of Cengkok [29]. The Malacostraca class was found the most because Malacostraca is a member of the phylum Crustacea. The phylum Crustacea is the most common zooplankton phylum found in marine areas because of its wide spread. It has excellent adaptability to various water conditions, even under extreme conditions [25]. In addition, crustaceans are zooplankton that are euryhaline or can adapt to a wide salinity range, so their abundance remains high at each sampling location and time [31]. Based on the number of classes found, the types of zooplankton in Bojonegara waters were more than those found on Majakerta beach [32], but less than those found on Cengkok beach [29].

Zooplankton abundance in Bojonegara coastal waters ranged from $2.8x10^4$ to- $9.9x10^4$ ind/L. The highest zooplankton abundance was found at the estuary station, whereas the lowest was found at the river station. The highest quantity of zooplankton found at the estuary station follows previous research conducted by [32], [33], and [34], where this could happen because the estuary area has sufficient nutrients to support phytoplankton growth, which would then be followed by the development of zooplankton. High abundance at the river mouth was not found along the Cengkok coast [29]. This zooplankton abundance value is higher than that of the zooplankton found in the Majakerta [32] and Bengawan Solo River [34]. Zooplankton were also relatively high during the rainy season, similar to the results of a previous study [35]. Similar conditions were also found in research conducted in the waters of Majakerta. Optimal phytoplankton growth will certainly support zooplankton growth directly, because zooplankton are the primary consumers and phytoplankton eaters. In addition, zooplankton abundance is highly influenced by phytoplankton abundance; therefore, zooplankton migrate to areas where there are many phytoplankton.

The zooplankton diversity index in the Bojonegara coastal waters ranged from 1.15-1.43, classified as moderate diversity. The zooplankton diversity index value is closely related to the availability of food, namely, the availability of phytoplankton. This shows that the aquatic ecosystem in the Bojonegara coastal waters provides a diverse supply of phytoplankton to fulfill the needs of various zooplankton classes [36]. The zooplankton evenness index ranged from 0.71-0.77, which indicating high species evenness. This shows that each class of zooplankton was evenly distributed, and the water was in good condition.

The zooplankton dominance index ranges from 0.28-0.43, so it indicating that zooplankton in Bojonegara coastal waters have low dominance or no class domination. Based on this statement, it can be concluded that the waters were in excellent and stable conditions because, if there was pollution, only a few types of zooplankton would survive. Hence, the dominance index value tends to be high and vice versa [37]. The values of diversity, evenness, and dominance indices of zooplankton in Bojonegara are different from the results of studies on Serang Beach (Yogyakarta) [12], Tangerang coast [38], Majakerta beach (Indramayu) [33], and Cengkok beach (Banten Bay) [29], with a wider range. However, this condition is different from the diversity and dominance index values, which have a better value in the waters of Mempawah (West Kalimantan) [11]. A comparison of the biological indices of the beaches is presented in Table 3.

Table 3. Comparison of biological indices in several water areas

	Tangerang [1)]	**Mempawah (Pontianak)** [2)]	**Majakerta (Indramayu)** [3)]	**Serang (Yogyakarta)** [4)]
H'	0.69-1.50	2.05-3.04	0.35-1.91	1.217-1.755
E	0.27-0.65	0.63-0.84	0.50-0.95	0.746-0.948
D	0.36-0.73	0.09-0.18	0.17-0.80	0.127-0.310

	Cengkok (Serang) [5)]	**Bojonegara** [6)]
H'	0.7-1.6	1.15-1.43
E	0.6-0.8	0.71-0.77
D	0.3-0.6	0.28-0.43

Remarks: [1)][38], [2)][11], [3)][33], [4)] [12]. [5)][29], [6)][Research results]

The PCA biplot shows the relationship between zooplankton abundance and physicochemical water parameters in the Bojonegara coastal waters of Banten Bay. Through the PCA biplot, it was observed that zooplankton abundance has a close relationship with specific parameters, namely temperature and nitrate. Temperature is a significant environmental factor for meroplankton or plankton groups that only live as plankton in the early stages, for example, in the egg and larval stages, and turn into nekton or benthos when they mature. One example is the bivalve class. Bivalves in the larval stage are closely related to temperature because temperature is the main factor determining their survival rate and growth. According to [39], optimum temperature can increase the success rate of Bivalvia larvae in completing the metamorphosis cycle. Zooplankton abundance, which is closely related to nitrate, can occur because nitrate is an important nutrient for phytoplankton growth, followed by zooplankton development. It plays an essential role in the formation of phytoplankton cells [30], which is the leading food source for zooplankton.

The similarity index shows the level of similarity at each data sampling station based on the zooplankton abundance. Through dendrogram similarity, it can be seen that the highest similarity was found at Station 2 with 5, followed by Station 3 with 4. The high similarity indicates that the zooplankton community structures at these two stations were similar. High similarity can occur in waters with similar characteristics, in this case, between estuary stations and other estuary stations and between sea stations and other sea stations. In addition, data collection locations that are close will increase the possibility of a similarity index between stations [40], for example, at Stations 3 and 4 because the two stations are located close.

4 Conclusion

Ten zooplankton classes were identified: Malacostraca (16 genera), Rotifera (6 genera), Ciliata (4 genera), Rhizopoda (2 genera), and one genus each from Bivalvia, Gastropoda,

Polychaeta, Sagittoidea, Sarcodina, and Urochordata. Malacostraca and Ciliata were the dominant zooplankton classes in the water. The highest zooplankton abundance was observed at estuary stations during the rainy season. The diversity index of zooplankton was categorized as moderate, the evenness index as high, and the dominance index as low, indicating no species dominance. These index values suggest that Bojonegara coastal waters are experiencing light environmental stress. Water parameters, such as salinity, pH, dissolved oxygen (DO), nitrate, and phosphate, were within the required standards for zooplankton growth. However, the temperature, transparency, and turbidity did not meet the standards at some sampling sites and times. The zooplankton community in the Bojonegara coastal waters of Banten Bay is mainly influenced by temperature and nitrate levels.

References

1. L.P.P. Dewanti, I.D.N.N Putra, E. Faiqoh, JMAS **4**(2), 324-335 (2018)
2. M. Uttieri, Y. Carotenuto, I.D. Capua, V. Roncalli, J. Mar. Sci. Eng., **11**(10), 1875–1875, (2023)
3. W. O. Wan Maznah, N. Kassim, Z. Din, Sains Malays., **50**(6), 1577–1588, (2021)
4. [BPS] Badan Pusat Statistik Kabupaten Serang, Serang (ID), BPS Kab. Serang (2019)
5. Z. Wang, R. Li, F. Wu, C. Feng, C. Ye, C. Yan, Mar. Pollut. Bull., **114**, 397-407, (2017)
6. R.F. Heneghan, J.D. Everett, J.L. Blanchard, P. Sykes, A.J. Richardson, Nat. Clim. Change, **13**(5), 470–477, (2023)
7. G. Papantoniou, Y. Cladas, V. Ketsilis-Rinis, Z. Vaitsi, N. Fragopoulu, Estuar. Coast. Shelf Sci., **245**,106985, (2020)
8. L. Ratnarajah, R. Abu-Alhaija, A. Atkinson, S. Batten, N.J. Bax, K.S. Bernard, G. Canonico, A. Cornils, J.D. Everett, M. Grigoratou, N.H.A. Ishak, D.G. Johns, F. Lombard, E. Muxagata, C. Ostle, S. Pitois, A.J. Richardson, K. Schmidt, L. Stemmann, L. Yebra, Nat. Commun., **14**(1), 564, (2023)
9. S. Lomartire, J.C. Marques, A.M.M. Gonçalves, Ecol. Indic., **129**, 107867–107867, (2021)
10. G.E.P. Murphy, T.N. Romanuk, B. Worm, Ecol. Evol., **10**(4), 2170-2181, (2020)
11. X. Sun, H. Zhang, Z. Wang, T. Huang, W. Tian, H. Hai, Microorganisms, **11**(7), 1638–1638, (2023)
12. M. Décima, Mar. Ecol. Prog. Ser., **692**, 23–42, (2022)
13. I. Minggawati, Ziraa'ah, **39**(2), 81-85, (2014)
14. Y. Yuan, M. Jiang, X. Liu, H. Yu, M.L. Otte, C. Ma, Y.G. Her, Ecol. Indic., **91**, 1-12, (2018)
15. I. Yamaji, Illustration of the Marine Plankton of Japan, Osaka (JP), Hiokusha Publishing.co.ltd, (1979)
16. T. Mizuno, Illustration of the Freshwater Plankton of Japan, Osaka (JP), Hoikusha Publishing.co.ltd., (1979)
17. [APHA] American Public Health Association, Standard Methods for the Examination of Water and Wastewater. 21st ed., Ohio (US), (2012)
18. J. Haberman, M. Haldna, J. Limnol., **73**(2), (2014)
19. K.L. Bhavani, G. Venkatesh, P. Reddy, T.J. Chari, D.R. Kumar, IJISRT, **9**(10), 1610–1614, (2024)

20. S.R. Karmakar, M.B. Hossain, M.M. Sarker, A.-A.U. Nur, A. Habib, B.A.Paray, M.K. Al-Sadoon, A. Gulnaz, T. Arai, Diversity, **14**(9), 755–755, (2022)
21. D.H. Ismunarti, IJMS, **18**(1), 14-19, (2013)
22. L. Legendre, P. Legendre, Numerical Ecology, Elvesier Scientific Publishing Company, (1983)
23. I. Faturohman, Sunarto, I. Nurruhwati, JPK, **8**(1), 115-122, (2016)
24. M. Wati, N. Irawati, Indrayani, Jurnal Manajemen Sumberdaya Perairan, **4**(1), 61-73, (2019)
25. K. Paiki, J.D. Kalor, E. Indrayani, L. Dimara, Maspari J. **10**(2), 199-206, (2018)
26. J. Weinstock, L.G. Vargas, R. Collin, J. Mar. Sci. Eng., **10**(3), 427–427, (2022)
27. B.K. Yudhatama, S. Redjeki, C.A. Suryono, J. Mar. Res., **8**(3), 322-327, (2019)
28. D. Sulistiowati, R.H.R. Tanjung, D. Lantang, Jurnal Biologi Papua. **8**(2), 79-96, (2016)
29. E.M. Syafitri, E.S. Srimariana, Sulistiono, IOP Conf. Ser.: Earth Environ. Sci., **744**, 012088, (2021)
30. K. Paiki, J.D. Kalor, J. fish. mar. sci., **1**(2), 65-71, (2017)
31. S.B.A. Razak, Z. Sharip, Ecol. Process, **8**(44), (2019)
32. H. Widyarini, N.T.M. Pratiwi, Sulistiono, JITKT, **9**(1), 91-103, (2017)
33. M. Junaidi, Nurliah, F. Azhar, J. Biol. Tropis., **18**(2), 159-169, (2018)
34. R. Syafarina, R. Widodo, Sulistiono, N.T.M. Pertiwi, Biospecies. **11**(1): 19-36, (2018)
35. H. Pratiwi, A. Damar, Sulistiono, Biodiversitas, **19**(6), 2104-2110, (2018)
36. K.S. Tan, J. Ransangan, Estuar. Coast. Shelf Sci., **194**,16-29, (2017)
37. Sumiati, M. Kasim, Halili, JMSP, **5**(1), 37-49, (2020)
38. N.T.M. Pratiwi, Ardhito, D.Y. Wulandari, A. Iswantari, Procedia Environ. Sci., **33**, 470-477, (2016)
39. I. Hossain, M.B. Hossain, Acta Ecol. Sin., **41**(6), 611–619, (2021)
40. P. Wirabumi, Sudarsono, Suhartini, Jurnal Prodi Biologi, **6**(3), 174-184, (2017)

Smart temperature and vibration monitoring device for small-scale fishing vessel engines

Yuniar Endri Priharanto[1,2*], *Indra Jaya*[1], *Ayi Rahmat*[1], *Medria Kusuma Dewi Hardhienata*[1], *Donwill Panggabean*[3]

[1]IPB University, Bogor, Indonesia
[2]Politeknik Kelautan dan Perikanan Dumai, Dumai, Indonesia
[3]Universitas Terbuka, Jakarta, Indonesia

Abstract. Monitoring engine condition parameters is crucial for ensuring efficiency and reliability of small-scale fishing vessels. However, existing monitoring device and their application remains limited and they are not extensively utilized for small-scale fishing vessels. This study aims to design and develops a real-time temperature and vibration monitoring device for small-scale fishing vessel propulsion. The proposed device integrates temperature and accelerometer sensors, with ESP32 serving as the main processor. Blynk was used as a real-time monitoring platform. The prototype was tested on a single-cylinder diesel engine by comparing its monitoring results with a standard measurement tool. A trial monitoring the parameters of a single-cylinder diesel engine was conducted for 30 min to assess accuracy by the Mean Absolute Percentage Error (MAPE) value. The results of the trial indicated that the device could effectively record temperature and engine vibration data with a MAPE value below 20%. This level of accuracy indicates the effectiveness of monitoring the performance parameters. The proposed monitoring device can offer notifications regarding the conditions of the small-scale fishing vessel propulsion engines.

1. Introduction

Degradation of engine performance can lead to significant operational failures and increased maintenance costs in fishing vessels. Therefore, it is crucial to select maintenance strategies that align with the characteristics of the engine and system conditions [1,2]. The maintenance and repair of industrial and marine systems have received significant attention over the past few decades owing to their high capital investment and demand for high availability. Determining an optimal maintenance policy that balances the costs of preventive maintenance and system failure is essential. Various efforts are continually made to collect and analyse data from past experiences to prevent risky situations and enhance system reliability, as well as to develop effective preventive maintenance planning [3]. Effective maintenance and condition monitoring are vital because 75% to 96% of maritime accidents are caused by human errors at three levels of behavior: knowledge-based, rule-based, and

* Corresponding author: joeniarendri@apps.ipb.ac.id

skill-based [4]. In addition, the complexity of ship engine maintenance activities onboard, where seafarers perform tasks under various maritime environmental conditions (such as extreme weather, ship movements, noise, and vibrations) and operational conditions (such as excessive workload and stress), adds to this challenge [5]. In such conditions, the ability of humans to manually obtain engine performance data is limited compared to automated systems. The large volume of data and complexity of engine parameters require fast and accurate analysis, which can exceed the cognitive capacity of human operators. Therefore, it is important to consider human-machine interaction to enhance machine safety [6].

Several studies have addressed the challenges of monitoring engine performance, emphasizing the importance of machine condition monitoring using Condition-Based Maintenance (CBM) methods. These methods integrate temperature, pressure, torque, and fuel flow data to enhance the engine reliability and efficiency [7]. Temperature monitoring is crucial to prevent engine damage caused by extreme exhaust gas temperatures. This technique employs advanced sensors to detect thermal overload conditions, ensuring that the engine operating temperatures remain within safe limits, thereby reducing the risk of wear and damage [8]. The exhaust gas temperature in internal combustion engines is also measured to evaluate measurement errors caused by radiation, conduction, and the dynamic response of thermocouples or platinum resistance thermometers when placed in the exhaust gas flow. This allows for a more accurate temperature estimation, contributing to a better engine performance evaluation and supporting the monitoring of ship operational performance and efficiency [9,10]. In addition to engine temperature monitoring, methods for monitoring engine performance using vibration signals have also been explored. Methods for diagnosing valve clearances in piston combustion engines using vibration signals are conducted by employing triaxial accelerometers mounted on the cylinder head of a 4-stroke 6-cylinder diesel engine [11], and misfire detection in engines using sound quality metrics from generated noise [12]. Internal MEMS (Micro Electro Mechanical Systems) accelerometers are also used to measure rotor vibration directly without requiring additional sensors on the stator [13].

The development of engine monitoring using sensors involves integrating Artificial Neural Networks, including artificial neural networks (ANN), K-nearest neighbors (KNN), Fast Fourier Transform (FFT), Short-Time Fourier Transform (STFT), and Support Vector Machines (SVM), to model engine performance and detect faults. This integration enables more responsive and timely maintenance, minimizing the risk of operational failures, with accuracy rates ranging from 94% to 100% [7,11,12,14–16].

Another technology that has been utilized in machine condition monitoring is IoT, which can be employed to develop mathematical models based on monitoring data for diesel generator performance analysis, as well as fuzzy logic-based decision support systems to provide real-time warnings in critical situations [17]. The use of IoT can enhance maintenance efficiency and effectiveness and support predictive maintenance with higher reliability [16,17]. The integration of these technologies holds significant potential for improving machine reliability and operational efficiency as well as supporting more effective predictive maintenance.

Nevertheless, research has been conducted on the development of engine condition monitoring systems utilizing temperature and vibration sensors connected to IoT technology for small-scale fishing vessels under 5 GT and powered by single-cylinder diesel engine with less than 12 HP remains limited. Specifically, studies on the effective and efficient integration of multi-sensor systems for small fishing boats are sparse, despite the potential of such systems to combine temperature and vibration sensors into a single device that is easy to install and use on small vessels. Furthermore, infrastructure limitations in maritime environments often hinder the adoption of IoT technology and automation systems, highlighting the need for innovation in developing energy-efficient devices capable of autonomous operation. Therefore, the aim of this study is to develop an automatic monitoring system for small-scale fishing boat engines using temperature and vibration sensors to record data in real time. This system aims to provide a more efficient solution for monitoring engine conditions in small fishing vessels, reducing the risk of damage, and enhancing operational efficiency. Identifying potential risks and their causes can help determine critical components, thereby enabling the implementation of mitigation measures to lower risk levels.

2. Material and Method

This study used two approaches. The first approach uses cyber-physical instrumentation. This involves designing an IoT-based real-time engine monitoring system capable of recording engine temperature and vibration. By leveraging the IoT technology, the system can continuously collect data from machines and transfer it directly to the monitoring platform.

The second approach utilizes Artificial Intelligence, specifically Artificial Neural Networks (ANN) [18]. In this approach, the collected machine data are used to analyse the condition of the machine. The ANN processes these data and produces accurate machine-condition predictions based on patterns identified from the received temperature and vibration data. This approach aids in the early detection of potential problems, allowing for more effective preventative maintenance [18,19].

The combination of these two approaches facilitates the development of systems that not only accurately monitor machine conditions in real time, but also provide in-depth analysis to support better and timely maintenance decisions.

The detailed research methodology applied in this study is outlined in the flowchart presented in Figure 1.

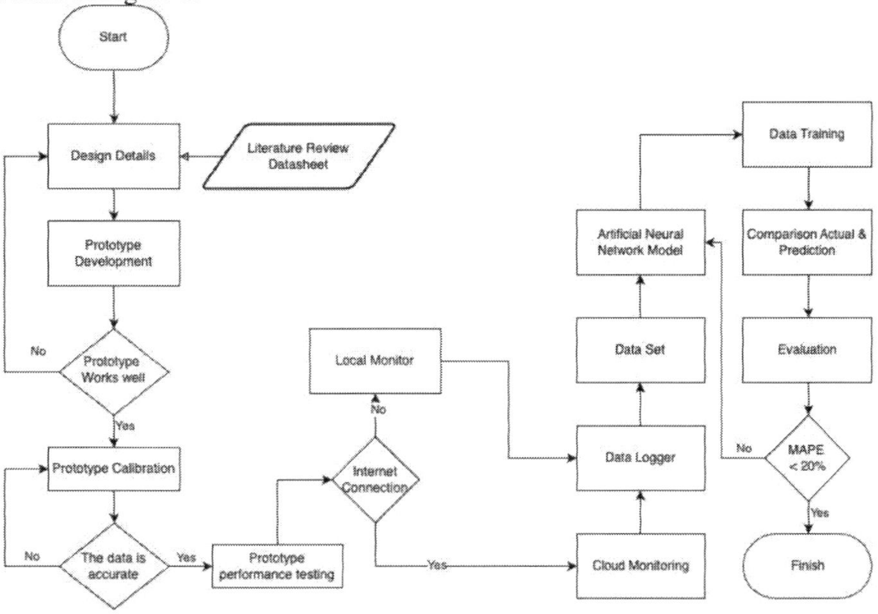

Fig 1. Flowchart Diagram of Research Methodology

In the detailed design phase, component selection was performed, the system was designed, and all necessary technical requirements were specified by referring to literature reviews and component datasheets. This step was conducted to ensure that the prototype functioned correctly. The next phase involves prototype development, which includes assembling the components, developing the required software, and integrating the system into an internet network. The testing phase was conducted to verify that the prototype operated as expected. If the prototype does not perform as anticipated, further development will be undertaken to make improvements. Once the prototype operates satisfactorily, calibration is performed to ensure that all the sensors and systems within the prototype operate with the desired accuracy. The evaluation of the generated data is essential to confirm that the data produced are accurate.

During the prototype performance testing phase, system testing was conducted to ensure that the prototype functioned properly in a single-cylinder diesel engine. Simultaneously, Internet connectivity was tested to allow for online data monitoring. In the absence of an Internet connection, the prototype stored the data locally. The data from the prototype were then transmitted to a cloud-based monitoring system, enabling more comprehensive and real-time data analysis. The recorded data were subsequently compiled into a complete dataset, which was utilized for the analysis and model training processes.

In the second approach, Artificial Neural Networks (ANNs) were used for analyse the collected dataset. ANNs were chosen for their ability to identify patterns and make predictions based on data. The collected data were used to train an ANN model. This training process aims to enhance the accuracy of the model predictions. The ANN model utilized 512 neurons and applied the rectified linear unit (ReLU) activation function. The network was trained for 1000 epochs. In this study, 80% of the data were used for training. After the model was trained, the predicted machine conditions were compared with actual data. The purpose of this comparison was to evaluate the performance of the model and to ensure prediction accuracy. Finally, an evaluation phase was conducted to assess the overall process, including the prototype performance, data accuracy, and model prediction effectiveness. This evaluation is crucial for determining whether all research objectives have been met [7,20].

To evaluate the accuracy and suitability of the ANN model and prototype, performance indicators such as the Mean Absolute Percentage Error (MAPE) are used to assess the accuracy of both the model and the prototype.

$$MAPE = \frac{1}{n}\sum_{t=1}^{n}\frac{|F_t - A_t|}{|A_t|} \times 100 \quad (1)$$

Where F_t is the measurement value of the prototype or the predicted value, A_t is the measurement value of the measuring instrument in period t or the actual value, and n is the number of measurement/data periods [21,22]. A smaller Mean Absolute Percentage Error (MAPE) value indicates better model performance; a reduced MAPE indicates that the prototype measurement results closely correspond to the actual values. A model is considered effective if its MAPE is less than 20% [22,23].

Figure 2 illustrates the hardware configuration and data flow process of the engine-monitoring system. The system utilizes a 5-volt battery as the power source to activate the ESP32. Two types of sensors were employed: an MLX90614 temperature sensor and a GY-521 vibration sensor. All data collected by these sensors were subsequently transmitted to the ESP32 TTGO T-SIM8000H microcontroller, which features GPRS communication capabilities. This microcontroller serves as the system processor, and manages and processes the data received from the sensors. The processed data are then sent to a data logger or a storage module. This data logger module is responsible for storing all the temperature and vibration data. In addition to local data storage, the system was equipped with cloud-based monitoring capabilities using the Blynk platform. This enables the real-time monitoring of temperature and vibration through devices such as laptops or smartphones.

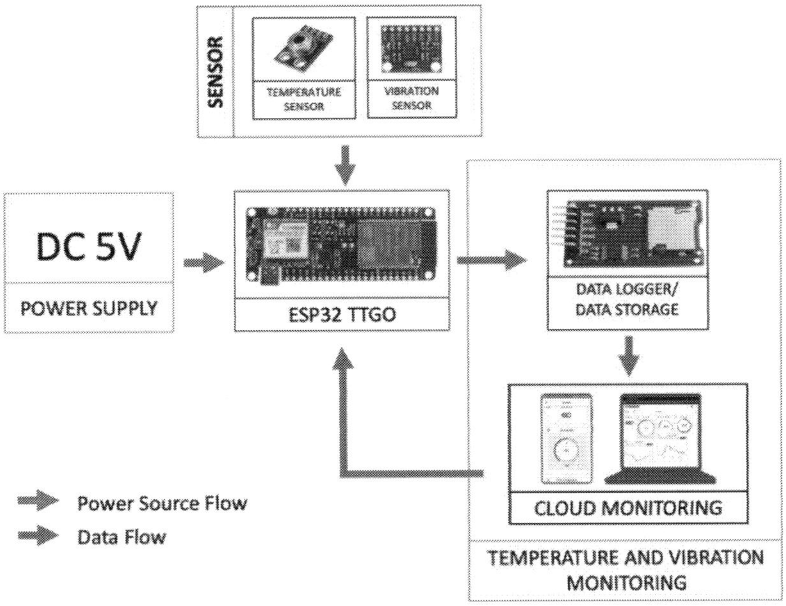

Fig 2. Hardware Block Diagram

3. Result and Discussion

The system was constructed by connecting the sensor components and other elements to the ESP32, following the wiring diagram illustrated in Figure 3. The entire system was assembled within a black enclosure to facilitate the operational processes. Subsequently, tests were conducted on a single-cylinder diesel engine operated under no-load conditions to verify and ensure the reliability and consistency of the measurements. During the testing phase, engine parameters such as temperature and vibration were periodically measured using a vibration-measuring device (vibration test) and infrared thermometer.

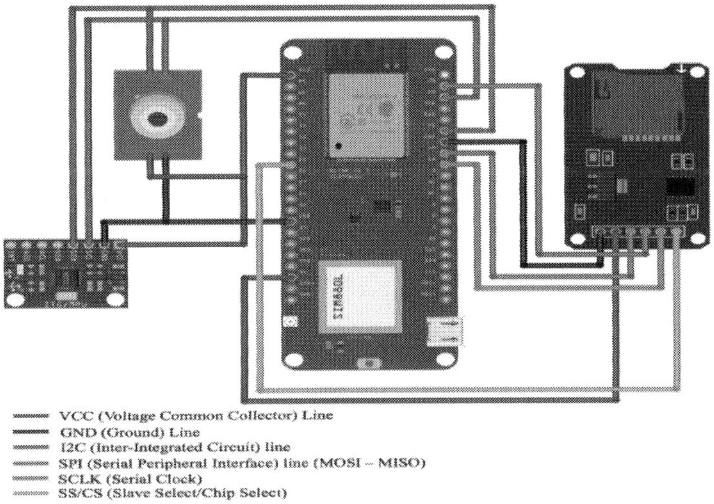

Fig 3. Wiring Diagram for Smart Temperature and Vibration Monitoring System (Schematic Made in Fritzing)

Figure 4 illustrates a prototype of a smart temperature and vibration device used for monitoring temperature and vibration parameters in small-scale fishing vessel engines. This system employs an Internet connection via GPRS signals, enabling real-time data transmission to the Blynk platform for immediate engine condition monitoring and historical data review.

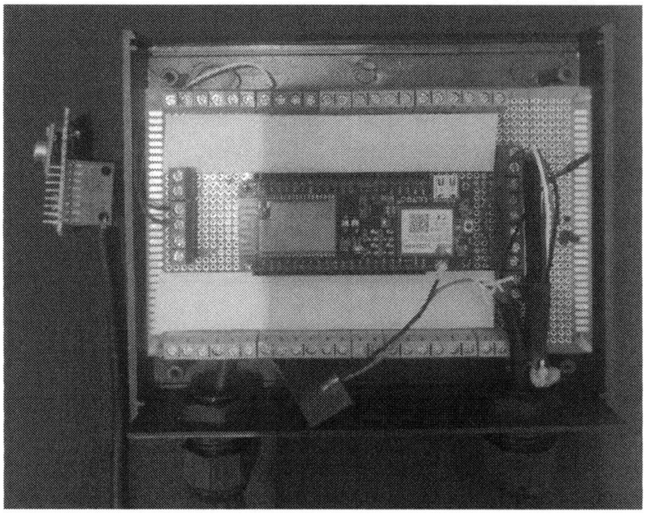

Fig 4. Prototype of Smart Temperature and Vibration Monitoring Device

During the development phase of the cyber-physical instrument, the Blynk platform was used to connect the monitoring devices to an online dashboard. This dashboard is designed to display measurement parameters, such as temperature and vibration sensor readings, enabling the real-time monitoring of engine parameters. This monitoring can be accessed through various devices, including computers, laptops, and smartphones, thereby enhancing the flexibility of engine parameter monitoring. The dashboard of this platform is shown in Fig. 5.

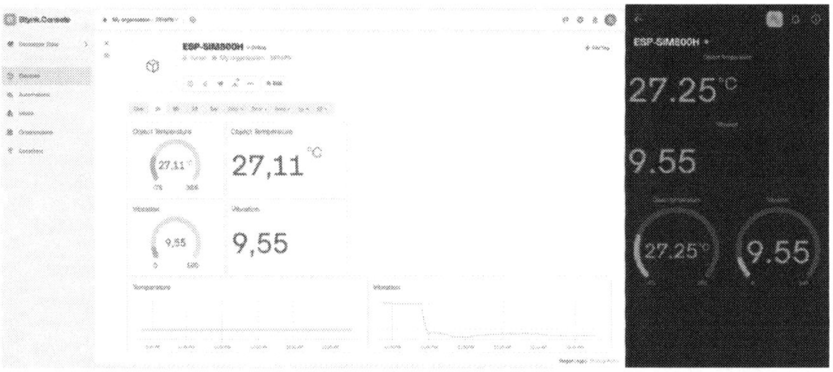

(a) and (b).

Fig 5. (a) Web Based Dashboard (b) Mobile Based Dashboard

The temperature and vibration sensors were tested using a prototype monitored under 'online' conditions on a Blynk platform dashboard. Figure 4 shows that the system operates effectively in an online state, with the temperature and vibration parameters being readable on the dashboard. The use of Blynk as an interface across various devices highlights its ease of access for monitoring machine conditions. As shown in Figure 5. A, the application display on a computer or laptop facilitates machine monitoring through the panel, with measured temperature readings of 27.11°C and vibration readings of 9.55 m/s². Figure 5.b shows the display when accessed via a smartphone. This monitoring system exhibits flexibility in machine monitoring across different devices, thereby expanding the scope of machine observation.

Figure 6 illustrates the results of prototype testing on a single-cylinder diesel engine used as a propulsion system for a small-scale fishing vessel. In this test, the measurements obtained from the developed monitoring prototype were compared with those obtained using a vibration-testing device and infrared thermometer. This comparison yielded Absolute Percentage Error (APE) values, which were then averaged to calculate the Mean Absolute Percentage Error (MAPE).

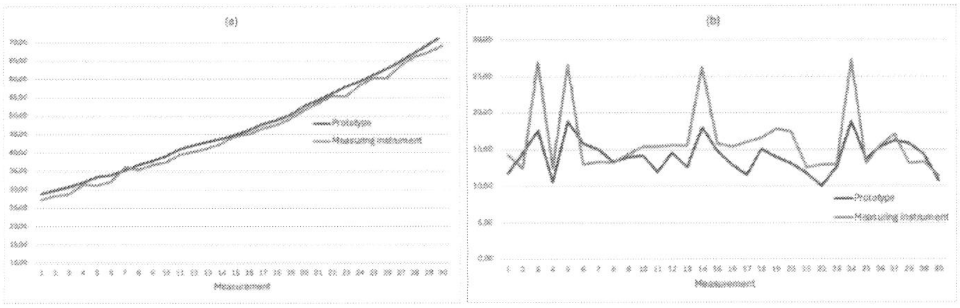

Fig 6. Graphics of Prototype Performance Test Results: (a) temperature and (b) vibration

For temperature measurements, the Mean Absolute Percentage Error (MAPE) obtained is 3.2%. This indicates that the prototype demonstrated a relatively high level of accuracy in temperature measurements, with an average absolute percentage error of 3.2%. Meanwhile, For vibration measurements, the MAPE value reached 18.6%. The vibration data exhibited more fluctuations than the temperature data. This is partly due to the nature of the vibration measurements, which can vary significantly in response to minor changes in the conditions. Therefore, it is evident that the prototype still exhibits variability in vibration measurement accuracy, with an average absolute percentage error of 18.6%.

There is a discrepancy between the prototype and the measurement instrument concerning the vibration measurements. The Absolute Percentage Error (APE) for vibration remained below 20%, with a Mean Absolute Percentage Error (MAPE) calculated at 18.6%. This indicates that although the prototype is capable of capturing general vibration trends, there is a greater level of error compared to temperature measurements.

ANN Model Development

The dataset utilized was derived from temperature and vibration measurements collected using a prototype device that served as the input layer. The machine condition, which was assessed based on the measured parameters, was used as the output. The accurate labelling of the output layer and partitioning of the dataset into training and testing sets ensured the accuracy of the model.

The details of the input layer dataset in the ANN analysis were temperature (°C) and vibration (m/s²). These inputs represent the temperature and vibration readings from sensors installed on the engine. The Output Layer (labels) consists of the machine condition, which is the target variable to be predicted by the ANN. The categories used for the machine's condition are normal (1) if the engine operates within normal parameters, warning (2) if the engine shows signs of potential issues that may require attention, and critical (3) if the engine is experiencing significant problems and may require immediate action. This dataset was utilized by the ANN to learn the patterns of the input and output data for prediction purposes. The dataset was divided into two parts, 80% for the training set and 20% for the testing set. The Training Set was used to train the ANN, including a balanced representation of normal, warning, and critical conditions. The Testing Set was used to evaluate the ANN's performance on data that the model had not previously encountered.

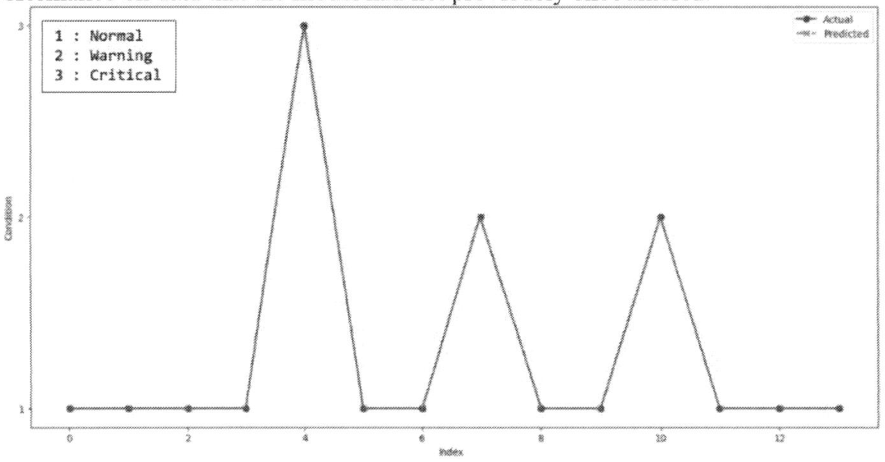

Fig 7. Comparison of Actual and Predicted Values

During the training phase, the Artificial Neural Network (ANN) received temperature and vibration data from the training set. The objective was to adjust the weights to minimize the error between the predictions and actual machine condition labels. The training process spanned 1000 iterations (or epochs), with each epoch representing a complete pass through the training set.

After undergoing 1000 epochs of training, the ANN achieved a 100% accuracy rate on the test data, as illustrated in Figure 7. This is evident from the line chart, which displays the results from the test set with two lines: a blue line representing the actual machine condition values, and an orange line representing the predicted values generated by the ANN. The closer these lines are to each other, the more accurately ANN predicts the condition of the machine. Therefore, this graph shows that the ANN provides predictions that closely align with the actual values, demonstrating the excellent performance quality of the developed model. Consequently, the ANN achieved a Mean Absolute Percentage Error (MAPE) of 0%, indicating that its predictions were highly accurate.

Achieving a Mean Absolute Percentage Error (MAPE) of 0%, this Artificial Neural Network (ANN) demonstrates an exceptionally high level of accuracy in its predictions. This indicates that the model does not produce significant errors in its estimations, where high accuracy is essential for preventing engine damage and ensuring safe and efficient operation. These results open up opportunities for the further application of ANN in machine condition monitoring systems, given their capability to effectively interpret and process sensor data.

4. Conclusion

A cyber-physical system for engine temperature and vibration monitoring was developed using the ESP32 TTGO and Blynk platforms. Prototype testing demonstrated good performance, with MAPE values below 20%. This indicates a satisfactory level of accuracy in monitoring and measuring the supporting parameters of the engine performance. The ANN Model Training Result shows an accuracy of 100% and Error Value (MAPE) of 0%, indicating that the model has been trained well and performs very well in predicting machine conditions.

References

1. Y. E. Priharanto, R. I. Yaqin, G. Marjianto, J. P. Siahaan, M. Z. L. Abrori. J. Fail. Anal. Prev. **23**, 822–836 (2023).
2. M. H. Abidi, M. K. Mohammed, H. Alkhalefah, Sustain. **14**, (2022).
3. K. Cicek, H. H. Turan, Y. I. Topcu, M. N. Searslan, 2010 2nd Int. Conf. Eng. Syst. Manag. Appl. ICESMA 2010 (2010).
4. A. Galieriková, Transp. Res. Procedia **40**, 1319–1326 (2019).
5. R. Islam, M. Anantharaman, F. Khan, V. Garaniya, TransNav **14**, 43–47 (2020).
6. N. Hasanspahić, S. Vujičić, V. Frančić, L. Čampara, J. Mar. Sci. Eng. **9**, 1–16 (2021).
7. O. C. Basurko, Z. Uriondo, Appl. Therm. Eng. **80**, 404–412 (2015).
8. S. K. Nanda, B. Jia, A. Smallbone, A. P. Roskilly, Energies **10**, (2017).
9. N. Papaioannou, F. Leach, M. Davy, J. Eng. Gas Turbines Power **142**, (2020).
10. O. S. Markelova, Proceedings of the 2022 Conference of Russian Young Researchers in Electrical and Electronic Engineering, ElConRus 2022.
11. M. Tabaszewski, G. M. Szymański, Eksploat. i Niezawodn. **22**, 331–339 (2020).
12. S. Singh, S. Potala, A. R. Mohanty, Proc. Inst. Mech. Eng. Part D J. Automob. Eng. **233**, 3112–3124 (2019).
13. S. Jiménez, M. O. T. Cole, P. S. Keogh, J. Sound Vib. **377**, 58–75 (2016).
14. S. M. Ramteke, H. Chelladurai, M. Amarnath, J. Vib. Eng. Technol. **10**, 175–192 (2022).
15. B. F. Yildirim, B. A. Mercangoz, Eurasian Econ. Rev. (2020).
16. I. Koene, V. Klar, R. Viitala, HardwareX **7**, 1–15 (2020).
17. A. G. Mohapatra, A. Mohanty, N. R. Pradhan, S. N. Mohanty, D. Gupta, M. Alharbi, A. Alkhayyat, A. Khanna, Alexandria Eng. J. **76**, 525–541 (2023).
18. C. Liu, C. Zhao, Y. Wang, H. Wang, Sensors **23**, (2023).
19. A. N. Bhatt, N. Shrivastava, Arch. Comput. Methods Eng. **29**, 897–919 (2022).
20. G. Ahmad, S. Imran, M. Farooq, A. N. Shah, Z. Anwar, A. U. Rehman, M. Imran, Sustain. **15**, (2023).
21. P. C. Chang, Y. W. Wang, Expert Syst. Appl. **30**, 715–726 (2006).
22. L. F. Rahman, M. Marufuzzaman, L. Alam, M. A. Bari, U. R. Sumaila, L. M. Sidek, Sustain. **13**, 1–14 (2021).
23. N. T. Ngo, A. D. Pham, T. T. H. Truong, N. S. Truong, N. T. Huynh, Sci. Rep. **12**, 1–24 (2022).

Mapping the habitat of demersal fish resources to determine suitable fishing gear

Rauzatul Nazzla[1]*, *Indra* Jaya[2], *Jonson Lumban* Gaol[2], *Yeni* Herdiyeni[3], and *Donwill* Panggabean[4]

[1]Postgraduate of Marine Technology, Department of Marine Science and Technology, Faculty of Fisheries and Marine Science, Jl. Agatis Gedung Marine Center Babakan Dramaga-Bogor
[2]Department of Marine Science and Technology, Faculty of Fisheries and Marine Science, Jl. Agatis Gedung Marine Center Babakan Dramaga-Bogor
[3]Department of Computer Science, Faculty of Mathematics and Natural Science, Jl. Agatis Gedung Babakan Dramaga-Bogor
[4]Department of Fisheries Management, Postgraduate School, Jl. Pondok Cabe, Tangerang Selatan-Banten

Abstract. Proper use of fishing gear is important in fishing operations. This study aimed to determine a suitable type of fishing gear to be implemented in seven coastal districts in Bengkulu Province based on the results of mapping and analysis of demersal fish habitats. The data used included: 1) bathymetry, 2) substrate type, 3) water brightness, and 4) demersal fish data from to 2019-2022. The methods applied involved classification and descriptive categories. Bathymetry and substrate type were classified and analyzed using mapping tools, while demersal fish catch data were analyzed using statisticals tools. The output was produced as thematic maps and infographics to illustrate the distribution of the three parameters. Interpretations were used to describe the relationship between demersal fish habitat and fishing gear type. The results showed that in the waters of Mukomuko, North Bengkulu, Central Bengkulu, Bengkulu City, Seluma, South Bengkulu and Kaur, demersal fish habitats dominated by Pampus argentus, Trachinotus carolinus, Lutjanidae sp., Epinephelus sp., and Ariidae sp. Suitable types of fishing gear include the Hand Line, Cash Net, Set Net, Bottom Long Line, Seine Net and Ocean-drift Gillnet.

1. Introduction

1.1 Background

Demersal fish is types of fish live most of their lives at the bottom or near the bottom of the water. Making formation of schools that are not too large, moving not too far, and relatively less movement [1], they occupy sea floors and lake beds, which usually consist of mud, sand, gravel, or rocks [2]. They divided into two main types: strictly benthic fish and benthopelagic

* Corresponding author: rauzatulnazzla@apps.ipb.ac.id

fish. Benthic fish are denser than water and have negative buoyancy so they can rest on the sea floor. Benthopelagic fish have neutral buoyancy that can float just above the sea floor. They either live and wait as ambush predators, at times covering themselves with sand or otherwise camouflaging themselves, or move actively above the bottom in search for food. Most demersal fish are benthopelagic [2].

Catches from demersal fisheries contribute significantly for human consumption to developing countries [3] and worldwide [4] They make up a large proportion of the marine harvest in industry and have stimulated an increase in fishing activities. There are two management principles: 1) equal access between small and large fishermen by regulating routes and limiting fishing capacity, 2) the protection of biodiversity through the selectivity of fishing gear and the development of eco-friendly fishing gear. The deployment of fishing gear in Fishing Lanes in the WPPNRI and the High Seas is adjusted to the requirements of fishing gear, including 1) properties, 2) capacity, 3) selectivity level, type, and size, 4) fishing area, and 5) water characteristics [5]

Fishing operations are all activities carried out by a person or group of people at sea using various types of fishing gear with the aim of making a profit. There are several classifications of fishing gear types in terms of point of view, purpose, and water conditions. The basic principle of classification is based on fish catch [6]. In practice, the problem that arises is the mismatch between fishing gear specifications and the target species. It leads to overfishing and environment damage.

In order to achieve a sustainable blue economy vision, it involves good management supported by appropriate decision-making. Information on the availability of fish resources is very important as an indicator to describe the current conditions and potential of fisheries in a water area [7].

1.2 The aim of research

This study aims are

1. Produce fish-catch maps and fishing gear based on territorial boundaries; bathymetry, and sediment type
2. To examine the possible relationship of habitat profile to estimate suitable fishing gear in the study area.
3. To recommend the most suitable fishing gear for demersal fisheries in the study area

2. Methods

This study uses fishery production data based on fish species equipped with the type of fishing gear taken from statistical data provided by the Indonesian Ministry of Maritime Affairs and Fisheries (KKP) [8] data series from 2019 to 2022. Fisheries production is classified according to fish species and fishing gear. The assumptions used were estimating demersal fish habitats based on the existence substrate types and bathymetry. Data on the presence of bathymetry and substrate type were used as a basis for habitat estimates by creating partitions that indicate demersal fish movement areas.

The method includes techniques such as data clustering and combining data with spatial information. The data will be provided in a spatial distribution containing the distribution of the mean fish catch in each regency, distribution of sediment type, and depth distribution. By combining these data, we can determine suitable fishing gear for the study area. Fish production raw data contains fish catch data divided by species and fishing gear. Fish species data were sorted and selected based on dominance, represented by a higher total catch number and gentle fluctuation [9]. Each data type needs to produce a mean value in order to calculate the Catch Per Unit Effort (CPUE) and effectivity. For oceanographic data, such as sediment type and water depth, need to be computed by using an interpolation technique. Sediment type data were collected from DKP (2017). The samples were converted into scores and classified into five categories. Water depth data sourced from Dishidros is provided image format and need to be converted to digital format. All dataset must be stored in spatial attributes. Both interpolation results of oceanographic data were combined with fish catch and fishing gear results to recommend suitable fishing gear use. The workflow of this study is illustrated in Figure 1.

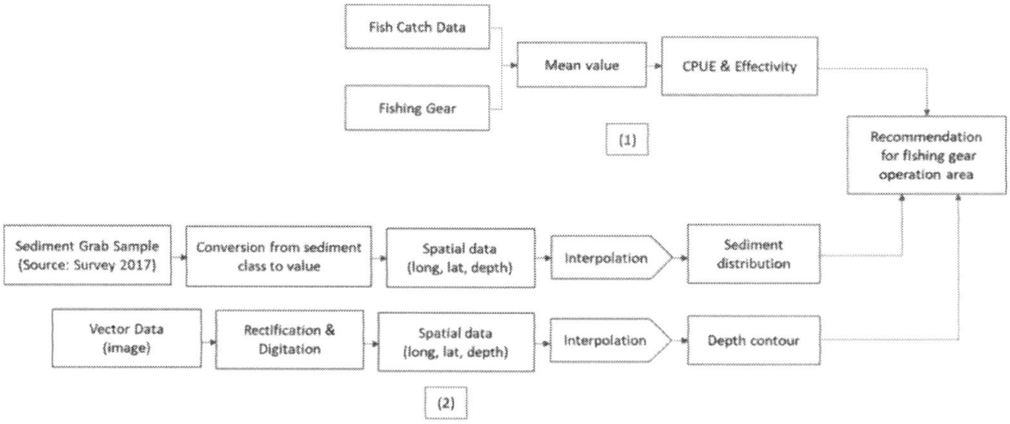

Figure 1. Research flow

Spatial analysis is the process of manipulating spatial information to extract new information and meaning from original data. Interpolation is a spatial analysis tool for calculating feature statistics and conducting geoprocessing activities. This technique uses a deterministic mathematical function to assume a closer value, which is more related than a further value, called Inverse Distance Weighting (IDW) interpolation. The IDW formula is as follows:

$$FZ_x = \sum_{i-1}^{n} \quad W_i Z_i \tag{1a}$$

$$W_i = \frac{\left(\frac{1}{d_i}\right)^p =}{\sum_{j=1}^{n} \left(\frac{1}{d_j}\right)^p} \tag{1b}$$

$$\sum_{i=1}^{n} \quad W_i = 1, p > 1 \tag{1c}$$

Fish catch data were sorted based on fish species as well as fishing gear data, and each was calculated using the formula shown below.

$$mean = \frac{\sum_{i=1}^{n} \quad catch}{n} \tag{2}$$

Fish catch and fishing gear data related to Catch Per Unit Effort. The CPUE is calculated by dividing the catch of each fishing trip by the number of hours fished during that trip. In this study, we assume that the effort is an overnight trip of 24 h.

$$CPUE = \frac{Total\ catch}{Total\ Effort} \tag{3}$$

Effectivity is efficient, suggesting an acting or a potential for action or use in such a way as to avoid loss or waste of energy in affecting, producing, or functioning.

(4)

The CPUE is the main evaluation tool for fishing gear productivity. This study intended to make a recommendation for suitable fishing gear based on the national standard Permen-KP No. 71 2016 about Fishing Gear Deployment [10], but it is too general, so we need another parameter to specify the most suitable fishing gear in Bengkulu.

3. Results

The CPUE results from a number of assumptions that can be made in various ways: (i) the fish vulnerability to the fishing gear [11], (ii) the spatial distribution of the stock which determines fishing effort distribution [12]. The distribution of fishing effort will be randomly distributed if the stock is randomly spaced; otherwise, the fisherman will search in a non-random way, functioning as an ideal predator; Every fishing operation is autonomous; otherwise, the process would be Markovian, with the fisherman repeating his activity in more abundant locations. The proportionality between CPUE and the actual circumstances in which CPUE was most likely to stay high while abundance decreased [13].

Figure 1 shows two bar charts represents catch volume (blue) in tons and five types of dominant fishing gear (orange) in units. Visually, the information that can be obtained is that high fishing productivity is indicated by the high blue and low orange bars. The resulting values differed because they were influenced by the effectiveness of each fishing gear. Under ideal conditions, catch volume increases as the number of fishing gear units increases because each fishing gear unit contributes to catching fish. The ideal condition is when fish resources are still abundant and there is no overfishing. If the number of fishing gear operations exceeds the optimal water capacity, the catch per fishing gear unit (catch per unit effort/CPUE) will decrease. For example, the use of trawls in Mukomuko is higher, but the catch obtained does not reach the same value as that of the ocean-drift gillnet. This can be caused by increased competition between fishing gear units, causing overfishing, but resources have been reduced.

https://doi.org/10.1051/bioconf/202516805002

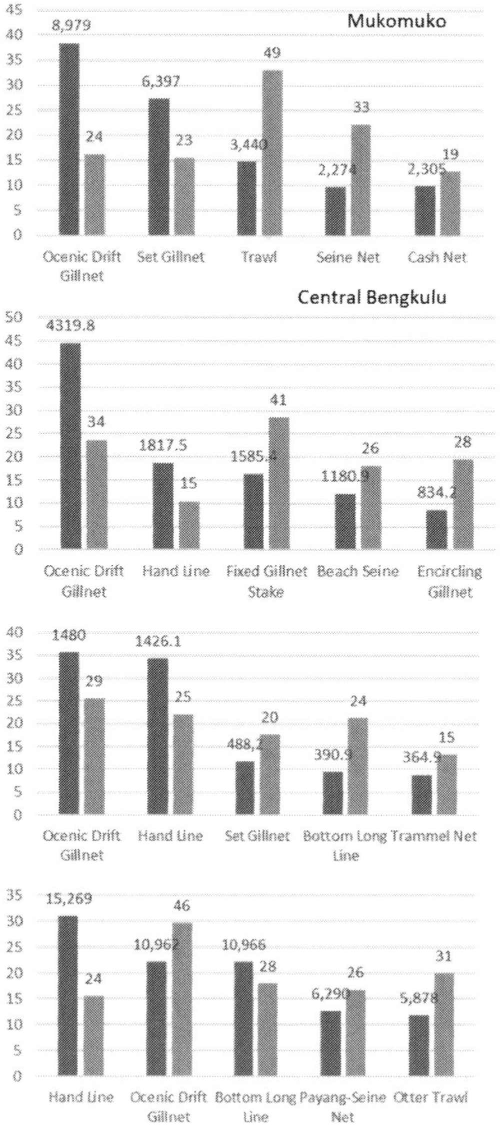

North Bengkulu

Bengkulu City

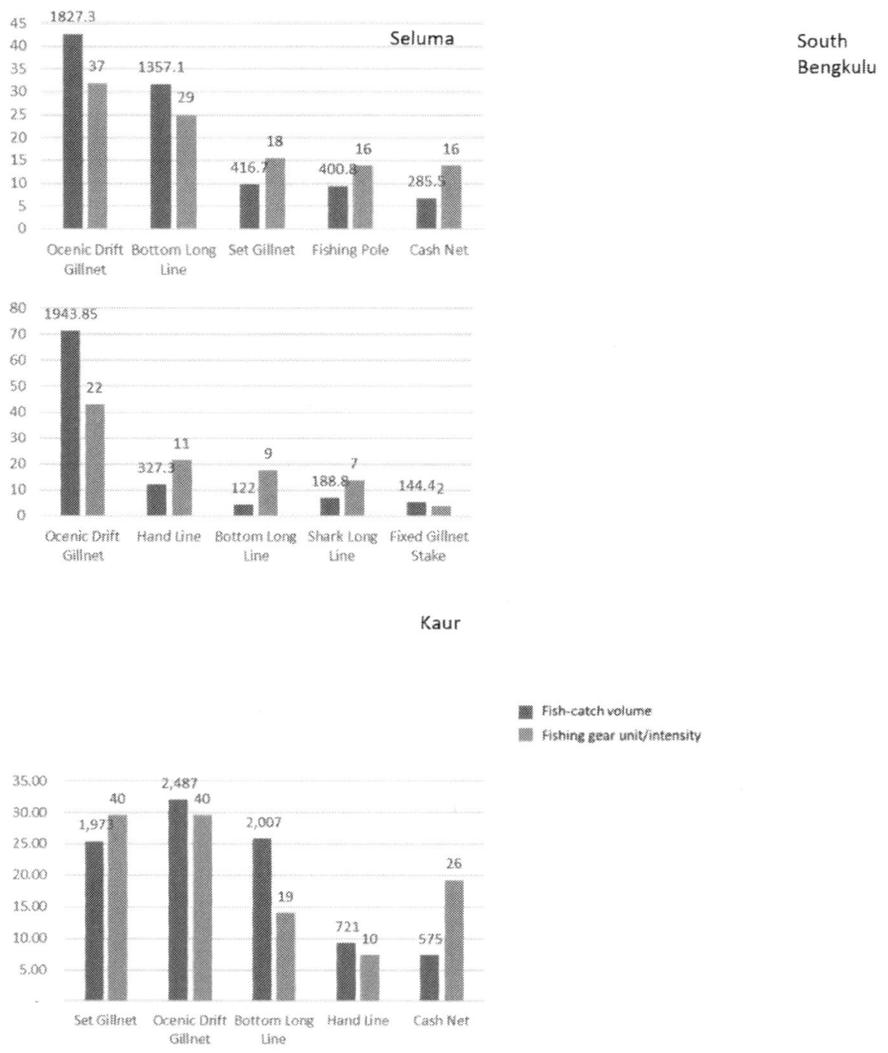

Figure 1. Bar diagram of Fish-catch Volume and Fishing Gear Unit

Catch-per-unit effort (CPUE) methods [14][15][6][17][18] can be applied to calculate the absolute abundance of closed populations when removal occurs one after the other. Catch-per-unit-effort (cpue) is frequently used as a source of stock size estimation. This approach is particularly helpful when there is a linear relationship through the origin between the catch and effort [19][20][21]. Ocean-drift gillnets had the highest catch, with a mean catch of 6,100 tons. If we look at the CPUE, Hand line has the highest value of 8.82 tons/year but the lowest value of effectiveness compared between fish catch number and fishing gear utility. Fixed gillnets are highly effective in terms of the number of catches produced with an average of 11 fishing gear units. The 5 (five) most commonly used fishing gears in Bengkulu waters are listed in Table 1

Table 1. CPUE and Effectivity

No	Fishing Gear type	Catch Volume (ton)	Fishing gear (unit)	CPUE	Effectivity
1	Ocean-drift Gillnet	6100	58	4.38	20.25
2	Hand Line	4497.75	21	8.82	-60.50
3	Bottom Long Line	3754.75	27	5.74	-4.48
4	Set Gillnet	4043.5	25	6.72	-21.43
5	Fixed Gillnet Stake	396.25	11	1.53	72.04
	total	18792.25	142		

Changes in catch per unit effort were inferred to signify changes in the true abundance of the target species. A decreasing CPUE indicates overexploitation, whereas an unchanging CPUE indicates sustainable harvesting [22]. As the cumulative catch increases, CPUE decreases linearly [20].

As show in Table 2, the average fish catch based on the five dominant demersal fish species in the seven coastal districts. It is consisted of 11 dominant fish species that are always available, with relatively stable fluctuations during 2019-2022. The demersal fish species are Pomfret, Ariid catfish, Yellowfin mojarras, Leaftail croaker, Toli shad, Doublespot queenfish, Pompano, snapper fish, Mullet, Grouper fish, and soles.

Table 2. Mean Fish Catch by Species 2019-2022

No	Regency	Mean Volume of Fish Species (kg)				
1	Mukomuko	Pomfret	Ariid catfish	Yellowfin mojarras	Leaftail croaker	Toli shad
		995,168.5	781,316	542,291.75	522,050	354,535
2	North Bengkulu	Doublespotted -queenfish	Pomfret	Pompano	Toli shad	Yellowfin mojarras
		503,345.25	492,777	434,350	407,482.5	349,443.25
3	Central Bengkulu	Snapper fish	Pompano	Pomfret	Mullet	Toli shad
		192,120.8	161,057.5	132,900	105,331.8	76,552.5
4	Bengkulu City	Pomfret	Grouper fish	Snapper fish	Ariid catfish	Pompano
		3,153,345	2,687,601.75	1,932,421	1,452,437.5	1,273,446
5	Seluma	Grouper fish	Pompano	Snapper fish	Mullet	Doublespotted queenfish
		197,027.5	121,065	120,677.5	82,630.75	55,925.5
6	South Bengkulu	Ariid catfish	Snapper fish	Grouper fish	Leaftail croaker	Sole
		315,866.8	264,114.25	191,495	187,077.5	89,498.25
7	Kaur	Pompano	Ariid catfish	Yellowfin mojarras	Mullet	Doublespotted queenfish

		351,224.5	273,385.5	171,972.3	160,990.3	124,252.5

The average fish catch based on three types of fishing gear is most widely used in each of the seven coastal districts. The types of fishing gear most often used in the waters of the Bengkulu Province are drift nets, fixed nets, trawls, handlines, gill nets, and bottom longlines. Table 3 is a list of catches based on fishing gear:

Table 3. Mean Fish Catch by Fishing Gear 2019-2022

No	Regency	Mean Fish Catch by Fishing Gear (kg)		
1	Mukomuko	Oceanic Drift Gillnet	Set Gillnet	Trawl
		93,535	69,533.75	17,550
2	North Bengkulu	Oceanic Drift Gillnet	Hand Line	Fish Gillnet Stake
		1,079,939	454,367.5	396,361.3
3	Central Bengkulu	Oceanic Drift Gillnet	Hand Line	Set Gillnet
		369,992.8	356,527.5	122,049.3
4	Bengkulu City	Hand Line	Oceanic Drift Gillnet	Bottom Long Line
		3,817,259	2,740,494	2,741,623
5	Seluma	Oceanic Drift Gillnet	Bottom Long Line	Set Gillnet
		456,825.8	339,287	104,179
6	South Bengkulu	Oceanic Drift Gillnet	Hand Line	Bottom Long Line
		485,962.5	81,833.75	30,505
7	Kaur	Set Gillnet	Oceanic Drift Gillnet	Bottom Long Line
		493,178.8	621,810.8	501,808.3

The relationship between water depth, bottom substrate, and fishing gear is very close and mutually influential. The depth and type of substrate are crucial in choosing the appropriate fishing gear since they affect both the target fish's habitat and the fishing gear's efficiency. The results from two parameters for determining the right fishing gear in this study are presented in Table 4.

Table 4. Contribute Parameters for Fishing Gear Determination

Regency	**Fishing Area (Nm)**	**Fishing Gear (based on deployment and effectivity)**	**Depth range (m)**	**Substrate**
Mukomuko	0-4	Set Gillnet	5-50	Sandy mud
	4-12	Ocean-drift Gillnet	10-785	Sandy mud
North Bengkulu	0-4	Encircling Gillnet	5-15	Muddy sand
	4-12	Hand Line	10-220	Muddy sand
Central Bengkulu	0-4	Hand Line	0-20	Gravel-mud sand
	4-12	Ocean-drift Gillnet	20-125	Gravel-mud sand
	0-4	Bottom Long Line	5-85	Gravel Sand

https://doi.org/10.1051/bioconf/202516805002

Regency	Fishing Area (Nm)	Fishing Gear (based on deployment and effectivity)	Depth range (m)	Substrate
Bengkulu City	4-12	Hand Line	15-285	Gravel Sand
Seluma	0-4	Bottom Long Line	5-105	Gravel Sand
	4-12	Ocean-drift Gillnet	20-270	Gravel Sand
South Bengkulu	0-4	Fixed Gillnet Stake	5-25	Gravel Sand
	4-12	Ocean-drift Gillnet	15-530	Gravel Sand
Kaur	0-4	Bottom Long Line	5-205	Sand
	4-12	Hand Line	25-1350	Sand

As shown in Figure 3a, depth was divided into five classes, and Figure 3b, substrate type was divided into five classes. The depth in the 12 nautical mile area ranges from 0 to 1400 m from minimum to maximum. The substrate type that was successfully obtained and scored consisted of sand, gravelly sand, sand-gravel-mud, sandy mud, and muddy sand. This justification is based on statistical data [8] and national standards [10].

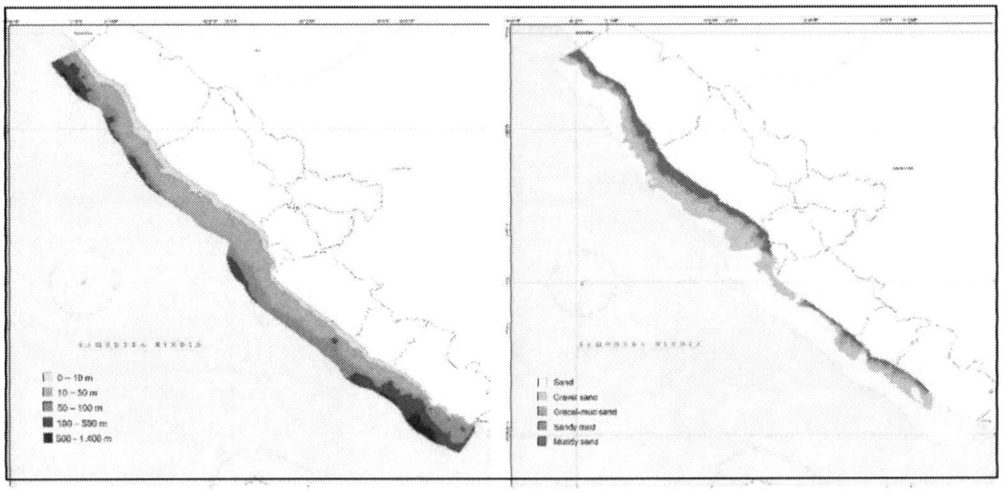

Figure 3. Map of (a) Bathymetry and (b) Sediment Type

The fishing area can be mapped based on the catch results, as shown in the two images of the distribution of demersal fish. Figure 4a shows the distribution of the average value of demersal fish catch in each district. Figure 4b shows the dominant fishery resource. The sediment type is a characteristic of the fish habitat. Based on this, we examined two categories to predict the distribution of fish habitats.

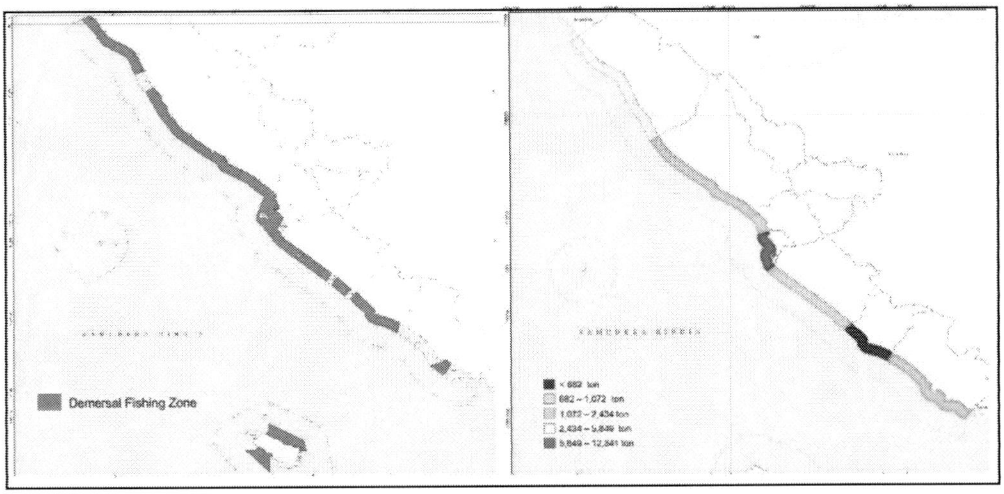

Figure 4. Map of (a) fisheries zones (source: RZWP3K, 2020) (b) fish catch distribution

A map that recommends fishing gear suitable for the substrate type in conjunction with KKP [10] [5] can be created based on the overlay findings between the depth variables and sediment type. Specifically, Fishing Lanes and Placement of Fishing Gear in the Fisheries Management Area of the Republic of Indonesia and Placement of Fishing Gear and Fishing Aids in the Fisheries Management Area of the Republic of Indonesia and the High Seas, as well as the Arrangement of Fishing Andons.

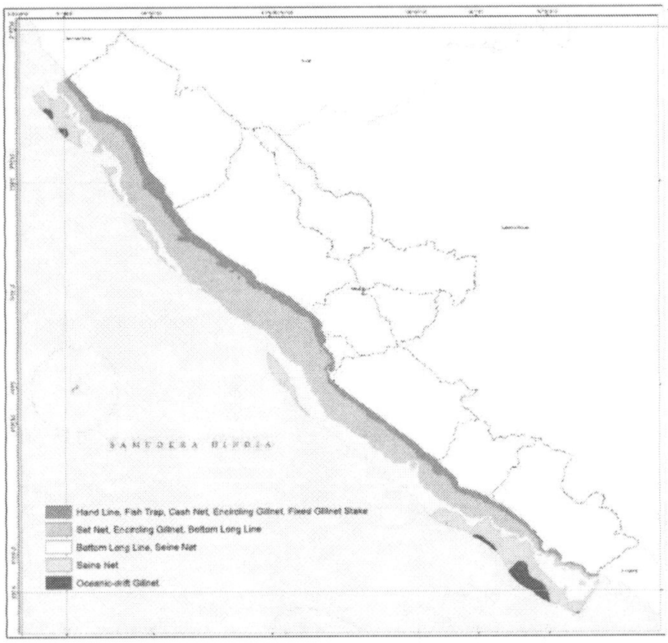

Figure 5. Map of Recommended Fishing Gear Use Based on Depth and Substrate Type

Based on the results shown in Figure 5, there are five recommended lanes for fishing gear use based on sediment type and depth. The coastal area from the coastline to 4 Nm is a cruising area for vessels with a tonnage of less than 5 GT, indicating that this area should be dominated by traditional fishermen who use traditional fishing gear.

4. Discussion and Conclusion

In coastal areas, passive fishing gear is often found in the form of handlines, cash nets, encircling gillnets, and fishing traps [23]. Meanwhile area of 4-12 Nm dominated by bottom longlines, set nets, encircling gillnets, seine nets, fixed gillnets, and ocean-drift gillnets. The type of fishing gear known to have been used in fishing operations is not only special demersal fishing gear, but also pelagic gear. This refers to demersal species frequently occur in mid-water and pelagic species occur close to the seabed, so that 'demersal' species are frequently caught in 'pelagic' fisheries and 'pelagic' species in demersal fisheries [4].

Fish behavior is generally influenced by food availability [24], reproductive cycles [25], and shelter [26]. The number of species at higher trophic levels may be affected by the modifications made to fishing gear, which may also have an impact on benthic macroinvertebrate ecosystems [27][28][29][30] found that excessive sedimentation created anoxic conditions with a high concentration of nutrients in the sediment of a tropical estuary lagoon with a high density of artisanal fishing gear.

The relationship between the right fishing gear and the type of substrate is related to the type of seabed, which is the habitat and roaming area of the demersal fish. Different seabed characteristics require different treatment methods. Water areas that have coral reef ecosystems are sensitive areas so that in carrying out fishing efforts it is necessary to pay attention to the applicable rules related to the working principles of the fishing gear used.

This study successfully produced fish-catch volumes based on region and sediment type, habitat estimation based on bathymetry and sediment type, and determination of fishing gear based on bathymetry and habitat. The largest catch of demersal fish is landed in the city of Bengkulu, followed by Mukomuko, North Bengkulu, Kaur, and Seluma, and the least in South Bengkulu. The most abundant species in each district were Pampus argentus, Scomberoides lysan, Lutjanidae sp., Epinephelus sp., Ariidae sp., and Trachinotus carolinus. The sediment type was classified into five classes: sandy mud, muddy sand, gravel-mud sand, gravel-sand, and sand.

Each district has a different type of substrate, and the species of demersal fish caught is also different, even though the fishing gear used tends to be almost the same. The seven districts use ocean-drift gillnets, set gilnets, bottom longlines, and hand lines, but there is a large variation in the species caught. Based on the type of substrate, depth, and type of fish obtained, Mukomuko, North Bengkulu, Bengkulu, Central Bengkulu, Bengkulu City, Seluma, South Bengkulu, and Kaur districts were recommended to use handlines, fish traps, cash nets, encircling gillnets, and fixed gillnet stakes at depths of 0-10 m. At depths of 10-50 m, set nets, encircling gillnets, and bottom longlines can be used. At depths of 50-100 m, bottom longlines and seine nets can be used, as well as at depths of 100m-500m. However, for depths >500 m, ocean-drift gillnets are generally used. The recommendation of a suitable fishing gear is determined based on the 5 (five) highest productivity or most efficient fishing gear and is also adjusted to national standards.

References

1. Aoyama, T. 1973. The Demersal Fish and Fisharies ff The South China Sea. IPFC/SCS/DEV/73/3. Rome
2. W.C. Carl. Coastal fish - Fish of the open sea floor. Te Ara - the Encyclopedia of New Zealand. (2009)
3. I.C. Stobutzki, G.T. Silvestre, A. Abu Talib, A. Krongprom, M. Supongpan, P. Khemakorn, N. Armada,L.R. Garces. Decline of demersal coastal fisheries resources in

three developing Asian countries. Fisheries Research. Volume 78, Issues 2–3, May 2006, Pages 130-142. (2006)

4. K. Brander. Demersal Species Fisheries. Encyclopedia of Ocean Sciences (Second Edition) (2001)
5. KKP. Peraturan Menteri Kelautan dan Perikanan Nomor 18 Tahun 2021 tentang Penempatan Alat Penangkapan Ikan Dan Alat Bantu Penangkapan Ikan Di Wilayah Pengelolaan Perikanan Negara Republik Indonesia Dan Laut Lepas Serta Penataan Andon Penangkapan Ikan. (2021)
6. Sudirman. *Metode Penangkapan Ikan: Modul 1 Klasifikasi Alat dan Metode Penangkapan Ikan*. Perpustakaan UT. (2018)
7. D. Panggabean & R. Nazzla. Determining Swimming Layer and Pelagic Fish Distribution in the Western Flores Sea Using Acoustic Detection. Marine Fisheries. **11**, 2, November 2020. page: 213-228. (2022)
8. KKP. Statistika KKP https://statistik.kkp.go.id/home.php?m=prod_ikan_laut_kab#panel-footer. (2022)
9. Depdagri. Directorate general regional development. circular letter No 050.05/2910/III/BANDA. (1999).
10. KKP. Menteri Kelautan Dan Perikanan Republik Indonesia Nomor 71/Permen-Kp/2016. *Jalur Penangkapan Ikan Dan Penempatan Alat Penangkapan Ikan Di Wilayah Pengelolaan Perikanan Negara Republik Indone*sia. (2016)
11. M.N. Maunder, J.R. Sibert, A. Fonteneau, J. Hampton, P. Kleiber, S.J. Harley. Interpreting catch per unit effort data to assess the status of individual stocks and communities. ICES Journal of Marine Science, **63**, 8, 2006, Pages 1373–1385. (2006)
12. J.E. Paloheimo, and L.M. Dickie. Abundance and fishing success. Journal du Conseil. International pour l'. Exploration de la Mer, **155**(1), p. 152-163. (1964)
13. S.J. Harley, R.A. Myers. & A. Dunn. Is catch-per-unit-effort proportional to abundance? Canadian Journal of Fisheries and Aquatic Sciences, **58**, 9, p. 1760-1772. (2001)
14. P.H. Leslie & D.H.S. Davis. An attempt to determine the absolute number of rats on a given area. J. Anim. Bcol. **8**:94-113. (1939)
15. W.E. Ricker. Handbook of Computations for Biological Statistics of Fish Populations. The Fisheries Research Board of Canada Under the Control of the Honourable the Minister of Fisheries. Ottawa. Bulletin 119. (1958)
16. J.R. Skalski, D.S. Robson, & C.L. Matsuzaki. Competing probabilistic models for catch-effort relationships in wildlife censuses. Ecological Modelling **19**:299-307. (1983)
17. J. Laake. Catch-Per-Unit-Effort Models: An Application to an Elk Population in Colorado. Chapter. Springer Nature. pp 44–55. (1992)
18. R.A Lancia, J.W. Bishir, M.C. Conner, C.S. Rosenberry. Use of catch-effort to estimate population size.Wildl Soc Bull. 1996; **24**: 731–737.
19. J.A. Gulland. Catch per unit of effort as a measure of abundance. Rapport et process-verbaux des réunions du Conseil International pour l'Exploration de la Mer, **155**(1), p. 8-14. (1964).
20. W.E. Ricker. Computation and interpretation of biological statistics of fish populations. Bull. Fish. Res. Board Can. **191**: 1–382. 1975.
21. A.C. Lima, C.E.C. Freitas, Maabuabara, M. Petrere & V.S. Batista. On the standardization of the fishing effort. Acta Amazonica, **30**, 1, p. 167-169. 2000.

22. P.E. Puertas & R.E. Bodmer "Hunting effort as a tool for community-based wildlife management in Amazonia". Columbia University Press. https://doi.org/10.7312/silv12782-008. (2004).
23. S.G. Kim. W.I. Lee. M.Y. Seok. The estimation of derelict fishing gear in the coastal waters of South Korea: Trap and gill-net fisheries. Marine Policy. **46**:119–122. (2014)
24. Pratiwy & Rosidah. Exploring the intricate relationship between foodavailability and feeding behavior in fish larvae: A review. International Journal of Fisheries and Aquatic Studies 2023; **11,** 4: 01-04. DOI: https://doi.org/10.22271/fish.2023.v11.i4a.2816
25. A.J. Khursigara, T. J. Hamilton, Fish Behavior. Encyclopedia of Fish Physiology (2nd Edition). (2024)
26. M. Gesto & A. Jokumsen. Effects of simple shelters on growth performance and welfare of rainbow trout juveniles. Aquaculture. **551**,737930. (2022)
27. M. Marconi, M. Sarti, & F. Marincioni. Sustainability assessment of traditional fisheries in Cau Hai lagoon (South China Sea). Marine Environmental Research. **70**, 3–4, pp 253-263. (2010)
28. J.H. Grabowski., M. Bachman., C. Demarest., S. Eayrs., B.P. Harris., V. Malkoski., D. Packer & D. Stevenson. Assessing the Vulnerability of Marine Benthos to Fishing Gear Impacts. Reviews in Fisheries Science & Aquaculture. **22**, 2. pp 142-155. (2014)
29. T. Sinso, S. Sa-nguansil, A. Buranapratheprat, K. Wangkulangkul. Distribution of artisanal fishing gears with impacts on sedimentation and benthic macroinvertebrate assemblages in Songkhla Lake, the largest lagoon in Thailand. Marine Environmental Research. **190**, September 2023, 106126. (2023)
30. M. Rueda & O. Defeo. Linking fishery management and conservation in a tropical estuarine lagoon: biological and physical effects of an artisanal fishing gear. Estuarine, Coastal and Shelf Science. 56(5–6). pages 935-942. (2003)

Coastline change and coastal habitat loss: A case study in Trabzon City, Southeast Black Sea, Türkiye

Neira Purwanty Ismail[1*], *Coşkun* Erüz[2]

[1]Department of Fisheries Technology Engineering, Sürmene Faculty of Marine Sciences, Karadeniz Technical University, Trabzon 61530, Türkiye
[2]Department of Marine Science and Technology Engineering, Sürmene Faculty of Marine Sciences, Karadeniz Technical University, Trabzon 61530, Türkiye

Abstract. The Black Sea is the largest semi-enclosed waterbody, with anoxic deep waters, and suffers from ecological degradation due to human activities, including eutrophication, overfishing, pollution, military conflicts, and land reclamation. For instance, in the Southeastern Black Sea region of Turkiye, rapidly growing coastal cities strain the narrow coastal zone. Land reclamation is frequently becoming a go-to solution for local governments seeking expansion. This study investigated the impact of human intervention on the coastline and its associated habitats in Trabzon city, Turkiye, between the 1960s and 2015. This study examined coastline, and habitat changes in the study area by comparing remote sensing data and Geographic Information Technology (GIS) for comparison. Additionally, coastal bathymetry data were used to create a 3D model of the coastal habitats. Analysis of this approximately 20 km coastline revealed a significant loss of coastal habitat of roughly 305.06 ha - over the 50-year period. This loss encompasses shallow water habitats up to 26 m deep and most natural sandy habitats. Ground-truthing confirmed that reclaimed coastal land has been converted for various land uses, including roads, car parks, sports, and recreation facilities. This study contributes to the policymakers in decision making for coastal development practices in the region.

1 Introduction

Coastal areas are recognized as some of the most productive and biologically diverse ecosystems on Earth. These regions are among the most dynamic on the planet, constantly evolving owing to the interplay between land and sea [1]. Coastal zones are unique and fragile environments in which land and marine processes interact and change [2]. Coastal changes can occur naturally because of geological and hydrodynamic processes. However, human activities, such as coastal intervention, land reclamation, construction, and sand mining can significantly disrupt these natural processes and accelerate coastal changes [3]. Coastal

* Corresponding author: neira.ismail@gmail.com

changes can have both beneficial and detrimental effects. Artificial coastlines disrupt natural processes and causeerosion and sedimentation [4]. In addition, coastal interventions can alter habitats and affect biodiversity [5].

The Black Sea, which has a unique and ecologically sensitive marine ecosystem, is facing increasing anthropogenic pressure [6,7]. As the world's largest semi-enclosed water body, it is critical for the regional climate and biodiversity [8]. However, decades of human activity, such as eutrophication, overfishing, pollution, and military conflicts, have led to significant degradation of the marine environment [9]. The Black Sea coastal zone is particularly vulnerable to human impact. Rapid urbanization and population growth in countries, including Turkiye, have increased the pressure on these fragile coastal areas [10]. Land reclamation has become a common practice for supporting the expansion of urban areas and infrastructure. While offering short-term solutions, this approach often has detrimental effects on coastal ecosystems [11]. In the southeastern Black Sea region, human-induced coastal change has been particularly pronounced with rising urbanization, tourism, and industrialization, further intensifying these pressures [12]. A study by Erüz et al. [13] found that coastal road construction in the southeastern Black Sea led to a significant loss of natural coastlines. This highlights the vulnerability of the region's unique ecosystem.

The southeastern Black Sea in Turkiye, including Trabzon city, is an example. The narrow coastal zone is undergoing rapid development, with land reclamation projects transforming the shoreline at an alarming rate [12]. Despite the increasing awareness of the negative impacts of coastal development, comprehensive studies quantifying habitat loss and its ecological consequences in this region remain limited [13].

Both natural and anthropogenic coastal changes, including the impact of human intervention and habitat loss, can be analyzed using remote sensing techniques [14,15]. This involves analysis of aerial photographs, satellite imagery, and other data collected at different times. Direct GPS measurements, along with printed and digital maps, aerial photographs, satellite images, and time-series imagery, can be processed using Geographic Information Systems (GIS) to analyze coastline changes, natural and anthropogenic area and volume changes, and associated habitat loss [11,16]. Remote sensing data are especially effective for coastal monitoring in countries with extensive coastlines, such as Turkey [17], and offer a cost-effective method for collecting spatial and temporal data [18].

To bridge this knowledge gap, this study investigated the impact of human intervention on the coastline and associated habitats in Trabzon city between the 1960s and 2015. Using remote sensing and Geographic Information System (GIS) techniques, we aimed to (1) quantify the extent of coastal habitat loss due to land reclamation, (2) analyze changes in habitat types, and (3) identify the implications of these changes in the marine ecosystem.

To develop efficient conservation and management plans, it is essential to understand the pace and trends of coastal habitat loss in Trabzon city. The results of this study will aid in the general understanding of how human activity affects coastal ecosystems and offer useful information for policymakers and managers of coastal zones.

2 Methodology

2.1 Study area

The study area encompasses approximately 20 km of coastline in Trabzon city, Southeast Black Sea, Turkiye (Fig. 1). Trabzon is one of the largest cities in Turkey's southeastern Black Sea region, covering approximately 4,865 square kilometers and has a growing population of 335,628 people, making it the third most populous city in the region [19]. The region's economy is driven significantly by tourism, fishing, and maritime activities, with the sea serving as a crucial hub for trade and transportation [20]. Trabzon has been a key seaport for centuries, linking various ports in the Black Sea and coastal cities. Its strategic location is further enhanced by the vast hinterland accessible through a robust road network extending into central Anatolia and reaching as far as the Caucasus and Iran [21].

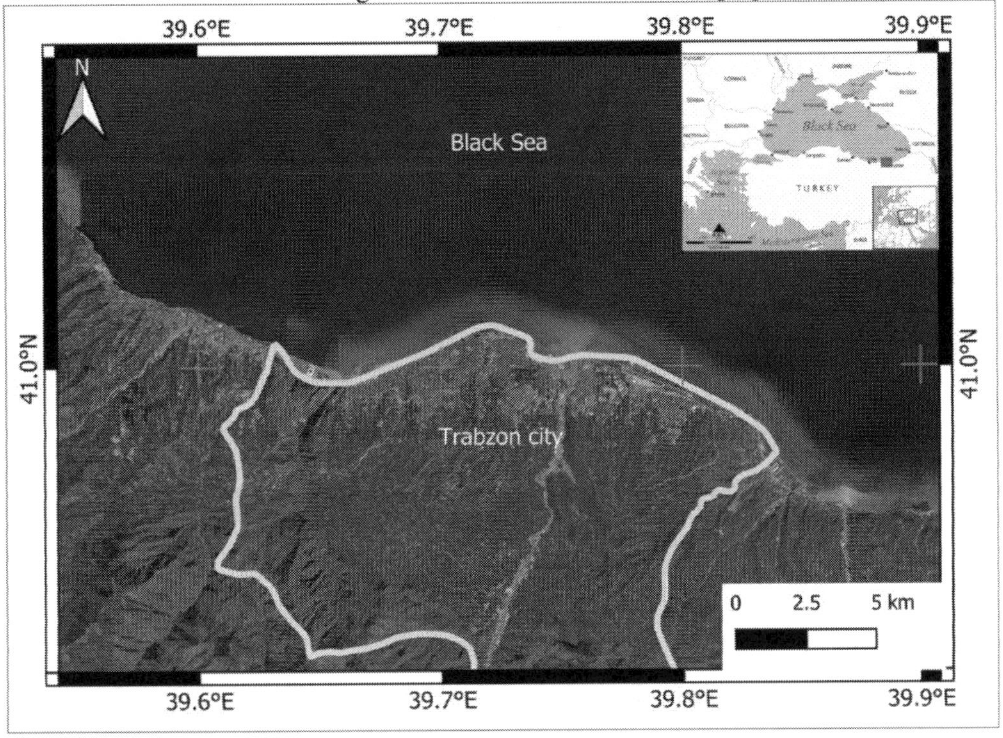

Fig. 1. Study area in Trabzon city coast, Turkiye

Trabzon city features a mountainous landscape with steep slopes that descends directly into the Black Sea [22]. While this rugged terrain has historically restricted coastal development, recent decades have seen rapid urbanization and industrialization, intensifying the pressure on fragile coastal ecosystems [12,13]. The climate of the region, characterized by heavy rainfall and frequent fog, drives intense erosion and sediment transport, shaping the dynamic coastline [23]. Additionally, the presence of diverse coastal habitats such as rocky shores, sandy beaches, rocky cliffs, and estuaries highlights the ecological richness of the area.

This study analyzed coastal data from 1957 to 2015, focusing on this timeframe because of the significant increase in coastal interventions, such as development and construction, along the Eastern Black Sea coast, including Trabzon, between 1960 and 2015 [13,24].

2.2 Dataset

To monitor changes in coastal areas, this study utilized multi-temporal remote sensing datasets and geographic information system (GIS) techniques. A time-series analysis was conducted to track changes in the coastline over time. By comparing data from different periods, changes in land cover (waterbody and land), shoreline position, and coastal features were identified and assessed [25].

To monitor coastline changes over a fifty-five-year period, historical maps derived from 1:25.000 scale aerial photographs taken in the 1960s were used as baseline coastal data. Old maps were obtained from the General Directorate of Mapping, Republic of Turkiye. The physical maps were scanned and converted into a digital format for coastline digitization. The coastal data for 2015 derived from satellite imagery with the same spatial resolution represent the recent conditions obtained from the KTÜ GIS Lab (derived from WorldView-1 satellite). To assess hydrographic changes along the Trabzon coast and determine changes in coastal water, a 1:100,000 scale bathymetric chart of the Southeast Black Sea produced in 2014 by KTÜ Sürmene Marine Science Faculty as part of the Turkish TÜBİTAK MAM "Our Seas Monitoring Project" was used. All data were processed using the WGS84 datum and UTM Zone 37 North projection.

2.3 Coastline Change Analysis

Changes in coastal areas that are subject to constant transformation can be monitored both spatially and temporally. To determine changes in coastal areas, specific regions must be monitored and documented over time. By comparing the historical and current data of the same coastal areas, it was possible to determine the spatial and temporal changes in the coastlines.

To assess temporal changes in the coastline and associated habitats, remote sensing data and Geographic Information System (GIS) techniques were utilized. Multi-temporal data from the 1960s to 2015 provided a synoptic view of the study area, allowing the identification and delineation of coastal features. Digitation and classification techniques through visual interpretation were used to extract information on the coastline position, land, and coastal water. Subsequently, GIS analysis was applied to quantify the changes in coastal areas, habitat types, and spatial distributions over time. Additionally, bathymetric data were incorporated to create a three-dimensional model of the coastal environment, providing a deeper understanding of habitat loss and its implications, as shown in Figure 2.

An overlay method was used to analyze coastline change by comparing the 1960 and 2015 coastlines. This approach identified areas where the coastline advanced seaward, indicating sedimentation, and areas that retreated inland, indicating erosion or abrasion. The primary cause of these changes in this study was human activity, specifically, extensive coastal reclamation. The rate of coastal change was calculated by measuring coastline displacement. A visual representation of the methodology used in this study is shown in Figure 3.

All image processing, geospatial analysis, and data visualization were performed using the open-source software tool QGIS 3.28.15-Firenze [26].

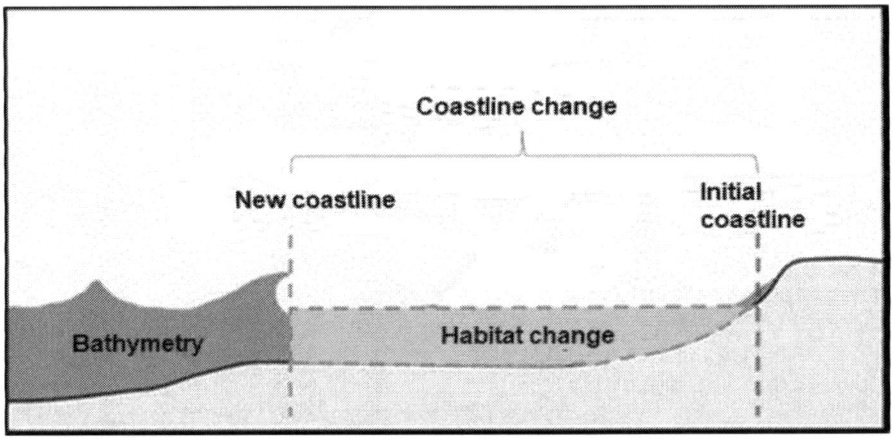

Fig. 2. Coastline and habitat change analysis illustrations.

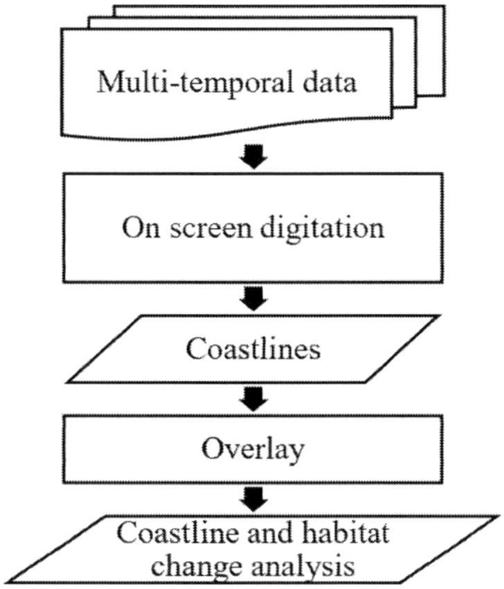

Fig. 3. Coastline and habitat change analysis workflow

3 Results and Discussion

Trabzon city boasts a 20 km coastline, the second longest in the province. Coastal reclamation in Trabzon city began in the 1970s, and continued intermittently until the 2000s. Trabzon city continues to undertake coastal reclamation projects.

The results revealed that between 1960 and 2015, approximately 305.06 hectares of land were gained through coastal reclamation in Trabzon city, resulting in a loss of 304.49 hectares of ecologically significant coastal habitats, as shown in Figure 4. Additionally, 11,775,362 m^3 of habitat or water bodies were destroyed during this period. Reclamation projects in the central district primarily affected the sandy (63.40%) and rocky (36.6%) coastal habitats (Figure 5). Most of these sandy habitats later replaced by concrete structure due to land reclamation along the city's coastline. Analysis and field observations revealed that Akyazı and Moloz underwent the most significant changes in the district of Trabzon. In these areas, reclamation extended 530-550 meters from the initial coastline and reached depths up to 26 m.

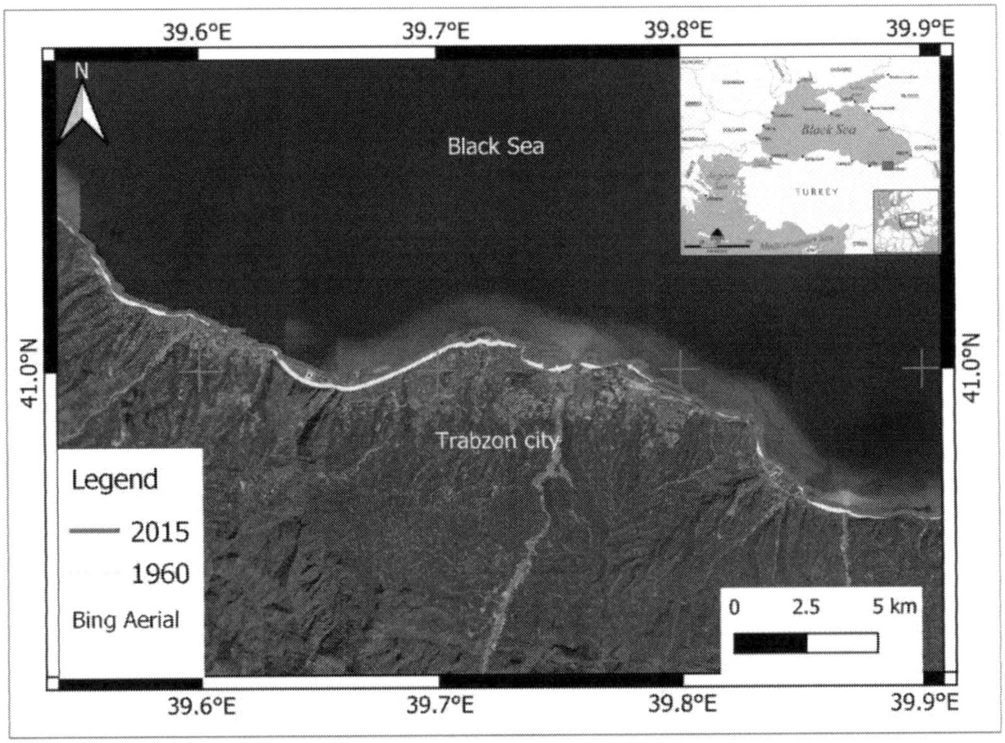

Fig. 4. Coastline changes in Trabzon city from 1960s to 2015

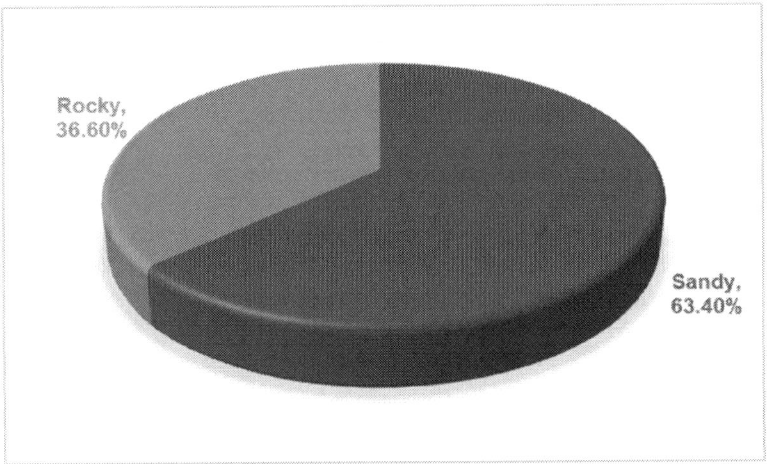

Fig. 5. Initial coastal habitat composition in Trabzon city

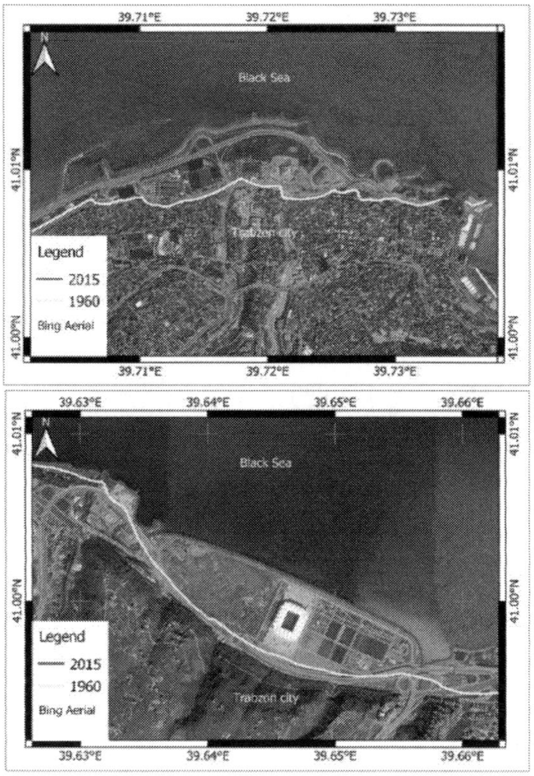

Fig. 6. Coastal area changes in Trabzon city Moloz (left) and Akyazı (right).

Analysis of the Trabzon coastline revealed substantial loss of coastal habitat between 1960 and 2015. Shallow water habitats and sandy shores were converted into anthropogenic land use. The rate of coastal erosion varied across the study area, with some sections experiencing an accelerated shoreline retreat. The 3D bathymetric model indicated a significant reduction in water depth due to land reclamation activities, impacting marine ecosystems, and hydrodynamic processes. Ground-truthing confirmed that reclaimed land has been converted into various infrastructures, including roads, parking areas, and recreational facilities, highlighting the extent of human intervention in the coastal zone. These reclaimed areas are now used for roads, buildings, green spaces, playgrounds, and recreational facilities. In addition to reclamation, the coastline of the district was affected by the construction of three ports/fishing shelters and 24 coastal protection groins.

Moloz is the primary location in central Trabzon and has undergone significant and continuous transformation, with new reclamation projects undertaken in every planning phase (Figure 6, upper). Field investigations and documentation revealed that the reclaimed area now accommodates highways, rest areas, parks, walking paths, sports facilities, markets, petrol stations, municipal parking lots, fire stations, storage areas, and other facilities. In addition, a portion of the former storage area remains vacant.

A Field survey and coastal analysis of the Akyazı area revealed that in addition to a previously constructed highway, approximately 78.5 hectares of the reclaimed land in the eastern part has been developed into a football stadium and sports facilities. Meanwhile, the western part of the area has yet to be finalized in terms of its development plan (Figure 6, lower).

In Trabzon, the Gülcemal project was underway in 2016, targeting an area of 800,000 square meters (80 ha) between Beşirli and Kemerkaya. Although the exact location, extent, and nature of this new reclamation project, similar to the Akyazı project, will be determined upon project completion, its stated purpose is to create new social and cultural spaces for the city, provide public access to the sea, and attract both local and foreign tourists.

Coastal planning and reclamation activities in Trabzon have significantly altered the coastline and hydrographic features of the region. The construction of reclamation structures and coastal infrastructure has led to the transformation of natural sandy and rocky habitats, resulting in near-complete loss of natural sandy beaches and shallow sandy habitats. Consequently, a large portion of Trabzon's natural sandy habitats have been converted into rocky habitats. As reported by Ge et al. [5] and Bianchini et al. [27], such physical coastal interventions not only modify topographic features, hydrology, and water circulation systems, but also exacerbate coastal erosion, increase land-based pollution, alter sediment transport and deposition processes, and cause shallowing in harbor areas and shipping routes. Additionally, the ecosystems and marine life in reclaimed and filled areas have been significantly damaged or destroyed.

The shallow coastal waters of the Black Sea are crucial for maintaining the function of the coastal ecosystems in the region. Because of the anoxic, hydrogen sulfide-rich deep waters that comprise over 87% of the Black Sea, only the top 10% of the water column can support life (Figure 7). Consequently, marine organisms in the Black Sea rely on oxygenated shallow surface waters for their survival. Başçınar et al. [28] found that the abundance and diversity of benthic organisms in the Eastern Black Sea significantly decreased below a depth of 35 m. Extensive reclamation along the Trabzon coast has further endangered these benthic organisms and their habitats, deepening coastal waters in reclaimed areas to 5-25 meters and further destroying the limited shallow-water habitat.

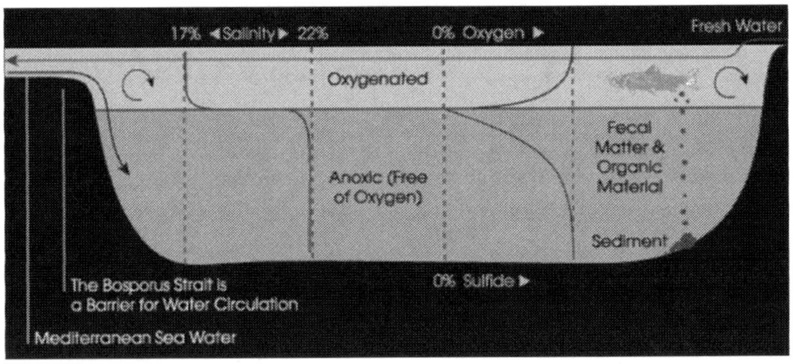

Fig. 7. Illustration of oxic and anoxic layers of the Black Sea ecosystem [29]

The substantial loss of coastal habitats in Trabzon city, as evidenced in this study, is a critical concern. The conversion of diverse coastal ecosystems into artificial ecosystems has far-reaching ecological consequences including habitat loss, biodiversity decline, and erosion. The reduction of shallow water habitats is particularly alarming, as these areas support crucial ecological functions, such as nursery grounds for fish and other marine organisms. Moreover, the alteration of coastal topography through land reclamation can modify hydrodynamic processes, thereby impacting water circulation, sediment transport, and wave energy dissipation. These changes can trigger cascading effects in marine ecosystems. Therefore, it is imperative to adopt sustainable coastal management strategies that prioritize habitat conservation, restoration, and protection. This requires enforcing strict land reclamation regulations, promoting development approaches, and investing in coastal ecosystem-based management plans. Additionally, the presence of diverse coastal habitats such as rocky shores, sandy beaches, and submerged vegetation underscores the ecological richness of the area. Coastal interventions have altered natural coastal environments.

4 Conclusion

This study demonstrated the significant impact of human-induced coastal modification on the Black Sea ecosystem. Specifically, the rapid expansion of Southeast Black Sea region of Türkiye has led to a significant loss of coastal habitats owing to land reclamation. The conversion of approximately 305.06 hectares of diverse coastal environments, including shallow-water habitats and natural sandy shores, into artificial structures for urban development underscores the urgent need for comprehensive coastal management strategies. These findings highlight the detrimental effects of short-sighted coastal planning and emphasize the importance of balancing economic growth with ecological preservation. To mitigate future habitat loss and protect the unique characteristics of the Black Sea, it is imperative to implement sustainable coastal development policies that prioritize ecosystem conservation and restoration.

References

1. F. Asensio-Montesinos, R. Molina, G. Anfuso, G. Manno, and C. L. Ro, J. Mar. Sci Eg. **12**, 11 (2024)
2. M. de Andres, J. M. B. Munos, J. G. Onetti, and L. D. C. Zuniga, Ocean & Coastal Management, **231**, 106402 (2023)
3. Y. Zhang, R. Chen, and Y. Wang, Land Use Policy **91**, 104370 (2020)
4. M. Martín-Antón, V. Negro, J. M. Del Campo, J. S. López-Gutiérrez, and M. D. Esteban, https://doi.org/10.2112/SI75-133.1 **75**, 667 (2016)
5. Y. Ge and J.-Y. Zhang, Energy Procedia, **5** (2011)
6. N. Serpetti, C. Piroddi, E. Akoglu, E. Garcia-Gorriz, S. Miladinova, and D. Macias, Plos One, 20,1 (2025)
7. L. Lazar, A. Spanu, L. Boicenco, A. Oros, N. Damir, E. Bisinicu, V. Abaza, A. Filimon, G. Harcota, O. Marin, E. Panteaü F. Timofte, O. Vlas, and S. Korpinen, Frontiers in Marine Scince, **11** (2024)
8. T. Oğuz, in *Black Sea Marine Environment: The Turkish Shelf* (2017), pp. 1–13
9. K. Seyhan, Ö. Dürrani, L. Papadaki, E. Akinsete, Ş. Atasaral, K. Özşeker, H. Akpınar, E. Kurtuluş, R. E. Mazlum, P. Koundouri, and A. Stanica, Frontiers in Marine Science, **12** (2025)
10. A. Güneroğlu, Ocean Coast Manag **118**, 225 (2015)
11. Y. Zhang, R. Chen, and Y. Wang, Land Use Policy, **91**, 104370 (2020)
12. N. P. Ismail and C. Erüz, in *III. International Architectural Sciences and Applications Symposium*, edited by A. Gül, Ö. Demirel, S. Seydoşoğlu, and F. Zucaro (IKSAD Publishing House, Naples, Italy, 2023), pp. 591–601
13. C. Erüz and N. P. Ismail, in *International Conference on Environmental Science and Technology* (2015)
14. B. El Mahrad, A. Newton, J. D. Icely, I. Kacimi, S. Abalansa, and M. Snoussi, Remote Sens (Basel) **12**, 2313 (2020)
15. O. Chukwuka, I. Emeka, Y. Ibrahim, Y. Fenetahun, Y. Yuan, and W. Yongdong, Heliyon, **9**, 11 (2023)
16. L. Sui, J. Wang, X. Yang, and Z. Wang, Sustainability **12**, 3242 (2020)

17. F. Temiz, S. S. Durduran, IOP Conf. Ser: Earth Environ. Sci., **44**, 042033 (2016)
18. D. Dominici and S. Zollini, Journal of Marine Science and Engineering 2020, Vol. 8, 498 **8**, 498 (2020)
19. TÜİK, data.tuik.gov.tr (2022)
20. The economy and trade in Trabzon https://avrasya.edu.tr/the-economy-and-trade-in-trabzon/ (2024)
21. S. Kose, and D. M. Sekban, Transportation Safety and Environment, **4**, 2 (2022)
22. S. Taşçıoğlu, Online Journal of Art and Design, **11**, 4 (2023).
23. T.C. Sanayi ve Teknoloji Bakanlığı, http://www.doka.org.tr (2025)
24. E. M. Alpak, D. G. Özkan, and T. Düzenli, Journal of Balkan Ecology **19**, 291 (2016)
25. M. E. Fagan and R. S. DeFries, Reference Module in Life Sciences (2022)
26. QGIS Development Team, http://www.qgis.org (2025)
27. A. Bianchini, F. Cento, M. Pellegrini, and C. Saccani, Journal of Envionrmnetal Management, **248**, (2019)
28. N. S. Bascinar, A. M. Gozler, C. Şahin, C. Eruz, L. Tolun, E. Agirbas, M. Mantikci, S. Serdar, and B. Verep, Iran J Fish Sci **13**, 719 (2014)
29. L. Neretin, M. E. Böttcher, B. B. Jörgensen, I. I. Volkov, and H. Lüschen, Geochemistry of the Earth´s Surface 331 (1999)

Benthic habitats mapping in Pari Island, Kepulauan Seribu, Indonesia using drone and sentinel-2B imagery with object based image analysis method

James P. Panjaitan[1*], *Ghani* Mahendratama[1] *Vincentius* Siregar[1]

[1]Department of Marine Science and Technology, Faculty of Fisheries and Marine Science, IPB University, Bogor, 16880, Indonesia

Abstract. Benthic habitats are living places and shelters for various types of aquatic organisms, including seagrass, seaweed, algae, dead corals, and living corals with different types of substrates, such as rubble, sand, and mud. This habitat has the potential to change, so it is necessary to monitor it regularly by mapping shallow-water habitats using remote sensing technology such as satellite imagery and drones. This study aims to classify and map benthic habitats in the shallow waters of Pari Island using drone and satellite images using the object-based image analysis (OBIA) classification method. A total of 1052 aerial photographs with a resolution of 3.6 cm / pixel and Sentinel-2B images with a spatial resolution of 10 m were processed using the support vector machine (SVM) algorithm. Both images produced six classes: live corals, macroalgae, seagrass, dead corals with algae, sand, and rubble. The overall accuracy test results for drone imagery and satellite imagery were 81.88% and 69.57%, respectively. Based on the overall accuracy, the drone images proved to be better than Sentinel-2 imagery.

1 Introduction

Indonesia is a country with a coral reef area reaching 39,583 km2 with 18% in the form of world coral reef cover and 45.7% in the form of coral triangle area [4]. The amount of potential possessed must be optimized and utilized on an ongoing basis. One way to accomplish thisis to utilize remote sensing technology to inventory basic habitat cover [5]. A shallow sea water habitat is a place to live and shelter various types of aquatic organisms, including seagrass, seaweed, algae, dead coral, and living corals with different types of substrates, such as coral fragments, sand, and mud [11].

Pari Island is one of the islands located in the waters of the Thousand Islands. This water area is situated in the DKI Jakarta region, outside Jakarta Bay. Geographically, this water area plays an important and significant role in regional economic growth through the development of marine industries such as sea transportation services and tourism. The Pari

* Corresponding author: jppanjaitan@apps.ipb.ac.id

Island group consists of Pari Island, Kongsi Island, Tengah Island, Burung Island, and Tikus Island, which are all surrounded by an atoll reef. The water conditions in the area outside the atoll are deep, while the water inside the atoll is relatively shallow, and in some places, there are lagoons [18].

Remote sensing technology has been widely used for the detection of benthic habitats in Indonesia because this method works faster and with sufficient accuracy. In addition to satellite imagery, currently the increasing use of remote sensing using drone technology has been widely used [23]. High-resolution aerial photography images produced with drones can be repeated in time series at the same location so that studies can be conducted on the impact of changes on the closure of coral reef ecosystems associated with certain phenomena [22]. Drones produce image data with higher spatial resolution, are more economical, avoid cloud cover, provide more flexible data acquisition while maintaining high accuracy potential, and real-time. Digital classification consists of two approaches: pixel and object-based classification. Object-Based Image Analysis (OBIA) is a object-based remote sensing image analysis method that can detect several objects with certain similarities. OBIA classifies satellite images based on spectral values and optimizes the spatial features of images according to interpretation elements, such as texture, shape, and other parameters [19]. A wide variety of algorithms can be used in the image classification process. The SVM algorithm is an algorithm whose classification method uses a hypothetical space in the form of a linear function in a feature space. The SVM algorithm has been proven to produce better accuracy than other classification algorithms, such as Decision Tree (DT), Bayes, and K-Nearest Neighbor (KNN), in image-based classification.

Benthic habitat mapping needs to be done regularly because this habitat has the potential to change due to certain factors, such as habitat utilization, which causes pressure on shallow sea-bottom habitats [21]. In general, shallow water areas on small islands are vulnerable to climate change, which results in potential habitat destruction, changes in natural processes in ecosystems, and pollution [20].

2 Methods

2.1 Study Site

This study was conducted from October 2023 to May 2024. The data used are from field observations carried out on Pari Island, Kepulauan Seribu. The research sites are shown in Figure 1.

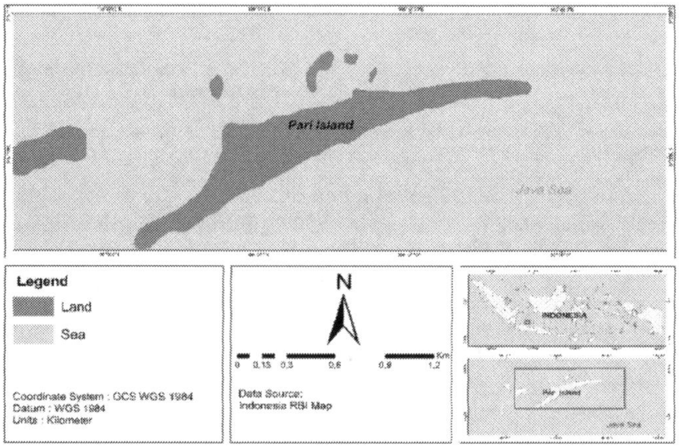

Fig. 1. Study area at Pari Island.

2.2 Data Collection

Field observations were conducted by taking aerial photos using drones and observing benthic habitats by taking the coordinates and documentation of objects. The process of taking aerial photos consists of creating a flight plan using a drone deployment application. A flight area plan that includes three flight plans with flight altitude settings of 120 mfrontlap 70%, sidelap 70%, and drone flight speed set at 12 m/s and camera viewing angle -90°. Drone flights were carried out at 16.00-17.00) [9]. There were 242 points of benthic habitat divided into two different inputs: 104 points for classification and 138 points for accuracy tests.

2.3 Data Processing and Analysis

2.3.1 Photogrammetry

The process of processing aerial photos until the results of obtaining mosaic images includes the following processes: 1) input aerial photos, 2) aligned photos, 3) building a dense cloud, 4) building 3D polygon model, 5) building texture, 6) building orthophotos, and 7) exporting aerial photo mosaic results. Photos taken using drones already have coordinate points and are in JPEG format. The drone design track is shown in Figure 2.

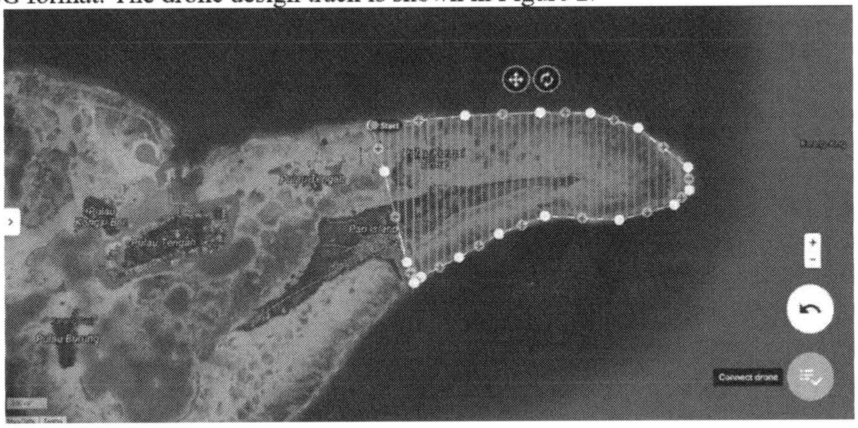

Fig. 2. Drone track design at Pari Island.

2.3.2 Image Preprocessing

Preprocessing is the initial process of repairing an image to remove noise. Preprocessing removes parts that are not required in the input image for the next process [15]. The image preprocessing stage began with cropping and composite imagery. Cropping on the image serves to determine exactly which part of the image will be used as an area of the object to be processed, so that it can be separated from areas that are not needed in the next processing. The composite band is the process of composite (merging) three bands to obtain red, green, and blue colors for the image [13]. Atmospheric correction was then carried out to eliminate errors caused by atmospheric influences on the image [8]. Dark object subtraction (DOS) was used. DOS is an empirical method of atmospheric correction for satellite imagery, which assumes that reflectance from dark objects includes a substantial component of atmospheric scattering.

2.3.3 Water Column Correction

Water column correction is performed to improve image quality by red (3)
the water column so that the appearance of objects is clearer, making it e
contained in the image [1]. The information obtained in the previous image was mixed with reflections from several components in the water. Water column correction improves the accuracy of digital image classification performed [6]. In drone imagery, water column correction is not carried out because drone data collection is carried out at low tide such that the appearance of the object is clearly visible. The method used to correct the influence of the water column is the Depth Invariant Index (DII) using the Lyzenga algorithm (1981)[10] with the following equation:

$$Y = ln\, ln\,(Li)\ - \left[\left(\frac{ki}{kj}\right) \times ln\,(Lj)\right]$$

With,

$$\frac{ki}{kj} = a + \sqrt{a^2 + 1}$$

$$a = \frac{(\sigma_{ii} - \sigma_{jj})}{(2\sigma_{ij})}$$

Where:

Y:	Depth-invariant index.
Li:	Reflectance value of blue band
Lj:	Reflectance value of green band
ki/kj:	Attenuation ratio
a:	Coefficient of attenuation
σ_{ii}:	Variance band blue
σ_{jj}:	Variance band green
σ_{ij}:	Covariance of bands blue and green

2.3.4 Segmentation (4)

Segmentation is the concept of building objects or segmentation from pixels into segments or objects that have the same properties. The segmentation algorithm used was multi resolution segmentation (MRS) [12]. In this study, three main segmentation parameters were selected:scale, shape, and compactness, with different variation values to obtain optimal segmentation parameters for land cover classification [7].

2.3.5 Level Classification

Image classification is carried out using contextual editing method (1)
application of machine learning algorithms at level 2. Machine lea
mathematical algorithms that use data as a way of learning and produce predictions or decisions [16]. Level 1 classification (reef level) is used to assign classes with a certain threshold value so that it becomes the desired object class, while leve
used classifier with the Supported Vector Machine (SVM) algorithm. (2)
classes that are classified which is life coral, macroalgae, seagrass, rubble, sand, and dead coral with algae. SVM algorithm equation are:

$$f(x) = \Sigma_{i\epsilon S}\lambda_i y_i K(x_i x) + w_0 \quad (5)$$

Where:

K	= Kernel function	(6)
y_i dan x_i	= Represent training sample	
λ_i	= Lagrange multiplier	
S	= Part of the training sample corresponds to the multiplier.	(7)
w_0	= Hyperplane parameter	

2.3.6 Accuracy Test

Accuracy tests were performed to determine the optimal results of each machine learning algorithm used and to determine the results of mapping shallow marine habitats. A confusion matrix was used as the accuracy test formula. The confusion matrix is one of the accuracy testing methods used to determine the overall accuracy value of the cla: results by comparing the classification result classes with data training (can be calculated using the following equation [2]: (8)

$$Producer\ accuracy = \frac{n_{jj}}{n_{+j}} \times 100\%$$

$$User\ accuracy = \frac{n_{ii}}{n_{i+}} \times 100\%$$

$$Overall\ accuracy = \frac{\sum_{i=1}^{k} n_{ii}}{n} \times 100\%$$

The value of the kappa coefficient describes the condition of whether or not the accuracy of the data being compared is used to determine the closeness or agreement of the results of testing two datasets. The kappa coefficient value has a range of up to 1.00, with value categories referring to Congalton and Green (2019)[2] The equation and range of kappa coefficient values can be expressed as follows:

$$K = \frac{\sum_{i=1}^{k} n_{ii} - \sum_{i=1}^{k} n_{i+}n_{i+i}}{n^2 - \sum_{i=1}^{k} n_{i+}n_{i+i}}$$

Where
K = kappa coefficient
k = number of rows in the matrix
n = number of observations
nii = number of observations in the i column and i row
njj = number of observations in the j column and j row
ni+ = total marginal row i
n+i = total marginal column i
n+j = total column

3 Result and Discussion

3.1 Ortophoto

Drone flights were carried out three times in the North, South and East regions, and 1052 aerial photographs were obtained. Aerial photographs of the three regions were then processed using orthophotos to produce images of Pari Island. An aerial photograph map is shown in Figure 3.

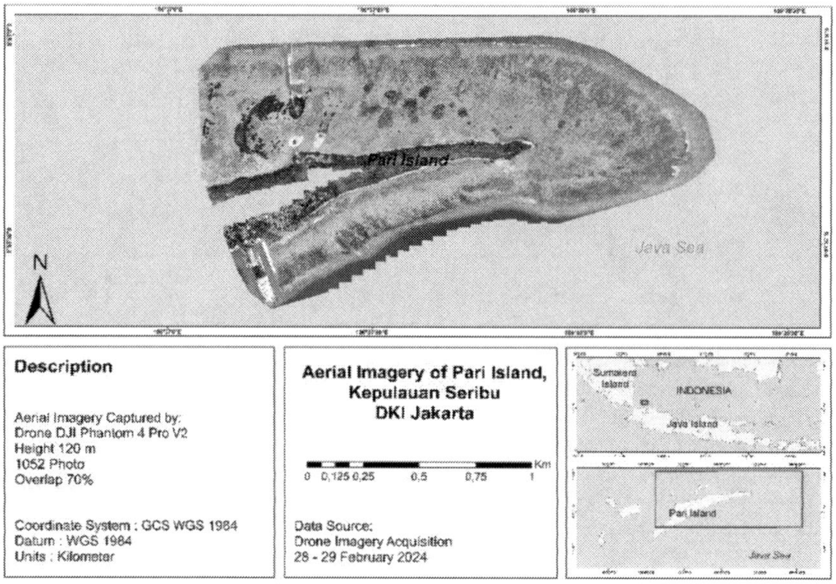

Fig. 3. Aerial imagery using drone at Pari Island.

The aerial photography acquisition process acquired an area of 161 ha with a flight time of ±30 min per flight plan and a spatial resolution of 3.6 cm/px. In the aerial photo image of Pari Island, there was a difference in brightness caused by the angle of incidence of the sun and the ability of the camera sensor to focus on an object. In addition, other weather conditions such as clouds and wind also affect the results of aerial photography and the duration of aerial photo acquisition. The orthophoto results were used to segment the benthic habitat.

3.2 Image Preprocessing

3.2.1 Atmospheric Correction

Atmospheric corrections were made to the satellite images using the DOS. The DOS searches each band for the darkest pixel value. Scattering was eliminated by subtracting the value of each pixel in the band [14]. Satellite images before and after correction show visual differences, where the results of images from the application of atmospheric correction are sharper and brighter compared to images before correction. The differences in satellite imagery before and after correction are shown in Figures 4 and 5.

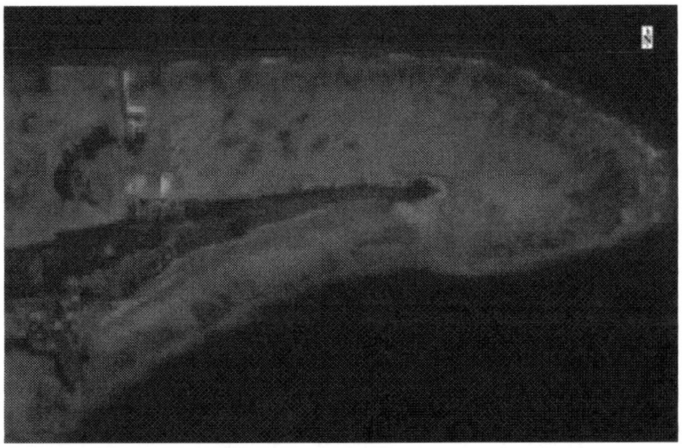

Fig. 4. Sentinel-2B image before atmospheric correction.

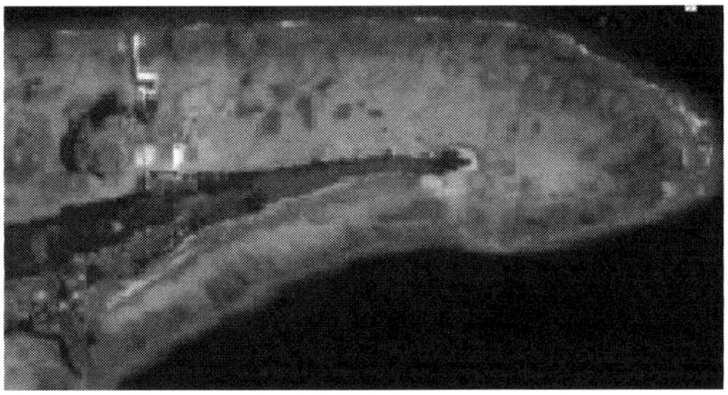

Fig. 5. Sentinel-2B image after atmospheric correction.

3.2.2 Water Column Correction

Water column correction is applied to improve the image quality so that objects at the bottom of shallow water can be better distinguished. The Sentinel-2B imagery acquired on February 29, 2024, was taken by satellite at 2:55 p.m. and the receding time of Pari Island waters was at around 4:00 p.m., making it necessary to correct the water column for the Sentinel-2B imagery. Image transformation is carried out using a combination of blue and green canals, which are excellent bands for penetrating visible light energy into the water column and are highly dependent on water conditions (attenuation coefficient) [17]. Lyzenga (1981)[10] stated that obtaining an appropriate linear correction requires training areas on the same bottomwater substrate at different depths. In this study, training areas were in sand habitats with different depths to extract their pixel values and search for variant values of each band and covariance values between bands. The seabed information can be determined from the attenuation coefficient between channels i and j (ki/kj).

Band	Attenuation Ratio (ki/kj)
Blue – Green	0.361738

3.3 Segmentation and Classification

Level 1 segmentation of individual drones and satellite imagery resulted in land, shallow water, and deep-sea classes. The segmentation process is performed using a multi resolution segmentation algorithm with shape parameters valued at 0.1, compactness valued at 0.5 (default), and scale using a value of 250 for drone imagery and a scale of 50 for satellite imagery. Level 2 segmentation is carried out by resegmentation of the shallow water class. Each drone image and imagery produced six classes of benthic habitats, namely corals, seagrass, sand, macroalgae, DCA, and rubble. For drone imagery, segmentation was performed with a shape parameter value of 0.1, compactness of 0.5 (default), and scale of 50 and 10, respectively, because benthic habitats have a smaller area, and the scale value of level 2 segmentation must be smaller than level 1 to be more focused and avoid under-segmentation.

The results of level 1 segmentation (reef level) are then classified into three classes: land, shallow sea, and deep sea, and the shallow sea area is used for level 2 segmentation. The benthic habitat classification scheme in the Pari Island area is classified into six classes: seagrass, rubble, sand, macroalgae, dead coral with algae, and coral. In determining the benthic habitat classification scheme with six classes used, there are no standard provisions in Indonesia, so the determination of these six classes uses ad hoc classification, namely, the determination of classes with what is appropriate found in field conditions. Benthic habitat classification was performed based on the results of shallow sea segmentation, where the segment was trained using 104 field observation points. Classification of benthic habitat distribution in aerial photographic imagery using the support vector machine (SVM) algorithm. The results of benthic habitat classification are shown in Figures 6 and 7, and the overall area of each benthic habitat class is shown in Table 5.

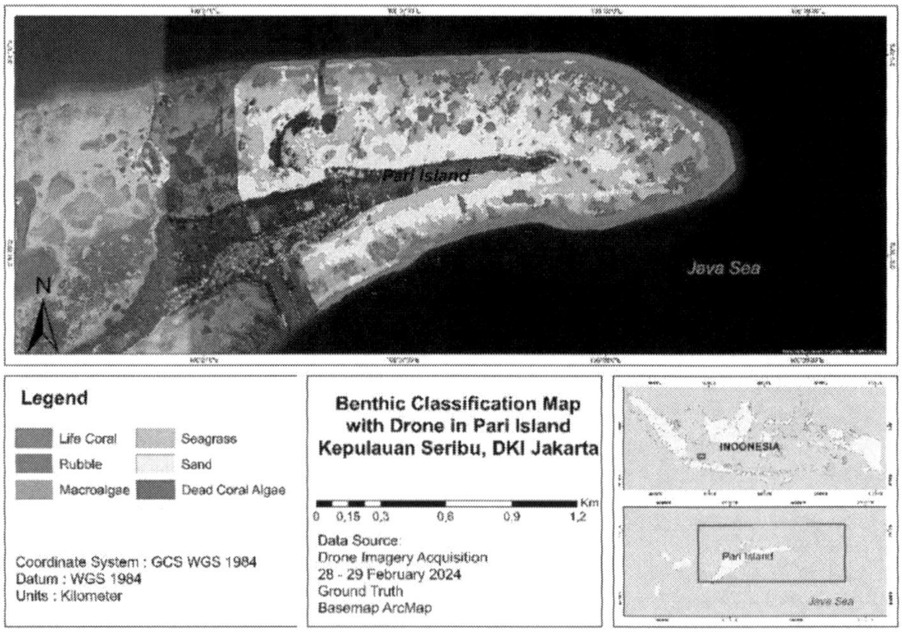

Fig. 6. Benthic map with drone at Pari Island.

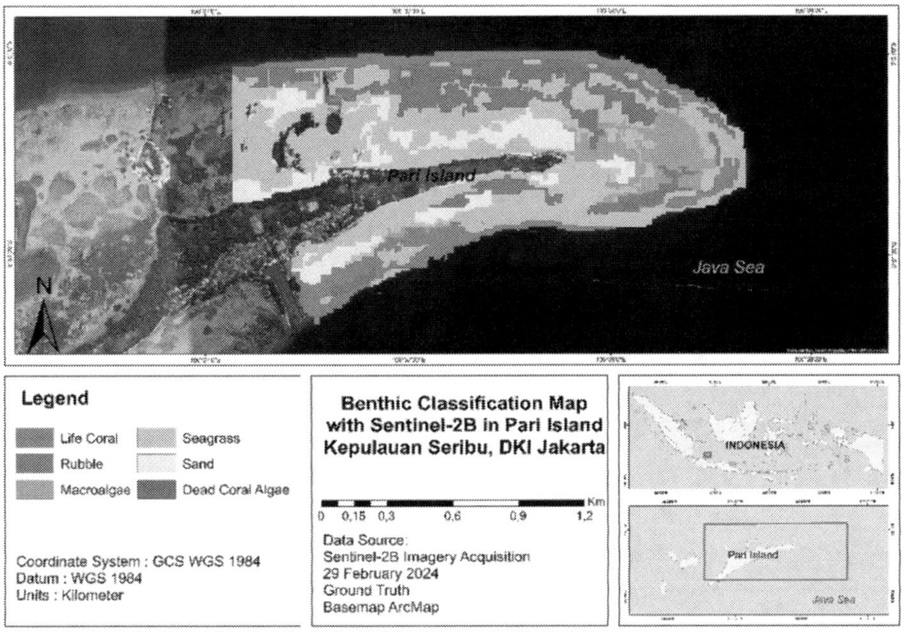

Fig. 7. Benthic map with Sentinel-2B at Pari Island.

The benthic habitats found on Pari Island are quite diverse and evenly distributed in shallow water areas. There were differences in the area of the benthic habitat in each image due to differences in the spatial resolution of each image used. Significant differences were observed in the results of the object-based classification using Sentinel-2B imagery. This is due to the lower resolution of Sentinel-2B imagery compared to drone imagery to perform object-based classification in classifying a relatively small spatial area, namely, Pari Island. The low spatial resolution of Sentinel-2B images affects the segmentation process, which causes satellite images to experience errors in the form of under-segmentation. In the results of drone images, the distribution pattern formed is very complex, and is influenced by the data source used (drone imagery) having a high spatial resolution of 3.6 cm/pixel so that it is able to classify objects on Pari Island quite well.

3.4 Accuracy Test

An accuracy test was conducted to determine the accuracy of the data from the classification process determined through 138 observation points from a total of 242 observation points. Accuracy tests were performed using the confusion matrix method, including overall accuracy (OA), producer accuracy (PA), and user accuracy (UA).

Drone Imagery Result									
		DCA	LC	SG	M	S	R	Total	UA (%)
Field Points	DCA	15		4				19	78.95
	LC		17	2			1	20	85.00
	SG	2	1	33		3		39	84.62
	M	1			10	2		13	76.92
	S	2		3		24		29	82.76
	R		3		1		14	18	77.78
	Total	20	21	42	11	29	15	138	
	PA (%)	75.00	80.95	78.57	90.91	82.76	93.33		
	OA (%)	81.88							
	Kappa	0.80							

Satellite Imagery Result									
		DCA	LC	SG	M	S	R	Total	UA (%)
Field Points	DCA	12		2	3	1	1	19	63.16
	LC		11	1	1		2	20	73.33
	SG	2	5	44	6	2	5	64	68.75
	M	1	1	2	10	1		15	66.67
	S			4	1	15		20	75.00
	R	1					4	5	80.00
	Total	16	17	53	21	19	12	138	
	PA (%)	75.00	64.71	83.02	47.62	78.95	33.33		
	OA (%)	69.57							
	Kappa	0.68							

Based on Table 7 and 8 it can be seen that the OA value for drone imagery is 81.88% and for satellite imagery is 69.57%. Compared to previous studies by Ari Anggoro (2017), which achieved 75% overall accuracy by using Worldview 2, benthic habitat mapping using drones showed higher spatial resolution and overall accuracy. The kappa coefficient for drone imagery, with a value of 0.80 is included in the very good category, and the kappa coefficient for satellite imagery with a value of 0.68 is included in the good category. There is a difference in OA values from the results of the drone and Sentinel-2B image classification owing to differences in the spatial resolution of the images. The spatial resolution of drone imagery in one pixel is 3.6 cm/pixel while in Sentinel-2B imagery, it is 10 m. According to Firansyah *et al.* (2019), the higher the spatial resolution of the image data, the better the accuracy strength. Referring to SNI 7716:2011, the value of the acceptable overall accuracy test results for mapping shallow seawater habitats was ≥ 60% [9] . Compared to previous studies by Ari Anggoro (2017), which achieved 75% overall accuracy by using Worldview 2, benthic habitat mapping using drones showed higher spatial resolution and overall accuracy.

4 Conclusion

Object-based benthic habitat mapping on Pari Island using drone imagery and Sentinel-2B satellite imagery resulted in a scheme of six classes: seagrass, macroalgae, rubble, sand, coral, and dead coral with algae. Drone and satellite imagery produced OA values of 81.88% and 69.57%, respectively, with kappa coefficient values of 0.80 (very good) and 0.68 (good) respectively. The results of the kappa accuracy and coefficient test indicate that the results of drone image classification are better than those of satellite images, and the spatial resolution of the image affects the accuracy value. Further research is expected on the collection of coordinate point data using geodetic GPS so that it can produce classification and test values with better accuracy.

References

[1] Aldin F, Prasetyo Y, Helmi M. 2020. Studi pemetaan habitat dasar perairan laut dangkal berdasarkan analisis digital menggunakan citra Pleiades multispektral di Perairan Pulau Menjangan Besar, Kepulauan Karimunjawa, Jawa Tengah. *Jurnal Geodesi Undip*. 9(1): 77-86.

[2] Congalton RG, Green K. 2019. *Assessing the Accuracy of Remotely Sensed Data*. New York (NY): CRC Press.

[3] Firansyah S, Gaol J, Susilo SB. 2019. Perbandingan klasifikasi SVM dan *decision tree* untuk pemetaan mangrove berbasis objek menggunakan citra satelit Sentinel-2B di Gili Sulat, Lombok Timur. *Journal of Natural Resources and Environmental Management*. 9 (3): 746 – 757.

[4] Giyanto, Winardi, Kusmanto E, Salmin, Soeroyo, Manuputty A. 2014. Panduan Monitoring Kesehatan Karang. Jakarta (ID): CRITC COREMAP LIPI.

[5] Hafizt M, Iswaril MY, Prayudha B. 2017. Kajian Metode Klasifikasi citra landsat-8 untuk pemetaan habitat bentik di Kepulauan Padaido, Papua. *Oseanologi dan Limnologi di Indonesia*. 2(1): 1-13.

[6] Ilyas TP, Nababan B, Maduppa H, Kushardono D. 2020. Pemetaan ekosistem lamun dengan dan tanpa koreksi kolom air di Perairan Pulau Pajenekang, Sulawesi Selatan. *Jurnal Ilmu dan Teknologi Kelautan Tropis*. 12(1): 9-23.

[7] Koman WAF, Izza AS, Candraningtyas D. 2022. Perbandingan parameter segmentasi OBIA dalam klasifikasi tutupan lahan. *Jurnal Paradigma: Jurnal Multidisipliner Mahasiswa Pascasarjana Indonesia*. 3(2): 104-113.

[8] Kristianingsih L, Wijaya A, Sukmono A. 2016. Analisis pengaruh koreksi atmosfer terhadap estimasi kandungan klorofil-a menggunakan citra Landsat 8. *Jurnal Geodesi Undip*. 5(4): 56–64.

[9] LIPI. 2014. Panduan Teknis Pemetaan Habitat Dasar Perairan Laut Dangkal. Jakarta (ID): CRITC COREMAP LIPI.

[10] Lyzenga DR. 1981. Passive remote sensing techniques for mapping water depth and bottom features. *Applied Optics*. 17: 379-383.

[11] Mastu LOK, Nababan B, Panjaitan JP. 2018. Pemetaan habitat bentik berbasis objek menggunakan citra sentinel-2 di Perairan Pulau Wangi-Wangi Kabupaten Wakatobi. *Jurnal Ilmu dan Teknologi Kelautan Tropis*. 10(20): 381-396.

[12] Prabowo NW, Siregar VP, Agus SB. 2018. Klasifikasi habitat bentik berbasis objek dengan algoritma *support vector machines* dan *decision tree* menggunakan citra multispektral SPOT-7 di Pulau Harapan dan Pulau Kelapa. *Jurnal Ilmu dan Teknologi Kelautan Tropis*. 10(1):123-134.

[13] Putra RD, Napitupulu HS, Nugraha AH, Suhana MP, Ritonga AR, Sari TEY. 2022. Pemetaan luasan hutan mangrove dengan menggunakan citra satelit di Pulau Mapur, Provinsi Kepulauan Riau. *Jurnal Kelautan Tropis*. 25(1): 20-30.

[14] Rahman A, Siregar VP, Panjaitan JP. 2020. Estimasi kedalaman perairan dangkal menggunakan citra satelit multispektral Sentinel-2A. *Jurnal Segara*. 16(3): 151-162.

[15] Rohpandi D, Sugiharti A, Winara GA. 2015. Aplikasi pengolahan citra dalam pengenalan pola huruf ngalagena menggunakan MATLAB. Konferensi Nasional Sistem dan Informatika 2015.

[16] Roihan A, Sunarya PA, Rafika AS. 2020. Pemanfaatan *machine learning* dalam berbagai bidang: review paper. *Indonesian Journal on Computer and Information Technology*. 5(1): 75-82.

[17] Sabilah AA. 2021. Performa algoritma *maximum likelihood, support vector machine,* dan logika *fuzzy* dalam memetakan sebaran lamun menggunakan citra multi-sensor. [tesis]. Bogor (ID): Institut Pertanian Bogor

[18] Salim H, Ahmad L. 2013. Karakteristik pantai Gugusan Pulau Pari. *Jurnal Region* 5(1): 1-11.

[19] Saputra R, Gaol JL, Agus SB. 2021. Studi perubahan tutupan lahan mangrove berbasis objek (OBIA) menggunakan citra satelit di Pulau Dompak Provinsi Kepulauan Riau. *J. Ilmu dan Teknologi Kelautan Tropis*. 13(1): 39-55.

[20] Sari CA, Syah AF, Prayudha B, Salatahol A. 2020. Pemetaan habitat bentik menggunakan citra satelit sentinel-2A di Pulau Liki, Papua. *Jurnal Penginderaan Jauh*. 17(1): 33-42.

[21] Siregar VP, Agus SB, Sunuddin A, Subarno T, Aziizah NN. 2020. Analisis perubahan habitat dasar perairan dangkal menggunakan citra satelit resolusi tinggi di Karang Lebar, Kepulauan Seribu. *Jurnal Ilmu dan Teknologi Kelautan Tropis*. 12(1): 37- 51.

[22] Sugara A, Siregar VP, Agus SB. 2020. Klasifikasi habitat bentik perairan dangkal dari citra worldview-2 menggunakan data in-situ dan *drone*. *Jurnal Ilmu dan Teknologi Kelautan Tropis*. 12(1): 135-150.

[23] Thenu IM, Makailipessy MM. 2017. Pemetaan perubahan ekosistem wilayah pesisir Kecamatan Dullah Utara Kota Tual. *Jurnal Teknologi Perikanan dan Kelautan*. 8(1): 39-48.

The relation between volume backscattering strength (SV) and density of *Enhalus acoroides* at Wangi-Wangi Island, Wakatobi Regency

Dianti[1], *Sri* Pujiyati[1*], *Steven* Solikin[1]

[1]Departement of Marine Science and Technology, Faculty of Fisheries and Marine Sciences, IPB University (Bogor Agricultural University), 16680 Bogor, Indonesia

Abstract: Seagrass is a fundamental vegetation that can optimize the function of coastal areas, to obtain information about the survival of seagrass. Thus an effective and efficient seagrass monitoring method is required. Seagrass density measurements can be performed using in-situ observation and hydroacoustic methods. The purpose of this study was to analyze the oceanographic parameters, SV values, and their correlation with the density and height of *E. acoroides*, as well as oceanographic parameters at Wangi-Wangi Island, Wakatobi Regency. Stationary data were acquired at five stations through acoustic clustering and in-situ diving. The density values at each station were 32 ind/m^2, 33 ind/m^2, 29 ind/m^2, 35 ind/m^2 and 37 ind/m^2. The range of SV values for *Enhalus acoroides* was (-50,19) – (-37,41) dB. The average SV values at each station were -40,76 dB, -40,63 dB, -41,10 dB, -40,48 dB, and -40,46 dB. Logarithmic regression analysis of density against SV values showed a strong relationship R = 0,96. PCA with a cumulative percentage of 92,7% from the biplot correlated most closely with the SV values of densities and seagrass heights. The project's implications and contributions include advancing seagrass monitoring techniques using acoustic methods for improved oceanographic parameter analysis.

1 Introduction

Seagrass, a flowering plant (Angiospermae), anchors itself with rhizomes, leaves, and roots [3], and reproduce through pollination and shoot growth [2]. Its roots and rhizomes stabilize coastlines against wave action, thriving in sandy, muddy, and sometimes rocky substrates, where root penetration varies [14]. Ecologically, seagrass serves as nursery, feeding, and spawning grounds [17], which are crucial for coastal functions and human well-being [10], an interact with abiotic (substrate, water) and biotic (flora, fauna) components (Supriyadi et al., 2018). From the intertidal to subtidal zones, seagrass thrives in submerged high-salinity waters [3].

Indonesia boasts 15 seagrass species, including seven : *Cymodocea rotundata*, *Halodule uninervis*, *Thalassodendron ciliatum*, *Syringodium isoetifolium*, *Thalassia hemprichii*,

* Corresponding author: sripu@apps.ipb.ac.id

Enhalus acoroides, and *Halophila ovalis*. *Enhalus acoroides*, including *Enhalus acoroides*, are documented on Wangi-Wangi Island, Wakatobi Regency [6]. *Enhalus acoroides*, distributed uniformly in intertidal and shallow waters on sandy/muddy substrates, is the focus off SV analysis in this study, across various research stations.

Hydroacoustics is a method for monitoring and mapping the seabed, assessing the substrate, and underwater vegetation using echo signals [14]. This technology categorizes survey data and this; widely applied to in study fish and plankton distributions. Hydroacoustic devices can detect underwater objects by transmitting sound waves from a transducer through a water column. The reflected waves were converted to electrical signals, recorded, amplified, and displayed. The backscattering strength (SV), which indicate the intensity of the reflected signals, is crucial. Instruments such as the Simrad EK-15 echo sounder used here are efficient and offer high accuracy in delineating water body characteristics based on SV values.

Wangi-Wangi Island, part of the Wakatobi Archipelago in Southeast Sulawesi Province, has healthy seagrass potential due to minimal industrial activity and eco-friendly fishing practices. Most Wakatobi research focuses on coral reefs, but seagrass and coral reefs play crucial roles in coastal ecosystems. Despite extensive seagrass beds, little in-situ and hydroacoustic research has been conducted. Thus, further research analyzing seagrass in this area is of great interest.

This study aimed to analyze oceanographic parameters, the volume backscattering strength (SV) values of *Enhalus acoroides* seagrass, and the relationship between SV values and the density and height of *Enhalus acoroides* seagrass.

2 Materials and Methods

This study was conducted on May 3rd, 2023. The research location was Pulau Wangi, Wakatobi Regency, Southeast Sulawesi Province (Figure 1). This is a coastal area with seagrass vegetation at a depth of 2.9 meters. Five stationary stations were used for the research, including in-situ seagrass data acquisition and acoustic, oceanographic, and substrate data.

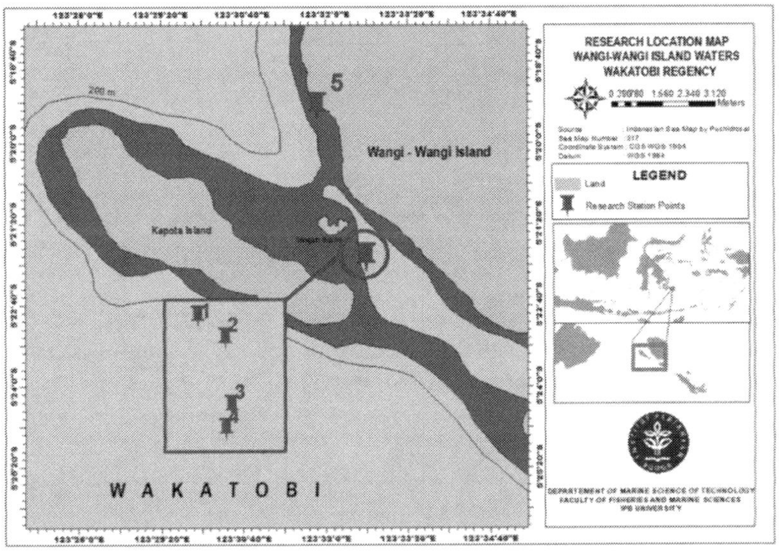

Fig. 1. Research Location Data Acquisition at Wangi – Wangi Island Waters

The equipment used for data acquisition wass aSimrad EK-15 instrument, Garmin Montana GPS, laptop, boat, refractometer, water quality checker, 1x1 m transect, PVC Pipe Frame, Underwater Camera, diving equipment, buoy and weights, fabric meter, underwater recording device, plastic clips, and labels. The software used for data processing included Echoview 4.0, Ms Office, and other necessary software packages.

2.1 Data Acquisition and Processing Procedure

The process begins with the acquisition and processing of oceanographic parameters data, in-situ *Enhalus acoroides* seagrass, and the processing of acoustic data.

2.1.1 Oceanographic Parameters Data

Oceanographic data were obtained from recordings using a refractometer and water quality checker. The oceanographic data includes temperature, salinity, and pH, which are used to help calculate and display the visualizations of the results we have obtained, and facilitate the analysis of oceanographic data in Wangi-Wangi Island, Wakatobi Regency.

2.1.2 In-Situ Enhalus Acoroides Seagrass

In situ data collection involves identifying seagrass types and measuring density, and height using 1x1 m random transects at each station. Density was determined by counting seagrass stands within the transect, whereas height was measured from the rhizome to the leaf tip. Substrate information was directly observed in each transect/station and documented using an underwater mobile phone camera. Visual observations were used to validate the acoustic data obtained later. Diving transects at observation stations precede acoustic data collection [11]. The variables included substrate type, seagrass density (ind/m^2), and average height (m). Seagrass density is the number of individuals per unit area, calculated using Equation (1) as follows [9] :

$$D = \frac{N}{A} \quad (1)$$

Explanation :
D = species density (stands/1m^2)
N = total number of species stands (individuals)
A = sampled area (m^2)

The average height of theseagrass was calculated using at sampling method with a sample size of 10. The average seagrass height was calculated the following formula:

$$\underline{Ti} = \frac{\Sigma Ti}{n} \quad (2)$$

Explanation :
$\underline{Ti}$ = Average height of seagrass obtained from the number of samples (m)
ΣT = Total seagrass height obtained from number of samples (m)
n= Number of seagrass samples (individuals)

2.1.3 Acoustic of Seagrass

Acoustic data were obtained by directing the transducer at each seagrass and substrate target at frequency of 200 kHz. The transducer was placed on a frame composed of PVC pipes with a diameter of 0.75 inches. The transducer was positioned perpendicular to the target (90°), and recordings were made for 5 min The 5-minute seagrass data collection was deemed sufficient because it showed the presence of detected seagrass. This was supported by the condition of the waters during the sounding, which tended to be calm so that. The noise in the data is very small, so the seagrass can be detected well. resulting in the acoustic data in (. raw) format. These data were processed using the Echoview Demo software with a threshold of -70 to -35 dB to detect *Enhalus acoroides* seagrass based on previous research references, producing seagrass SV values. The volume backscattering strength (SV) value can be explained as the ratio of the intensities reflected by a group of single targets. SV can describe a group of targets in a water column. Data processing to obtain the SV value using Echoview 4.0 software can be performed using the following equation [11]:

$$SV = 10 \; log \; log \; Sv \tag{3}$$

$$Sv = 10^{(\frac{SV}{10})} \tag{4}$$

$$\underline{Sv} = \frac{\Sigma Sv}{n} \tag{5}$$

$$\underline{SV} = 10 \log \underline{Sv} \tag{6}$$

Explanation :
SV = Volume Backscattering Strength
Sv = Volume Backscattering Strength linier
n = Number of data

2.2 Data Analysis Procedure

2.2.1 The Relation Between Volume Backscattering Strength (SV) and the Density Value of Enhalus acoroides Seagrass Using Logarithmic Regression Analysis

The relationship between SV and the density value of *Enhalus acoroides* seagrass, using logarithmic regression analysis, expresses the functional relationship of the predetermined variables in mathematical form. The logarithmic regression analysis useds two variables The first variable is the independent variable, which is the density value, while the dependent variable is the SV value of *the Enhalus acoroides* seagrass. The relationship between the two variables is visualized in the form of a graph. This graph produces a regression coefficient (R^2) that indicates the percentage of correlation between two variables. The strength of the correlation between these variables can be determined by examining the correlation coefficient (R) values in Table 1.

Table 1 Coefficient Correlation Value (R) [6]

R Value	Correlation
R ≥ 0,00 - < 0,20	Very low
R ≥ 0,20 - < 0,40	Low
R ≥ 0,40 - < 0,70	Moderate
R ≥ 0,70 - < 0,90	Strong
R ≥ 0,90 - ≤ 1,00	Very Strong

2.2.2 The Relation Between Volume Backscattering Strength (SV), Density, Height of Enhalus acoroides Seagrass, and Oceanographic Parameters using Principal Component Analysis (PCA)

The SV values, density, and height of *Enhalus acoroides* seagrass, along with oceanographic parameters, were analyzed using Principal Component Analysis (PCA). PCA is a statistical method that highlights variations in variables or parameters and; extracts the most significant information from at dataset (Rahimah et al., 2020). Due to differing measurement units, data normalization (0 to 1 range) is crucial for precise relationship interpretation and for reducing bias in forming principal components [8]. The principal components were derived using the cumulative variance proportion, eigenvalues from covariance/correlation matrices, and scree plots illustrating the variance/eigenvalues across components. The reduced data were visualized using a biplot graph showing the two principal component axes.

3 Results and Discussion

3.1 Oceanographic Parameters Data

Oceanographic data measurements were obtained during high-tide conditions (Table 2). Sea surface temperatures ranged from 31.30 to 33.50 °C, which is optimal for *Enhalus acoroides* seagrass growth (28.00 - 34.00 °C). The salinity was stable at 34.00 ppt across all stations, with *Enhalus acoroides* showing good tolerance. pH levels ranged from 6.80 to 6.84, within the normal range (6.81 - 8.06) supporting seagrass and organism growth. Weather conditions varied: clear skies at stations 1, 4, and 5 and; light rain at stations 2 and 3.

Table 2 Resullt of Acquisition oceanographic parameters

Parameters	**Station**				
	1	**2**	**3**	**4**	**5**
Temperature (°C)	33,50	33,50	33,50	31,30	31,30
Salinity (ppt)	34,00	34,00	34,00	34,00	34,00
pH	6,84	6,84	6,84	6,80	6,80

3.2 In Situ Data of Seagrass

The seagrass community found diversity on Wangi-Wangi Island, including seven species: *Cymodocea rotundata, Halodule uninervis, Thalassodendron ciliatum, Syringodium isoetifolium, Thalassia hemprichii, Enhalus acoroides, and Halophila ovalis* [6]. This study focuses on *Enhalus acoroides* because of its dominance, high density, and uniform distribution across all transects (Table 3). Unlike the others, *Cymodocea rotundata* and I have a low density and short seagrass height (0-0.015m), resulting in overlapping SV values with seabed SV, precluding further analysis.

Table 3 Species of Seagrass Found at Each Observation Station

Spesies	**Station**				
	1	2	3	4	5
Enhalus acoroides	✓	✓	✓	✓	✓
Cymodocea rotundata	-	✓	✓	-	✓
Thallasia hemprichii	✓	-	✓	✓	-

Explained:

✓ = founding at the research location

- = Unfounded at research location

The study was conducted Wangi-Wangi Island, and with data are ummarized in (Table 4). The observation stations showed varying densities of dense seagrasses Visual observations indicate homogeneous seabed conditions. Substrate information was directly observed in each transect/station and documented using an underwater mobile phone camera. The Results from the substrate with fine sand, sandy mud, and sandy substrates, average a depth of 2.9 meters across the five stationary stations..

Table 4 Result of Acquisition data of *Enhalus acoroides* with diving

Station/Transect	**Density** (ind/m^2)	**Average High (m)**	**Type of Substrat**
1	32	0,79	fine sand
2	33	0,91	sandy mud
3	29	1,19	sandy mud
4	35	0,81	sandy mud
5	37	1,23	sand

Enhalus acoroides seagrass is a large species with leaf lengths exceeding 1 meter [11] (Figure 2). This aligns with the diving data in Table 7, showing the seagrass dense green color of seagrass and its ability to adapt to various substrates such as mud, sand, and coral sand submerged in water. Therefore, they were found across all stations [6]. Seagrass density analysis revealed values ranging from 29 inds/m^2 to 37 inds/m^2 across transects 1-5. Station 5 had the highest density, whereas station 3 had the lowest density. Differences in density are attributed to natural factors, such as topography and substrate variations at each station, which influence seagrass adaptation patterns. Density is also influenced by seasonal changes, wave energy, substrate organic content, and other environmental factors (Short & Coles, 2001). Average seagrass height measurements ranged from 0.79 to 1.23 meters from rhizome to leaf tip in the study area..

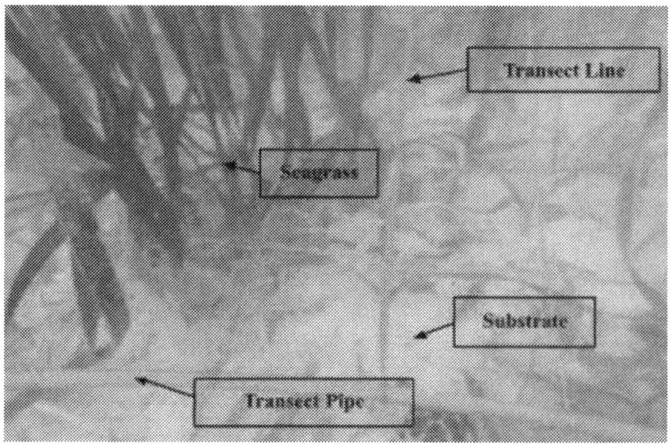

Fig. 2. *Enhalus acoroides* founding at research location

The type of seabed substrate is crucial for understanding the growth and distribution of seagrasses. According to [14], suitable substrate conditions are a primary factor influencing seagrass development and growth and are characterized by adequate sediment content. Thinner seabed substrates can destabilize seagrass, whereas thicker substrates support more vigorous growth.

3.3 Acoustic Instrument Calibration (Simrad EK-15)

An acoustic instrument calibration was conducted to measure the standard Target Strength (TS) using a 35 mm diameter tungsten sphere at a frequency of 200 kHz. The calibration TS results were verified against the theoretical TS measurements of the sphere. This calibration value determines the accuracy of the instrument, considering factors such as the sound propagation speed, transmitter and receiver transducers, and noise levels. The acoustic calibration results are show in Figure 3.

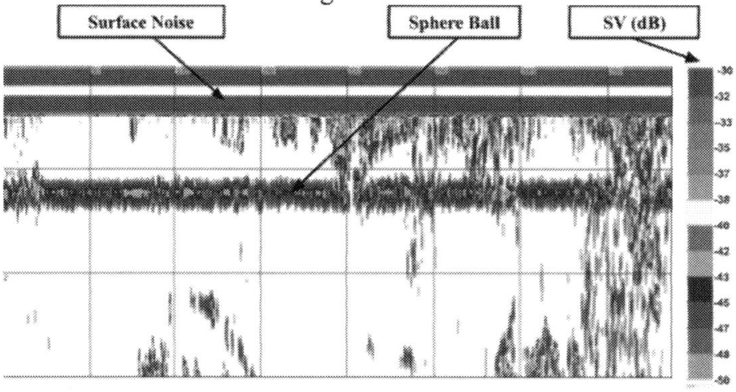

Fig. 3. Calibration Simrad Ek-15 with Sphare ball

Calibration of the acoustic instruments was performed using a 35 mm tungsten sphere at 200 kHz. The measured standard TS value was -43.94 dB, while the theoretical TS for the Simrad EK-15 is -45.96 dB (Elson et al. 2022). This results in a 2.013 dB correction factor for the volume backscattering strength (SV) of *Enhalus acoroides* seagrass. Calibration with the Echoview software showed an absorption coefficient of 8.8789 x 10^-3 dB/m and a sound speed of 1544.38 m/s

3.4 Seagrass Acoustic Data

Hydroacoustic technology enables the retrieval of information from the water column to the seabed, including organism migration, seabed structure, plankton, and underwater vegetation, such as seagrass (Ma'mun et al., 2013). The Simrad EK-15 displays the survey results in the *. raw format, which weare then processed using the Echoview 4.0 Demo software to generate echogram displays.

3.4.1 Echogram

An echogram, derived from echosounder recordings; visually depicts the time, echo strength, and echo return times [4]. The x-axis shows the ping count, while the y-axis represents the water depth (m), measured 1 m below the transducer from the water surface. Each ping's interpretation of the echogram displays the volume backscattering strength (SV) in decibels (dB). Figure 4 illustrates the echogram of the eagrass, where color variations indicate different dB ranges corresponding to the detected objects. Depths detected on the echogram extend to 1.8m, with false echoes attributed to noise from surface reflections caused by waves

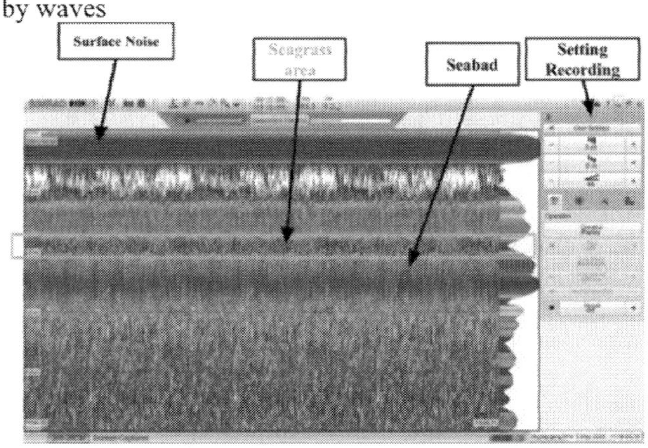

Fig. 4. Echogram result acqusition acoustic data

3.4.2 SV Value of Enhalus acoroides

The processed acoustic survey data for seagrass spanned 5 min. The SV values represent the backscattering strength obtained from the echoes reflected by objects; sourced from the transducer. The SV values for *Enhalus acoroides* seagrass at each station are compared in Table 5.

Table 5 Comparison of Statistical Parameters of Volume Backscattering Strength (SV) for *Enhalus acoroides* Across All Stations

Parameters	Station				
	1	2	3	4	5
ESDU	41	45	35	46	32
SV Minimum (dB)	-50,19	-43,94	-47,57	-44,75	-44,40
SV Maximum (dB)	-37,71	-38,63	-37,60	-38	-37,41
$\underline{SV}$ (dB)	-40,76	-40,63	-41,10	-40,48	-40,46
Standard Deviation	0,94	0,79	0,99	1,56	1,34

Observations of the volume backscattering strength (SV) for *Enhalus acoroides* seagrass compared across the five stations revealed distinct variations. The lowest minimum SV value, -50.19 dB, occurreds at Station 1, indicative of softer seagrass morphology owing to flexible and soft leaves adapted to fine sand substrate conditions at this station. This aligns with Ole's [13] findings that the substrate type significantly affects seagrass root penetration. Conversely, the highest maximum SV value, -37.41 dB, was observed at Station 5, which was attributed to both the morphology and harder sand substrate, challenging seagrass root penetration and supporting diverse attached biota on seagrass leaves (Figure 5a).

Stations 2, 3, and 4 share a similar muddy sand substrate yet exhibit varied SV values: Station 2 ranges from -47.57 to -38.63 dB ($S\overline{V}$ = -40.63 dB), Station 3 from -47.57 to -37.60 dB ($S\overline{V}$ = -41.10 dB), and Station 4 from -44.75 to -38 dB ($S\overline{V}$ = -40.48 dB). These differences may stem from seagrass morphology, particularly leaf color and texture, with Stations 2 and 3 showing younger, smoother leaves, resulting in lower SV compared to Station 4. Overall, a comparison of SV values across all stations in Table 4 reveals notable standard deviation differences: station 1 ± 0.94, station 2 ± 0.79, station 3 ± 0.99, station 4 ± 1.56, and station 5 ± 1.34. Station 2 demonstrateds relatively accurate data acquisition with a low standard deviation, indicating good measurement precision [13].

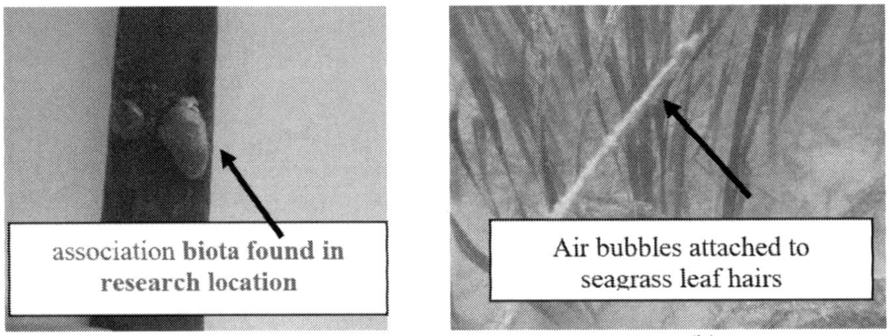

(a) (b)

Fig. 5. Associated biota (a) and air bubbles (b) attached to the hair/nodes of seagrass leaves

Additionally, current flow plays a crucial role, with nutrient-rich currents likely to enhance seagrass fertility. Stations 1 to 4, located in closed current areas, may experience suboptimal nutrient intake compared to Station 5, benefiting from open currents and nutrient accumulation from nearby docks. Thus, Station 5 exhibiteds higher SV values, reflecting superior seagrass growth fertility. Air bubbles, which are ideal reflectors, notably impact hydroacoustic recordings near the surface, potentially introducing noise, including bubbles adhering to fine *Enhalus acoroides* seagrass hairs (Figure 5b), further amplifying SV values [15].

3.4.3 Frequency Distribution of Enhalus acoroides

The frequency of occurrence of *Enhalus acoroides* seagrass data (Figures 6, 7, 8 9, 10 and 11) revealed significant variations across the study stations. Station 1 (Figure 6) exhibits a rising frequency distribution, peaking at 12 occurrences at -40 dB, with data ranging from -43.10 to -38.84 dB and showing slight variability. Station 2 (Figure 7) displays a sharply peaked frequency distribution, with the highest frequency at 11 occurrences at -40 dB, spanning -42.25 to -38.70 dB with minimal variability. Station 3 (Figure 8) shows a moderate distribution with a peak frequency of seven occurrences at -41 dB, ranging from -43.08 to -39.48 dB with even occurrence trends. Station 4 (Figure 9) presents a rising frequency distribution, peaking at 15 occurrences at -40 dB, spanning -42.82 to -37.55 dB with scattered and varied occurrences. Station 5 (Figure 10) has a broader distribution shape compared to other stations, with the highest frequency of 6 occurrences at -40 dB, ranging from -43.41 to -37.32 dB, showing a flat, even, and highly variable occurrence trend. Figure 11 combines these frequency distributions, indicating diverse occurrences despite sharing the same SV scale range, reflecting the variability in the presence of acoustically detected seagrasses

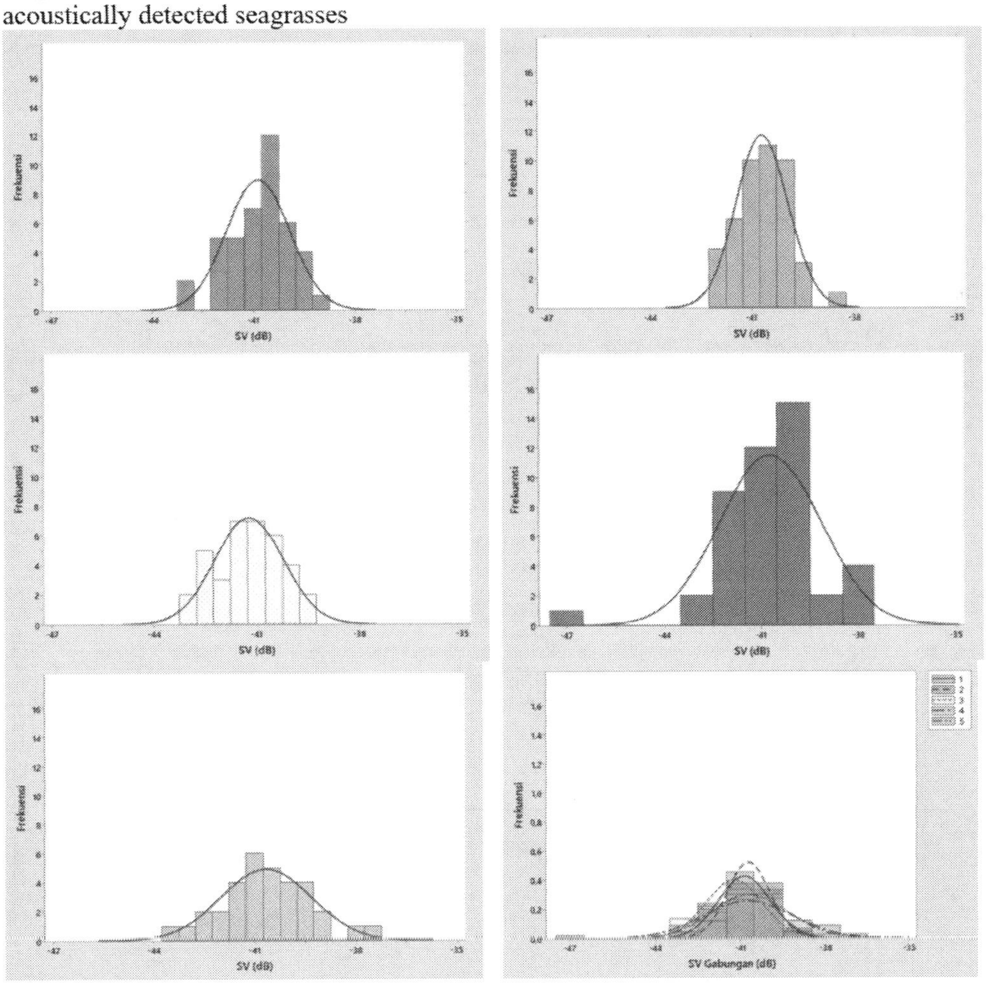

3.5 The Relation Between SV Values, Density of Enhalus acoroides, and Other Parameters

Based on the obtained data, further analysis was conducted to examine the relationship between SV values and seagrass density, as well as between SV values and other parameters.

3.5.1 The Relation Between SV and Density Values of Enhalus acoroides Using Logarithmic Regression Analysis

It looks like you are summarizing the results of a logarithmic regression analysis in your study, focusing on the relationship between seagrass density and the average backscattering strength (SV) of *Enhalus acoroides* seagrass. The table (Table 6) illustrates that lower seagrass density corresponds to lower average SV values, as depicted in Figure 12. The analysis reveals a positive correlation (symbolized by σ) between SV values and seagrass density. The logarithmic regression equation derived is $(\overline{SV}) = 2.7445\ln(\sigma) - 50.289$, with a regression coefficient (R2) of 0.9407, indicating that 94% of SV variability is explained by seagrass density, while 6% is influenced by other factors. The correlation coefficient (R) of 0.96 falls within the range of $R \geq 0.90 - \leq 1.00$, signifying a very strong/high relationship between seagrass density and $(\overline{SV})$ values. This strong relationship suggests that higher seagrass density results in higher acoustic echo amplitudes. In-situ data acquisition by scuba divers ensures a 94% accuracy in acoustic detection. The backscattering of seagrass depends on leaf canopy and echo sounder frequency, consistent with findings by [7], highlighting the frequency influence on received echo levels in decibels.

Table 6 Result of volume backscattering strength (SV) value and density of seagrass

Transect/Stasion	$\underline{SV}$ (dB)	Density (Ind/m²)
1	-40,76	32
2	-40,63	33
3	-41,10	29
4	-40,48	35
5	-40,46	37

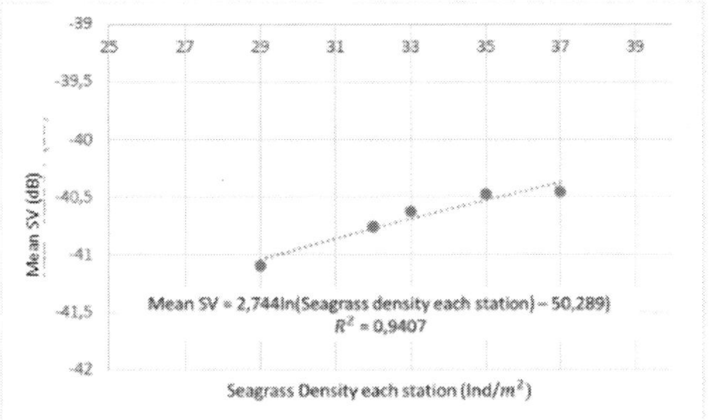

Fig. 12. Logarithmic regression of seagrass SV at all stations with density

3.5.2 The Relation Between Volume Backscattering Strength (SV), Density, Height of Enhalus acoroides Seagrass, and Oceanographic Parameters using Principal Component Analysis (PCA)

PCA analysis included SV, density, and height of *Enhalus acoroides* seagrass, along with oceanographic parameters (temperature, salinity, and pH). Two main components, F1 and F2 (eigenvalues > 1), collectively explained 92.7% of the variance, meeting the PCA criterion for reducing components while retaining data variability (Supranto, 2010). The biplot (Figure 13) shows the SV, density, height, and oceanographic parameters on the F1 and F2 axes. Factor 1 (69% variance) was positively correlated with temperature, salinity, and pH; Factor 2 (23.7% variance) was positively correlated with SV and density, and negatively correlated with seagrass height. Positive and negative correlations we are shown by line proximity, illustrating variable associations (Ma'mun et al., 2019)

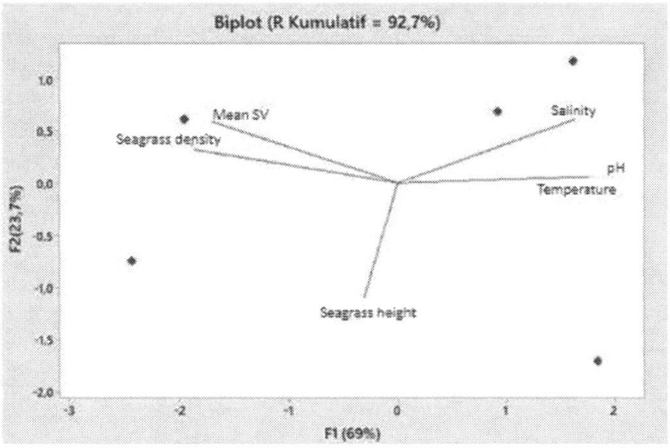

Fig. 13. Biplot of PCA analysis results of seagrass SV values at each station with seagrass density and height values, as well as oceanographic parameters on axis (F1) and axis 2 (F2).

The PCA analysis across five stations (S1-S5) on Wangi-Wangi Island (Figure 14) revealeds correlations among variables, as shown in Figure . The figure illustrates the grouping of stations into zones characterized by different variable relationships. Stations S4 and S5 correlated with SV, density, and seagrass height. Stations S1, S2, and S3 correlated with salinity, temperature, and pH. Clear correlations were observed, particularly between SV and density values, evident in the station zoning. For instance, S4 and S5 show similar

SV (-40.48 dB and -40.46 dB, respectively) and density (35 ind/m² and 37 ind/m²), indicating proximity between these and SV values, resulting in a more dispersed zone in the figure. This analysis highlights positive and negative correlations among SV, density, seagrass height, and oceanographic parameters, demonstrating the varied environmental influences on *Enhalus acoroides* seagrass growth on sand, mud, and coral sand substrates [6]. Similarly, S1 and S2 exhibit close values, S1 at SV -40.76 dB and density of 32 ind/m², and S2 at SV -40.63 dB and density of 33 ind/m². In contrast, S3 exhibited a lower density.

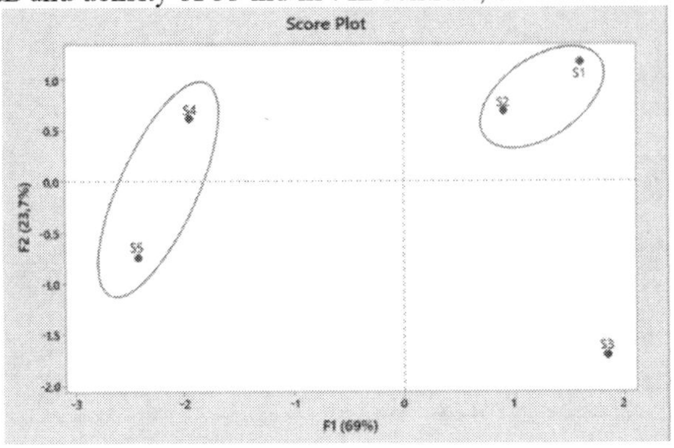

Fig. 14. Score Result of PCA Analys on five stasiun in Wangi – Wangi Island, Wakatobi Regency

4. Conclusion

The conclusion of this study indicates that the oceanographic parameters in Pulau Wangi – Wangi and Wakatobi Regency are suitable for the growth of *Enhalus acoroides* seagrass. The temperature ranges from approximately 31.30 to 33.50 °C, salinity is around 34.00 ppt, and pH levels range from 6.80 to 6.84. The SV values of seagrass detected at all research stations range from -50.19 dB to -37.41 dB, with seagrass density ranging from 29 ind/m2 to 37 ind/m2, and an average seagrass height of 0.79 to 1.23 m. Logarithmic regression analysis demonstrateds a strong positive correlation (R2 = 0.96) between SV and seagrass density. Principal Component Analysis (PCA) confirmeds the dominance of seagrass density and height over SV, with Axis 1 (F1) and axis 2 (F2) explaining 92.7% of the variability. The correlation matrix indicates the positive and negative contributions to the variability.

References

1. L. Elson, HM. Manik, T. Hestirianoto, S. Pujiyati, J. Ilmu dan Teknologi Kelautan Tropis. **14,** 1 15-29 (2022)
2. F. Feryatun, B. Hendarto, N. Widyorini. J. of Management of Aquatic Resources. **1,** 1 44-50 (2012)
3. B.A.J. Gosari, A. Haris. J. Ilmu Kelautan dan Perikanan. **22,** 3 156-162 (2012)
4. B. Hamuna, L. Dimara, S. Pujiyati, N.M.N. Natih, J. Kelautan. **11,** 1 31-37 (2018)
5. E.L. Hazen, J.K. Horne, JoMS. **60,** 1 555-562 (2011)
6. N. Ikhsan, N.P. Zamani, D. Soedharma, J. Teknologi Perikanan dan Kelautan. **10,** 1 27-38 (2019)
7. H.M. Manik, D. Apdillah, Pertanika Journal of Science and Technology. **28,** 2 421-429

(2020)

8. A.A Mansor, S. Shamsul, S. Abdullah, N.C. Dom, N.N. Napi, Ahmed, M. Ismail, Materials Science and Engineering. 1144 012055 (2020).
9. A. Minerva, F. Purwati, A. Suryanto, Diponegoro Journal of Maquares, Management of Aquatic Resources. **3,** 4 88-94 (2014).
10. A. Mulyaningsih, F. Sahami, S.N. Hamzah, Universitas Negeri Gorontalo (2015).
11. A. Ningtyas. [Skripsi] Departemen Ilmu dan Teknologi Kelautan, Institut Pertanian Bogor (2014)
12. F.T.A. Nuswantari. [Skripsi] Departemen Ilmu dan Teknologi Kelautan, Institut Pertanian Bogor (2023).
13. L. Ole. [Thesis] Sekolah Pascasarjana, Institut Pertanian Bogor (2011)
14. S. Pujiyati, B. Hamuna, L. Dimara, N.M.N Natih, J. Kelautan Nasional. **15,** 3 165-174 (2020)
15. S. Pujiyati, M.A Maulana, A. Rahmat, M.H.S Alajuri, J. Kelautan Nasional. **18,** 1 13-22 (2023)
16. I. Rahimah, E.S.Y. Siregar, T. Heriyanto, V.P. Siregar, Marine and Fisheries Science Research. 695 (2020)
17. M.J. Rugebregt, Widyariset. **1,** 1 79-86 (2015).
18. I.H. Supriyadi, R. Rositasari, M.Y. Iswari, J. Segara. **14,** 1 1-10 (2018).

Analysis of changes in shallow water benthic habitat area using sentinel-2 imagery on the Ujung Genteng coast, Sukabumi, West Java, Indonesia

James Parlindungan Panjaitan[1*] · *Jasmine Fikrinda Qirani*[1], and *Syamsul Bahri* Agus[1]

[1]Department of Marine Science and Technology, Faculty of Fisheries and Marine Sciences, IPB University, 16680 Bogor, Indonesia

Abstract. Ujung Genteng is a remarkable place known for its rich marine diversity and provide various marine biota habitats. However, uncontrolled human activity and natural factors can cause changes in benthic habitats in shallow marine waters. Remote sensing technology can be used to monitor these changes through spatial and temporal analyses. This study aimed to map benthic habitats in Ujung Genteng waters and determine their accuracy using a pixel-based classification. The results were highly accurate, with an overall accuracy of 70,49% and excellent image interpretation accuracy, with a kappa coefficient value of 0,619. By analyzing Sentinel-2 satellite image data from 2017, 2020, and 2023, we observed changes in the area of each benthic habitat. The most significant real change was the addition of 8,92 hectares of sand class area and the reduction of 22,89 hectares of macroalgae class area.

1 Introduction

The benthic habitat of shallow marine waters is part of the coastal area which consists of three tropical ecosystems: coral reefs, seagrasses, and mangroves [16]. A benthic habitat is a place of life for several types of organisms that are composed of living corals, dead corals, seagrasses, algae, and seaweed with a substrate of sand, mud, and coral fragments. Coral reef and seagrass ecosystems are very important constituents of shallow marine waters as spawning grounds, feeding grounds, shelter for marine biota, protection from waves, sediment stabilizers, water purifiers, carbon sinks, sources of industrial and pharmaceutical materials, and tourism [18].

* Corresponding author: jppanjaitan@apps.ipb.ac.id

Ujung Genteng is located in the southern coastal area of Sukabumi Regency, West Java. This area has a variety of tourist attractions and a high diversity of marine conditions, including coral reef, seagrass, and mangrove ecosystems, which are a place of life for various marine animals from the Echinodermata phylum [25]. Behind the diversity of its marine conditions, this region faces directly to the Indian Ocean with sloping coastal characteristics with a topography of 0-25 m above sea level, making it an area prone to being affected by tsunami wave sweeps [22]. This is supported by a statement from Marabessy *et al.* (2018) [15] that coastal areas have the potential to be affected by habitat destruction due to changes in natural ecosystem processes and pollution.

Benthic habitats have the potential to change as a result of human activities and the uncontrolled utilization of natural resources and natural factors. Remote sensing technology is the best approach to view coastal areas and shallow marine waters with large area coverage [16]. The use of remote sensing technology for benthic habitat change detection has been widely carried out because the availability of data can be accessed easily and can save time, energy, and cost [20]. Several satellite images can be used to map benthic habitat, one of which is Sentinel. Sentinel is a satellite image that has the advantage of a high spatial resolution reaching 10 m and open source. Sentinel-2 performs well in scientific habitat monitoring, as its high spatial resolution can well define important features in benthic habitats, such as seagrasses and coral reefs [12]. This study aimed to map the benthic habitat and determine its accuracy around the Ujung Genteng Coast using the maximum likelihood classification method, and examine changes in benthic habitat spatially in 2017, 2020, and 2023.

2 Methods

2.1 Time and Location

This study was conducted from June - October 2023. Field observations were conducted on the coast of Ujung Genteng (Fig. 1.). Data processing was conducted at the Mapping and Spatial Modelling Laboratory of the Department of Marine Science and Technology, Faculty of Fisheries and Marine Science at IPB University.

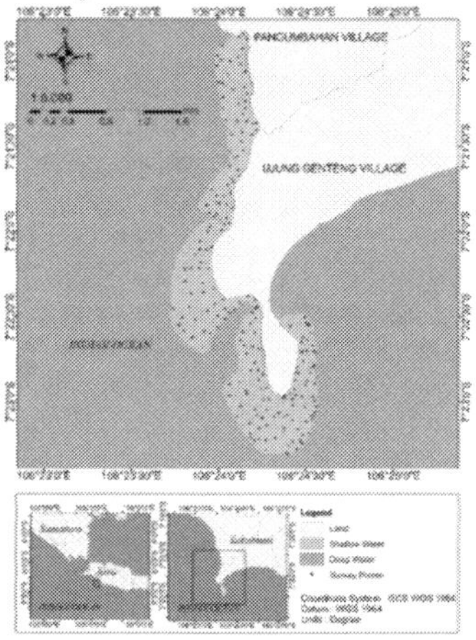

Fig. 1. Study site on the coast of Ujung Genteng, Sukabumi

2.2 Tools and Materials

The tools used in this research were laptops equipped with ArcGIS 10.8, QGIS 3.16, Microsoft Excel 2019, ENVI 5.3, and Microsoft Word 2019. Tools used in data collection include GPS / Mobile topographer, snorkeling equipment, and camera. The materials used were Sentinel-2A image data acquired on May 24, 2023 downloaded on the website (scihub.copernicus.eu/dhus/#/home), Sentinel-2B image data acquired on May 24, 2020, and Sentinel-2A image data acquired on May 25, 2017 downloaded on the website (https://explore.creodias.eu).

2.3 Research Procedurs

2.3.1 Image Pre-processing

The Sentinel-2 image used in this study does not go through a geometric correction process because it is a geometrically corrected and radiometrically calibrated, 1C level data product in the form of TOA reflectance (Top of Atmosphere), so the next correction process is atmospheric correction to convert TOA reflectance to at surface reflectance or BOA reflectance (Bottom of Atmosphere) using the Dark Object Reflectance (DOS) method [19]. The aim is to reduce noise from the atmosphere and make it clearer to recognize objects when interpreted. The corrected pixels are displayed as reflectance values. The reflectance value that will be generated has a range of 0-1.0 [3]. Masking and cropping aim to limit the image analysis area by removing the effects of land areas, deep water, and other unnecessary objects.

2.3.2 Image Transformation

Image transformation is applied to reduce the effect of disturbances in the water column. It combines the blue-green visible (true color) channel [9, 10]. The correction was applied using the Lyzenga transformation (Lyzenga, 1981) with the following equation:

$$\mathrm{DII} = \ln(\mathrm{RBi}) - \left((\mathrm{ki}/\mathrm{kj}) \times \ln(\mathrm{RBj})\right)$$

Description:
DII = Invariant underwater depth index (Lyzenga)
RBi = Short wavelength band (low water attenuation)
RBj = Longer wavelength band (Higher water attenuation)
ki/kj = ratio of attenuation coefficients of compared channels
σi = variance of band I (shorter-wavelength channel)
σj = variance of band j (longer-wavelength channel)
σij = covariance of bands i and j,
ki/kj is obtained by the formula: $ki/kj = a + \sqrt{(a^2 + 1)}$ while $a = \frac{\sigma i - \sigma j}{2 \times \sigma ij}$

2.3.3 Field Survey

Field data collection was carried out using the random sampling method while still paying attention to the representation of each class. Observations were made by recording coordinate points, benthic habitats, and documenting them. In this study, 202 sample points were collected from shallow marine habitats. The points were divided into two categories: 70% (141 points) for the input in the image classification process and 30% (61 points) for the accuracy test process of the image classification [23].

2.3.4 Image Classification

Image classification uses the maximum likelihood method of supervised classification. This method compares and takes into account the average value of the diversity between classes and channels. The input of this classification is in the form of sample points (ground truth) that already have benthic habitat attribute data on each class obtained during field surveys.

2.4 Data Analysis

2.4.1 Mapping Accuracy Analysis

The accuracy test of the classified image is carried out by creating an error matrix/confusion matrix (Table 1.) consisting of the overall accuracy (OA), producer accuracy (PA), user accuracy (UA), and kappa coefficient (Table 2.), using the following equation:

$$\text{Producer accuracy} = \frac{X_{ii}}{X_{+i}} \text{ x } 100\%$$

$$\text{User accuracy} = \frac{X_{ii}}{X_{i+}} \text{ x } 100\%$$

$$\text{Overall accuracy} = \frac{\sum_i^k X_{ii}}{N} \text{ x } 100\%$$

$$\text{Kappa coefficient} = \frac{N \sum_{i=1}^k X_{ii} - \sum_{i=1}^k (X_{i+} X_{+i})}{N^2 - \sum_{i=1}^k (X_{i+} X_{+i})}$$

Table 1. Confusion Matrix

Classification Image	Field Survey 1	2	3	Total
1	X_{11}	X_{12}	X_{13}	X_{1+}
2	X_{21}	X_{22}	X_{23}	X_{2+}
3	X_{31}	X_{32}	X_{33}	X_{3+}
Total	X_{+1}	X_{+2}	X_{+3}	N
PA (%)	X_{11}/X_{+1}	X_{22}/X_{+2}	X_{33}/X_{+3}	
UA (%)	X_{11}/X_{1+}	X_{22}/X_{2+}	X_{33}/X_{3+}	
OA (%)		$(X_{11}+X_{22}+X_{33})/N$		

Description:
N = total number of pixels used for accuracy test
X_{ii} = the diagonal value of the matrix of i row and I column
X_{+i} = number of pixels in column i
X_{i+} = number of pixels in row i
k = number of rows in the confusion matrix (number of classes)

Table 2. Kappa Agreement [24]

Kappa Coefficient Value	Strength of Agreement
< 0,00	Poor
0,00-0,20	Slight
0,21-0,40	Fair
0,41-0,60	Moderate
0,61-0,80	Substantial
0,81-1,00	Almost Perfect

2.4.2 Detection of Benthic Habitat Change

Benthic habitat change detection using the post-classification method. This is a pixel-based method. Two images were classified, and each classification class was compared using logical equations to detect habitat changes from one class to another.

3 Result and Discussions

3.1 Image Processing

The difference in the appearance of the sentinel-2 image before image pre-processing and after image pre-processing, which is cropping and atmospheric correction using the DOS method, is shown in Fig. 2. This method aims to improve the radiometric value, namely, the pixel value.

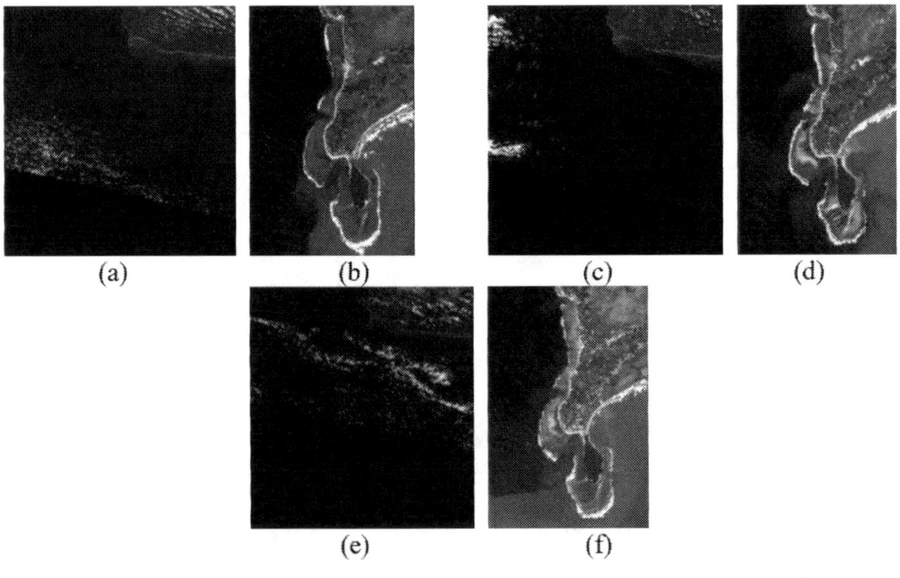

Fig. 2. Comparison of image appearance (a) acquired in 2017 (b) acquired in 2017 after image cropping and atmospheric correction (c) acquired in 2020 (d) acquired in 2020 after image cropping and atmospheric correction (e) acquired in 2023 (f) acquired in 2023 after image cropping and atmospheric correction

Image difference before and after pre-processing the image are shown in the form of a histogram (Fig. 3.)

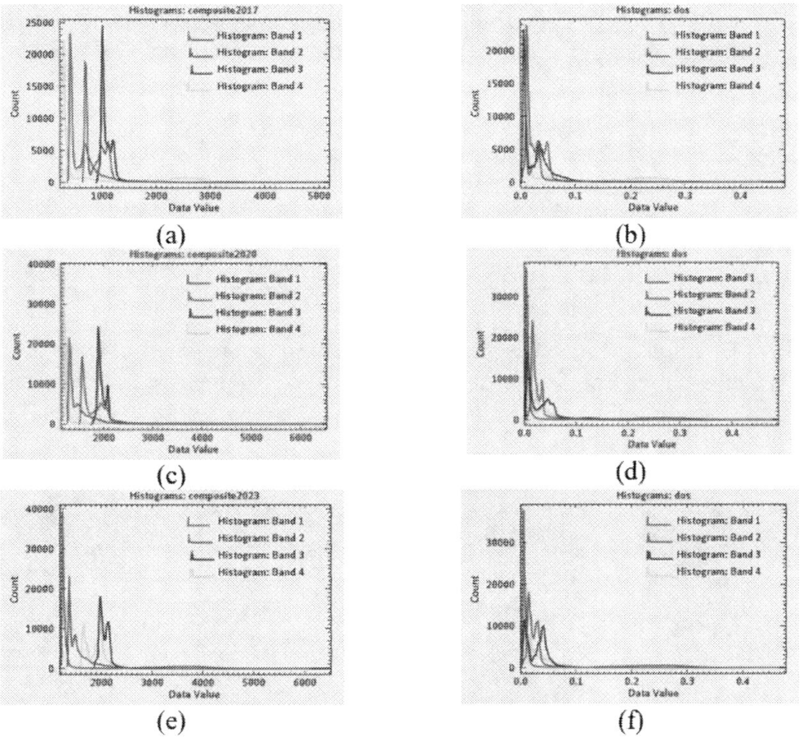

Fig. 3. Image histogram (a) acquired in 2017 before atmospheric correction (b) acquired in 2017 after atmospheric correction (c) acquired in 2020 before atmospheric correction (d) acquired in 2020 after atmospheric correction (e) acquired in 2023 before atmospheric correction (f) acquired in 2023 after atmospheric correction

The x-axis represents reflectance values, whereas the y-axis represents frequency values. Uncorrected images have values of up to thousands, whereas corrected images have values less than 1 or close to 0. This is in accordance with the theory that the range of reflectance values is 0-1.0 [2]. The reflectance value is the ratio of the light energy reflected by the surface to the light energy received. This value cannot be greater than one because an object cannot reflect more light than it receives. Therefore, 0 means that all light is absorbed (dark object), and 1 means that all light is reflected (light object).

The image composite combines RGB bands (true color composite) because it has a good penetration rate into clear waters [3-5]. The image cropping and masking process was carried out to focus on the coverage area only so that the data size was smaller and easier for further processing. There is a clear difference where the atmospheric-corrected image is brighter and the objects are more observable. The masking process was performed by manually digitizing the images of land and deep water areas with shallow water in shapefile data.

Image transformation using water column correction, an algorithm developed by Lyzenga known as Depth Invariant Index (DII). using a combination of blue-green bands. Input data by taking several pixel samples in the image in the form of Region of Interest (ROI) polygons in sandy areas at different depths to equalise the spectral values of the same object at different depths. Here are the results (Table 3.):

Table 3. Lyzenga algorithm equation

	Band	Algoritm
Sentinel 2A acquired in 2023	B2B3	alog(B2)-[(1.444345249)*alog(B3)]
Sentinel 2B acquired in 2020		alog(B2)-[(1.472743101)*alog(B3)]
Sentinel 2A acquired in 2017		alog(B2)-[(1.633754722)*alog(B3)]

Ki or Kj is a variable value of signal weakening due to water column absorption [2]. The attenuation coefficient ratio (ki/kj) was used for further calculations using Lyzenga's algorithm. The results of the water column correction and masking processes are shown in Fig. 4.

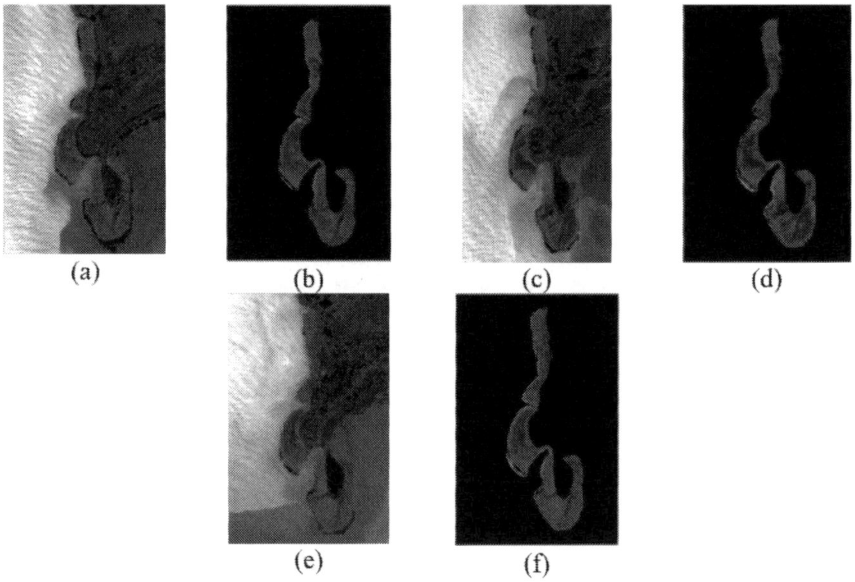

Fig. 4. Image appearance (a) DII correction in 2017 (b) masking in 2017 (c) DII correction in 2020 (d) masking in 2020 (e) DII correction in 2023 (f) masking in 2023

Images that have been corrected by DII approach will be different and produce new visual images. The difference is that the disappearance of the water column where there is no longer a color gradation for depth differences because the water column correction has been carried out. In addition, bottom water objects will look clearer and more visible due to the Lyzenga transformation [2].

3.2 Benthic Habitat Classification

Image classification is done in a supervised method using field data. The working principle of this method is to group image pixels into several classes based on the training area used as a classification reference. Maximum likelihood uses the calculation of the probability of a pixel to enter a particular class, by looking at the probability threshold of the pixel value so that the pixel can be classified [1, 11].

The image classification results obtained five benthic habitat classes based on the reference from Dwinasari (2017) [7]. These include sand (yellow), macroalgae (light green), seagrass (purple), coral (brown) and dead coral algae (dark green) which can be seen in the benthic habitat map for the three years in Fig. 5.

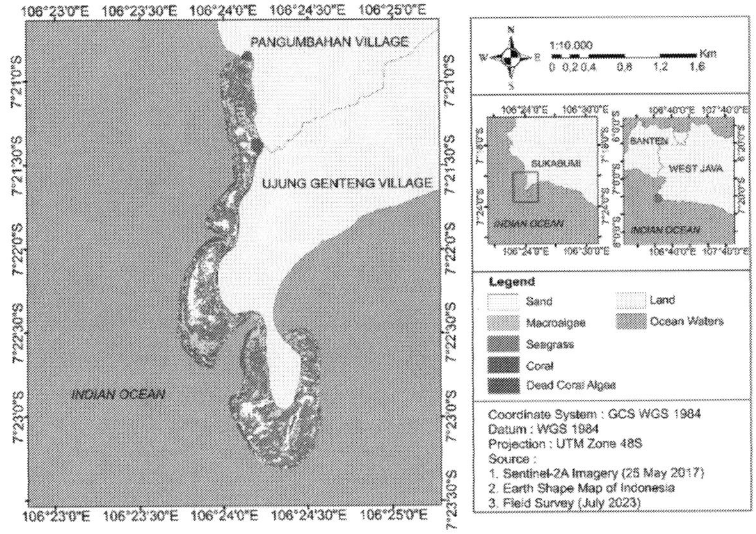

(a)

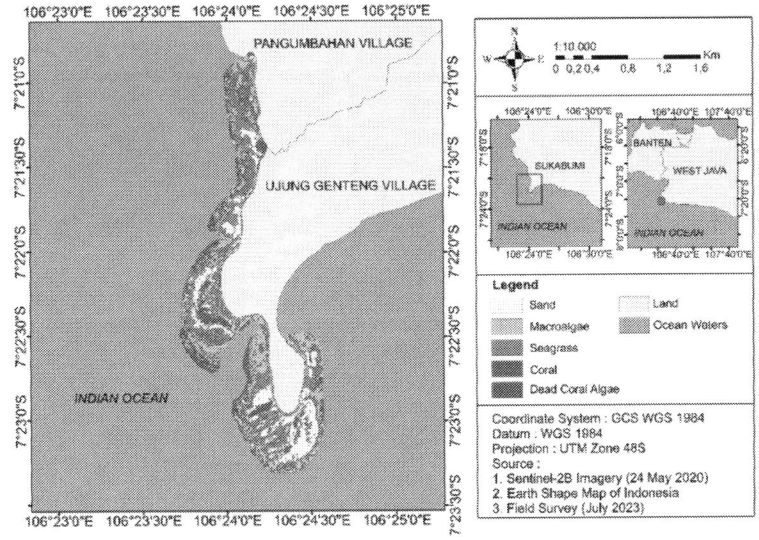

(b)

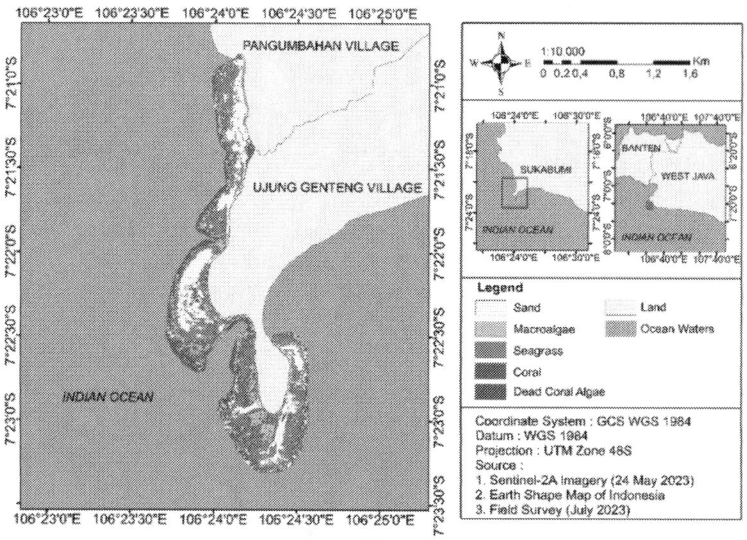

(c)

Fig. 5. Map of benthic habitats classification in Ujung Genteng Coast (a) in 2017 (b) in 2020 (c) in 2023

Overall, the benthic habitat area and changes in its area in 2017, 2020, and 2023 on the Ujung Genteng coast are shown in Table 4.

Table 4. Benthic habitat class area and changes in area in Ujung Genteng coast

Benthic Habitat Classes	Area (Ha)			Change in Area Class (Ha)		
	In 2017	In 2020	In 2023	2017-2020	2020-2023	Total 2017-2023
DCA	17,14	27,43	11,46	10,29	-15,97	-5,68
K	14,48	11,17	19,07	-3,31	7,90	4,59
L	82,78	71,42	81,54	-11,36	10,12	-1,24
MA	57,31	53,58	34,42	-3,73	-19,16	**-22,89**
P	50,48	58,66	75,76	8,18	17,10	**8,92**

Description: Dead coral with algae (DCA), mixed dead and live coral (K), seagrass (L), macroalgae (MA), and sand (P).

The seagrass class always dominates the benthic habitat of shallow marine waters in Ujung Genteng every year but decreases in area from 2017 to 2020 and increases in area from 2020 to 2023. The total change in the seagrass class area is a decrease of 1.24 Ha. The seagrass class is distributed along the coast of Ujung Genteng. In this area, thalassia sp. and enhalus sp. seagrasses are commonly found because they generally grow well in the reef flat zone [21]. The DCA class also decreased in total area by 5.68 Ha, had increased in area from 2017 to 2020, but decreased in area from 2020 to 2023. The DCA class was widely distributed in the reef crest area near the shoreline. Algae are attached to hard substrates such as rocks, pebbles, shellfish, plants, and dead corals. The macroalgae class showed the largest decrease in area, which was 22.89 Ha. It has decreased in area each year. Macroalgae found at the site are red algae, green algae, and brown algae species commonly found because of their ability to adapt to their mucus layer so that they can overcome water shortages and have a high level of tolerance to tides for a long time [13].

In addition to decreasing the area, some benthic habitat classes also increased, such as sand and coral classes. The sand class had the largest total additional area of 8.92 Ha. Based on the benthic habitat maps in 2017 and 2020, the sand class dominates in the western and southern areas, but in 2023, it spreads to the north. The coral class increased in the total area by 4.59 Ha. There was a decrease in area from 2017 to 2020, but there was an increase in area from 2020 to 2023. The meaning of coral class here is a combination of live and dead corals. The coral class is mostly found in the reef crest zone near breaking waves.

The main factors that are thought to be the cause of benthic habitat changes at that location are oceanographic conditions and increasing human activities such as land use and tourism activities. According to Nuraeni (2016) [17] the number of tourists visiting Ujung Genteng Beach has increased annually. The number of fishing boats that move and stop around Ujung Genteng Beach can also trigger the destruction of benthic habitats, particularly macroalgae. The growth and development of macroalgae are strongly supported by substrate conditions dominated by sand and coral fragments (rubble) [6]. The sand class experienced an increase in area due to the sand substrate that was easily lifted and deposited, covering seagrass due to fishing boat activities. Oceanographic conditions, such as water currents, tides, temperature, waves, salinity, water quality conditions, climatic conditions, and the entry of watersheds into the sea also affect the survival of shallow marine benthic habitats [27]. In addition, image mismatches in the classification of benthic habitat classes can because of to the absence of field data from 2017 to 2020 at the study site.

3.3 Classification Accuracy

The number of samples used for the accuracy test was 30% of the total number of samples taken (61 points) (Table. 5).

Table 5. Confusion Matrix

		Field Data					
		DCA	K	L	MA	P	Total
Classification Image	DCA	5	0	1	1	0	7
	K	2	5	0	1	0	8
	L	1	1	13	2	2	19
	MA	1	1	0	7	1	10
	P	0	0	3	1	13	17
	Total	9	7	17	12	16	61
	Producers Accuracy (%)	55,56	71,43	76,47	58,33	81,25	
	Users Accuracy (%)	71,43	62,50	68,42	70,00	76,47	
	Overall Accuracy (%)	70,49					
	Koefisien Kappa	0,619					

Description: Dead coral with algae (DCA), mixed dead and live coral (K), seagrass (L), macroalgae (MA), and sand (P).

The range of PA values in this study was 55,56-81,25%. The largest PA value belongs to the sand class, with a value of 81,25%, whereas the smallest PA value belongs to the DCA class, with a value of 55,56%. The range of UA values in this study was 62,50-76,47%. The sand class has a value 76,47%, while the coral class has the smallest UA value 62,50%. The sand class had the highest level of accuracy for both PA and UA values. As shown in Table 5., from the 16 sand class field data, 13 points were classified as sand, only two points were classified as seagrass, and one point as macroalgae. This indicates that the sand class was well-classified.

In this study, the OA value was 70,49%, which is in accordance with the SNI 7716:2011 standard in the Guidelines for Processing Remote Sensing Data for Coral Reef Ecosystems by LAPAN (2015) [14], which is a benthic habitat map that is acceptable if the overall accuracy is above 60%. The kappa coefficient for this study was 0,619. This shows that the image classification results have an accuracy level with strength of agreement in the good category (substantial).

3.4 Benthic Habitat Change Analysis

Combining the classification data from the two years resulted in a new classification class of 25 classification classes that had changed to other classes. A benthic habitat change map is shown in Fig. 6.

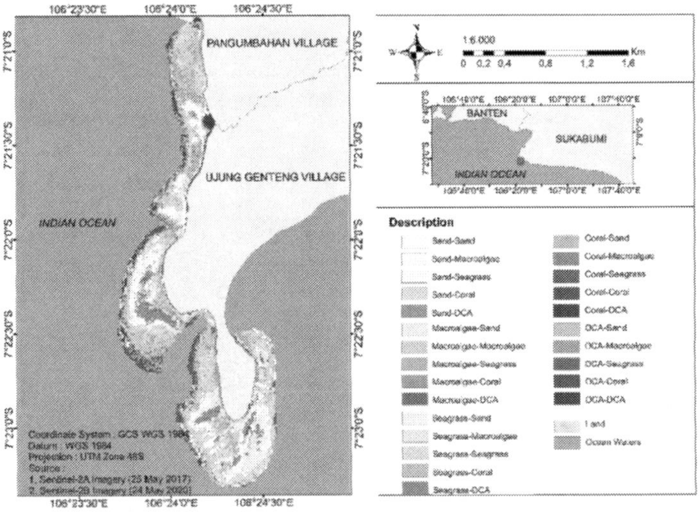

(a)

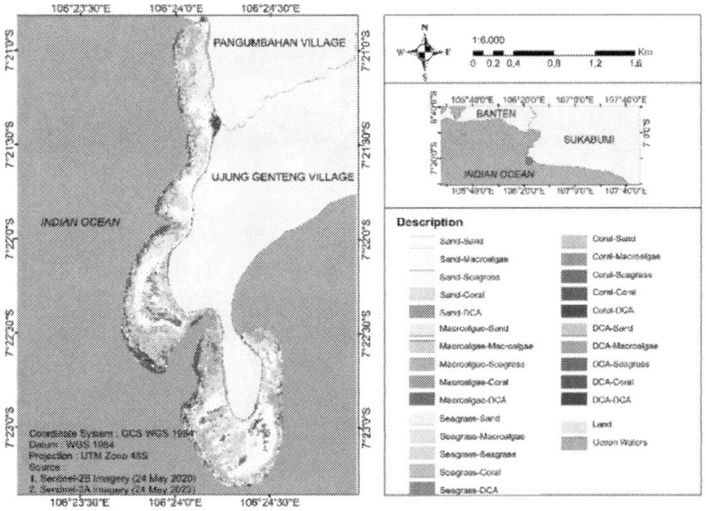

(b)

Fig. 6. Thematic map of benthic habitat change in (a) 2017-2020 (b) 2020-2023 in Ujung Genteng coastal area.

Table 6. Changes in benthic habitat composition in 2017-2020

		Transformed into (2020) -Ha				
		DCA	K	L	MA	P
Initial Class (2017)	DCA	6,390	1,890	1,820	1,780	4,870
	K	2,180	4,960	1,940	3,910	1,260
	L	6,430	1,020	33,340	19,900	21,980
	MA	3,960	1,770	19,580	22,510	9,380
	P	8,280	1,260	14,550	5,220	21,030
	Total Area (2020)	27,240	10,900	71,230	53,320	58,520

Table 7. Changes in benthic habitat composition in 2020-2023

		Transformed into (2023) -Ha				
		DCA	K	L	MA	P
Initial Class (2020)	DCA	4,900	3,160	5,710	1,870	11,560
	K	0,630	6,280	1,340	1,740	0,900
	L	1,760	2,350	30,240	11,840	25,040
	MA	1,400	5,040	21,380	13,210	12,280
	P	2,650	1,780	22,780	5,620	25,720
	Total Area (2023)	10,710	18,610	81,450	34,280	74,600

Description: Dead coral with algae (DCA), mixed dead and live coral (K), seagrass (L), macroalgae (MA), and sand (P).

Tables 6 and 7 list the changes in benthic habitat composition in 2017-2020 and 2020-2023. There are several classes whose results are dubious and quite difficult to define (impossible change), such as the sand class that changes to the DCA and coral classes, and vice versa [26]. This can occur because of the absence of field data in 2017 and 2020 as initial reference data, causing the possibility of incorrect information in the classification. In addition, it can be related to the level of accuracy, with an error rate of approximately 30% obtained from the accuracy test.

4 Conclusion

The results of benthic habitat mapping of shallow waters in the Ujung Genteng coastal area revealed five benthic habitat classes namely dead coral covered with algae (DCA), a mixture of dead coral and live coral (Coral), Seagrass, Macroalgae, and Sand. The overall accuracy of benthic habitat mapping at the study site was 70,49%, and the accuracy of interpretation of image classification was in the good category (substantial) with a kappa coefficient value of 0,619. Spatial temporal analysis conducted using the Sentinel-2A satellite can display changes in the area of each benthic habitat. The largest total change occurred with the addition of a sand class area of 8,92 ha and a reduction in macroalgae class area of 22,89 ha. Oceanographic conditions and increasing human activity are the causes of benthic habitat changes. In addition, the absence of field data in 2017 and 2020 may have caused mismatches in the classification of benthic habitat classes.

References

1. A. Adiwijoyo, P. Danoedoro, Jurnal Bumi Indonesia, **3 (4)**, 1-9 (2014)
2. F. Aldin, Y. Prasetyo, M. Helmi, Jurnal Geodesi Undip, **9 (1)**, 77-85 (2020)
3. L. O. Alifatri, B. Prayudha, K. Anggraini, Jurnal Ilmu Pertanian Indonesia (JIPI), **27 (3)**, 372-384 (2022)
4. I. D. M. K. P. Astaman, I. W. G. A. Karang, I. G. Hendrawan, K. T. Setiawan, Journal of Marine and Aquatic Sciences, **7 (2)**, 184-195 (2021)
5. Badan Informasi Geospasial, *Pedoman Teknis Pengumpulan dan Pengolahan Data Geospasial Habitat Dasar Perairan Laut Dangkal* (BIG, 2014)
6. R. Dwimayasanti, D. Kurnianto, Oseanologi dan Limnologi di Indonesia, **3 (1)**, 39-48 (2018)
7. W. Dwinasari, Analisis spasial secara geomorfologi untuk habitat bentik menggunakan citra pleiades di sebagian perairan Taman Nasional Karimunjawa, thesis, Universitas Muhammadiyah Surakarta (2017)
8. European Space Agency, *Sentinel-2 UserHandbook. ESA Standard Document UserHandbook* (ESA, 2015)
9. E. A. Giofandi, Y. Safitri, A. Eduardi, Jurnal Kelautan, **12 (2)**, 165-174 (2019)
10. Guntur, D. Prasetyo, Wawan, *Pemetaan Terumbu Karang*, (Ghalia Indonesia, 2012)
11. A. N. R. W. Hamdir, H. B. S. Sigit, Jurnal Bumi Indonesia, **3(4)**, 1-6 (2014)
12. J. D. Hedley, C. M. Roelfsema, I. Chollett, A. R. Harborne, S. F. Heron, S. Weeks, W. J. Skirving, A. E. Strong, C. M. Eakin, T. R. L. Christensen, V. Ticzon, S. Bejerano, P. J. Mumby, Remote Sens, **8**, 118-157 (2016)
13. O. Johan, Erlania, I. N. Radiarta, Jurnal Riset Akuakultur, **10 (4)**, 609-618 (2015)

14. Lembaga Penerbangan dan Antariksa Nasional, *Pedoman Pengolahan Data Pengindraan untuk Ekosistem Terumbu Karang* (Pusat Pemanfaatan Pengindraan Jauh, LAPAN, 2015)
15. I. Marabessy, A. Fahrudin, Z. Imran, S. B. Agus, Journal of Regional and Rural Development Planning, **2 (1)**, 1-22 (2018)
16. Mukrimin, L. O. M. Y. Haya, A. Takwir, Sapa Laut, **6 (1)**, 63-74 (2021)
17. L. Nuraeni, Kemenarikan daya tarik wisata Kawasan Pantai Ujung Genteng Kecamatan Ciracap Kabupaten Sukabumi, thesis, UPI Bandung, Fakultas Pendidikan Ilmu Pengetahuan Sosial (2016)
18. C. D. Prawoto, Hartono, Jurnal Bumi Indonesia, **7 (3)**, 2-8 (2018)
19. L. N. Purhartanto, P. Danoedoro, P. Wicaksono, Jurnal Nasional Teknologi Terapan, **3 (1)**, 63-86 (2019)
20. A. D. Purwanto, K. T. Setiawan, Jurnal Kelautan, **12 (2)**, 141-153 (2019)
21. F. R. Putri, M. C. Putri, P. O. Rinda, R. Ridwana, Nandi, Jurnal Geografi, **12 (1)**, 19-26 (2023)
22. N. D. Rezita, Sobirin, Supriatna, Jurnal Geosains Terapan, **2(4)**, 8-12 (2016)
23. W. Siska, Widiatmaka, Y. Setiawan, S. H. Adi, Tata Loka, **24 (1)**, 74-83 (2022)
24. Y. Susanti, Syafrudin, M. Helmi, BIOEDUKASI: Jurnal Pendidikan Biologi, **13 (1)**, 23-30 (2020)
25. Z. I. E. C. Triacha, M. P. Pertiwi, R. T. Rostikawati, Jurnal Ilmu Dasar, **22 (1)**, 9-18 (2021)
26. N. Wahiddin, V. P. Siregar, B. Nababan, I. Jaya, S. Wouthuyzen, Jurnal Ilmu dan Teknologi Kelautan Tropis, **6 (2)**, 507-524 (2014)
27. Y. Wahyudin, Bonorowo Wetlands, **1 (1)**, 37-50 (2011)

Temporal analysis of mangrove carbon stock using Sentinel-2 data on Menjangan Island, West Bali National Park, Indonesia

Ranti Dwimarsa Salsabila[2], *James Parlindungan Panjaitan*[1*]

[1]Department of Marine Science and Technology, Faculty of Fisheries and Marine Sciences, IPB University, 16680 Bogor, Indonesia
[2]Undergraduate Student of Department of Marine Science and Technology, Faculty of Fisheries and Marine Sciences, IPB University, Bogor 16680, Indonesia

Abstract. Mangrove forests are vital for their ecological function as the largest carbon storage systems in tropical regions, making a significant contribution to reducing carbon emissions. The mangrove areas in Menjangan Island part of the West Bali National Park (TNBB), are protected conservation zones that face various threats. This study aims to evaluate the changes in mangrove extent based on density, map actual biomass, estimate total carbon reserves, and analyze the relationship between changes in density and total carbon stock and sequestration in 2019 and 2023 using Sentinel-2 imagery. The NDVI-based classification data were categorized into three density classes by the Indonesian Forestry Ministerial Regulation (2012). Actual biomass values were derived using species-specific allometric methods with purposive random sampling. Biomass mapping was modeled with a simple linear regression equation using NDVI and actual biomass, resulting in the equation y = 1578,2x – 269,79 with an R^2 of 82,81%. The result showed decline in mangrove carbon reserves on Menjangan Island decreased from 124,075 x 10^3 tonsC in 2019 to 105,705 x 10^3 tonsC in 2023, indicating a positive correlation between mangrove area changes, total carbon reserves, and carbon sequestration.

1 Introduction

Mangrove forests are tidal zone plant communities that provide various benefits across multiple sectors [1]. One of their essential ecological functions is their ability to store the largest amount of carbon. Mangrove forests can store more carbon than most tropical forests and even more than all terrestrial forests [2]. This function can play a significant role in reducing carbon emissions and mitigating the impacts of climate change [3]. Carbon stock in vegetation represents the amount of stored carbon, indicating how much CO_2 the plants have absorbed. Through photosynthesis, mangroves perform carbon sequestration in the atmosphere and store carbon in the form of biomass and sediment [2]. Tree biomass consists of Above Ground Biomass (AGB), which includes leaves, stems, branches, and flowers, and Below Ground Biomass (BGB), which includes roots [4]. The crucial role of mangrove

* Corresponding author: jppanjaitan@apps.ipb.ac.id

ecosystems in maintaining environmental balance necessitates effective management and conservation efforts. A method for monitoring coastal ecosystem conditions combines Geographic Information Systems (GIS) and remote sensing technology utilizing satellite imagery with sensors sensitive to green vegetation, such as Sentinel-2. One of the critical data sets that can be analyzed using this method is the estimation of biomass and carbon in mangrove ecosystems. Accurate biomass assessment is traditionally performed using destructive and non-destructive approaches. However, these methods are constrained by limited spatial coverage and extended processing times. Therefore, remote sensing-based approaches are essential for estimating mangrove biomass on a broader scale with greater efficiency [5]. Through imagery data, it is possible to estimate carbon by analysing of vegetation index values and biomass data from field surveys, processed through regression analysis [6].

The mangrove ecosystem in West Bali National Park (TNBB) is one of the largest mangrove areas on the island of Bali. Although it is part of a protected national park, this ecosystem remains vulnerable to degradation caused by both natural and anthropogenic factors. Despite its ecological significance, research on the dynamics of this mangrove ecosystem, particularly in terms of carbon stock estimation, remains limited. To address this gap, this study aims to evaluate changes in mangrove extent based on density, map actual biomass, estimate total carbon stocks, and analyze the relationship between density changes and total carbon stock and sequestration between 2019 and 2023 using Sentinel-2 imagery.

2 Methodology

2.1 Location

This study was conducted from August 8th to 11th, 2023. The data used were obtained from field observations conducted on Menjangan Island, located in West Bali National Park. The research sites are shown in Figure 1.

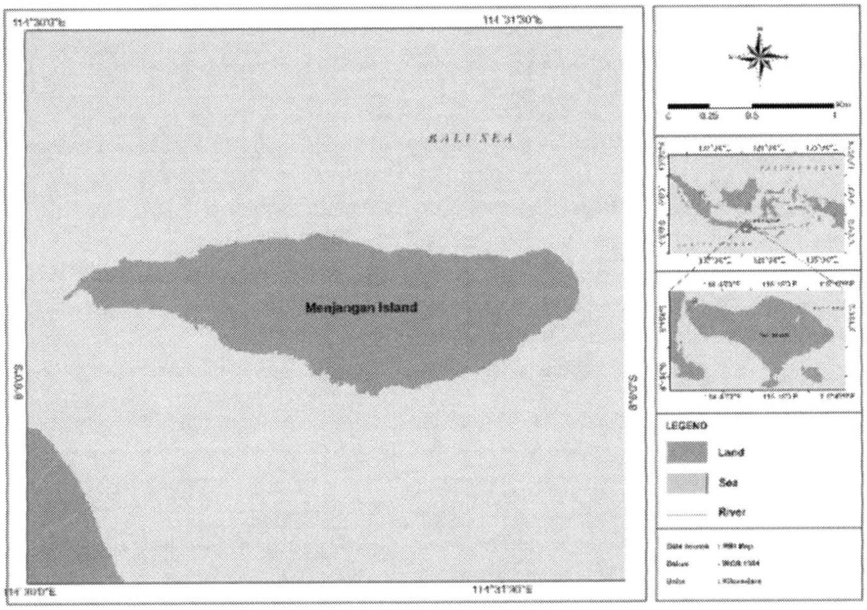

Fig. 1. Map of research locations on Menjangan Island, West Bali National Park

2.2 Data source

Field data collection during the ground check survey was essential for gathering primary information on mangrove species and tree girth. Sampling points were determined using purposive random sampling techniques, considering accessibility and the density of mangrove forest [7]. Data collection involved establishing 10m x 10m quadrat transects in the predetermined mangrove ecosystem areas, with a total of 9 plots. GPS coordinates were recorded at each plot to verify the satellite imaginary data. Within the transect, species identification and girth measurement at breast height (GBH) were conducted using measuring tapes. Girth data were then converted to diameter at breast height (DBH), and processed using allometric equations to calculate mangrove tree biomass. The study utilized Sentinel-2 imagery acquired on July 25, 2019, and September 27, 2023, as well as field survey data collected from August 08-11, 2023. The Sentinel-2A imagery, with a spatial resolution of 10 meters was suitable for spatial and temporal mapping of mangrove carbon stock [8].

2.3 Data processing

This study was conducted through several stages of data processing and analysis. The stages included field survey data collection, image data processing, accuracy assessment of images classification using the confusion matrix method, processing field data to calculate the total carbon stock, and analyzing the relationship between image data and field observation. The overall workflow the data processing and analysis is shown in Figure 2.

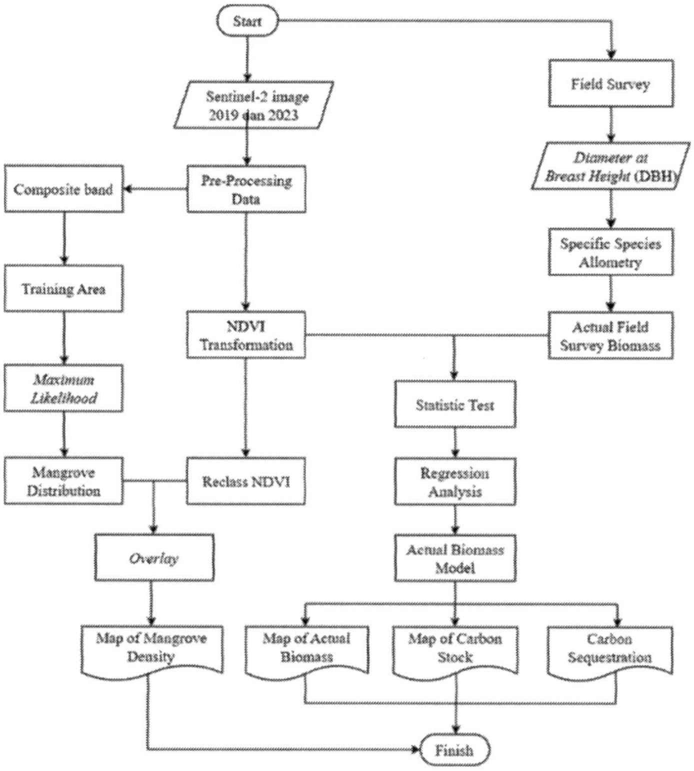

Fig. 2. Data processing flow chart.

2.3.1 Mapping of mangrove land cover

The process begins with the preprocessing of the image data, including image correction and cropping. Sentinel-2 satellite data has undergone radiometric and geometric corrections [9]. The images, classified as Level MSIL2A, were processed using the Sen2Cor algorithm and represent conditions at the Top of Atmosphere (ToA). Subsequently, image cropping was performed to focus on the study area, separating the specific region of interest from the larger image. Next, image classification was conducted to group pixels into specific classes based on predefined criteria, where each classified pixel represents an object with homogeneous characteristics [10]. A supervised classification approach using the Maximum Likelihood (MLH) algorithm was employed because of its ability to account for variance and correlation in spectral response patterns when classifying unknown pixels [11]. This probabilistic approach assumes that the spectral values of each class follow a normal distribution, allowing for more accurate classification by considering the probability that a given pixel belongs to a specific class. Field survey data in the form of coordinate points were used to define training areas within the study region, enabling the differentiation of mangrove and non-mangrove objects, and resulting in a detailed map of mangrove ecosystem distribution.

2.3.2 Analysis of Vegetation Index

The vegetation index utilized in this research was the Normalized Difference Vegetation Index (NDVI). This transformation is essential for calculating vegetation index values for each pixel, which are subsequently utilized in regression equations with mangrove carbon stock values [12]. The NDVI was selected for its simplicity and its sensitivity to chlorophyll, making it an effective indicator of active photosynthetic radiation [13]. The NDVI calculation from satellite imagery follows the following formula:

$$\text{NDVI} = \frac{NIR - Red}{NIR + Red} \tag{1}$$

Where NDVI represents the Normalized Difference Vegetation Index, where NIR is the spectral reflectance value in Sentinel-2 channel 8, and Red is the spectral reflectance value in Sentinel-2 channel 4.

The resulting NDVI values were classified based on mangrove density levels specified by the Indonesian Ministry of Forestry Regulation 2012. These density levels were categorized into three classes; sparse, moderate, and dense as shown in Table 1.

Table 1. Mangrove Density Categories [14]

No	Condition	Category	NDVI
1	Moderate	Sparse	$0{,}15 \leq x \leq 0{,}25$
2	Good	Moderate	$0{,}25 < x \leq 0{,}35$
3	Good	Dense	$0{,}35 < x \leq 1{,}00$

2.3.3 Actual Biomass Modeling

Estimation of actual biomass is performed using allometric methods, where diameter at Breast Height (DBH) served as the independent variable in equations that link DBH to actual biomass as the dependent variable. These allometric equations were used to calculate the actual biomass, including Above Ground Biomass (AGB) and Below Ground Biomass (BGB) of mangroves (Tables 2 and 3). Each mangrove species has different wood structure and density, thus species-specific allometric equations were utilized [15].

Table 2. Above Ground Biomass (AGB) allometric equation

No	Species	Allometric equation (kg)	Source
1	*Bruguiera gymnorrhiza*	$B = 0{,}0754\ (D)^{2{,}505*\rho}$	[16]
2	*Ceriops decandra*	$B = 0{,}251\ \rho\ (D)^{2{,}46}$	[17]
3	*Excoecaria agallocha*	$B = 0{,}251\ \rho\ (D)^{2{,}46}$	[17]
4	*Lumnitzera racemosa*	$B = 0{,}740\ (D)^{2{,}32}$	[18]
5	*Rhizophora apiculata*	$B = 0{,}235\ (D)^{2{,}42}$	[19]
6	*Rhizophora stylosa*	$B = 0{,}105\ (D)^{2{,}68}$	[20]
7	*Sonneratia alba*	$B = 0{,}3841\ \rho\ (D)^{2{,}101}$	[21]
8	*Xylocarpus granatum*	$B = 0{,}082\ (D)^{2{,}59}$	[20]

Table 3. Generalized allometric equation according to [17]

No	Biomass class (kg)	Allometric equation	DMAX (cm)
1	*Above Ground Biomassa (AGB)*	$B = 0{,}251\ \rho\ (D)^{2{,}46}$	49
2	*Below Ground Biomassa (BGB)*	$B = 0{,}199\ \rho^{0{,}899}\ (D)^{2{,}22}$	45

Wood densities values for each species were sourced from the International Centre for Research in Agroforestry (ICRAF) database at http://db.worldagroforestry.org/wd (Table 4). The actual biomass of individual trees was calculated and then aggregrated to determine the total biomass per plot by summing the biomass values of all trees within the plot.

Table 4. Mangrove Wood Densities (ICRAF)

No	Species	*Wood Density* (g/cm^3)
1	*Bruguiera gymnorrhiza*	0,8683
2	*Ceriops decandra*	0,7250
3	*Excoecaria agallocha*	04288
4	*Lumnitzera racemosa*	0,8325
5	*Rhizophora apiculata*	0,8814
6	*Rhizophora stylosa*	0,9400
7	*Sonneratia alba*	0,6443
8	*Xylocarpus granatum*	0,6721

Absorption is estimated by calculating biomass and carbon stock per hectare, which requires the conversion of biomass values [21]. Biomass per hectare was calculated using the following formula :

$$B_{ax} = \frac{Ba}{1000} \times \frac{10000}{L.Plot} \quad (2)$$

where Ba represent the estimated actual biomass (ton/ha), B_{ax} is the estimated biomass value per plot (kg), and LPlot is the area of the observation plot (m^2).

The calculated actual biomass values will be used for modeling through simple linear regression analysis to explore the relationship between carbon stocks and NDVI values derived from Sentinel-2 imagery. Carbon stock estimation can be estimated using the following equation [23]:

$$Cb = B_{ax} \times \%\ C\ organic \quad (3)$$

where Cb represents the carbon stock from biomass (ton/ha), B_{ax} is the actual biomass (ton/ha), and % C organic is the percentage value of carbon stock 0,47 [24].

The carbon stock identified from the biomass values was used to estimate carbon sequestration. Variations in in habitat types influence CO_2 absorption capacity due to differences in biomass and carbon stocks. The total CO_2 absorption value was calculated by multipliying the carbon value from the stored actual biomass by a factor of 3.67, which is the conversion constant for carbon to CO_2 [25]. This value was derived using the following equation:

$$\text{Absorbtion } CO_2 = \frac{\text{Mr } CO_2}{\text{Ar C}} \times \text{Cb} \quad (4)$$

where Mr CO_2 is the molecular weight of CO_2 (44), Ar C is the relative atomic weight of carbon (12), and Cb is the carbon stock from the biomass (ton/ha).

3 Result and Discussion

3.1 Mangrove Classification

Sentinel-2 satellite image data were classified using supervised classification with the Maximum Likelihood (MLH) algorithm, based on field survey data and mangrove distribution data in the FCC combination 8 (NIR)-4 (Red)-3 (Green). In thus study, masking was applied to distinguish mangrove areas from others land cover types, such as non-mangrove vegetation, resulting in a map of the mangrove ecosystem distribution at the study site. The MLH method for mangrove ecosystem classification and masking produced maps for Menjangan Island for 2019 and 2023. The total mangrove area on Menjangan Island was 9.56 ha in 2019, which decreased to 6.74 ha by 2023.

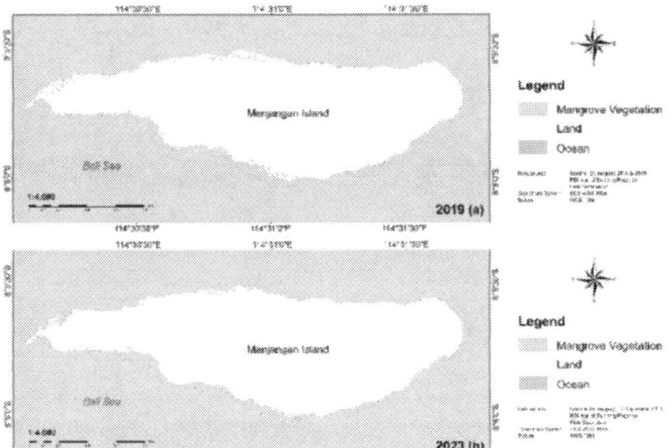

Fig. 3 Map of mangrove distribution on Menjangan Island in 2019 (a) and 2023 (b)

3.2 Vegetation Indices Classification

The Sentinel-2 satellite spectral values were tranformed into the Normalized Difference Vegetation Index (NDVI), which serve as parameter for monitoring vegetation conditions [26]. The NDVI values at the study site ranged from -0.347 to 0.749 in 2019 and -0.403 to 0.876 in 2023. The NDVI-transformed images were overlaid with result of supervised classification to produce maps of mangrove distribution and density at the study site. The

NDVI values for mangrove were reclassified into three density classes; dense, moderate, and sparse [13], as shown in Figure 4.

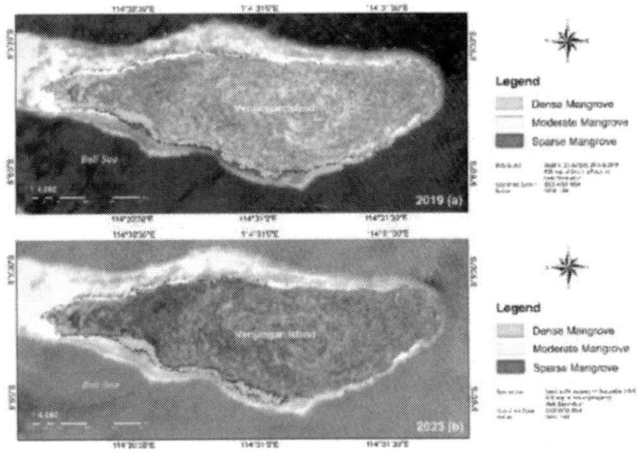

Fig. 4. Map of mangrove density distribution on Menjangan Island in 2019 (a) and 2023 (b)

Processing of mangrove density distribution data from NDVI transformation in the study area revealed three density classes: dense (green), moderate (yellow), and sparse (red). Analysis of data from 2019 and 2023 showed that dense density classes were predominated in the study area. The table below presents the area of each density class in the mangrove of Menjangan Island West Bali National Park, for 2019 and 2023. All density classes decreased in the study area from 2019 to 2023 (Table 5). The total decrease in density area from 2019 to 2023 was 2,8 ha. This decrease could be due to various factors such as coastal erosion, human activities such as land conversion, and natural factors [27].

Table 5. Area of mangrove density classes in the research area

Category	Area in 2019 (ha)	Area in 2023 (ha)	Changes (ha) 2019 – 2023
Dense	5,02	4,62	-0,4
Moderate	1,43	1,20	-0,2
Sparse	3,10	0,93	-2,2
Total	9,56	6,74	-2,8

3.3 Actual Biomass Modeling using Vegetation Indices

Field data on actual biomass values were processed to estimate biomass using Sentinel-2 satellite image data. A biomass estimation model was derived using a simple linear regression equation. The independent variable (x) is the field biomass, and the dependent variable (y) is the NDVI. Based on the results, the derived equation is as follows as seen in Figure 5

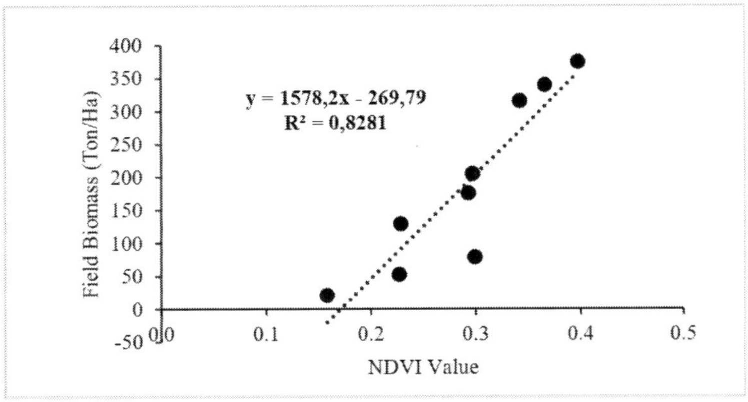

Fig. 5. Regression analysis of actual mangrove biomass values with NDVI values

The estimation of the biomass model from the image data resulted in a coefficient of determination (R^2) of 0,8281. This coefficient indicates a strong correlation between the actual biomass and NDVI values. The analysis showed that NDVI accounted for 82.81% of the variation in actual biomass per plot, with the remaining 17,19% being influenced by other factor not examined in this study. Although the model shows a strong correlation, it still has limitations, as 17.19% of the biomass variation remains unexplained. NDVI may not fully capture the structural differences in mangroves, such as canopy density and species composition, whereas satellite data can be affected by atmospheric conditions and resolution constraints. Additionally, field data uncertainties and sampling limitations may impact model accuracy, highlighting the need to integrate other indices and higher-resolution imagery for improved biomass estimation. Despite these limitations, plots with higher NDVI density categories showed higher AGB and BGB, indicating a direct relationship between NDVI values and actual biomass [27].

3.4 Distribution Maps of Actual Biomass and Carbon Stock

Simple linear regression analysis resulted in the equation Y = 1578,2x – 269,79. This equation illustrates the relationship between the actual biomass values and NDVI pixel values. This equation was applied to the 2019 and 2023 data to generate maps of actual biomass distribution in the mangrove vegetation at the study site.

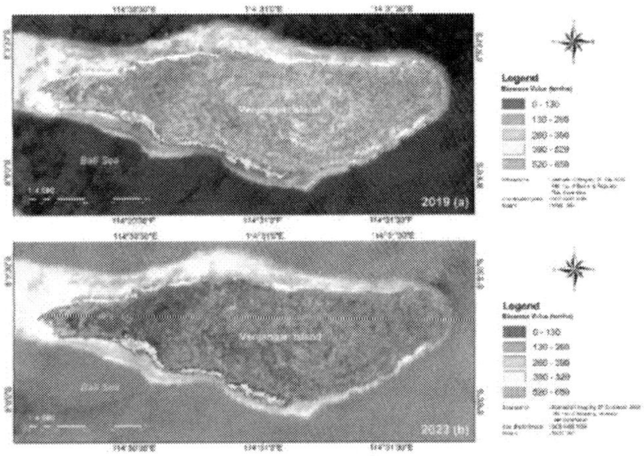

Fig. 6. Map of mangrove actual biomass estimation on Menjangan Island in 2019 (a) and 2023 (b)

The actual biomass values on Menjangan Island ranged from 0 to 631,51 tons/ha in 2019 and from 0 to 583,99 tons/ha in 2023. The total estimate of actual biomass in 2019 was 263,989 x 10^3 tons with a mangrove area of 9,56 ha and an average actual biomass of 211,962 tons/ha. However, these values decreased in 2023 for both the actual biomass estimates and mangrove area. In 2023, the total estimate of actual biomass was 224,903 x 10^3 tons with a mangrove area of 6,74 ha and an average actual biomass of 317,638 tons/ha.

Processed biomass data were converted into carbon stock estimates. The estimated carbon stock values on Menjangan Island in 2019 ranged from 0 to 296,812 tonsC/ha. In 2023, estimated carbon stock values on Menjangan Island ranged from 0 to 274,47 tonsC/ha. The total estimated carbon stock in 2019 was 124,075 x 10^3 tonsC with an average of 99,622 tonsC/ha. Similar to actual biomass, carbon stock estimates also decreased in 2023, with the total estimated carbon stock in the study area being 105,705 x 10^3 tonsC and an average of 149,290 tonsC/ha. Maps of carbon stock estimates are presented in Figure 7.

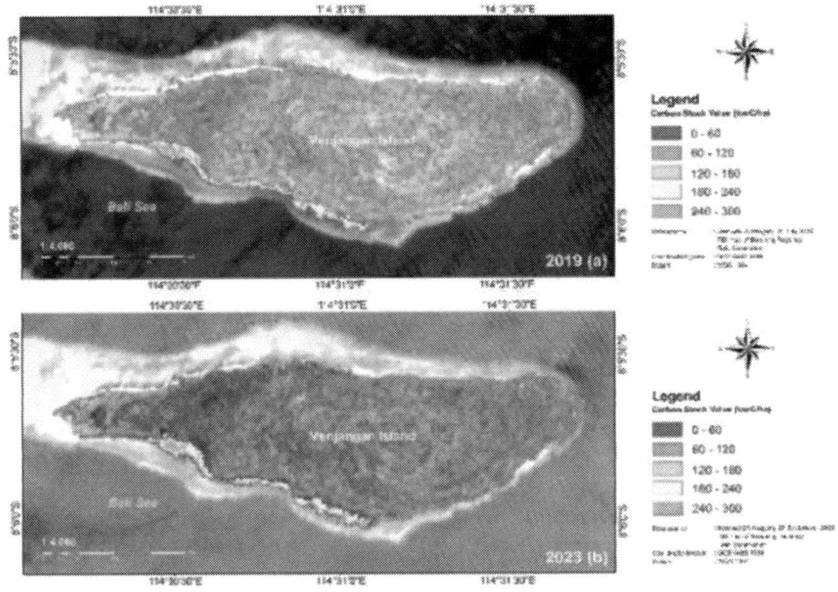

Fig. 7. Map of mangrove carbon stock estimation on Menjangan Island in 2019 (a) and 2023 (b)

3.5 Carbon Sequestration

Estimated carbon stock data for each mangrove density class can be used to assess the amount of carbon sequestration at the study site. Carbon sequestration (CO_2 absorption) refers to the process of transferring carbon from the atmosphere and storing it in reservoirs (carbon pockets). Based on data processing from 2019 to 2023, the dense mangrove density class exhibited the highest level of carbon absorption and stock, followed by the moderate density class, with the sparse density class showing the lowest values. The data analysis shows that a decrease in the estimated carbon stock value is accompanied by a corresponding decline for carbon sequestration. Therefore, the amount of carbon stock is directly proportional to the value of mangrove carbon sequestration [2]. The carbon sequestration values for each mangrove density class in 2019 and 2023 are shown in Table 6.

Table 2. Actual biomass, carbon stock, and carbon sequestration values (2019 – 2023)

Density classification	2019			2023			Changes (2019-2023)		
	Ba (10^3 ton)	Cb (10^3 tonC)	Sq,CO2 (10^3 tonCO2)	Ba (10^3 ton)	Cb (10^3 tonC)	Sq,CO2 (10^3 tonCO2)	Ba (10^3 ton)	Cb (10^3 tonC)	Sq,CO2 (10^3 tonCO2)
Dense	182	86	314	150	70	258	-32	-15	-56
Moderate	75	35	130	68	32	118	-7	-3	-12
Sparse	7	3	12	7	3	12	0.05	0.03	0.10
Total	264	124	455	225	106	388	-39	-18	-67

Note: Ba (Actual biomass) Cb (Carbon stock); Sq.CO2 (Sequestration carbon dioxide); - (Decrease); + (Increase)

Carbon sequestration in the dense density class in 2019 was 314 x 10^3 tonsCO2 and decreased by 56 x 10^3 tonsCO2 in 2023 to 258 x 10^3 tonsCO2. Similarly, the moderate density class recorded a sequestration value of 129 x 10^3 tonsCO2 in 2019, which decreased by 12 x 10^3 tons CO_2 to 118 x 10^3 tonsCO2 in 2023. In contrast, the sparse density class showed a slight increase, with sequestration values of 12 x 10^3 tonsCO2 in 2019 rising by 0,10 10^3 tons CO_2 to 11 x 10^3 tonsCO2 in 2023. Overall, the total carbon stock and carbon sequestration in the Menjangan Island area decreased from 2019 to 2023. This could be attributed to the positive relationship between density, carbon stock and sequestration. An increase in density results in higher carbon stock and carbon sequestration, whereas a decrease in density results in reduction in both.

4 Conclusion

The mangrove ecosystem experienced a decrease in area by 2.80 hectares from 9.56 hectares in 2019 to 6.74 hectares in 2023. Simple linear regression analysis was conducted to evaluate the correlation between the NDVI algorithm results and actual biomass values which showed a strong relationship, with a coefficient of determination of 0,8281. This modeling resulted in the mapping of actual biomass and carbon reserves at the study site for 2019 and 2023. The changes in area, total carbon reserves, and carbon sequestration in the mangrove areas of Menjangan Island, West Bali National Park exhibit a positive relationship.

References

1. L.O.A.F. Hasidu, A.F. Ibrahim, A. Prasetya, M. Maharani, A. Asmi, A. Agusriyan, A.A. Mubarak, S. Kamur, G.N. Kharisma, *JFSI* **5**, 60-71 (2021).
2. D. Alviana, R. Anggraini, J.R. Hidayati, I. Karlina, F. Lestari, D. Apdillah, A.D. Syakti, D. Sihite, *JKT* **26**, 464-472 (2023).
3. R.A.T. Nuraini, D. Pringgenies, C.A. Suryono, V.H. Adhari, *Bul. Oseanogr. Mar.* **10**, 180-188 (2021).
4. A.A.E. Andiani, I.W.G.A. Karang, I.N.G. Putra, I.W.E. Dharmawan, *JITKT* **13**, 483-496 (2021); J.P. Situmorang, S. Sugianto, D. Darusman, *AIJST* **5**, 126-139 (2016).
5. G. Pratiwi, B. Sasmito, N. Bashit, *Elipsoida: J. Geod. Geomat.* **4**(1), 27-33 (2021).
6. V.N. Husna, V.P. Siregar, S.B. Agus, T. Arifin, *JPSAL* **9**, 456-466 (2019).
7. D. Phiri, M. Simwanda, S. Salekin, V.R. Nyirenda, Y. Murayama, M. Ranagalage, *Remote Sens.* **12**(14), 1-35 (2020).
8. [ESA] European Space Agency, *Sentinel-2 User Handbook Revision 2*, Noordwijk: Esa Communication (2015).
9. A.F. Hanan, I. Pratikto, N. Soenardjo, *J. Mar. Res.* **9**(3), 271-280 (2020).
10. R. Septiani, I.P.A. Citra, A.S.A. Nugraha, *J. Geogr.* **16**(2), 90-96 (2019).

11. H. Frananda, H. Hartonso, R.H. Jatmiko, *Maj. Ilm. Globe* **17**(2), 113-123 (2015).
12. L.Z. Albasit, R. Pribadi, R.J.M.R. Pramesti, *J. Mar. Res.* **11**, 620-640 (2022).
13. Kementerian Kehutanan Republik Indonesia, *Peraturan Menteri Kehutanan RI Nomor P.23/Menhut-II/2009 tentang Tata Cara Penyusunan Rencana Teknik Rehabilitasi Hutan dan Lahan Daerah Aliran Sungai (RTk RHL-DAS)*, Jakarta: Kementerian Kehutanan Republik Indonesia (2012).
14. A. Malik, A.R. Jalil, A. Arifuddin, A. Syahmuddin, *Geogr. Environ. Sustain.* **13**(3), 32-38 (2020).
15. J.B. Kauffman, D.C. Donato, *(Vol. 86)*, Bogor (ID): CIFOR (2012).
16. A. Komiyama, S. Poungparn, S. Kato, *JTE* **21**, 471-477 (2005).
17. A. Kangkuso, J. Jamili, A. Septiana, R. Raya, I. Sahidin, U. Rianse, S. Rahim, A. Alfirman, S. Sharma, K. Nadaoka, *For. Sci. Technol.* **12**(1), 43-50 (2016).
18. J.E. Ong, W.K. Gong, C.H. Wong, *For. Ecol. Manag.* **188**(1-3), 395-408 (2004).
19. B.F. Clough, K. Scott, *For. Ecol. Manag.* **27**(2), 117-127 (1989).
20. J.B. Kauffman, T.G. Cole, *Wetlands* **30**(6), 1077-1084 (2010).
21. M.A. Rahmayanti, A. Jauhari, A. Fitriani, *J. Sylva Scienteae* **4**(5), 902-909 (2021).
22. [IPCC] The Intergovernmental Panel on Climate Change, *Supplement to the 2006 IPCC Guidelines for National Greenhouse Gas Inventories: Wetlands: Methodological Guidance on Lands with Wet and Drained Soils, And Constructed Wetlands For Wastewater Treatment* (2014).
23. S. Brown, A.E. Lugo, *Science* **223**(4642), 1290-1293 (1984).
24. A. Maruapey, I. Irnawati, *Median: J. Ilmu Eksakta* **11**(1), 26-38 (2019).
25. A.A. Romadoni, R. Ario, I. Pratikto, *J. Mar. Res.* **12**(1), 71-82 (2023).
26. S. Febrianto, H.A. Syafina, N. Latifah, M.R. Muskananfola, *J. Kelaut. Tropis* **25**(3), 369-377 (2022).
27. S. Pirasteh, D. Mafi-Gholami, H. Li, Z. Fang, A. Nouri-Kamari, B. Khorrami, *Int. J. Appl. Earth Obs. Geoinf.* **128**, 103769 (2024).

Distribution of fish target strength value in Menjangan Island waters, West Bali National Park

Imam Syahnanda Irawan[1*], *Sri* Pujiyati[2], *Syamsul* Bahri Agus[3], *Steven* Solikin[4]

[1]Marine Science and Technology Study Program, Faculty of Fisheries and Marine Science, Bogor 16680, Indonesia

[2]Department of Marine Science and Technology, Faculty of Fisheries and Marine Science, IPB University, Bogor 16680, Indonesia

Abstract. Research on the target strength (TS) value of fish is generally influenced by various internal and external factors, either from fish or signals. This study aims to identify the composition and analyze the target strength value of fish in four directional zones in the waters of Menjangan Island. The distribution of fish TS values in 4 zones at depths of 1.5 - 5.0 m ranged from -59.8 to -39.5 dB. The presence of fish was validated by dive data, which showed that the composition of fish was dominated by the Pomacentridae family, at approximately 54.5%. Results from the estimation of fish length using Foote's (1987) algorithm, ranged from 4.3 - 44.6 cm. Based on the zone, the North Zone had the highest number of fish detected at 2179 fish, while the West Zone had the lowest number at 1213 fish. The range of the TS values detected was dominated by the range of -61 to -58 dB and the least detected range was -40 to -37 dB. This indicates that the fish individuals detected around the depth near the surface of these waters tend to be small, and few were large, based on the results of the literature study.

1 Introduction

Menjangan Island is a popular tourist destinations within the West Bali National Park. This area has a variety of attractions in the field of tourism, with a stretch of coral reefs and various cultural tourism sites sacred to the Balinese Hindu community. Based on the Minister of Forestry Decree No. 493/Kpts-II/1995, West Bali National Park has a land area of 15587.89 ha and waters of 3.145 ha located at coordinates 8°05′20″S – 8°15′25″E and 114°25′00″S – 114° 56′30″E, and is administered in the territory of two regencies in Bali, namely Buleleng and Jembrana [1]. The water conditions at this location have many underwater caves, such that the habitat is rich in fish diversity, especially fish that live in coral reef ecosystems, and have several points with a sudden vertical decrease in depth (drop off) due to various factors such as geological changes and other underwater formations [2].

*Corresponding author: sripu@apps.ipb.ac.id

The word "fish" in this study refers to the interpretation of fish in general, which is defined as a vertebrate organism that lives in water and is taxonomically included in the phylum Chordata with the characteristics of having gills that are useful for taking dissolved oxygen from water and fins that are used for swimming [3]. The most common fish found in Menjangan Island waters is the reef fish group, where 16 fish families were found and dominated by the Pomacentridae or reef fish family, Chaetodontidae or indicator fish, and Acanthuridae or herbivorous fish [4]. Each type of fish has different physiological mechanisms, based on differences in habitat conditions and the surrounding environment [5].

The hydroacoustic method is an underwater detection method that uses a transducer to generate sound waves or echoes that propagate in a water medium, and then the signal is scattered or reflected back by the object [6]. This technology is highly effective for studying aquatic biota and the environment without causing significant damage or environmental changes [7]. Target strength (TS) is the ability of an object to reflect the sound that hits it [8]. If the TS value is small, then the object is estimated to be small; otherwise if the TS value is large, then the object is estimated to be large [9]. The factors that affect changes in TS values in fish are size, shape, behavior, presence of swim bladder, pulse incidence angle, wavelength, frequency, acoustic impedance, flesh, bones, scales, and fins [10]. This study aims to identify the composition and analyze the target strength value of fish in four directional zones in the waters of Menjangan Island. This research is important to fill the data gap of fish target strength (TS) values in Menjangan Island waters, provide basic information for conservation, and validate acoustic methods for identifying fish composition and distribution.

2 Method

2.1 Data acquisition

The survey was conducted during the Himiteka Expedition VIII by the Marine Science and Technology Student Association of IPB University on Wednesday, August 9th 2023, in the waters of Menjangan Island, West Bali National Park. This activity starts in the morning between 9.00 AM – 3.00 PM (WITA). Fish data acquisition was also carried out using the Underwater Visual Census (UVC) method because it can be used as supporting data to validate the presence of fish up to a depth of 5 m. Bathymetric information was obtained by downloading bathymetry images from the Batimetri Nasional (BATNAS) website page for the eastern part of Java Island and the western part of Bali Island. The selection of this bathymetry data source was based on the fact that this information is only a complement to the visuals on the research map. A map of the research locations is shown in Figure 1.

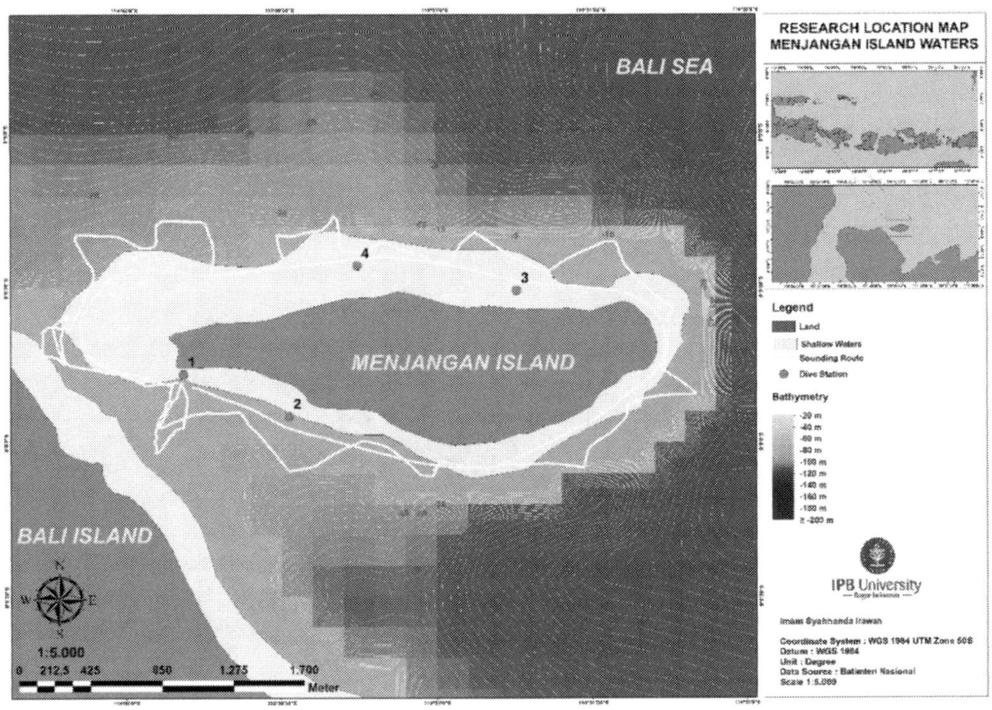

Fig 1. Research location map of Menjangan Island waters

The equipment used in the acoustic data acquisition process in the waters of Menjangan Island, West Bali National Park is the Scientific Echo Sounder Simrad EK15. Simrad EK15 is a single beam hydroacoustic Sound Navigation and Ranging (SONAR) instrument designed with a frequency of 200 kHz and a beam width of 26° for a variety of needs in aquatic environments. The full specifications of the Simrad EK15 are listed in Table 1 [11].

Table 1. Simrad EK15 Specifications (Simrad 2012)

Instrument Specifications	
Operational frequency	200 kHz
Transmission	Simultaneous or sequential ping rate up to 40 s Pulse durations 80 to 1240 µs Data rate 1,6 Mbps
Transceiver Unit	Maximum number in use 15 Output power 45 W IP rating, IP66
Data output formats	Raw data (EK60 format)
Third party post-processing	Echoview and SONAR5
Transducer	Type Single Beam Maximum installation depth 600 meters Beamwidth 26 degrees

The 200 kHz frequency was operated in the shallow area of the Menjangan Island Waters. The output data recorded by the echosounder were the amplitude value with *.raw extension. Echograms of acoustic recording results from Simrad EK15. The raw format was converted into *.uuu format using the SONAR5_pro software dongle borrowed from Syiah Kuala University, Banda Aceh. The supporting equipment used was GPS Garmin 585, diving equipment, and stationery. The GPS has 2 frequencies of 50/200 kHz, a maximum depth of

1500 ft or 457.2 m, and an accuracy level of about 0.2 m [12]. The Simrad EK15 settings related to the beaming activity are shown in Table 2.

Table 2. Simrad EK15 Settings

Instrument Settings	
Sound Speed (m/s)	1537,69
Transducer Gain (dB)	9,200
Transmission Frequency (kHz)	200
Transmit Power (W)	46,080
Beam Width (°)	26°
Pulse Length (ms)	0,320
Ping Interval (m/s)	500
Minimum Threshold (dB)	-100
Time Variable Gain (TVG)	*Target Mix*

The acoustic sounding process was carried out in one round for each of 2 different paths around the waters of Menjangan Island to fulfill the area representation of each observation zone. The transducer was installed under clear weather conditions and calm waves 1 m from the keel, to the right of the center of the ship, because it is assumed that the part of the ship is the most balanced and receives the least disturbance or noise. Tool settings related to the target to be detected or Time Variable Gain (TVG), ping interval, sound speed, gain, temperature, salinity, and depth need to be performed to maximize the tool performance and adjustment to field conditions through environmental parameters.

The fish data acquisition was conducted in four zones (West, South, North, and Northeast). The acquisition process aims to identify the diversity and size of fish in situ in Menjangan Island waters using the Underwater Visual Census (UVC) method. UVC is a method of directly observing fish underwater to determine their diversity, abundance and size. This method was performed by visual observation at a depth of 3 - 5 m for three replicates [13]. The dive direction at stations 1 and 4 moved westward, whereas at stations 2 and 3 it moved eastward. Each of the four stations had a transect length of approximately 50 m with a visibility of 2.5 m right and left of the diver. The observers dive along the transect while observing the fish.

2.2 Data Processing and Analysis

2.2.1 Acoustic Data

Data processing and analysis were conducted at the Laboratory of Acoustics, Instrumentation, and Marine Robotics (AIR) Division, Department of Marine Science and Technology (ITK), IPB University. Acoustic data in the form of echograms were obtained, and further analyzed to obtain information on the detection results of the distribution of target strength values of fish. This data processing and analysis was performed from a depth below the near field line to 5.0 m. The near field zone can be calculated using equations (1) and (2) as follows [14] :

$$\lambda = \frac{c}{f} \quad (1)$$

$$Nf = \frac{L^2}{\lambda} \quad (2)$$

Where Nf is the near field (m), L is the longest area of the transducer surface, λ is the wavelength (m), and c is the speed of sound in water (m/s). The acoustic data from the sounding results had a *.raw format and must first be converted to *uuu format. The conversion process of SONAR5_pro software produces two types of data, one of which is

40log(R) for single target backscatter. 40log(R) represents the single detected target information. The use of 40log(R) is related to how large the physical size of the detected target is, so that it can provide information about the size of the target reflected by the SONAR signal. SONAR signals can experience a decrease in intensity that is proportional to the change in the square of the distance from the target (R).

Data processing was performed using a minimum threshold of -60.0 dB and a maximum of -34.0 dB [15]. Processing and analysis of acoustic data using SONAR5_pro software by dividing the data by 50 pings in every 1 Elementary Sampling Distance Unit (ESDU) at a depth stratum of 1.5 - 5.0 m. The main reason for this is related to data quality. The more data that are divided, the more detailed the information. The number of detected targets was calculated using the SONAR5_pro software system. The information obtained is "Detected No. SED", or the number of single signals detected by the fish. This value is compiled to obtain the number of single fish targets. The TS value was calculated using the following equation [14] :

$$TS = 10log\frac{I_r}{I_i} \qquad (3)$$

Where TS is the target intensity in decibels, I_r is the intensity of the transmitted energy, and I_i is the intensity of the energy backscattered from the object. The TS value obtained from equation (1) was used to estimate the length of the fish using the following equation [16] :

$$TS = 20logL - A \qquad (4)$$

Where L is the length of the fish in centimeters and A is the constant TS value for 1 cm of fish length. Basically the "A" value for all types of fish is always different whether the fish has an open swim bladder (physotome), closed (physochlist), even those without a swim bladder (bladderless fish). Fish with an open swim bladder have an "A" value of 71.9, fish with a closed swim bladder have an "A" value of 67.5, fish without a swim bladder have an "A" value of 80, while the "A" value of 68.0 is used to determine the general length of fish that are not known specifically related to species and swim bladder conditions. It should be noted that all of these "A" values are derived from focusing on detecting, measuring and estimating the length of pelagic and demersal fish in situ using various acoustic system operations (single-beam, dual-beam, split-beam) integrated at a frequency of 38 kHz [16].

2.2.2 Fish Data

Fish data processing was carried out by calculating the fish composition to obtain data on the ratio between the number of individuals of each species to the number of individuals of all species in the waters of Menjangan Island. Reef fish composition can be calculated using the following equation [17] :

$$KJ = \frac{ni}{N} \times 100\% \qquad (5)$$

Where KJ is the fish species composition in percentage (%), n_i is the number of individuals of each i-th fish species, and N is the total number of individuals of all fish species. The results of these calculations were then tabulated in Microsoft Excel to produce the fish family composition data visualized in graphical form. The data obtained are associated with the results of acoustic data processing in the area adjacent to the dive point. The next step was to sort the fish data to determine the type of fish swim bladder. The length of fish obtained in situ is used in equation (2), and is based on validation data and comparison with the results of acoustic data processing.

3 Result and Discussion

3.1 Fish Data

The fish family composition in Menjangan Island at a depth of 3 – 5 m was dominated by reef fishes, where nine fish families were identified. The largest percentage of fish families is the Pomacentridae family at 54.5%, followed by Acanthuridae at 10%, Chaetodontidae at 5.7%, Labridae at 5.6%, Balistidae at 5.4%, Serranidae at 4.2%, Caesionidae at 3.5%, Lutjanidae at 3.5%, Zanclidae at 2.1%, and others at 5.4%. The fish family composition information was used to validate the presence of fish from the acoustic results. A graph of the percentage composition of fish families in Menjangan Island is shown in Figure 2.

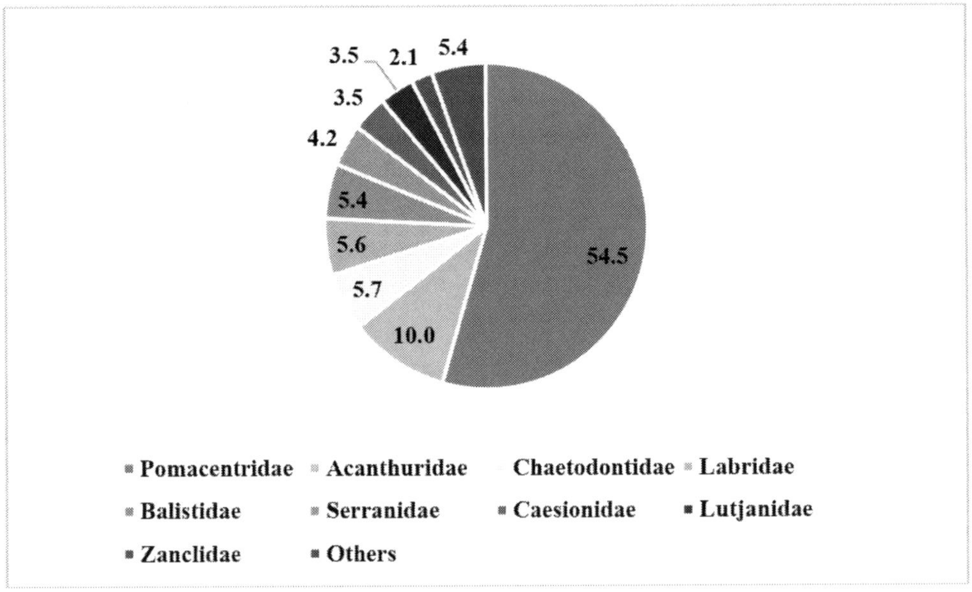

Fig 2. *Chart of fish family composition in Menjangan Island waters*

Fish can be divided into three types based on the shape of their swim bladder. Physostome or open swim bladder, the swim bladder has a connecting channels with the digestive tract, physoclist or closed swim bladder that does not have connecting channels with the digestive tract, and bladderless fish or without swim bladder and is usually a type of deep-sea fish. The swim bladder is an organ that functions to regulate the fish body when floating in water by adjusting the body volume and specific gravity to the depth of the water environment. Another function of the swim bladder is as a space to resonate in producing or receiving sound, so that sound waves can be reflected perfectly because it generally contains air [18]. Taxonomy of fish families distinguished by their swim bladder type are presented in Figure 3 [19].

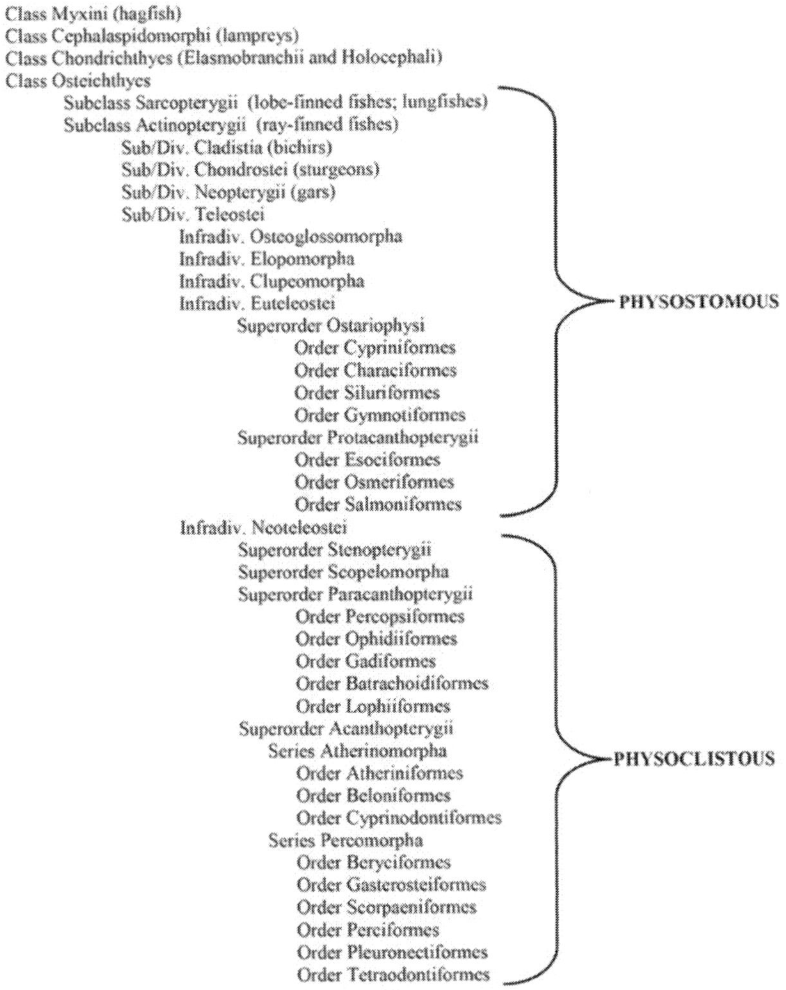

Fig 3. *Fish family taxonomy distinguished based on swimming bladder type*

Based on the identification results of the fish family composition, fish in Menjangan Island Waters can be declared as a group of fish that have closed swim bladder. All fish families belong to the order of true bony fish or Perciformes, which is a group of fish with closed swim bladder (physoclistous).

3.2 Acoustic Data

The distribution of TS (single) values in the depth stratum is from -59.8 to -39.5 dB. The estimated length of fish at this depth is from 4.3 - 44.6 cm and is thought to be dominated by the Pomacentridae or reef fish family. The South Zone consisted of 141 ESDUs, with the highest number of targets detected in the -61.0 to -58.0 dB value range of 899 targets and the least number detected in the -40.0 to -37.0 dB range of 13 targets. The blue line shows the relationship between the two variables in the form of polynomial regression of degree two. The blue dashed line shows the relationship between the two variables in the form of polynomial regression of degree two. This regression model was chosen because it has the highest coefficient of determination (R^2) with a correlation coefficient of r = 0.988 (very strongly correlated). A histogram graph of fish TS values againts the number of detected targets in the South Zone of the 1.5 to 5.0 m depth stratum is shown in Figure 4.

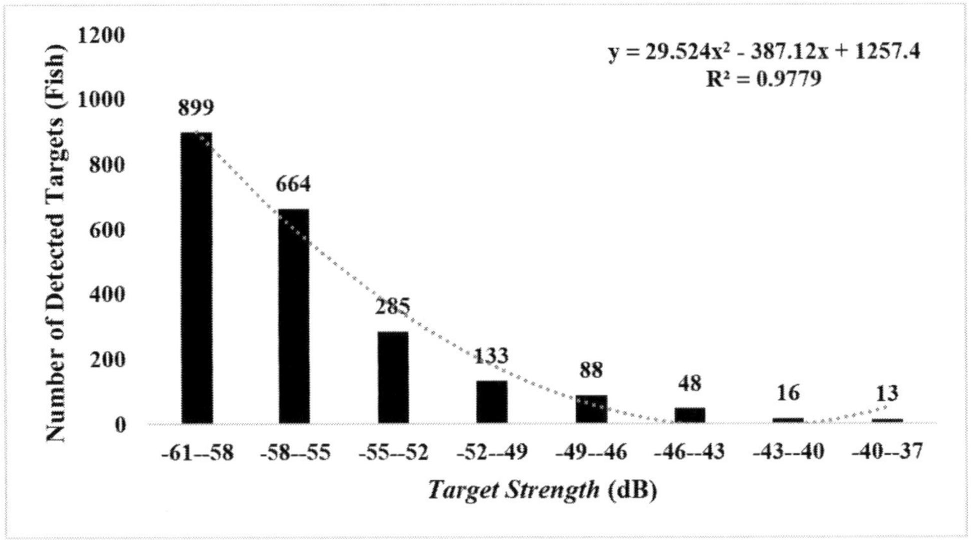

Fig 4. *Histogram of fish TS toward the number of detected targets at the South Zone*

The East Zone consists of 62 ESDUs. The highest number of detected targets was in the range of -61.0 to -58.0 dB values with 485 targets and the least number detected was in the range of -40.0 to -37.0 dB with 29 targets. The blue dashed line shows the relationship between the two variables in the form of a polynomial regression of degree two. This regression model was chosen because it had the highest coefficient of determination (R^2) with a correlation coefficient of $r = 0.986$ (very strongly correlated). A histogram graph of fish TS values with respect to the number of detected targets in the East Zone of the 1.5 - 5.0 m depth stratum is shown in Figure 5.

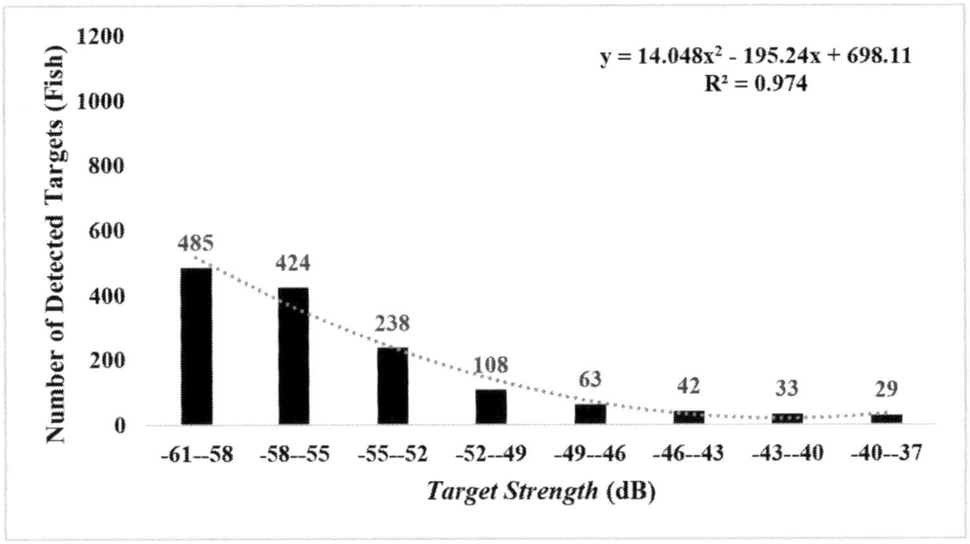

Fig 5. *Histogram of fish TS toward the number of detected targets in the East Zone*

The North Zone consists of 162 ESDUs. The highest number of targets detected was in the range of -61.0 to -58.0 dB values with 1034 targets and the least number detected were in the range of -40.0 to -37.0 dB with 16 targets. The blue dashed line shows the relationship between the two variables in the form of polynomial regression of degree two. This regression model was chosen because it had the highest coefficient of determination (R^2) with a correlation coefficient value of r = 0.983 (very strongly correlated). A histogram graph of fish TS values with respect to the number of detected targets at the North Zone of the 1.5 - 5.0 m depth stratum is shown in Figure 6.

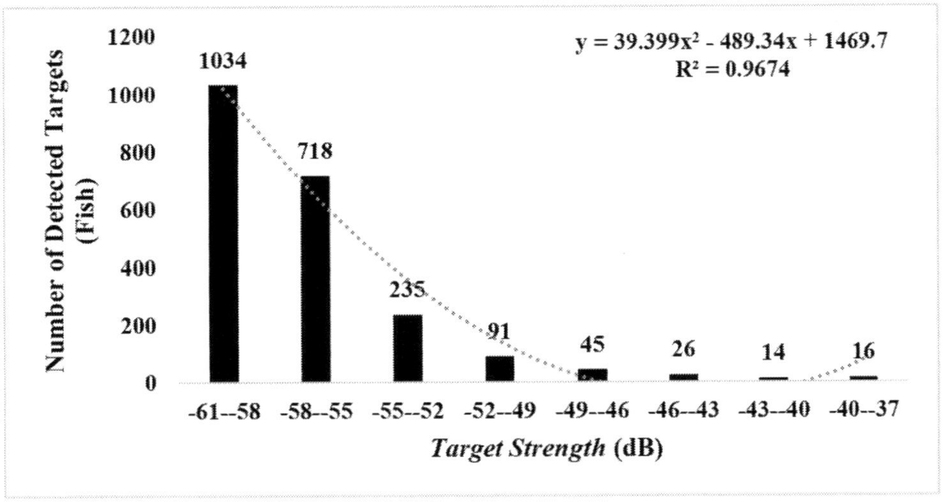

Fig 6. *Histogram of fish TS toward the number of detected targets in the North Zone*

The West Zone of 1.5 to 5.0 m depth consisted of 49 ESDUs and the most number of targets detected was in the range of -61.0 to -58.0 dB values with 453 targets and the least number detected was in the range of -40.0 to -37.0 dB with 13 targets. The blue dashed line shows the relationship between the two variables in the form of a polynomial regression of degree two. This regression model was chosen because it had the highest coefficient of determination (R^2) with a correlation coefficient of r = 0.984 (very strongly correlated). A histogram graph of fish TS with respect to toward the number of detected targets in the West Zone of the 1.5 - 5.0 m depth stratum is shown in Figure 7.

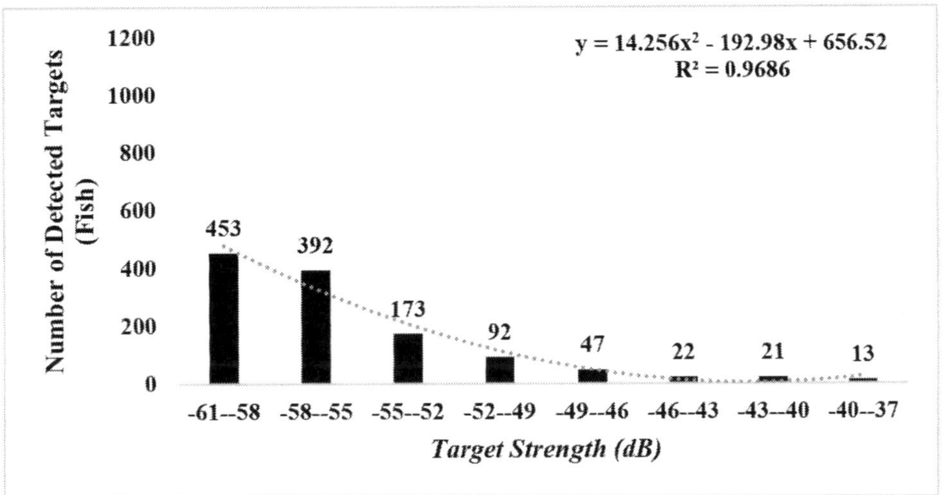

Fig 7. *Histogram of fish TS toward the number of detected targets in the West Zone*

The North Zone had the highest number of detected fish by zone at 2179 individual fish, followed by the South Zone at 2146 individual fish, the East Zone at 1422 individual fish, and the West Zone at 1213. Overall, the successful and dominant TS values detected in the 1.5 - 5.0 m depth stratum were in the range of -61 to -58 dB and the least detected were in the range of -40 to -37 dB. This indicates that this depth stratum is not only inhabited by small fish but also by some that are thought to be larger fish. Some fish species tend to rise to the surface when looking for food, therefore the presence of larger fish at certain depths is thought to be to find food by preying on smaller fish [20]. A graph of the distribution of fish TS values against the number of targets detected in all zones of the Menjangan Island waters in the 1.5 - 5.0 m depth stratum is shown in Figure 8.

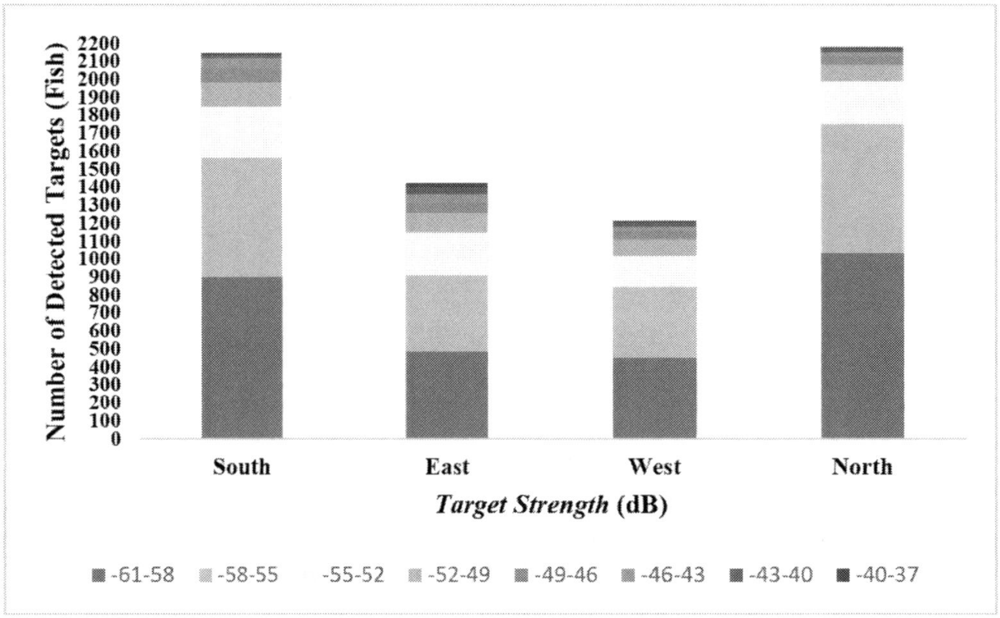

Fig 8. *Graph of distribution of fish TS values toward the number of detected targets in all water zones of Menjangan Island*

Fish length can be estimated by knowing the type of swim bladder, where the longer the fish size, the larger the swim bladder [21]. Physoclist-type fish can regulate buoyancy through a relatively slow gas-diffusion process, which can regulate the flow of gas in and out of the swim bladder. This causes the filling rate of the swim bubble and the resulting partial pressure to vary, to achieve a neutral level of buoyancy at greater depths [18]. The fish size can be distinguished in 3 groups, namely small (-70.0 to 60.0 dB), medium (-59.9 to -50.0 dB), and large (-49.9 to 30.0 dB), so it is known that the composition of individual fish detected in all depth stratum of 1.51 - 5.0 m in Menjangan Island waters belongs to the small to large group of fish [22].

4 Conclusion

Based on the results of processing and analyzing acoustic data of the 1.5 - 5.0 m depth stratum, the distribution of fish TS values is in the range of -59.8 to -39.5 dB. At this study site, the total number of detected targets was 6960 single individual fish, where TS values of -61.0 to -58.0 were the dominant TS value range with 2871 targets detected and TS values of -40 to -37 were the least detected TS value range with 71 targets detected in all zones. The North Zone had the highest number of detected targets by zone with 2179 individual fish, followed by the South Zone with 2146 single individual fish, the East Zone with 1422 single individual fish, and the West Zone with 1213 single individual fish. The estimated length of fish at this depth is from 4.3 - 44.6 cm and is estimated to be dominated by the Pomacentridae or reef fish family. At this depth, it is indicated that it is not only inhabited by small fish but also by large fish based on the results of the estimated length size.

Fish in these waters are not only schooling but also solitary. The main strength of this study is that, it was able to detect, estimate and quantitatively justify the specific number of solitary fish targets in this protected marine area. The main weakness of this research lies in the limited supporting data that can be used to validate the correctness of fish presence, given that fishing activities are strictly prohibited in these waters; therefore, an alternative approach in the form of a literature study was used.

Imam Syahnanda Irawan, as the main contributor expressed his deepest gratitude to the management of West Bali National Park for their cooperation in supporting the success of this research activity. The author's greatest appreciation goes to Mr. Syahrul Purnawan, S.Pi., M.Si as a lecturer at Syiah Kuala University Banda Aceh, Mrs. Sri Ratih Deswati, S.Pi, M.Si as Education Personnel at IPB University, Friansyah as the Chief Executive of Himiteka VIII Expedition who has allowed the author to use the data and all committees who have helped and facilitated the author during the research.

References

1. E. Sulistyadi, Zoo Indonesia, 25(2), pp. 142-159 (2017).
2. R. Hernowo, D. Djumanto, and N. Probosunu, Biota: Jurnal Ilmiah Ilmu-Ilmu Hayati, 18(1), pp. 11-20 (2013).
3. S.S. Fitrah, I. Dewiyanti, and T. Rizwan, Jurnal Ilmiah Kelautan dan Perikanan Unsyiah, 1(1), pp. 66-81 (2016).
4. M.I. Febrianti, F. Purwanti, and A. Hartoko, Jurnal Ilmu dan Teknologi Kelautan Tropis, 10(1), pp. 15-24 (2018).
5. C.K. Bangsa, Sugito, Zuhrawati, D. Razali, A. Nuzul, and Azhar, Jurnal Medika Veterinaria, 9(1), pp. 9-11 (2015).
6. S. Pujiyati, N.M.N. Natih, B. Hamuna, and L. Dimara, Journal of Applied Geospatial Information, 3(2), pp. 240-243 (2019).
7. F.W. Shabangu, J.C. Coetzee, I. Hampton, S.E. Kerwath, W.M. Wet, and A.L. Ochoa, Reflections on the State of Research and Development in the Marine and Maritime Sectors in South Africa, Pretoria: Council for Scientific and Industrial Research (2014).
8. M.Z. Lubis, Oceanography and Fisheries, 3(2), pp. 1- 6 (2017).
9. D.P. Anggraeni, F. Supriyadi, Jurnal Ilmiah Matematika dan Ilmu Pengetahuan Alam, 16(2), pp. 176-184 (2019).
10. M. Farhan, H.M. Manik, and T. Hestirianoto, Jurnal Teknologi Perikanan dan Kelautan, 13(2), pp. 175-186 (2022).
11. Simrad, Multi Purpose Scientific Echo Sounder Simrad EK15, Norway: Kongsberg Maritime (2012).
12. S. Marona, Junaidi, Dalrino, and E. Fernando, Andalas Civil Engineering Conference (2018).
13. I. Manuputty, Jurnal Ilmu Kelautan, 11(1), pp. 1-12 (2006).

14. E.J. Simmonds, D.N. Maclennan, Fisheries Acoustic: Theory and Practice 2nd ed, Oxford: Blackwell Science Ltd (2005).
15. D. Bakhtiar, Y.A. Ompusunggu, A. Anggoro, and S. Supiyati, *Jurnal Kelautan Tropis*. 27(1), pp. 51-58 (2024).
16. K.G. Foote, J. Acoust. Soc., 82, pp. 981-987 (1987).
17. A.E. Greenberg, Standard Methods for the Examination of Water and Waste water for 4th Edition, Washington (WA): American Public Health Association (1989).
18. A.I. Burhanuddin, Biologi Kelautan, Yogyakarta: Lily Publisher (2019).
19. R.S. Brown, A.H. Colotelo, B.D. Pflugrath, C.A. Boys, L.J. Baumgartner, Z.D. Deng, L.G.M. Silva, C.J. Brauner, M. Mallen-Copper, O. Phonekhampeng, G. Thorncraft, and D. Singhanouvuong, Fisheries, 39(3), pp. 108-122 (2014).
20. N. Aryani, Ikan dan Perubahan Lingkungan, Padang: Bung Hatta University Press (2014).
21. E. Ona, X. Zhao, I. Svellingen, and J.E. Fosseidengen, Proceedings of the Herring, pp. 461-487 (2001).
22. V. Samedy, M. Wach, J. Lobry, J. Selleslagh, M. Pierre, E. Josse, P. Boet, Fisheries Research, 172, 225-233 (2015).

Parametric study on effect of condensing and evaporating temperatures on cold storage performance and carbon footprint of traditional fishing vessels

Suhengki[1,2], *Henry Munandar Manik*[2*], Totok Hestirianoto[2], Susilohadi[3], Muhamad Yulianto[4,]

[1]Department of Mechanical, Institut Teknologi PLN, Jl. Lingkar Luar Barat Duri Kosambi, Cengkareng 11750 West Jakarta, Indonesia
[2]Department of Marine Science and Technology Jl Raya Dramaga Kampus IPB Dramaga Bogor 16680 West Java, Indonesia
[3]Department of Research Center For Geological Resources-BRIN, Jalan Dr Djunjunan No 236, Husen Sastra Negara, Kec Cicendo, Kota Bandung Jawa barat Indonesia.
[4]Departement of Mechanical and Biosystem Engineering Faculty of Agricultural Engineering and Technology, IPB University, Jl Raya Dramaga Kampus IPB Dramaga Bogor 16680 West Java, Indonesia

Abstract. This paper presents a study on cold storage applications' performance and carbon footprint on traditional fishing vessels. The cooling method used is Vapor Compression Refrigeration (VCR). However, VCR systems have issues related to CO_2 emissions. This study examines refrigerants' Global Warming Potential (GWP) to assess their environmental impact. The refrigerants analyzed in this study are R32, classified as a low-GWP refrigerant, and R404a, classified as a high-GWP refrigerant. The cold storage capacity is based on a 5 GT traditional fishing vessels, approximately 500 kg. A steady-state numerical model was developed using thermodynamic equations. The condensing temperature was varied at 30°C, 35°C, 40°C, and 45°C, while the evaporating temperature was varied at -15°C, -20°C, -25°C, and -30°C. The parameters evaluated include power consumption, cooling capacity, coefficient of performance, and CO_2 emissions. The CO_2 contribution was assessed based on the Life Cycle Climate Performance (LCCP) model developed by the International Institute of Refrigeration (IIR). The results indicate that R32 has a higher cooling capacity than R404a, while R404a contributes more to CO_2 emissions than R32. Based on these findings, R32 has significant potential to replace R404a in cold storage applications, particularly for traditional fishing vessels, to achieve better environmental sustainability.

1. Introduction

Indonesia is an archipelagic nation [1], with approximately 76% of its territory consisting of

* Corresponding author: henrymanik@apps.ipb.ac.id

the ocean [1]. However, the utilization of marine resources, particularly in traditional fishing, remains suboptimal. This is primarily because the storage method relies on ice boxes or slurries. These methods have limitations in terms of storage duration [2]; as a result, fish spoil quickly once the ice starts to melt as they become submerged in water. On the other hand, a widely known and commonly applied cooling method is the vapor compression refrigeration system [3]. This method compresses, condenses, expands, and evaporates a refrigerant in a closed cycle to remove the heat from a designated area. The vapor compression refrigeration system is based on the Carnot cycle and consists of four main components: compressor, condenser, expansion valve, and evaporator, as illustrated in Figure 1(a) [4]. A thermodynamic cycle can be observed in the P-h diagram shown in Figure 1(b) [5].
Process 1-2: Isentropic compression, where the working fluid is compressed, results in an increase in temperature and pressure.
Process 2-3: Heat rejection to the environment causes the refrigerant to transition from a superheated vapor to a liquid state.
Process 3-4: Expansion, where the refrigerant experiences a pressure drop and enters a mixed-phase state.
Process 4-1: Heat absorption from the storage space leads to a decrease in temperature to the desired level.
Despite their widespread use, vapor-compression refrigeration systems face a major challenge related to environmental concerns. These concerns primarily stem from high energy consumption and the use of refrigerants with high Global Warming Potential (GWP) values.

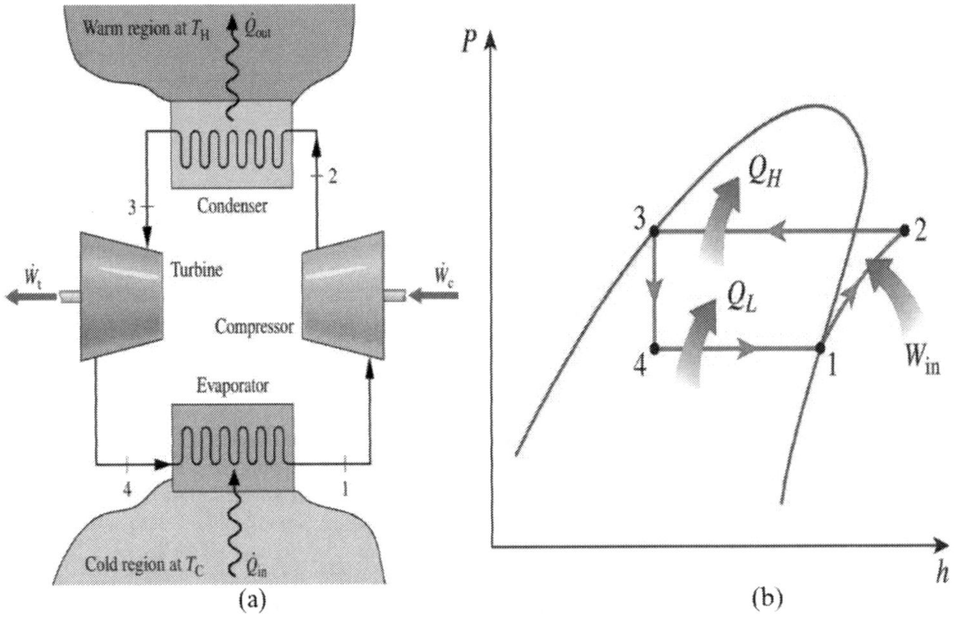

Fig. 1. Vapor compression refrigeration (a) Main component [4], (b) P-h diagram [5].

Several studies have been carried out to reduce energy consumption and CO2 emission contribution, with a focus on the use of various refrigerants both experimentally and through simulations, as shown in Table 1. Research indicates that R32, a low-GWP refrigerant (675), is increasingly being applied and recommended, especially for air conditioners and heat pump water heater applications [6]. The use of R32 for air conditioners and heat pumps also resulted in high COP (Coefficient of Performance) values. Based on Table 1, R404a is widely used as a refrigerant for cold storage applications because it delivers good performance, high availability, and greater cost-effectiveness than R507.

However, as can be seen in the GWP value, R404 had a fairly high GWP value of approximately 3943. The use of R404a for cold storage resulted in a COP value, as shown in Table 1.

Table 1. Research maping of varian refrigerant for refrigeration and heatpump

Refrigerant	Application	Information	Reference
R32	HPWH	COP is affected by mass charge	[7]
R32, R134a	HPWH	COP R32 is better. of R134a (simulation).	[8]
R32	HPWH	COP is worth 5 on climate tropical (simulation and experiment).	[9]
R455A, R468a, R290, R1270	Refrigerator	R1270 increases COP (33.26%) at the same condensing temperature	[10]
R290, R134a	Ice storage AC	The COP of both refrigerants is almost the same	[11]
R134a	Cold storage 3 GT ship	The COP value is 3.14 experimentally	[12]
R32, R410a	Air Conditioning	R32 COP 4% lower but the cooling capacity is greater	[13]
R32, R290	Air Conditioning	COP R32 is better	[14]
R32,R134a, R152a	Air Conditioning	COP R32 is 4.13-7.12 (simulation)	[15]
R404a	Cold storage	COP is worth 2.29 – 2.47	[16]
R32, R410a	Air Conditioning	COP R32 is better with rotary compressor	[17]
R507	Cold storage	COP between 3.43 – 3.7	[18]

In the design and testing of refrigeration systems, including cold storage, it is necessary to consider several parameters such as cooling load, cooling capacity, superheated temperature (TSH), subcooling temperature (TSC), condensation temperature, evaporation temperature, compressor power, coefficient of performance (COP). Evaporation and condensation temperatures are well-known parameters used as variables to evaluate the performance of refrigeration and heat pump systems [19, 20] because these two temperatures are related to the target temperature conditions and the environmental temperature at which cold storage is used.

Based on the explanation above, this research will contribute to understanding the influence of condensation and evaporation temperatures on the development of refrigeration systems for cold storage with low GWP (R32) and high GWP (R404a) as working fluids. The system performance parameters that will be evaluated based on changes in evaporation and condensation temperatures are the cooling capacity (Qe), compressor power (Wc), heat removed by the condenser (Qc), and coefficient of performance COP. This study also discusses the potential contribution of CO2 to the environment due to variations in evaporation and condensation temperatures. The CO2 contribution was calculated based on the life-cycle climate performance (LCCP) method.

2 Material and Methods

2.1 Material

The refrigerants used in this study are R32 and R404a. R32 is a refrigerant with high energy efficiency, low global warming potential (GWP of approximately 675), good cooling capacity, and suitable operating pressure for various temperature conditions; however, it is flammable and requires careful handling and installation. R404A is a mixed refrigerant frequently used in commercial refrigeration applications with good thermodynamic performance, but it has a high global warming potential (GWP of approximately 3943) and is being replaced by more environmentally friendly refrigerants. The properties of the refrigerants R32 and R404a are listed in Table 2. As shown in Table 2, R32 has a lower boiling point than R404a does. This indicates that R32 evaporates at a lower temperature than R404a, making it an effective refrigerant for low-temperature cooling applications. This boiling point also affects the design and operation of the cooling systems that use it because the ability of the refrigerant to absorb heat from the environment is directly proportional to this low temperature.

Table 2. Properties refrigerant R32 and R404a

No	Refrigerant	R32	R404a
1	Formula	CH_2F_2	CHF_2CF_3
2	Boiling Point (°C)	-51.6	-46,2
3	Critical Temperature (°C)	78,4	72,1
4	Critical Pressure (bar)	58,3	37,4
5	GWP	675	3943
6	ODP	0	0
7	Safety Group Classification	A2L	A1

2.2 Methods

The influence of evaporation and condensation temperatures was studied by simulation using software developed by the Department of Mechanical and Biosystems Engineering, IPB University. However, before carrying out the simulation, validation must first be performed using experimental data on the performance of R32 cold storage. This study was conducted based on the study matrix in Table 3.

Table 3. Matrix parameter study

Parametric Study	**Evaporation Temperature (°C)**	**Condensation Temperature (°C)**
1	-30	30
2	-30	35
3	-30	40

Parametric Study	Evaporation Temperature (°C)	Condensation Temperature (°C)
4	-30	45
5	-25	30
6	-25	35
7	-25	40
8	-25	45
9	-20	30
10	-20	35
11	-20	40
12	-20	45
13	-15	30
14	-15	35
15	-15	40
16	-15	45

The experiment was performed according to the experimental setup shown in Figure 2. Temperature and pressure data were measured using a K-type thermocouple and pressure transmitter installed at the inlet and outlet positions of each main component during cold storage. All temperature and pressure data were recorded every minute using a data acquisition (DAQ) system. The refrigerant mass flow rate data in the cold storage system were measured and recorded every minute by using a Coriolis flow meter. The power consumption data were measured and recorded every minute using a power meter. In the R32 cold-storage test, the refrigerant mass charge inserted into the cold-storage system was 220 g.

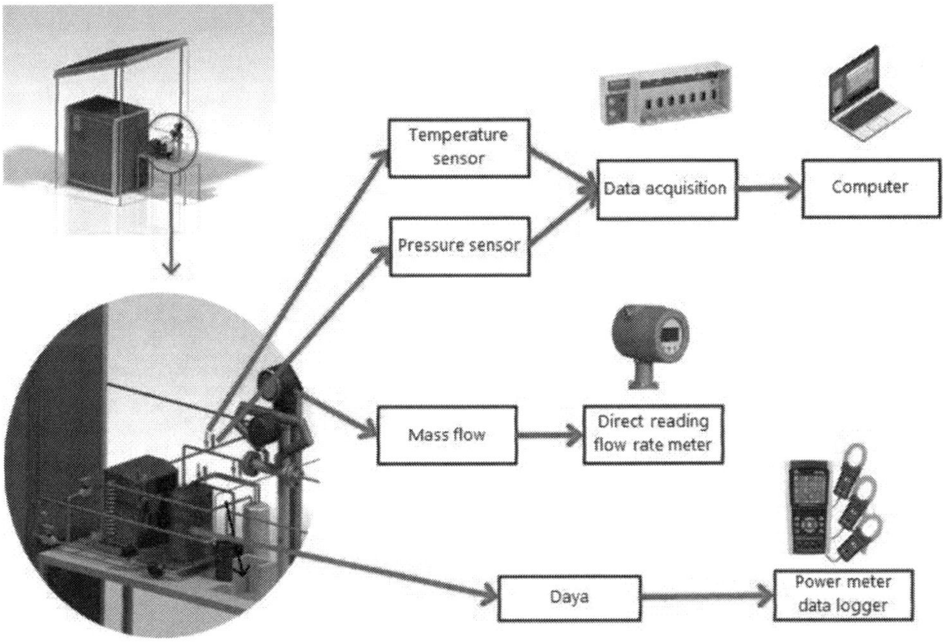

Fig. 3. Experimental set-up

The simulation used in this study was the Poci (Thermal Power Component Integrated) software developed by the Department of Mechanical and Biosystem Engineering IPB

University, as seen in Figure 3. The equations used in this software are based on thermodynamic equations applied to the refrigeration system's main components, such as the Compressor, Condenser, Expansion Valve, and Evaporator.

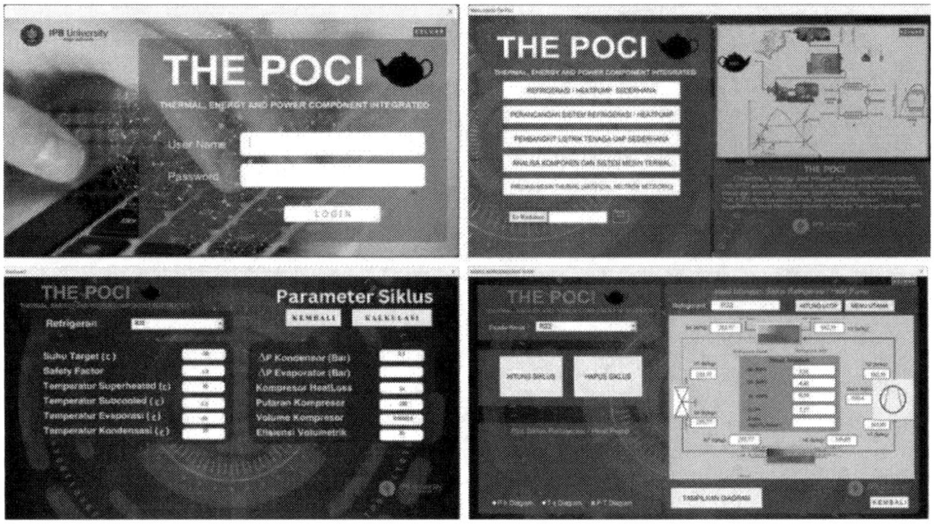

Fig. 4. The Poci Software

The equation that used in The Poci Software are listed below:

a. Cooling Capacity

Cooling capacity can be calculate by multiply the mass flowrate of refrigerant in the system ($\dot{m}$) with differences of inlet Evaporator (h_4) and Inlet Compressor (h_1) which is as mathematically can be expressed as below [4,5]:

$$Q_e = \dot{m}\,(h_1 - h_4) \tag{1}$$

b. Power of Compressor

The power of compressor (W_c) is calculated by multiply the mass flowrate of refrigerant in the system ($\dot{m}$) with the differences of outlet (h_2) and inlet of compressor (h_1). Based on that as mathematically can be expressed as below [4,5]:

$$W_c = \dot{m}\,(h_2 - h_1) \tag{2}$$

c. Heat rejection in the condenser

In this proses, the heat rejection(Q_c) can be calculate by multiply the mass flow rate of the refrigerant ($\dot{m}$) with the differences of enthalpy inlet (h_2) and outlet condenser (h_3). mathematically can be expressed as below [4,5]:

$$Q_c = \dot{m}\,(h_2 - h_3) \tag{3}$$

d. COP

COP is the energy efficiency level of the cooling system in converting electrical energy into cooling energy. The higher the COP value, the more efficient the cooling system in producing cooling. To calculate COP, the cooling capacity of a refrigeration system (Q_e) is divided by the compressor power (W_c) or if written in the formula as below [4,5]:

$$COP = \frac{Qe}{Wc} \tag{4}$$

e. Evaluation of CO_2 Contribution

Evaluation of the contribution of CO_2 to the environment is an evaluation to assess how much impact the cooling or air conditioning system has on carbon dioxide (CO_2) to atmosphere. The contribution value of CO_2 missions to the environment using the Life Cycle Climate Performance method can be calculated using the following formula :

$$LCCP = Direct + Indirect \quad (5)$$
$$Direct = C \times (L \times ALR + EOL) \times (GWP + Adp.GWP)/Q_E \quad (6)$$
$$Indirect = (L \times AEC \times EM + \sum(m + MM) + \sum(mr \times RM) + C \times (1 + L \times ALR) \times RFM + C \times (1 - EOL) \times RFD)/Q_E \quad (7)$$

To determine the deviation between the experimental and simulation values, calculations were carried out using the relative deviation method, as in the following equation:

$$e_{rel} = \left|\frac{Exp-Sim}{Rel}\right| * 100\% \quad (8)$$

3 Result and Discussion

3.1 Experimental Results and Validation

Figure 5 and Table 4 present the validation results by comparing the experimental data with the simulation results. Figure 5 shows the P-h diagram results obtained from both the experiment and simulation. Experimental conditions carried out by a condensation temperature of 32°C and an evaporation temperature of -28°C can produce lower side pressures of around 0.28 MPa (Sim) and 0.27 MPa (Exp). Meanwhile, the upper side produces pressure of 2.06 MPa (Sim) and 2.11 MPa (Exp). In Table 4, it can also be seen that the average deviation is 2.62%, which means that the model used in The Poci has excellent accuracy and can be used to predict cooling system performance.

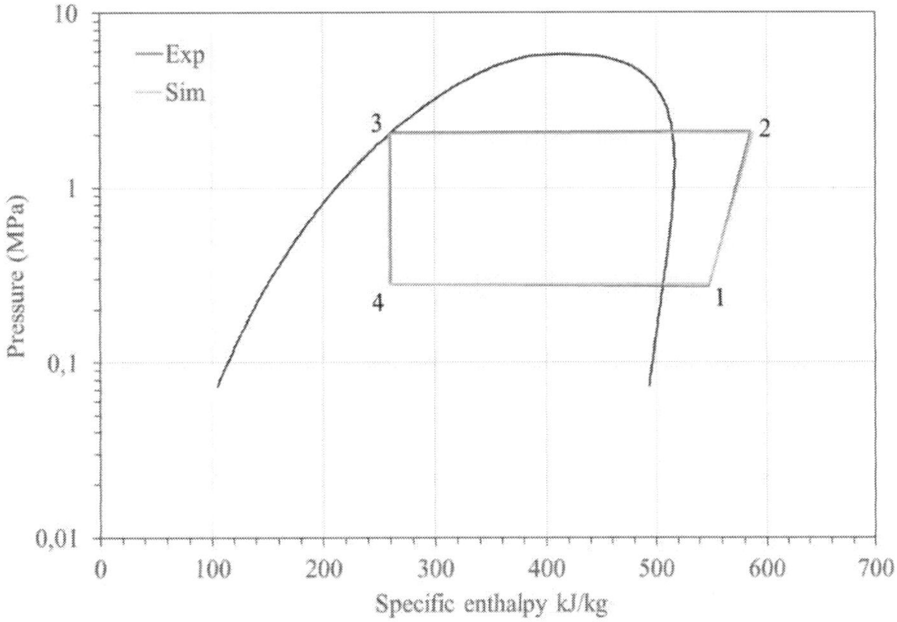

Fig. 5. Validation P-h Diagram

Table 4. Validation result between experiment and simulation of R32 Cold Storage

No	Evaluation Parameter	Experiment	Optimum conditions based on simulation calculations	Relative Eror (%)	Ave. Rel Error (%)
1	P1 (MPa)	0,27	0,28	2,64	2,62
2	P2 (MPa)	2,11	2,06	2,49	
3	P3 (MPa)	2,09	2,03	3,18	
4	P4(MPa)	0,27	0,28	0,18	
5	h1 (kJ/kg)	548,53	548,35	0,03	
6	h2 (kJ/kg)	585,39	587,89	0,42	
7	h3 (kJ/kg)	260,22	259,26	0,37	
8	h4 (kJ/kg)	261,09	259,26	0,7	
9	Mass Flow (kg/s)	0,018	0,017	6,52	
10	Comp. Power (Wc) (kW)	0,69	0,67	3,53	
11	Cooling Cap. (Qe) (kW)	5,23	4,92	5,96	
12	Heating Cap. (Qc) (kW)	7,50	7,31	5,53	
13	COP	7,50	7,31	2,53	

3.2 Study of Effect evaporation and condensation temperature on cold storage

Figure 6 shows the P-h diagram for an evaporation temperature of -30°C with variations in condensation temperature of 30°C, 35°C, 40°C, and 45°C for refrigerants R32 Figure 6 (a) and R404a Figure 6 (b). In both refrigerants, R32 and 404a have the same trend for the evaporation temperature remaining constant and increasing the condensation temperature, and the upper side pressure (condensation pressure) will increase. This is because of the principles of thermodynamics that regulate the relationship between the temperature and pressure of the refrigerant. Additionally, the gas pressure is directly proportional to the temperature if the volume remains constant. In the context of a refrigeration system, as the condensation temperature increases, the refrigerant pressure also increases because the system volume is relatively constant. In Figures 6 (a) and (b), it can also be seen that the pressure that occurs in R32 is higher than that in R404a. The compressor outlet pressure of R32 was higher than that of R404A, owing to the differences in the thermophysical properties of the two, where R32 has a lower boiling point, lower molecular mass, and a higher cooling capacity than R404a.

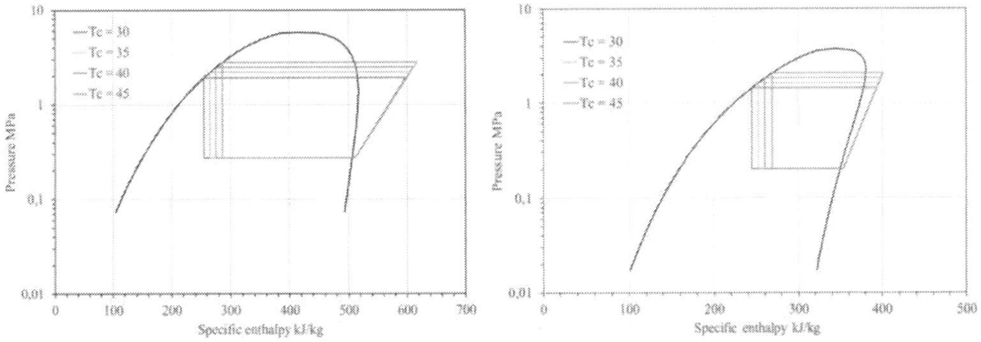

(a) (b)

Fig. 6. P-h Diagram changing the condensation at constant evaporation -30°C: (a). R32, (b). R404a

Figure 7 (a) and (b) can be seen the effect of changing the evaporation temperature at a condensation temperatures of 30°C for refrigerants R32 and R404a. A decrease in the evaporation temperature under a constant condensation temperature condition will cause a decrease in the pressure on the lower side (compressor inlet) for refrigerants R32 and R404a. This is because of the thermodynamic properties of the refrigerant and the basic principles of the refrigeration cycle, especially those related to the pressure and saturation temperature. During the refrigeration cycle, the refrigerant undergoes a phase change from liquid to gas in the evaporator. This process occurs at a certain interrelated temperature and pressure; at the saturation point, when the evaporation temperature decreases, the saturation pressure required to maintain the refrigerant in the gas phase also decreases.

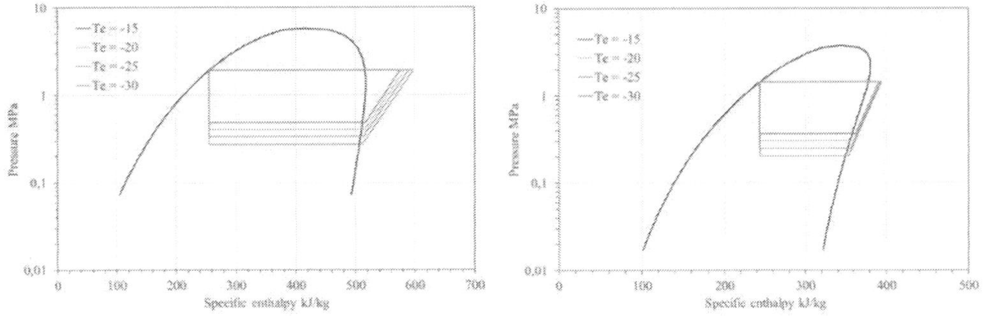

(a) (b)

Fig. 7. P-h Diagram Changing the evaporation at constant condensation 30°C: (a). R32, (b). R404a

The results of the study on the influence of evaporation and condensation temperatures in Figures 6 and 7 obtained results and analysis that were similar to the vapor compression refrigeration system using a variable speed compressor [19] and the adsorption refrigeration system, desalination cooling system [20].

Figure 8 (a) and (b) show the P-h diagrams of the best and worst conditions from the 16 case matrix studies of changes in evaporation and condensation temperatures. The results of the P-h diagram analysis for R32 produced the best conditions in study parameter 13 (Te = -15°C and Tc = 30°C) where in this condition the potential for the most optimum cooling capacity is around 261.7 kJ/kg with upper and lower pressure conditions of around 1 .93 MPa and 0.48 MPa. Meanwhile, the worst condition for R32 occurs in parametric study 4 (Te = -

30°C and Tc = 45°C) where in this condition the potential cooling capacity is 224.8 kJ/kg with upper and lower pressure conditions of 2.79 MPa and 0, 27 MPa. In R404a, the best and worst conditions are also the same, that is, in parametric studies 13 and 4 under conditions where the potential cooling capacity, top pressure, and bottom side are lower R32

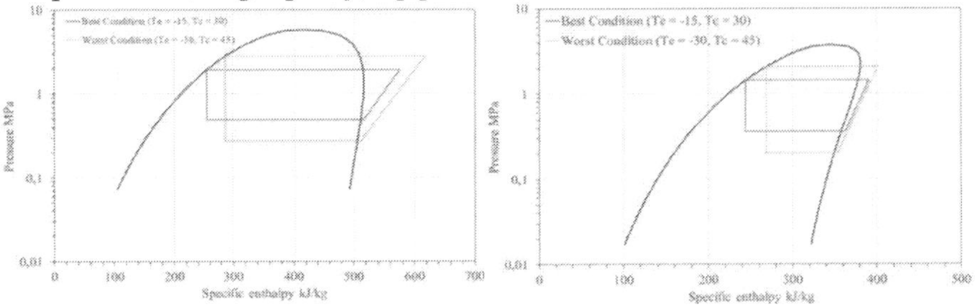

(a) (b)

Fig. 8. P-h Diagram for the best and worst condition: (a). R32, (b). R404a

Table 5 lists the performance of cold storage with a vapor compression refrigeration system that uses R32 and R404 refrigerants. Indonesian climate conditions are equal to the average environmental temperature of 32°C and for cooling fish with a target room temperature of -18°C; the condensation and evaporation temperatures that could possibly be applied are 30°C and -18°C. Under these conditions (Green Highlight), a possible value of compressor power, cooling capacity, and COP produced will be 0.62 kW and 1.85 kW for R32. Meanwhile, under the same conditions for R404a, the compressor power, cooling capacity, and COP were 0.4, 1.5, and 2.72, respectively. The power of the compressor, cooling capacity, and COP of R32 were better than those of R404a for cold storage applications. Meanwhile, the worst condition occurs if the evaporation temperature is -30°C and the condensation temperature is 45°C (red highlight). All the parametric study matrices that were carried out for R32 and R404a showed optimum performance when the evaporation temperature was -15°C and the condensation temperature was 30°C (light blue highlighted).

Table 5. Performance of cold storage using R32 and R404a

Parametric Study	Te (°C)	Tc (°C)	Evaluation							
			R32				R404a			
			Wc (kW)	Qc (kW)	Qe (kW)	COP	Wc (kW)	Qc (kW)	Qe (kW)	COP
1	-30	30	0,62	2,47	1,85	2,96	0,40	1,50	1,10	2,72
2	-30	35	0,67	2,45	1,78	2,64	0,43	1,45	1,02	2,37
3	-30	40	0,72	2,42	1,70	2,36	0,46	1,39	0,94	2,06
4	-30	45	0,77	2,39	1,62	2,12	0,48	1,33	0,85	1,78
5	-25	30	0,67	2,93	2,26	3,37	0,44	1,80	1,37	3,12
6	-25	35	0,73	2,90	2,17	2,99	0,47	1,74	1,27	2,70
7	-25	40	0,78	2,86	2,08	2,66	0,50	1,67	1,17	2,34
8	-25	45	0,84	2,82	1,99	2,37	0,53	1,59	1,07	2,02
9	-20	30	0,71	3,44	2,74	3,87	0,47	2,15	1,69	3,60
10	-20	35	0,77	3,41	2,63	3,40	0,51	2,07	1,57	3,11
11	-20	40	0,84	3,36	2,52	3,01	0,54	1,99	1,45	2,68
12	-20	45	0,90	3,32	2,41	2,67	0,58	1,90	1,33	2,30
13	-15	30	0,74	4,03	3,29	4,48	0,49	2,55	2,07	4,20
14	-15	35	0,81	3,98	3,17	3,90	0,54	2,46	1,93	3,59
15	-15	40	0,89	3,93	3,04	3,42	0,58	2,36	1,79	3,08
16	-15	45	0,96	3,87	2,90	3,01	0,62	2,25	1,64	2,64

Figure 9 shows the results of the CO_2 contribution using the Life Cycle Climate Performance (LCCP) method from 16 parametric studies for changes in the evaporation and condensation temperatures with a vapor compression system on both R32 and R404. From Figure 8, it can be seen that for conditions with an evaporation temperature of -30°C and a condensation temperature of 30°C, the CO_2 contribution produced for 10 years is 15334 $kgCO_{2e}$ or around 1533 $kgCO_{2e}$ per year for R32. Meanwhile, under the same conditions, R404a for 10 years produces 19082 $kgCO_{2e}$ or around 1908 $kgCO_{2e}$ per year. Using R32 in cold storage can reduce emissions contributions by 20% per year.

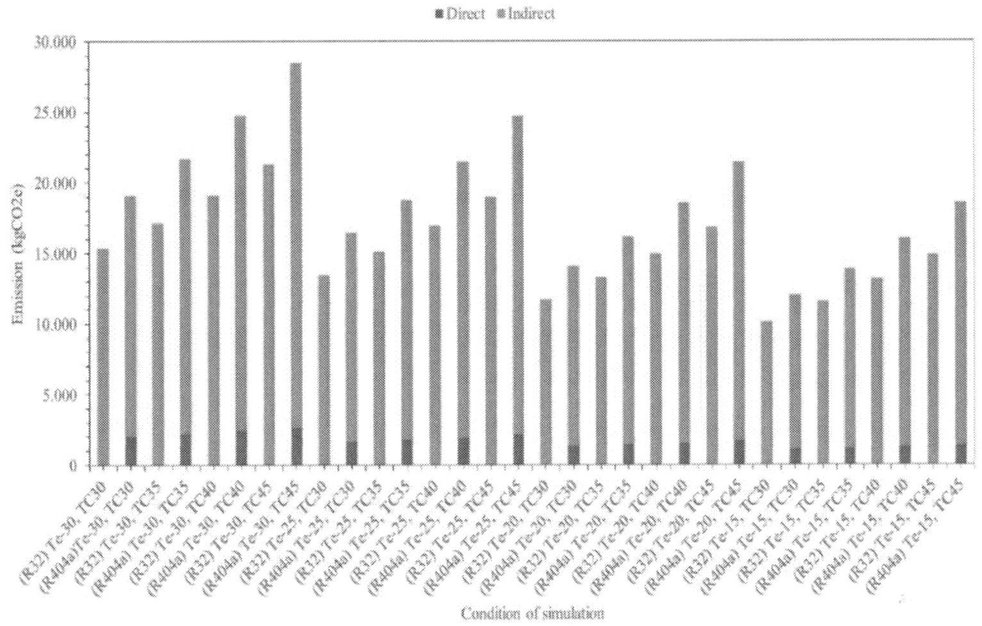

Fig. 9. Analysis results of LCCP

4. Conclusion

Cold storage performance studies were conducted using R32 and R404a refrigerants. The validation results comparing the experimental data and simulation outcomes obtained using Poci software demonstrated excellent accuracy, with an average error of 2.6%. The study results indicate that both R32 and R404a exhibit good performance when the evaporation temperature is -30°C and the condensation temperature is 30°C, respectively. The optimum conditions for both refrigerants are an evaporation temperature of -15°C and a condensation temperature of 30°C. The worst performance was observed when the evaporation temperature was approximately -30°C and the condensation temperature was approximately 45°C. Decreasing the evaporation temperature and increasing the condensation temperature increased the CO2 emissions. Over a 10-year period, R32 produces 15,334 kgCO2e or approximately 1,533 kgCO2e per year; under the same conditions, R404a produces 19,082 kgCO2e or approximately 1,908 kgCO2e per year. The use of R32 can reduce environmental CO2 emissions by 20% per year.

Thank you to the Directorate of Higher Education, Research, and Technology through Kedaireka Research Grant 2023 and PT Samudera Sinar Abadi Shipyard with contract number 31/E1/HK.02.02/2023 for funding this research.

References

1. Ramadan H., Cappenberg AD. *J Konversi Energi dan Manufaktur*. 5,2 (2018)
2. Arianto MF. *J Geogr*. 20,20(2020)
3. Mahmood, R. A., Ali, O. M., Al-Janabi, A., Al-Doori, G., & Noor, M. M. *Journal of Applied Mechanics and Engineering, 26*,3(2021)
4. Moran, M. J., Shapiro, H. N., & Bailey, M. B. *Fundamentals of Engineering Thermodynamics 9th Edition* (2018)
5. Cengel YA, Boles MA. *THERMODYNAMICS: AN ENGINEERING APPROACH, EIGHTH EDITION*. Ed ke-8. Volume ke-8. Stenquist B, (New York: McGrraw-Hill Education 2015)
6. Saito, K. *Latest heat pump technologies in Japan* (2017)
7. Yulianto, M., Giannetti, N., Miyaoka, Y., Ge, Z., Tsuchino, T., Urakawa, M., Taira, S., Jeong, J., & Saito, K. *AIP Conference Proceedings, 2836*,1. (2024)
8. Hadi, S. W., Wijaya, Y. T., Elysabeth, I. R., Aisyah, N., Putra, R. D. D., Ariyadi, H.M. *Advanced Engineering Letters*, 2,3 (2023)
9. Yulianto, M., Suzuki, T., Ge, Z., Tsuchino, T., Urakawa, M., Taira, S., Miyaoka, Y., Giannetti, N., Li, L., & Saito, K. *Sustainable Energy Technologies and Assessments, 49*(2022)
10. Cabello R., Sánchez D., Llopis R., Andreu-Nacher A., Calleja-Anta D. *Int J Refrig*. 137 (2022)
11. Zheng, H., Tian, G., Zhao, Y., Jin, C., Ju, F., & Wang, C. *Case Studies in Thermal Engineering, 36* (2022)
12. Fachruddin, F., Maimun, M., & Aliredjo, M. S. *International Journal of Advanced Engineering Research and Science, 8*, 2 (2021)
13. Ardita, I. N., Wirajati, I. G. A. B., Susila, I. D. M., & Sudirman, S. *Journal of Applied Mechanical Engineering and Green Technology, 2*,1 (2021)
14. Giriprasath, B. S., Chandramohan, R., Kumar, R. A., & Arun, N. K. *AIP Conference Proceedings, 2317* (2021)
15. Bhamidipati, A., Pendyala, S., & Prattipati, R. *Materials Today: Proceedings, 28* (2020)
16. Pattana, S., Tantakitti, C., & Wiratkasem, K. *Energy Reports, 6* (2020)
17. Shi, H., & Wu, J.. *IOP Conference Series: Materials Science and Engineering, 604*, 1 (2019)
18. Daging, I. K., Siahaan, J. P., Maimun, M., Perikanan, S. T., Hermawan, A., *November* (2018)
19. H. Al-Hassani, A., & R. Al-Badri, A.. *Engineering and Technology Journal, 39*, 1A (2021)
20. Youssef, P. G., Al-Dadah, R. K., Mahmoud, S. M., Dakkama, H. J., & Elsayed, A. *Energy Procedia, 75* (2015)

Nomenclature

Exp	Experiment
Sim	Simulation
COP	Coefficient of Performance
GWP	Global Warming Potential
$P1$	Pressure of Inlet Compressor (MPa)
$P2$	Pressure of Inlet Condenser (MPa)
$P3$	Pressure of Inlet Expansion Valve (MPa)
$P4$	Pressure of Inlet Evaporator (MPa)
$h1$	Enthalpy Inlet Compressor, $(kJ \cdot kg^{-1}. C^{-1})$
$h2$	Enthalpy Inlet Condenser, $(kJ \cdot kg^{-1}. C^{-1})$
$h3$	Enthalpy Inlet Expansion Valve, $(kJ \cdot kg^{-1}. C^{-1})$
$h4$	Enthalpy Inlet Evaporator, $(kJ \cdot kg^{-1}. C^{-1})$
$\dot{m}$	Mass flowrate refrigerant,$(kg \cdot s^{-1})$
ODP	Ozone Depletion Potential
Qc	Heat Rejection in condenser, (kW)
Qe	Cooling Capacity, (kW)
Wc	Work of Compressor, (kW)
Wp	Watt Peak

Study on design and construction of unmanned surface vehicle (USV) as a marine survey vehicle

Arif Baswantara[1*], *Muhammad Romdonul* Hakim[1], *Anas Noor* Firdaus[1], *Muhammad Riyono Edi* Prayitno[2], *Khaidan* Firmansyah[1], *Reinaldi Nur* Fasha[1]

[1]Pangandaran Polytechnic of Marine and Fisheries, Indonesia
[2]Karawang Polytechnic of Marine and Fisheries, Indonesia

Abstract. Marine surveys may provide hydrooceanographic, hydrobiological, and spatial information. Surveys often face the problem of limited professional resources and inadequate equipment. Continuous technological development requires further development in this field. This is also the reason for future research on the development of unmanned surface vehicles (USV) to support marine survey activities, especially underwater topography. The studies were conducted starting with design and construction, USV vehicle performance tests, control and motion system performance tests, and sensor performance tests. The USV design uses a catamaran ship model, where the USV has two hulls and two propulsion. Construction tests were performed using the inclining test and rolling test methods. Control and motion system tests were performed using the free running test and turning ability methods. The sensor test was performed by field sounding test. The results showed that the USV functioned well. All the systems were integrated as planned. Acoustic and GPS sensors can also be used for underwater topographic surveys. The deficiencies identified were related to the durability of the propulsion system and GPS sensor system, which can be further optimized.

1 Introduction

A survey is a discipline that includes all measurement methods and information collection related to the physical environment, information processing, and dissemination of various products produced. A hydrographic survey is a survey of coastal objects, lakes, rivers, seas, and other water bodies. More specifically, the marine survey is a survey activity carried out with the object of marine waters. Marine survey activities include hydrooceanography, hydrobiology, and spatial information. Conducting a survey according to certain categories requires professional resources and adequate equipment. Surveys, especially in Indonesia, often encounter problems of limited equipment availability, and with ongoing technological development, the ability to follow technological development must also be a concern. This certainly encourages the conducting of a study regarding the technological development of marine survey activities, and this also becomes the background for conducting a study

* Corresponding author: baswantara@pkpp.ac.id

regarding the design of an Unmanned Surface Vehicle (USV), which aims to assist marine survey activities in the future, especially for underwater topography survey activities.

An Unmanned Surface Vehicle (USV) is an unmanned vehicle that operates on the water surface and has a specific purpose. The USV is also known as an Autonomous Surface Vehicle (ASV) or surface vehicle that operates automatically using GPS to determine the direction and destination of its movement [1]. The concept of unmanned marine vehicles (UMV), including USV, was initially developed for military purposes, mine clearance, and battle-damage assessment. However, in recent decades, this technology has been used by civilians and various science missions, one of which has been used in environmental survey activities. An environmental survey that has become one of the USV developments is bathymetric survey activities [2][3].

The use of robots in bathymetric measurement must actually involve various considerations, including the quality of data, the number of data that can be processed, and system endurance when working close to 100% [4]. There are two kinds of unmanned systems that can be used in this activity: unmanned underwater vehicles (UUV) and USV, and the analysis of both showed that the USV is better when used for bathymetric survey activities. The parameters used for reference are speed/endurance, payload processing, payload sensor, survey coverage, navigation accuracy, low vehicle motion, real-time coms, and launch/recovery ease [5].

In this study, the activities carried out not only created the design of the USV vehicle but also designed the construction process. However, the USV testing was also performed. Testing was performed on the USV vehicle and USV maneuvers, and each sensor was used. This was carried out to achieve the objective of this study, which was to produce a USV design that can be utilized in marine survey activities in the future, especially water topography.

2 Method

2.1 Location, Time and Type of Study

This study was conducted in Bandung, Pangandaran, Indonesia. This study was conducted over 8 months, starting from design, construction, and testing. The study type was Research and Development (R&D). Based on [6]. There are four main characteristics in R&D: conducting a study or research to find findings related to the product that will be developed, developing the product according to the findings of previous studies or research, conducting field tests in real conditions where the product will be applied, and revising or improving the weaknesses found.

2.2 Boat Design and Construction

The USV vehicle was designed as a catamaran-type vehicle. The catamaran type is a multihull boat with two hulls or two pontoons, and this type of boat can trap fluid flow between the two hulls [7]. The USV vehicles are also designed with an asymmetric catamaran type, where the outer body of the boat is streamlined and the inner body of the boat is straight. The asymmetric catamaran was chosen because this type requires a lower thrust. Fluid acceleration in the convergent section involves a reduction in the quantity of motion of the water mass, thereby increasing the efficiency of the catamaran [8]. A detailed design of the boat is shown in Figure 1.

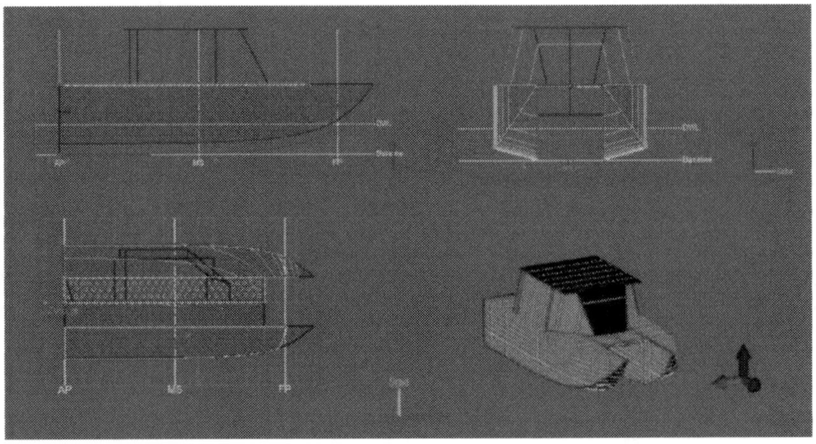

Fig 1. USV Boat Design

2.3 Control and Propulsion System

The USV control system is built wirelessly using control in the form of a Ky-023 joystick module, with a radio signal communication system using NRF24L01 on the transmitter and receiver. The transmitter was built as a remote control, whereas the receiver was built as a box placed on the deck of the USV vehicle. The signal receiver was directly connected to the USV drive system such that the USV movement could be controlled remotely.

The drive system of the USV uses two 22 mm Turbo Jet Thrusters and Brushless Motor 2845 with a speed controller in the form of a 13 V Electronic Speed Control (ESC). This machine works like a water jet pump; therefore, its movement produces a water spray.

2.4 Sensor and Electronic System

The acoustic sensor becomes a sensor is applied to the USV prototype. The acoustic sensor used had a transducer with a single beam and single frequency, working at a frequency of 115 kHz with a maximum range of 30m. This sensor can be used as an instrument to assist underwater topography mapping.

A GPS sensor is also added to the USV prototype. This sensor was used to determine the position of the USV when moving and as tracking points when the acoustic sensor sounded. The sensor used is Ublox Neo-6MV2 with a tracking and navigation sensitivity of -161 dBm, speed accuracy of 0.1 m/s, and horizontal position accuracy of 2.5m. This specification can change according to the amount of GPS satellite signals received by the sensor and the physical conditions of the sensor.

The electronic system of the USV was assembled separately. There are three main electronic circuits. First, the electronic circuit in the remote system is from the USV-control transmitter. Second, the circuit in the USV control receiver is connected to the USV drive. Third, the circuits in USV sensor systems are acoustic and GPS sensors. The third circuit consists of a data storage system using an SD Card module. The components of the electronic system are listed in Table 1. All electronic systems in the USV were powered by 9-volt and 12-volt DC power sources, with the control center as an Arduino microcontroller.

Table 1. Components in Electronic Systems

No	Components Name	Function
1	Ping Sonar Echosounder	Acoustic Sonar Sensor

No	Components Name	Function
2	GPS Neo-6MV2	USV Positioning System
3	NRF24101+	Control System Communication
4	Arduino Uno	Controls all electronic system functions
5	Remote Controller Kit	USV Motion Control

2.5 Boat Performance Test

The USV vehicle performance tests included load capacity and stability tests. A load capacity test was performed by placing a load in the middle of the boat until the lowest side of the boat was parallel to the water. The load was applied gradually at intervals of 1 kg. The parameters recorded were the weight of the load and the freeboard height. Freeboard is the vertical distance between the deck line and freeboard, leading down to the upper side of the load line measured in the middle of the boat. This is important because it is directly related to safety, production economy, and boat operation [9]. The measurements were performed 10 times. The testing method is illustrated in Figure 2.

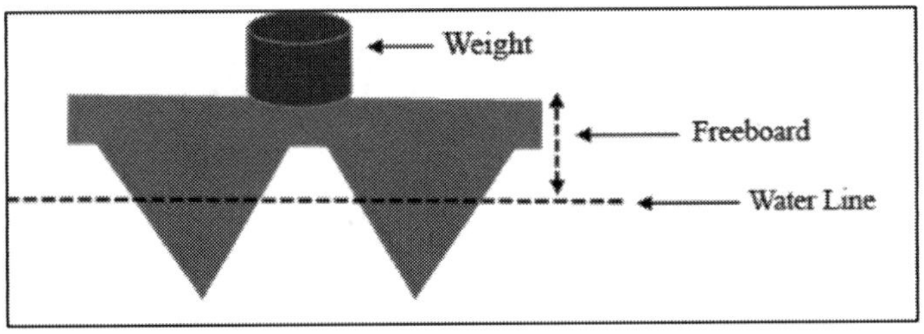

Fig 2. Load Capacity Test

A stability test was performed by inclination and rolling. Inclining was carried out by placing a load on one of the boat sides until the boat was capsized. The given load varies from 0 to 10 kg or 1 – 4 °. The parameters measured were the weight of the load and the angle of inclination formed between the metacentric line and vertical axis[10]. The inclination angle was measured by using a protractor. Measurements at each load weight were performed 10 times to obtain accurate results. Rolling was carried out by pressing the side of the boat until the boat tilted and then calculating the number of boat swings until the boat returned to an upright position. Testing was performed at a small inclination angle (10°) up to 60° with an interval of 10°. According to [11], the initial condition in stability is the ability of the boat to return to its original upright position when tilted at small angles (=60°). This initial stability stage is limited to discussions on transverse stability.

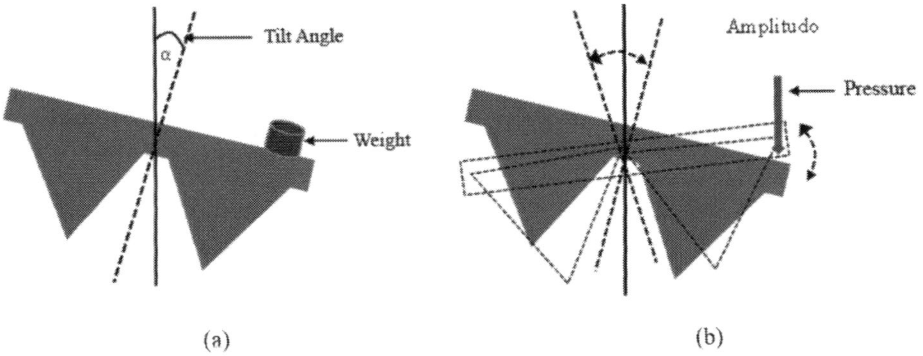

Fig 3. (a) Inclining Test, (b) Rolling Test

2.6 Control and Propulsion Performance Test

A USV control system test was conducted by testing the connectivity of the remote control. The remote control was tested by connecting the receiver and transmitter at a certain distance and carrying out a transfer file to generate data that was received to determine the connectivity between them. The distance used was between multiples of 50m and 500m with unobstructed conditions.

The USV motion test or maneuver test was carried out using two testing methods: open free running and turning ability tests [12][13]. Open-free running involved three testing steps. The first test was carried out in a straight line with an angle of 0°, the second test was moving towards the starboard with an initial angle of 0° to 90°, and the third test was moving towards the portside with an initial angle of 0° to 90°.

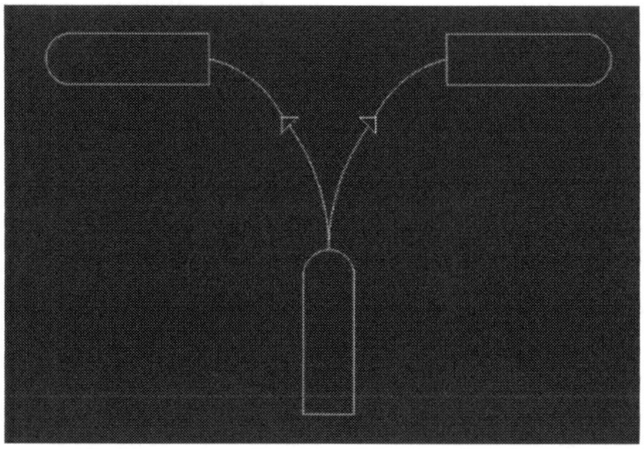

Fig 4. Open Free Running Test Illustration

The turning ability test was tested 3 rounds, starting with the control direction at -30° to the port side, and the engine was turned on at maximal speed until the boat formed a circle with a certain center diameter; when the boat was circling, the center point was marked as the center of the circle and had a circle diameter as the calculation [13][14]. An illustration of the turning ability test is shown in Figure 5.

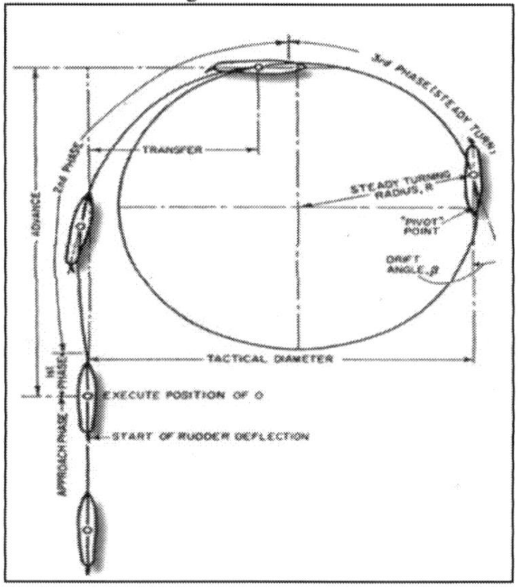

Fig 5. Turning Ability Test Illustration

2.7 Sensor Performance Test

Sensor performance tests were carried out simultaneously, where a GPS sensor and an acoustic sensor were operated to obtain the depth of the water that would then be made into the topographic map. Testing was carried out in the Cikidang Fish Landing Port, Pangandaran, and Bulaksetra River, Pangandaran. The testing was performed using a zigzag path with circular ends.

2.8 Data Analysis

Data analysis was performed based on the test results. The tests to be carried out and the analysis method to be used on the data obtained are listed in Table 2.

Table 2. Test Materials and Data Analysis

No	Test Material	Data obtained	Analysis used
1	USV Vehicle Performance Test	- USV Dimension - USV Stability	- The mean value of data obtained
2	USV Control System and USV Motion System Performance Tests	- Connectivity of transmitter and receiver - USV maneuver time	- Controller connection - Descriptive analysis
3	Sensor Performance Test	- Data on the depth - Coordinate position	- Depth analysis - Depth validation - Coordinate validation

3 Result and Discussion

3.1 USV Vehicle Performance Test

The USV vehicle was successfully manufactured as a catamaran-type boat using e-glass fibers or marine materials. The catamaran was manufactured as a type-B asymmetrical hull. The dimensions of the catamarans are listed in Table 3.

Table 3. Specification of USV Vehicle

No	Information	Size
1	Length of All (LoA)	148 cm
2	Width (B)	95 cm
3	Height	33 cm
4	Draft	12 cm
5	Maximum Load	80 kg
	Hull	
1	Interhull Distance (S)	30 cm
2	Width of Hull (Bl)	30,8 cm
	Deck	
1	Width of Deck	78,6 cm
2	Height of Deck	15 cm
3	Length of Deck (LDL)	122 cm
	Hatch	
1	Width of Hatch (L)	29 cm
2	Height of Hatch (D)	15 cm
3	Length of Hatch (P)	122 cm
	Superstructure	

No	Information	Size
1	Length of Structure	43 cm
2	Height of Structure	26 cm
3	Width of Structure	73 cm

The table shows that the catamaran vehicle had an overall length (LoA) of 148 cm, width (B) of 95 cm, and height of 33 cm. There were two hulls; each hull had a distance (S) of 30 cm, and the width of each hull (BI) was 30.8 cm. The catamaran deck is quite spacious, with a deck width of 78.6 cm and deck length of 122 cm.

The results of the load capacity test showed that the boat flooded the deck at a load weight of 80 kg with the load center in the middle of the boat. The load was gradually applied at intervals of 10 kg. The boat had an unladen draft measuring 12 cm. When the boat was given a load of 10 kg, the boat draft increased by 2.23 cm to 14.23 cm. The increase in the draft from a load of 30 kg to 40 kg was only 1.69 cm. An increase in draft height decreased with an increase in load, so the change was only 0.72 cm when the boat added a load from 70 kg to 80 kg. An increase in boat draft when given a load with an interval of 10 kg is shown in Figure 6.

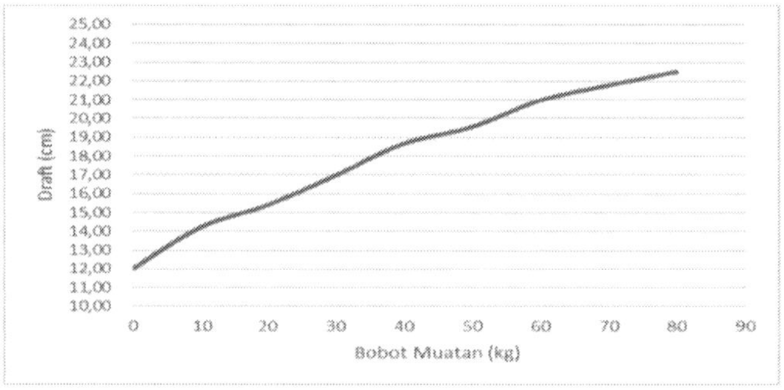

Fig 6. Load Capacity Test Result

Providing a load with the same weight interval caused the boat draft to increase with an increasingly small added value. This is because the shape of the boat hull is v-bottom. The amount of load required by a boat when the draft is increased by 1 cm is known as ton per centimeter (TPc). The magnitude of TPc in the v-bottom boat was smaller than that in the round-bottom or u-bottom boat. This shows that the loading capacity of a v-bottom boat with the same dimensions is lower than that of other or both types of boat hulls.

The results of the inclination test showed that the boat began to capsize when the load was 50 kg. When the boat was subjected to a load of 10 kg, it experienced a tilt of 14.9°. A load of 20 kg tilts the boat by 27.2 Â °. The boat was tilted by 34.3° when the load was 30 kg. A load weight of 40 kg resulted in a tilt angle of 44.8°. The final result was that when the boat was given a load of 50 kg, it began to capsize slowly. The change in tilt when adding a 10 kg interval load is shown in Figure 7.

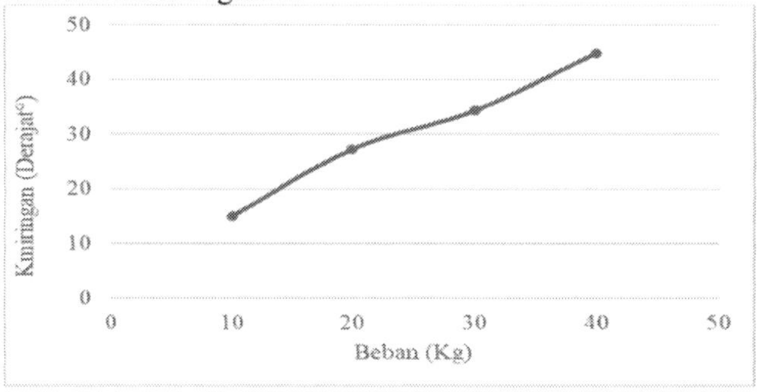

Fig 7. Inclining Test Result

The changes in the tilt were relatively high, with an average increase of 11°. This condition is caused by the use of large-volume loads in the form of sand-filled plastic bags. This type of load causes the center of gravity on the boat to move upward, making it easier for the boat to tilt. The center point of the downward force of gravity is called the gravity point or the center of gravity.

The rolling results showed that rolling testing with an initial angle of 10 °had an average of four swings in 1.7 seconds to return to stability. Moreover, in the rolling test with an initial angle of 20°, the boat had an average of five swings in 2.8 seconds to return to its initial position. This can occur because a boat with two hulls has good stability compared to a monohull boat. A catamaran boat still has good stability because, in a tilted position, to return to its initial position or stability, a catamaran boat does not take a longer time. The results of the rolling testing with initial angles of 10° and 20° are shown in Tables 4 and 5.

Table 4. Rolling Test Result with an angle of 10°

Repetition	Inclination (degree)	Rolling (times)	Time (second)
1	10	3	1,41
2	10	5	3
3	10	4	1,56
4	10	4	1,5
5	10	3	1,66
6	10	4	1,64
7	10	3	1,7
8	10	4	1,56
9	10	3	1,57
10	10	3	1,67
Average		4	1,7

Table 5. Rolling Test Result with an angle of 20°

Repetition	Inclination (degree)	Rolling (times)	Time (second)
1	20	4	2,20
2	20	8	6,00
3	20	6	3,00
4	20	5	2,64
5	20	5	2,53
6	20	4	2,28
7	20	5	2,45
8	20	4	2,35
9	20	5	2,36
10	20	5	2,40
Average		5	2,80

3.2 The Results of the USV Control System and Motion System Performance Test

In the connectivity test, to determine the performance of the USV control system, the frequency of NRF24L01 was set at two data rates: 250 kbps and 2 Mbps. A 250 kbps rate was used as the maximum range, and a 2 Mbps rate was used as the minimum range. The results of the NRF24L01 connectivity distance test at 250 kbps and 2 Mbps rates are listed in Table 6.

Table 6. Control Conectivity Test Result

Distance (m)	Rate	Signal	Result
50m	2 mbps	Min	Connect
	250 kbps	Max	Connect
100m	2 mbps	Min	Connect
	250 kbps	Max	Connect
150 m	2 mbps	Min	Connect
	250 kbps	Max	Connect
200m	**2 mbps**	**Min**	**Connect**
	250 kbps	Max	Connect
250m	2 mbps	Min	Disconnect
	250 kbps	Max	Connect
300m	2 mbps	Min	Disconnect
	250 kbps	Max	Connect
350m	2 mbps	Min	Disconnect
	250 kbps	Max	Connect
400m	2 mbps	Min	Disconnect
	250 kbps	Max	Connect
450m	2 mbps	Min	Disconnect
	250 kbps	Max	Connect
500m	2 mbps	Min	Disconnect
	250 kbps	**Max**	**Connect**

Based on the results obtained, it can be found that the maximum connection distance between the transmitter and receiver was 500 m with a frequency rate of 250 kbps. At a frequency rate of 2 Mbps, the connection maximum distance between the transmitter and receiver was only 200 m; therefore, the best frequency rate that could be used in the USV control system was 250 kbps. The frequency rate used affects the data exchange between the transmitter and receiver. Because the signal in this control system is one-way without two-way data exchange, the frequency rate used is not a problem at a low rate.

In the maneuver test, two types of tests were performed: free-running test and turning ability test. In the free-running test, the USV moved forward to the starboard and towards the port side. The calculated measurements were the speed and time per turn values, starting with the boat at 0° until it turned 90°. Testing was performed three times for each maneuver. The test results are shown in Table 7.

Table 7. Free Running Test Result

Repetition	Maneuver	Distance (cm)	Time (second)	Drift Angle (degree)	Speed (cm/s)
1	Straight	478	22	0°	21,7
2	Straight	408	22	0°	18.5
3	Straight	824	22	0°	37.5
Average	**Straight**	**570**	**22**		**25.9**
1	Starboard	411	22	30°	18.7
2	Starboard	511	22	30°	23.2
3	Starboard	363	22	30°	16.5
Average	**Starboard**	**428**	**22**		**19.4**
1	Portside	559	22	-30°	25.4
2	Portside	350	22	-30°	15.9
3	Portside	330	22	-30°	15
Average	**Portside**	**413**	**22**		**18.7**

The results of the free-running test showed that the USV could move straight forward at an average speed of 25.9 cm/second. This also depended on the power conditions used. The engine uses a DC electric current with a battery voltage of 12 V. During testing, the battery was in good condition and fully charged. During maneuvers towards the starboard and portside, the average speed of the USV was 19.4 cm/second and 18.7 cm/second, respectively. This almost identical speed shows that the drive system on the left and right sides of the USV was able to work properly. If there is a significant speed difference during the maneuver, one side of the USV drive system is not working properly.

In the turning ability test, the initial position of the USV at the reference point of 0° rotated up to 360° three times. The rotation diameter, circumference, time per turn, and speed were measured according to the maneuver. Data from the turning ability test results are listed in Table 8.

Table 8. Turning Ability Test Result

Path Num	Diameter (cm)	Circumference (cm)	Time per turn (second)	Speed (cm/s)
1	**148**	464,72	24,49	**18.9**
2	114	357,96	47,94	7,46
3	112,5	353,25	105,61	3,34

Based on the criteria of IMO 2022, calculations related to tactical diameter and advance were also carried out in the turning ability test. The results of the tactical diameter and advance calculations are shown in Table 9.

Table 9. Comparison of IMO Standard Criteria Values with Test Results

Num	Parameter	IMO Criteria	Test Result
1	Tactical Diameter	<5L (750 cm)	148cm
2	Advance	<4,5L (675 cm)	116,18 cm

According to the data in Table 9, the tactical diameter and advance tested at a speed of 18.9 cm/second obtained a value that could be included in the turning ability maneuver criteria. The results of tactical diameter were found below the 5 times length of the USV, with a value of 148 cm or 0.98 L, and advance was found below 4.5 times the length of the USV, with a value of 116.18 cm or 0.77 L.

3.3 The Results of the Sensor Performance Test

The sensors used in the USV test were GPS and acoustic sensors. The GPS sensor used was a Ublox Neo-6MV2. The acoustic sensor used was a Ping Sonar Echosounder from Blue Robotics. The GPS sensor and ping sonal echosounder were assembled into one instrument and placed on the USV vehicle. The instrument circuit containing the GPS sensor and the ping sonal echosounder is shown in Figure 8.

Fig. 8. Acoustic Instrument and GPS Sensor

A GPS sensor test was performed simultaneously with an acoustic sensor test. The GPS sensor Neo-6MV2 was tested by comparing it with GPS Garmin eTrex 10 and 585. The results of the test showed that compared with the GPS Garmin eTrex 10, the GPS sensor had an average difference of 15.4 meters, and compared with the GPS Garmin 585, it had an average difference of 55.9 meters.

The results of the test showed that the comparison between GPS Neo-6MV2 and Garmin eTrex 10, and Garmin 585 still had quite a large difference. This is because several factors can affect the GPS accuracy. According to [15], the accuracy of GPS is influenced by several factors, including the quality of the GPS receiver, number of satellites received, and location of satellites. The number of satellites read by GPS Neo-6MV2 ranged from 3 to 7 satellites. This is relatively small; therefore, it can affect the accuracy of the position read by the GPS.

The ping sonal echosounder used as an acoustic sensor can measure depth up to ±30 meters in water with a nearfield distance of 0.5 meters and a beam angle of 30°. The specifications of the sensors are listed in Table 10.

Table 10. Ping Sonar Echosounder Specification

Parameter	Specifications
Available Firmware Baud Rates	115200 bps (default), 9600 bps
Frequency	115 kHz
Beamwidth	30 degrees
Minimum Range	0.5 m
Maximum Range	30 m
Range Resolution	0.5% of range
Range Resolution at 30 m	15 cm

Parameter	Specifications
Range Resolution at 2 m	1 cm
Maximum Supply Voltage	5.5 volts
Minimum Suppy Voltage	4.5 volts
TTL Voltage Level	3.3 -5 volts

The use of ping sonar echosounders for measuring water depth has been well tested. However, development in providing coordinates and further data processing is very important to obtain good results [16], and this was performed in this study. When conducting sounding, the sensor takes depth data, and the GPS sensor takes data of latitude and longitude coordinates. During sounding, an LCD was used to monitor the data recorded by the sensor. The data were then calibrated and stored in a data logger, which entered the processing stage. Data calibration was performed using the tidal values in Pangandaran. Further processing of the data involved visualizing the data into a contour map and 3D view.

The acoustic sensor test was conducted in two locations: the Cikidang Fish Landing Port and Bulaksetra River. The visualization results of the sounding from the ping sonar echosounder are shown in Figure 9 and 10. Figure 9 is the visualization of a contour map and 3D view of the sounding location in Cikidang Fishing Port, Pangandaran, while Figure 10 is the visualization of a contour map and 3D view of the sounding location in Bulaksetra River, Pangandaran.

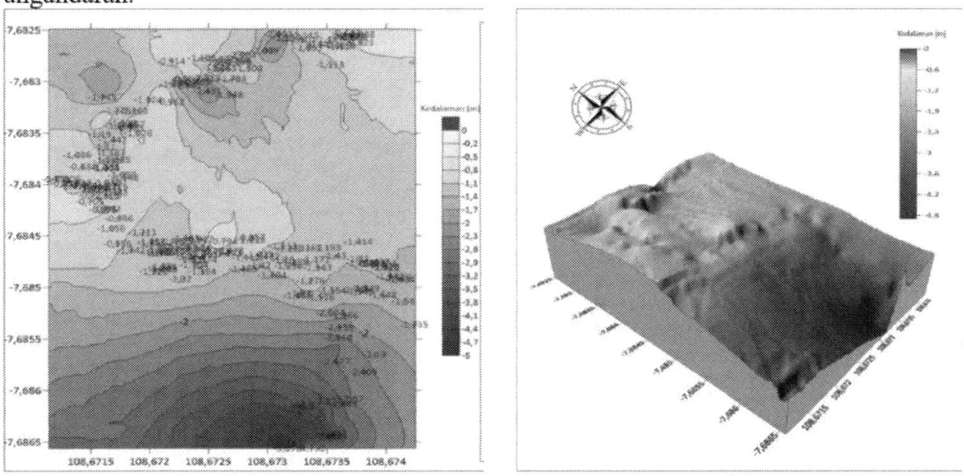

(a) (b)

Fig. 9. Visualization of (a) Contour Map and (b) 3D View from the Results of Sounding in Cikidang Fishing Port

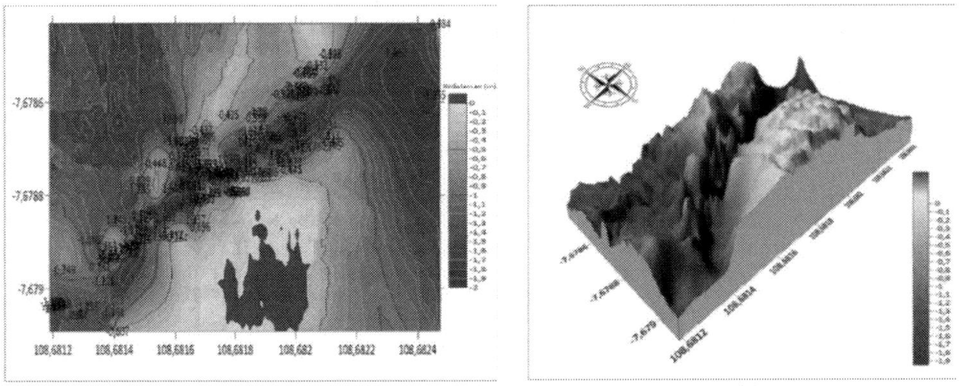

(a) (b)

Fig. 10. Visualization of (a) Contour Map and (b) 3D View from the Results of Sounding in Bulaksetra River

Based on the visualization results, the depth condition in the port area recorded that the deepest point was at a depth of 5.076 m and the shallowest point was at a depth of 0.562 m. In the Bulaksetra River, the deepest point was recorded at a depth of 2.49 meters, and the shallowest point was at a depth of 0.17 meters. Visualizations in Figures 9 and 10 were obtained by processing the sounding data corrected by the tide of 0.1 MSL.

4 Conclusion

Unmanned surface vehicles (USV) have been successfully built with success in USV vehicle design, USV control systems, USV vehicle tests, and USV sensor systems. Unmanned surface vehicles (USV) that have been built still encounter problems in the drive and GPS sensor systems, where the durability of the drive system used can be further improved and the accuracy of the GPS sensor used also needs to be further improved.

References

1. H. Niu, Y. Lu, A. Savvaris, A. Tsourdos, Ocean Engineering, **169**, 13 (2018)
2. NAP, *Thrusts of the time critical strike FNC program overview*, In 2002 assessment of the office of naval research's air and surface weapons technology program, (2002), Retrieved from https://nap.nationalacademies.org/read/10594/chapter/6
3. V. Schmidt, Best Practices Guide. International Hydrographic Review, **24**, 13 (2020)
4. M.L. Seto, A. Crawford, *Autonomous shallow water bathymetric measurements for environmental assessment and safe navigation using USVs*, in OCEAN 2015, 19-22 October 2015, Washington DC, USA (2015)
5. K. D. V. Ellenrieder, H. C. Henninger, S. Licht, *Dynamic Modelling and Control of a portable USV for Bathymetric Survey*, in Global Oceans 2020, 5-31 October 2020, Singapore-US. Gulf Coast (2020)
6. Organisation for Economic Co-operation and Development, Frascati Manual 2015 : Guidelines for Collecting and Reporting Data on Research and Experimental Development (OECD Publishing, Paris, 2015)
7. K. Ulgen, M. R. Dhanak, J. Mar. Sci. Eng, **10**(9), 1169 (2022)
8. M. A. Yengejeh, M. M. Amiri, H. Mehdigholi, M. S. Seif, O. Yaakob, J. Engineering for The Maritime Environment, 230 (2), 18 (2015)

9. I. Backalov, S. Rudakovic, J. FME Transactions, **45**, 1 (2017)
10. International Maritime Organization, International code on Intact Stability (IMO Publishing, 2008)
11. D. E. Setiawan, B. Hascaryo, F. Purwangka, V. Rumanti, S. Purbayanto, B. Wibowo, Int. J. of Marine Engineering Innovation and Research, **9** (2), 8 (2024)
12. P. Xu, Q. Cao, Y. Shen, M. Chen, Y. Ding, H. Cheng, J. Mar. Sci. Eng, **10** (12), 1899 (2022)
13. A. Sulisetyono, *The Simple Open Free Running Test for the Evaluation of Turning Ship Ability*, in ISOCEEN 2018, 8-10 August 2018, Surabaya, Indonesia (2018)
14. D. Kim, T. Tezdogan, A. Incecik, Applied Ocean Research, **123**, 103176 (2022)
15. P. Misra, P. Enge, Global Positioning System : Signals, Measurement and Performance (2nd ed)(Ganga-Jamuna Press, 2011)
16. T.F. Ilyas, F. Arkhan, R. Kurniawan, T.H. Budianto, G.B. Putra, *Thingsboard-based prototype design for measuring depth and pH of kulong waters*, in The 3rd ICoGEE 2021, 29-30 September 2021, Bangka, Indonesia (2021)

Shallow water habitat mapping in Bira Besar Island, Jakarta Province, Indonesia

Dimas Ridwan Maulana[1,2], *Fadhal* Abiyu Ridho[1,2], *Vincent* Austin Campbell[1,2], *Alfan* Fadhila Hijra Rabbani[1,2], *Haikal* Zidan Alvarizi[1,2], and *Riza* Aitiando Pasaribu[1*]

[1]Department of Marine Science and Technology, Faculty of Fisheries and Marine Sciences, IPB University, Bogor 16680, Indonesia

[2]Marine Mapping Club, Department of Marine Science and Technology, Faculty of Fisheries and Marine Sciences, IPB University, Bogor 16680, Indonesia

Abstract: Shallow water habitats are crucial for living and protective environments for various organisms. Mapping of these areas is necessary to understand the diversity of shallow water habitats around Bira Besar Island. The objective of this study was to map the shallow water habitat surrounding the Bira Besar Island. Remote sensing technology was applied using object-based image analysis (OBIA) on Sentinel-2A imagery acquired on May 28, 2024. The segmentation processed in OBIA uses a scale of 30 at level 1 and 10 at level 2, with the same value for both levels for compactness of 0.1 and shape of 0.5. The classification results indicated six habitat classes: sand, coral, macroalgae, seagrass, sand, and dead coral algae. Using water column correction with the Depth Invariant Index (DII), an overall accuracy of 82% and a kappa coefficient of 0.765 were achieved. The analysis showed that shallow water habitats around Bira Besar Island were dominated by dead coral algae (28.49 ha) and coral (24.88 ha), followed by sand (13.22 ha), rubble (9.69 Ha), macroalgae (1.43 ha), and seagrass (0.37 ha) as the smallest dominant habitat.

1. Introduction

Benthic habitats are natural habitats composed of various types of organisms, including algae, seagrass, life coral, and dead coral, with substrate types such as mud, sand, and coral fragment or rubble [1]. Shallow waters serve as a habitat for the growth of seagrass meadows and coral reef ecosystems, providing a habitat for a wide variety of highly productive tropical marine species. These ecosystems play a vital role in the ecological and economic aspects of coastal life. Their key functions include serving as fish habitats, facilitating the mixing of suspended materials, protecting shorelines from wave impacts, and contributing to the tourism industry [2]. In remote sensing, shallow waters are areas from the coastal boundary to a certain depth, which refers to the ability of solar energy to penetrate the water column. Satellite sensors can detect various habitats in shallow waters, including substrates, seagrasses, coral reefs, and algae [3]. This is because, for clear shallow waters, optical remote sensing methods are capable of sensing water depths between 15-30 m [4].

*Corresponding author: rizapasaribu@apps.ipb.ac.id

The Object-Based Image Analysis (OBIA) method for classifying shallow water habitats is becoming increasingly recognized in studies of benthic environments. Unlike pixel-based classification, which relies solely on spectral data at the pixel level, OBIA incorporates spatial information from homogeneous objects or image segments. This approach is based on both spectral and texture characteristics. The OBIA process involves several steps, starting with segmentation using specific parameters, followed by a classification phase. OBIA is preferred for its demonstrated high accuracy. For example, research that utilized the Support Vector Machine (SVM) algorithm achieved an accuracy of 75%. This highlights OBIA as a viable alternative for mapping geomorphological zones and benthic habitats [5].

In this study, remote sensing technology was used to monitor shallow water habitats. Research conducted on the coast of Bira Besar Island, Kepulauan Seribu, Jakarta employed Sentinel-2A image data and the OBIA method to map shallow water habitats. The waters surrounding Bira Besar Island, Kepulauan Seribu Regency, Jakarta Province, are home to seagrass meadows and coral reef ecosystems that support a diverse range of marine life in shallow waters [6]. Mapping the distribution of benthic habitats is crucial for the development and rehabilitation of Bira Besar Island, as well as for evaluating habitat availability for biodiversity on the island. Therefore, this research objective was to map the shallow water habitats surrounding Bira Besar Island.

2. Methodology

2.1. Location

Bira Besar Island is located in the northern part of the Jakarta Province and is surrounded by the Java Sea. This Island is part of the Kepulauan Seribu administrative regency in Jakarta Province. Research was conducted in 2024, with field observations taking place from May 24 to May 26. The data were collected during these field observations of the coastal waters surrounding Bira Besar Island.(Figure 1).

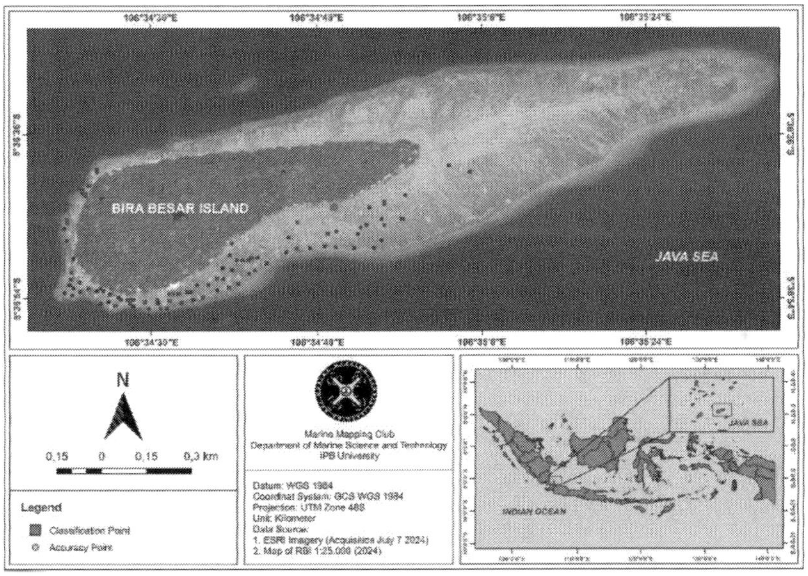

Fig. 1. Location of Field Survey at Bira Besar Island, Kepulauan Seribu, Jakarta Province

2.2. Data source

Field data collection included direct observation or ground truth habitat using 10 × 10 m transects and dominant objects recorded within the transect. Field observation sampling points were determined using a stratified random sampling method by taking random samples from each stratum of the heterogeneous population. In this study, stratified random sampling was performed by visually selecting transect samples in areas that showed varied spectral class variations based on the results of unsupervised classification to observe the distribution of benthic habitats [7]. The number of field samples collected was 163 throughout Bira Besar Island, which were divided into two groups, namely for classification and accuracy testing (see Figure 1). The purpose of this separation was to avoid imperfect accuracy in the accuracy test. The accuracy test data must be independent and not previously used. The split was performed randomly but considering the distribution and diversity of objects with almost the same number. Observation points were randomly assigned to the images before the field survey using the SVM algorithm classification method, followed by an accuracy test utilizing the confusion matrix. The data used were Sentinel-2A images acquired on May 28, 2024, and field survey data taken on May 25, 2024. The Sentinel-2A imagery had a spatial resolution of 10 meters, which allows it to effectively distinguish shallow-water habitats [8].

2.3. Data processing

The steps involved in image data processing include preprocessing images, such as image composite, geometric correction, atmospheric correction, and image transformation using the Depth Invariant Index (DII) method. Subsequently, the OBIA classification process proceeds, which includes image segmentation and the classification of all segmented results. An accuracy test was conducted using the confusion matrix method to determine the accuracy of the image classification compared to the field survey data.

2.3.1. Image preprocessing

The first step in improving image quality is to remove noise through image preprocessing [9]. The image preprocessing stage begins with image compositing and cropping, followed by image geometry correction. Geometric correction is the process of adjusting an image to eliminate distortion, rotation, and perspective differences. The primary goal of this process is to ensure that the orientation, projection, and additional information in the image align with existing map data. Geometric correction was performed to obtain a geographically accurate image [10]. The georeferencing method involves aligning the coordinate system of the satellite image with ground control points (GCP) on the surface. This is done by selecting coordinate points from the Indonesia Terrestrial Map (Peta Rupa Bumi Indonesia known as RBI) scale of 1:25.000 published by Indonesia Geospatial Agency (Badan Informasi Geospasial known as BIG) and matching them with the satellite images [11].

The accuracy level of the georeferencing process was measured based on the difference in coordinates between the transformed control points and the original control points. This difference is expressed in the form of root-mean-square errors (RMSE). Accuracy is determined by the Root Mean Square Error (RMSE). When two-dimensional mapping is being used, the coordinates of the test point (X, Y) and the locations of the points in the field must be considered [12]. The smaller the RMSE value, the higher the accuracy level of the georeferencing results. The RMSE in the geometric image correction measures the error by comparing the coordinates of the transformed GCP with their original coordinates. The precision of the transformation depends on the RMSE value; the smaller the RMSE value, the more accurate the transformation. The RMSE for each GCP was calculated using Equation [13].

$$RMS = \sqrt{(X_a - X_b)^2 + (Y_a - Y_b)^2 + (Z_a - Z_b)^2} \quad (1)$$

$$RMSE = \sqrt{\frac{\Sigma(X_t - \bar{X})^2}{n}} \quad (2)$$

Where,
X_a, Y_a, Z_a = Coordinate control-point measurement
X_b, Y_b, Z_b = Transformed coordinates.
X_t = RMS per point
$\bar{X}$ = Average RMS
n = total control-point data.

Atmospheric correction was then performed to convert the digital number values into reflectance values. This correction helps clarify the appearance of objects in the image by addressing inappropriate reflectance values caused by atmospheric influences, such as absorption and scattering [14]. The Dark Object Subtraction (DOS) method was used to improve the radiometric or pixel values in the image [15]. This method creates a corrected image with new values [16] using the following formula:

$$R_c = R_s - R_{si} \quad (3)$$

Where,
R_c = Object reflectance value was corrected
R_s = Object reflectance before correction
R_{si} = *(Mean R_w – (2*Standard Deviation R_w)*
R_w = spectral value, which is considered an offset.

2.3.2. Water column correction

Water column correction has been used to enhance image quality by reducing water column disturbances, which helps objects show more clearly and facilitates object detection [17]. The depth-invariant index (DII) with the Lyzenga algorithm 1981 [18] was used to adjust for the influence of the water column.

$$DII = \ln(Li) - \left[\left(\frac{ki}{kj}\right) \times \ln(Lj)\right] \quad (4)$$

With,

$$\frac{ki}{kj} = a + \sqrt{a^2 + 1} \quad (5)$$

$$a = \frac{(\sigma_{ii} - \sigma_{jj})}{(2\sigma_{jj})} \quad (6)$$

Where Li reflectance value of the blue band, Lj reflectance of the green band, ki/kj is the coefficient of attenuation, σ_{ii} variance band i, σ_{jj} variance band j, and σ_{ij} is the covariance pairs band i and j.

2.3.3. Object-based image classification

The process of creating objects or segments from pixels within an existing object is known as segmentation [19]. Benthic habitat mapping using the OBIA method is carried out by segmenting images (pixel level) into homogeneous segments/objects (object level) according to their parameters [20]. At this stage, a multiresolution segmentation algorithm is used to identify objects based on relative homogeneity, which combines the shape criteria and spectral values [21]. This algorithm is influenced by scale, shape, and compactness. An MRS scale value of 30 was used for level 1 classification, whereas an MRS scale value of 10 was used for level 2 classification. The compactness was set at 0,1 by default, and the shape value was 0,5. Because the rule-set parameter setting is devoid of some clauses, it is optimized for the repeat test procedure and yields optimal outcomes [22].

Classification of Sentinel-2A images was performed using the contextual editing method at level 1 and the Support Vector Machine (SVM) algorithm at level 2. The observation of objects at the research site refers to the principle of closing the dominant shallow water habitat during direct visual analysis [23]. There are two levels in the classification scheme (Figure 2): the reef level (classification level 1) and shallow water habitat class (classification level 2). The classification scheme depicted in Figure 2 was developed to systematically categorize the various types of shallow water habitats using data from field surveys and 10-meter quadratic transects taken at each research site observation point.

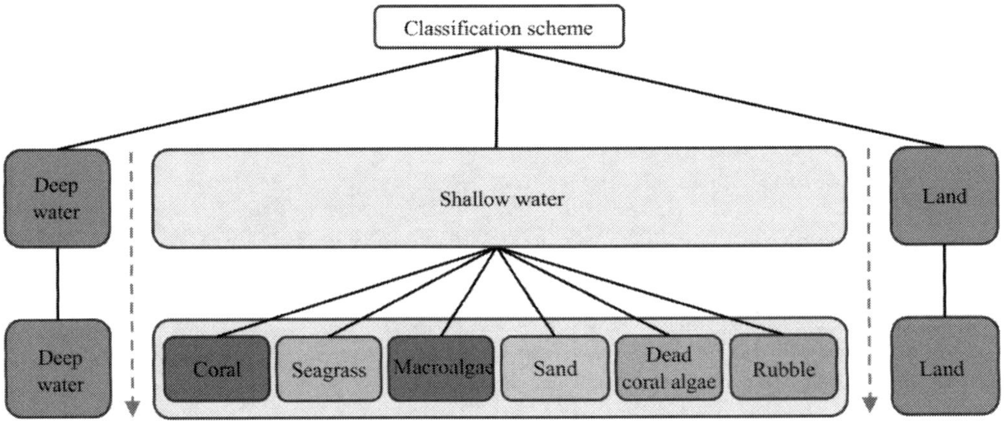

Fig. 2. Classification scheme

In level 1 classification, we used the assigned class algorithm to determine the threshold value for producing objects that belong to the target class. Trials were conducted to determine the ideal value before the threshold value was determined. Regarding level 2, we used the Support Vector Machine (SVM) algorithm as a classifier. The formula of the SVM algorithm is expressed as follows [24]:

$$f(x) = \Sigma_{i \epsilon S} \lambda_i y_i K(x_i x) + w_0 \quad (7)$$

Description: K is the kernel function, y_i and x_i represent the training samples, λ_i is the Lagrange multiplier, S is a part of the training samples corresponding to non-zero Lagrange multipliers, and w_0 is the hyperplane parameter [25].

Accuracy test. An accuracy test was applied to determine the level of conformity of the satellite image classification results with actual data obtained in the field or references [4]. The accuracy test was carried out using a confusion matrix that consisted of the overall accuracy (OA), producer accuracy (PA), user accuracy (UA), and Kappa coefficient [26]. The mathematical accuracy is expressed as follows:

$$Producer\ accuracy = \frac{n_{jj}}{n_{+j}} \times 100\% \quad (8)$$

$$User's\ accuracy = \frac{n_{ii}}{n_{i+}} \times 100\% \quad (9)$$

$$Overall\ accuracy = \frac{\sum_{i=1}^{k} n_{ii}}{n} \times 100\% \quad (10)$$

The kappa coefficient measures the agreement between two data points to be tested for accuracy. Kappa values were calculated from the error matrices of the classification results using Sentinel-2A imagery and the DII transformation. If the two Kappa values were independent and for two significantly different error matrices, the Kappa test was used to statistically compare each classification result [4]. The accuracy of the classification outcomes from the processed image data and field circumstances is shown by this value, which indicates whether the condition is strong or not. The kappa coefficient was calculated using the following formula [27]:

$$K = \frac{\sum_{i=1}^{k} n_{ii} - \sum_{i=1}^{k} n_i + n_{i+i}}{n^2 - \sum_{i=1}^{k} n_i + n_{i+i}} \quad (11)$$

Where,
K = kappa coefficient
k = number of rows in the matrix
n = number of observations
n_{ii} = number of observations in columns i and i
n_{jj} = number of observations in columns j and j
n_{i+} = total marginal row i
n_{+i} = total marginal column i
n_{+j} = total column.

3. Result and discussion

3.1. Image correction

3.1.1. Geometric correction

This research produced a digital map with Sentinel-2A satellite imagery that has been used for georeferencing processing using ArcGIS 10.8 software. The georeferencing processing was carried out with 35 points GCP and a total RMS Error of 0.0770844 m was obtained, as shown in Figure 3.

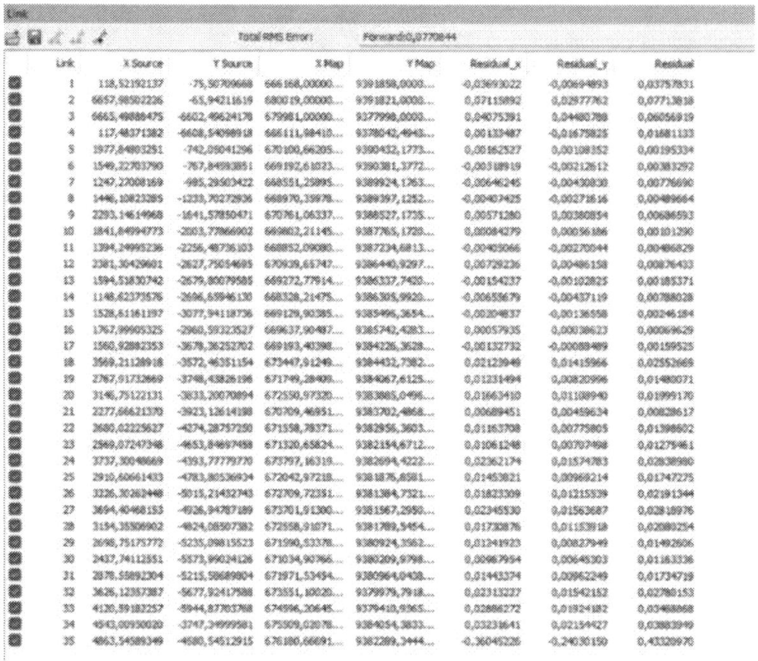
Link

Total RMS Error: Forward:0,0770844

Link	X Source	Y Source	X Map	Y Map	Residual_x	Residual_y	Residual
1	118,52192137	-75,50709668	666168,00000...	9391858,0000...	-0,03693022	-0,00694893	0,03757831
2	6657,98502226	-65,94211619	680019,00000...	9391821,0000...	0,07115892	0,02977762	0,07713818
3	6665,49888475	-6602,49624178	679981,00000...	9377998,0000...	0,04075391	0,04480788	0,06056919
4	117,48371382	-6608,54098918	666111,98410...	9378042,4943...	0,00133487	-0,01675825	0,01681133
5	1977,84803251	-742,05041296	670100,66205...	9390432,1773...	0,00162527	0,00108352	0,00195334
6	1549,22703790	-767,84593851	669192,61023...	9390381,3772...	-0,00318919	-0,00212612	0,00383292
7	1247,27008169	-985,29503422	668551,25895...	9389924,1263...	-0,00646245	-0,00430830	0,00776690
8	1446,10823285	-1233,70272936	668970,35978...	9389397,1252...	-0,00407425	-0,00271616	0,00489664
9	2293,14614968	-1641,57850471	670761,06337...	9388527,1735...	0,00571280	0,00380854	0,00686593
10	1841,84994773	-2003,77866902	669802,21145...	9387765,1720...	0,00084279	0,00056186	0,00101290
11	1394,24995236	-2256,48736103	668852,09080...	9387234,6813...	-0,00405066	-0,00270044	0,00486829
12	2381,30429601	-2627,75054695	670939,65747...	9386440,9297...	0,00729236	0,00486158	0,00876433
13	1594,51830742	-2679,80079585	669272,77914...	9386337,7420...	-0,00154237	-0,00102825	0,00185371
14	1148,62373576	-2696,65946130	668328,21475...	9386305,9920...	-0,00655679	-0,00437119	0,00788028
15	1528,61161197	-3077,94118736	669129,90385...	9385496,3654...	-0,00204837	-0,00136558	0,00246184
16	1767,99905325	-2960,59323527	669637,90487...	9385742,4283...	0,00057935	0,00038623	0,00069629
17	1560,92882353	-3678,36252702	669193,40398...	9384226,3628...	-0,00132732	-0,00088489	0,00159525
18	3569,21128918	-3572,46351154	673447,91249...	9384432,7382...	0,02123948	0,01415966	0,02552669
19	2767,91732869	-3748,43826196	671749,28409...	9384067,6125...	0,01231494	0,00820996	0,01480071
20	3146,75122131	-3833,20070894	672550,97320...	9383885,0496...	0,01663410	0,01108940	0,01999170
21	2277,66621370	-3923,12614198	670709,46951...	9383702,4868...	0,00689451	0,00459634	0,00828617
22	3680,02225627	-4274,28757250	671558,78371...	9382956,3603...	0,01163708	0,00775805	0,01398602
23	2569,07247348	-4653,84697498	671120,65824...	9382154,6712...	0,01061248	0,00707498	0,01275461
24	3737,30048669	-4393,77779770	673797,16319...	9382694,4222...	0,02362174	0,01574783	0,02838980
25	2910,60661433	-4783,80536934	672042,97218...	9381876,8581...	0,01453821	0,00969214	0,01747275
26	3226,30262448	-5015,21432743	672709,72351...	9381384,7521...	0,01823309	0,01215539	0,02191344
27	3694,40468153	-4926,94787189	673701,91300...	9381567,2950...	0,02345530	0,01563687	0,02818976
28	3154,35506902	-4824,08507382	672558,91071...	9381789,5454...	0,01730876	0,01153918	0,02080254
29	2698,75175772	-5235,09815523	671590,53378...	9380924,3562...	0,01241923	0,00827949	0,01492606
30	2437,74112551	-5573,99024126	671034,90766...	9380209,9798...	0,00967954	0,00645303	0,01163336
31	2878,55892304	-5215,58689804	671971,53494...	9380964,0408...	0,01443374	0,00962249	0,01734719
32	3626,12357387	-5677,92417588	673551,10020...	9379979,7918...	0,02313227	0,01542152	0,02780153
33	4120,39182257	-5944,87703768	674596,20645...	9379410,9365...	0,02886272	0,01924182	0,03468868
34	4543,00930020	-3747,34999581	675509,02078...	9384054,3833...	0,03231641	0,02154427	0,03883949
35	4863,54589349	-4580,54512915	676180,66691...	9382289,3444...	-0,36045226	-0,24030150	0,43320970

Fig. 3. RMSE Results

ArcGIS software can determine that the RMSE values obtained from the affine polynomial equation are quite good for 2D transformations, which have an influence on translation, rotation, and scale enlargement. The RMSE value obtained was 0.0770844 m, which was good for the georeferencing process. A low RMSE value indicates that the georeferencing results are accurate [10].

3.1.2. Atmospheric correction

After being corrected with atmospheric correction, Sentinel-2A satellite images showed a clear difference with sharper visuals and improved contrast. This indicates that the atmospheric correction successfully enhanced the visual quality, as depicted in Figure 4. Figure 5 illustrates a histogram showing that after correction, the pixel values approached zero with a stable frequency, indicating a decrease in the reflectance values.

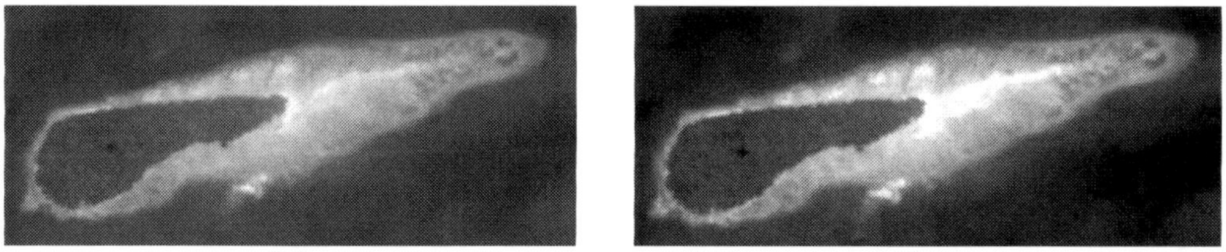

(a) (b)

Fig. 4. The visual image results show (a) acquisition before atmospheric correction and (b) acquisition after the atmospheric correction process.

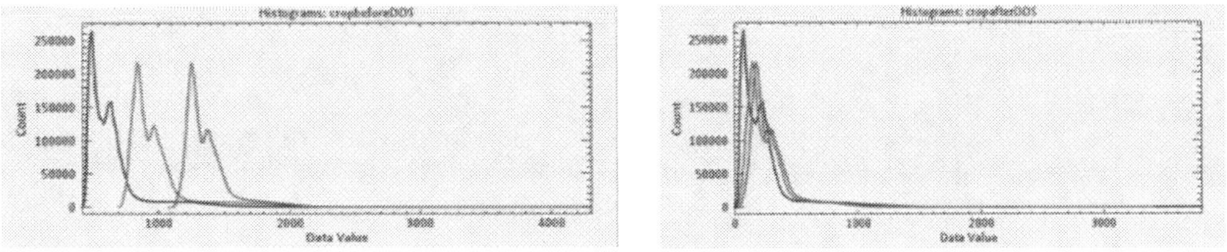

(a) (b)

Fig. 5. The histogram of the image presents (a) data captured prior to atmospheric correction and (b) data obtained following the atmospheric correction process.

3.1.3. Water column correction

The process of water column correction modifies the image by utilizing a blend of blue and green channels. This approach effectively enhances the penetration of visible light into the water column, with its performance largely dependent on water conditions, particularly the attenuation coefficient. Information about the seabed can be derived from the attenuation ratio between channels i and j (ki/kj). This coefficient represents the decline in light intensity as depth increases, meaning that a greater attenuation coefficient indicates higher water clarity.

The computation of attenuation ratio (ki/kj) values is derived from training areas across band pairs, as detailed in Table 2. These values serve as input parameters for level 2 object-based image classification utilizing the SVM algorithm. When ki or kj values are lower, the detection of shallow water habitats becomes more effective. Conversely, higher ki and kj values limit energy penetration to the seabed, reducing reflectance from underwater features and habitats. The results applied to the images before and after the water column correction are shown in Figure 6.

Table 2. Attenuation ratio of Sentinel-2A image

Band	Attenuation ratio (Ki/Kj)
Blue - Green	0.67803
Blue - Red	0.35518
Green - Red	0.52523

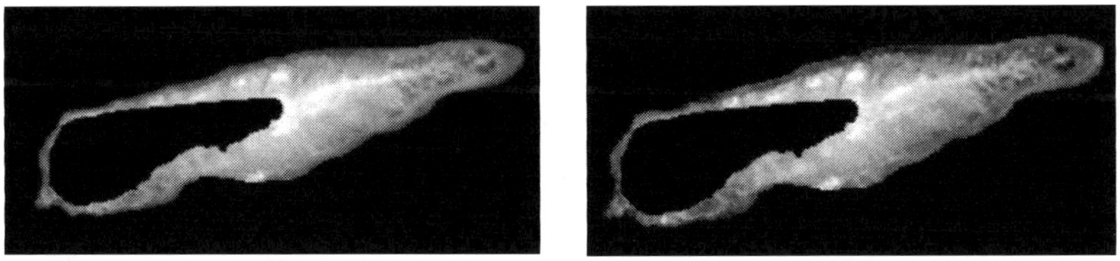

(a) (b)

Fig. 6. Image water column correction, (a) acquisition before water column correction and (b) acquisition after water column correction process.

3.2. Segmentation and Classification Level

After performing Image Correction on the images, the next initial step in OBIA was the segmentation process. To determine the optimal accuracy value for generating image data according to its resolution, trials and errors were performed to determine the segmentation scale value [14]. The segmentation scale values applied at the first and second levels were 30 and 10, respectively. These values were deemed suitable as they align with the classified objects that have been generated.

The parameters used included compactness, which was set to a default value of 0.1, and a shape value of 0.5. The purpose of segmentation at the first level was to group pixels into unified segments, whereas classification and subsequent accuracy assessments were conducted at the second segmentation level. There are no standard provisions for standard parameter values in object-based classification [1].

In the first level of classification (Reef level), segmentation was carried out to divide the area into three distinct categories: Land, Shallow Water, and Deep Water (refer to Figure 7). Meanwhile, the second level of classification (Shallow Water Habitat Class) was derived from the segmentation results of the first level and served as the boundary for the study area focusing on shallow water habitats. At this stage, the classification process was conducted using a supervised method with the application of an SVM algorithm. The DII method was applied for water column correction to enhance image quality by minimizing interference from the water column. Factors like water depth, which influence the reflectance values of seabed habitats or objects and exhibit an exponential relationship with depth, were effectively reduced through this process. Thematic layers were created using classification scheme data based on the ROI data from direct field observations. For the level 2 classification process, the input features consisted of layer values (mean and standard deviation) from all visible light bands, along with the composition of DII band pairs. The classification results depicting habitat distribution in shallow water before applying water column correction can be seen in Figure 8a, while Figure 8b presents the results after the correction was implemented.

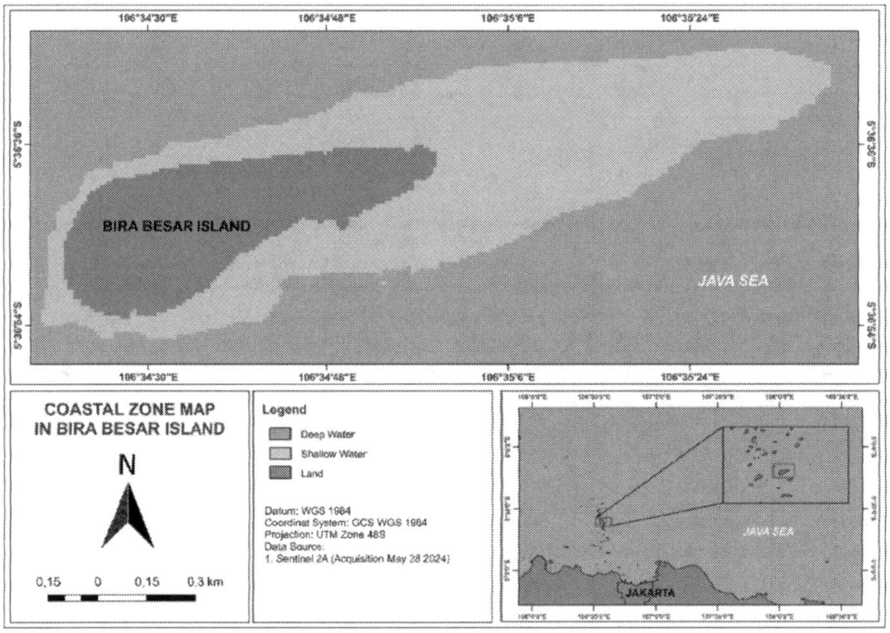

Fig. 7. Coastal zone or water zone distribution map based on Sentinel-2A level 1 segmentation.

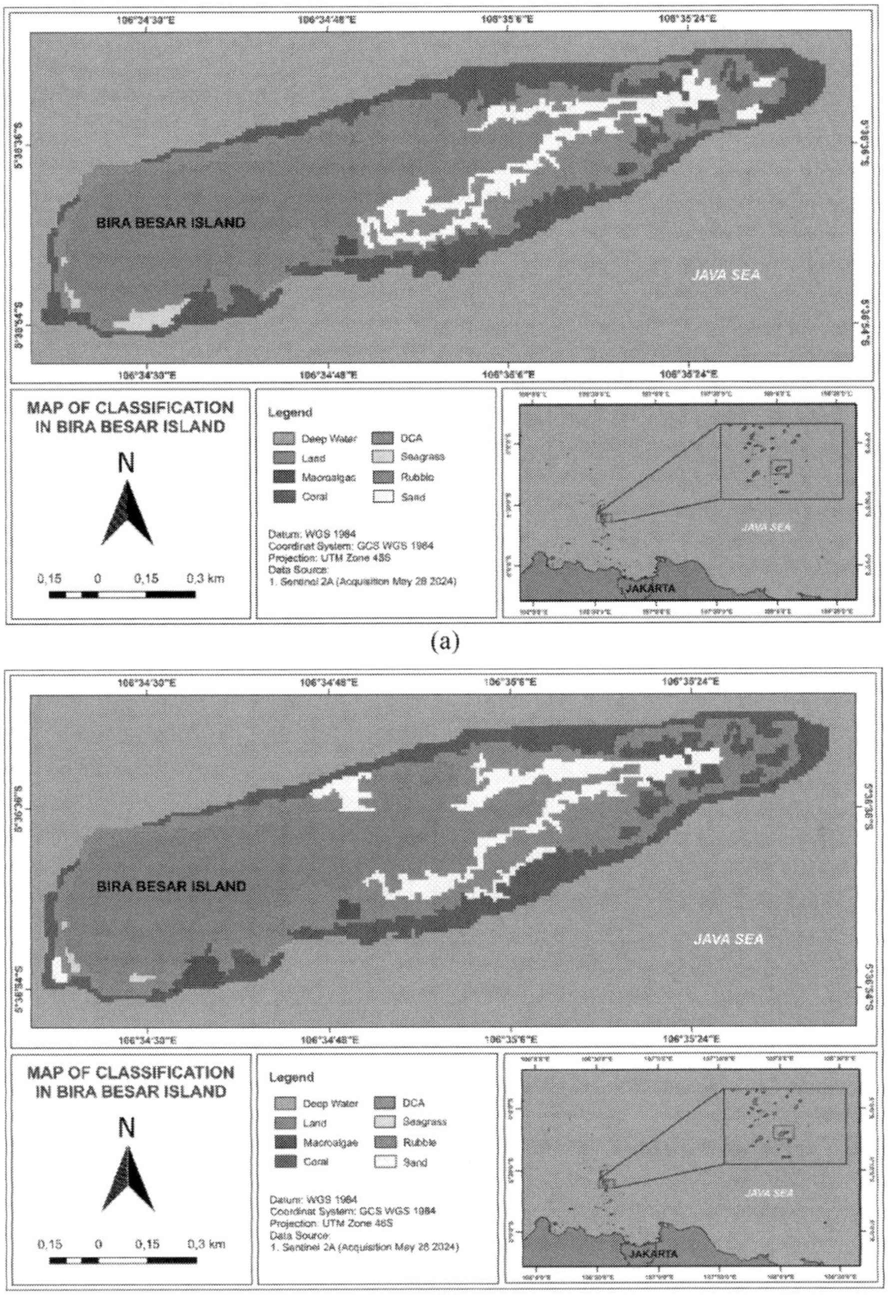

(a)

(b)

Fig. 8. Classification map in Bira Besar Island (a) before water column correction (b) after water column correction.

The shallow water habitats around Bira Besar Island exhibit significant heterogeneity and are evenly distributed across the waters. In the classification without applying water column correction (DII), the most dominant category was DCA, covering 27.7 hectares, followed by the coral category at 26.91 hectares. Sand occupied 12.46 hectares, Rubble covered 8.97 hectares, Seagrass extended across 1.26 hectares, and Macroalgae accounted for 0.36 hectares. After applying water column correction (DII), DCA remained the dominant category, expanding to 28.49 hectares, while the coral category decreased to 24.88 hectares. Sand increased to 13.22 hectares, Rubble to 9.69 hectares, Macroalgae to 1.43 hectares, and Seagrass to 0.37 hectares. Differences in area were particularly noticeable in the Macroalgae and Seagrass categories when comparing results before and after water column correction. A detailed breakdown of the total shallow water habitat area on Bira Besar Island is presented in Table 3.

Table 3. Area of shallow water habitat class on Bira Besar Island.

Class	Without DII		With DII	
	Area (Ha)	Percentage (%)	Area (Ha)	Percentage (%)
Sand	12.46	16.04	13.22	16.93
Coral	26.91	34.65	24.88	31.86
Macroalgae	0.36	0.46	1.43	1.83
Seagrass	1.26	1.62	0.37	0.47
Rubble	8.97	11.55	9.69	12.41
DCA	27.7	35.67	28.49	36.49

3.3. Accuracy test

The last step in the classification process using the OBIA method involves an accuracy assessment to ensure the maps produced are both reliable and practical. This evaluation was carried out using 40% of the total field data points. A comparison of two different accuracy assessment approaches was conducted to determine the most effective method for applying OBIA. The results of the image accuracy assessment, both before and after applying water column correction, are presented in Tables 4 and 5. The overall accuracy (OA) without DII was 65.6%. This value indicates suboptimal results in achieving a good mapping accuracy. However, after correcting the water column using the DII, the Overall Accuracy increased to 82%. This result demonstrates good performance, aligning with a previous research statement that OA > 60% is ideal for mapping shallow-water habitats [28].

The accuracy assessment, evaluated through the confusion matrix, resulted in a kappa coefficient of 0.589 for images without DII correction and 0.765 for images that underwent DII correction. The value for images without DII indicates moderate performance, whereas images corrected using DII show that the algorithm's performance falls within the good category [29]. This accuracy test proved that, to classify shallow water habitat mapping using the OBIA method, it would be better to perform water column correction before performing the segmentation. The accuracy test results of the mapping show that the overall accuracy of the DII transformation classification is improved. Thus, it is possible to conclude that the use of water column correction by the DII method is capable of improving the accuracy of the images in mapping shallow-water benthic habitats [30].

The limitation of this research is that there is no fixed calculation or standard test method for determining the value of scale, compactness, and shape for level 1 and 2 segmentation in the OBIA method. However, this limitation does not reduce the quality of data obtained for mapping shallow-water benthic habitats in the Bira Besar Island waters. A good results performed proved by accuracy test using OBIA method to map shallow water benthic habitat in Bira Besar Island waters showed this method potential to be applied on other area especially in the small island areas.

Table 4. Sentinel-2A image classification (without DII) accuracy test results

		Field Points						
		DCA	C	SG	MA	S	R	Total
Image Classification	DCA	15	0	1	3	1	1	21
	C	1	5	0	3	0	0	9
	SG	0	0	3	0	1	0	4
	MA	0	0	0	3	0	0	3
	S	6	0	0	1	12	0	19
	R	0	0	0	0	3	2	5
Total		22	5	4	10	17	3	61
	Producers Accuracy (%)	68.2	100.0	75.0	30.0	70.6	66.7	
	User's Accuracy (%)	71.4	55.6	75.0	100.0	63.2	40.0	
	Overall Accuracy (%)	65.6						
	Kappa coefficient	0.589						

Table 5. Sentinel-2A image classification (with DII) accuracy test results

		Field Points						
		DCA	C	SG	MA	S	R	Total
Image Classification	DCA	21	0	1	3	3	0	28
	C	1	5	0	2	0	0	8
	SG	0	0	1	0	0	0	1
	MA	0	0	0	5	0	0	5
	S	0	0	0	0	16	0	16
	R	1	0	0	0	0	2	3
Total		23	5	2	10	19	2	61
	Producers Accuracy (%)	91.3	100.0	50.0	50.0	84.2	100.0	
	User's Accuracy (%)	75.0	62.5	100.0	100.0	100.0	66.7	
	Overall Accuracy (%)	82.0						
	Kappa coefficient	0.765						

4. Conclusion

Using the OBIA and DII algorithms to map shallow water habitats on Bira Besar Island showed good results. The implementation of the DII algorithm must be performed before the OBIA is applied in the creation of a shallow water habitat map. The accuracy results showed a Kappa Coefficient value of 82% and 0.765 when the DII algorithm was used. The shallow water habitats in the waters surrounding Bira Besar Island were categorized into six distinct types: sand, coral, macroalgae, seagrass, rubble, and dead coral algae. The analysis indicates that shallow water habitats around Bira Besar Island mainly comprise dead coral algae (28.49

ha), coral (24.88 ha), sand (13.22 ha), rubble (9.69 ha), macroalgae (1.43 ha), and seagrass (0.37 ha) as the smallest dominant habitat areas.

Our sincere appreciation goes to the Department of Marine Science and Technology, Faculty of Fisheries and Marine Sciences, IPB University, for providing the necessary facilities and infrastructure that contributed to the success of this study. Our appreciation also goes to the Marine Science and Technology Student Association (HIMITEKA) for their support during the field survey, at the same time as the Marine Mapping Club II Expedition. Finally, we thank the Nusantara Geosains Institut (NGI) for their significant contribution and financial support throughout the research project. Without the assistance and support of these organizations, this research would not have proceeded smoothly.

References

[1] La Ode, K. Mastu, Bisman Nababan, James P. Panjaitan, *Jurnal Ilmu dan Teknologi Kelautan Tropis (JITKT)*, 10, 381-396 (2018).

[2] I Dewa Made, K. P. Astaman, I Wayan Gede A. Karang, I Gede Hendrawan, Kuncoro T. Setiawan, *Journal of Marine and Aquatic Sciences*, 7, 184-195 (2021).

[3] Vincentius, P. Siregar, Syamsul B. Agus, Adriani Sunuddin, Tarlan Subarno, Nunung N. Aziizah, *Jurnal Ilmu dan Teknologi Kelautan Tropis (JITKT)*, 12, 37-51 (2020).

[4] Peter, J. Mumby, William Skirving, Alan E. Strong, John T. Hardy, Ellsworth F. LeDrew, Eric J. Hochberg, Rick P. Stumpf, Laura T. David, *Marine Pollution Bulletin*, 48, 219-228 (2004).

[5] Ratna, Komala, "Keanekaragaman Teripang pada Ekosistem Lamun dan Terumbu Karang di Pulau Bira Besar, Kepulauan Seribu, Jakarta" in Prosiding Seminar Nasional Masyarakat Biodiversitas Indonesia, 4 Mei 2024, Surakarta, Indonesia (2015).

[6] Ari, Anggoro, Vincentius P. Siregar, Syamsul B. Agus, *Jurnal Penginderaan Jauh (JPJ)*, 12, 1-12 (2015).

[7] Favian, M.G. Putra, Nurul Khakhim, *Jurnal Bumi Indonesia (JBI)*, 3, 2-12 (2013).

[8] John, D. Hedley, Chris Roelfsema, Vittorio Brando, Claudia Giardino, Tiit Kutser, Stuart Phinn, Peter J. Mumby, Omar Barrilero, Jean Laporte, Benjamin Koetz, *Remote Sensing of Environment (RSE)*, 216, 598-614 (2018).

[9] Dani, Rohpandi, Asep Sugiharto, Giri A. Winara, "Aplikasi Pengolahan Citra dalam Pengenalan Pola Huruf Ngalagena Menggunakan MATLAB", in Prosiding Konferensi Nasional Sistem dan Informatika, 9-10 Oktober 2016, Bali, Indonesia (2015).

[10] Maulana prabandaru. *JIG (Jurnal Ilmiah Geomatika)*, 2(1): 73-84 (2022).

[11] Monica Nilasari, Bandi Sasmito, Adi Sukmono, *Jurnal Geodesi Undip (JGU)*, 6, 97-105 (2017).

[12] Pangeran S. Panjaitan, Jance M. Supit, *INTAN Jurnal penelitian Tambang (IJPT)*, 4, 121-125 (2021).

[13] Hayati, N., Institut Teknologi Sepuluh Nopember, Press (2012).

[14] El Syifa, Putri, Anita Widiasari, Rizal A. Karim, Lili Somantri, Riki Ridwana, *Jurnal Pendidikan Geografi Undiksha (JPGU)*, 9, 133-143 (2021).

[15] Lilik, Kristianingsih, Arwan P. Wijaya, Abdi Sukmono, *Jurnal Geodesi Undip (JGU)*, 5, 56-64 (2016).

[16] R.P. Dewa, P. Danoedoro, "The Effect of Image Radiometric Correction on The Accuracy of Vegetation Canopy Density Estimate Using Several Landsat-8 OLI's Vegetation Indices: A Case Study of Wonosari Area, Indonesia", in Prosiding IOP

Conference Series: Earth and Environmental Science, 25-26 Oktober 2016, Bogor, Indonesia (2017).

[17] Aldin, Faisal, Yudo Prasteyo, Muhammad Helmi, *Jurnal Geomatika Undip (JGU)*, 9, 77-86 (2020).

[18] David, R. Lyzenga, *Applied Optics*, 17, 379-383 (1978).

[19] Kumar, Navulur, Edisi Pertama, CRC Press (2006).

[20] Citra, A. Sari, Achmad F. Syah, *Rekayasa*, 14, 114-120 (2021).

[21] Ari, Setiani, Yudo Prasetyo, Sawitri Subiyanto, *Jurnal Geodesi Undip (JGU)*, 5, 112-121 (2016).

[22] Robin, Saputra, Jonson L. Gaol, Syamsul B. Agus, *Jurnal Ilmu dan Teknologi Kelautan Tropis (JITKT)*, 13, 39-55 (2021).

[23] Vincentius, Siregar, *Jurnal Ilmu dan Teknologi Kelautan Tropis (JITKT)*, 2, 19-30 (2010).

[24] Tzotsos A., Argialas D., "Support Vector Machine Classification for Object-Based Image Analysis", in chapter of Object-Based Image Analysis, Lecture Notes in Geoinformation and Cartography, Springer, Berlin, Heidelberg (2008).

[25] Giorgos, Mountrakis, Jungho Im, Caesar Ogole, *ISPRS Journal of Photogrammetry*, 66, 247–259 (2011).

[26] Congalton, Russel G., Kass Green, CRC Press, Taylor & Francis Group (2019).

[27] Congalton, Russel G., Kass Green, CRC Press (2008).

[28] Edmund, P. Green, Peter J. Mumby, Alasdair J. Edwards, Christopher D. Clark, UNESCO, Paris, France (2000).

[29] J. Richard Landis, Gary G. Koch, *Biometrics*, 33, 159-174 (1977).

[30] I Dewa Made Krisna Putra Astaman, I Wayan Gede Astawa Karang, I Gede Hendrawan, Kuncoro Teguh Setiawann, *Journal of Marine and Aquatic Sciences (JMAS)*, 7, 184-195 (2021)

Spatial distribution of tidal flooding in the coastal areas of Demak Regency, Central Java

Bayu Adnan[1], *Risti* Endriani Arhatin[1*], *Jonson* Lumban-Gaol[1] and *Setyo* Budi Susilo[1]

[1]Department of Marine Science and Technology, IPB University, Bogor 16680, Indonesia

Abstract: The melting of polar ice caps due to climate change has resulted in sea level rise that threatens coastal areas, such as the Demak Regency. The impact of tidal flooding is, exacerbated by land subsidence. This study aimed to map and predict the distribution of tidal flooding in coastal Demak and its impact on land use. The methods used included Differential Interferometry Synthetic Aperture Radar (DInSAR) for Land Subsidence (LS), supervised classification for land use, admiralty for tides, data visualization for elevation, linear regression analysis for SLR, meteorological data analysis for wind and rainfall, and spatial analysis for tidal flood inundation mapping. The results showed an LS rate of 2,74 cm/year, highest tide height of 61,93 cm, average elevation of 1,893 m above sea level, SLR rate of 0,435 cm/year, wind speed of 0,5-5,7 m/s blowing from the northwest, and average daily rainfall of 7,476 mm. The tidal flood inundation area in 2023 reached 8.954,6 hectares and is predicted to reach 11.394,5 hectares in 2031. Tidal flooding affects mangroves (523,6 hectares), ponds (3.857,9 hectares), settlements (1.095,4 hectares), industries (3.453,00 hectares), and rice fields (24,7 hectares).

1 Introduction

Global warming is a global challenge with widespread impacts, including shifting climate patterns and rising global temperatures that contribute to sea-level rise (SLR). If this trend continues, significant land areas may be inundated[1]. Changes in ocean currents and density further drive SLR, affecting coastal regions globally owing to increased sea temperatures[2]. The current global SLR rate is 3.25 mm/year, with projections indicating acceleration this century, although unevenly distributed[3][4][5]. SLR has intensified coastal disasters, making flooding the costliest issue in these areas[5][6].

Coastal regions, which account for 44% of the global population within 150 km of shorelines, are crucial to the world economy, but face challenges such as tidal flooding[7]. In Indonesia, tidal flooding results from high tides in low-lying coastal areas and is exacerbated by factors such as land subsidence, wind, waves, and rainfall[8][9]. One of the areas that was significantly affected was the Demak Regency.

The Demak Regency in Central Java, neighboring Semarang City, faces worsening tidal flooding due to land subsidence caused by groundwater exploitation and urban development, along with coastal erosion and SLR[10][11]. Although studies have examined tidal flood

* Corresponding author: risti@apps.ipb.ac.id

distribution in Demak, the impact of land subsidence on flooding and land use remains underexplored[12][13]. Spatial analysis is essential for mapping inundation, assessing land-use impact, and informing mitigation planning for tidal flood disasters in this region. This study aims to map the distribution and extent of tidal flood inundation in 2023, predict its occurrence in 2031, and analyze its impact on land use in the coastal area of Demak Regency, Central Java, using spatial analysis.

1.1 Study site

This study was conducted on the coast of Demak Regency, Central Java, focusing on the four affected sub-districts. These sub-districts are Sayung, Karangtengah, Bonang, and Wedung (Figure 1). These four sub-districts were chosen because they are directly adjacent to the sea and often experience tidal floods.

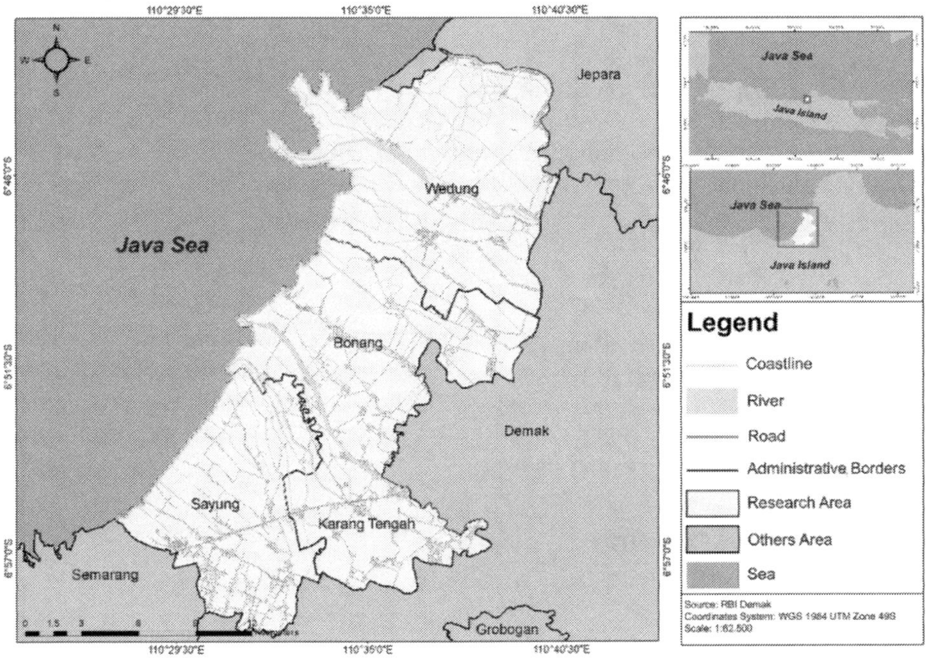

Fig. 1. Study site

2 Materials and Methods

The equipment used in this study included an Asus Zenbook 14 laptop with an AMD Ryzen 5 5500u processor, 8GB RAM, and an Nvidia MX450 graphics card equipped with software such as ArcMap 10.8, QGIS 3.23.3, ENVI 5.3, SNAP, JupyterLab, Microsoft Excel, and Microsoft Word. The materials used in this study included Sentinel-2A Level 1C imagery (dataspace.copernicus.eu), Sentinel-1A Single Look Complex (SLC) data (asf.alaska.edu), daily Sea Level Anomaly (SLA) data (cds.climate.copernicus.eu), duration and height of tidal inundation data (field data), hourly tidal data (srgi.big.go.id), National Digital Elevation Model (DEMNAS) data (tanahair.indonesia.go.id), wind data (climate.copernicus.eu), rainfall data (bmkg.go.id), and the Indonesian Rupabumi Map (RBI) of Demak Regency (tanahair.indonesia.go.id).

2.1 Satellite Image Data Processing

This study employs a multidisciplinary approach using geospatial and statistical methods to analyze land subsidence, sea level rise, tidal flooding, and their impact on coastal regions. Sentinel-1A SAR data (2018 and 2023) were processed using Differential Interferometry Synthetic Aperture Radar (DInSAR) in SNAP software with SNAPHU extensions[14][15]. Key steps included the coregistration of master (2018) and slave (2023) images, coherence estimation[16], topographic phase removal, and phase unwrapping. The DEMNAS data (2018) were integrated to update elevation maps with high geometric accuracy[17].

The land use classification used Sentinel-2A Level-1C imagery (2023), preprocessed with atmospheric correction using the Dark Object Subtraction (DOS) method, and geometric correction with 70 Ground Control Points[18]. The maximum Likelihood classification produced six land use categories—mangrove, agriculture, aquaculture, settlements, industry, and rice fields—validated using accuracy metrics including overall accuracy, user accuracy, producer accuracy, and kappa coefficients[19][20].

Hourly tidal data (September 2023 – February 2024) were analyzed for harmonic components to derive the tidal types, mean sea level (MSL), and highest high-water levels (HHWL)[21]. Historical sea-level anomaly (SLA) data (1993–2023) were analyzed via linear regression to identify sea-level rise (SLR) trends[22]. Field surveys and interviews provided data on the duration and depth of tidal flooding, which was validated by on-site documentation.

Rainfall data (November 2023) from the BMKG were processed to calculate daily and decadal averages, while wind data (November 2023) were classified using a modified Beaufort Scale and visualized with windrose diagrams[23]. The spatial analysis integrated elevation, land use, subsidence, and SLR trends to map tidal flooding for 2023 and predict scenarios for 2031, assuming linear trends in subsidence and SLR. Predictions were generated using ArcMap's Raster Calculator by applying conditional logic[24]. This comprehensive geospatial analysis highlights tidal flood risks and offers actionable insights into sustainable coastal management in the Demak Regency.

3 Result and Discussion

3.1 Land Subsidence

DInSAR processing results show that the Demak coastal area has a land subsidence rate ranging from +2 cm/year (indicating uplift or land rising) to -7.5 cm/year (indicating subsidence or land sinking). On average, the area experiences a land subsidence rate of 2.74 cm/year, as shown in Figure 2. Areas closer to the coast, industrial areas, and settlements experienced higher land subsidence rates than other areas (Figure 3). This is because the soil layer continues to compress closer to the coast, resulting in a greater land subsidence rate [25]. In industrial and residential areas, excessive groundwater withdrawal increases the land subsidence rate because the soil that was originally filled with water becomes empty and compact [26]. Building loads also affect the land subsidence rate, although not as much as the effect of massive groundwater withdrawal[27].

According to research conducted by Lumban-Gaol et al. (2024), the land subsidence pattern from DInSAR processing using Sentinel-1A SAR images is similar to direct measurements; therefore, it can be used as the main data. Land subsidence is one of the primary causes of tidal flooding. When the land surface drops, it becomes lower than sea level. As a result, it is easier for seawater to enter the land. This process causes tidal flooding, as seawater at the coastal boundary eventually overflows and floods in lower areas[28].

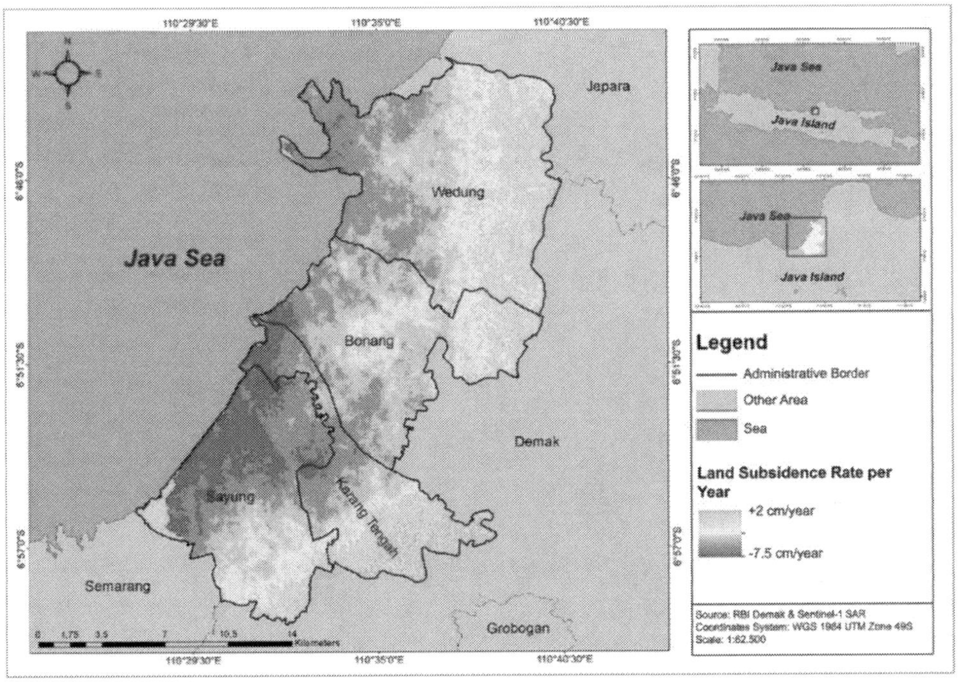

Fig. 2. Map of subsidence rate in Demak coastal area

3.2 Land Use

The results of land use processing using the maximum likelihood (MLH)-guided classification method resulted in six classes: mangroves, plantations, ponds, settlements, industries, and rice fields. These six classes are in accordance with the land cover classification standards issued by the National Standardization Agency (BSN)[29]. Land use data were used to map land use affected by tidal flood inundation. A land-use map is shown in Figure 3.

Sorted from the largest, the classes were rice fields (15,127 ha), industry (6,720 ha), settlements (6,167 ha), ponds (6,037 ha), mangroves (700 ha), and plantations (343 ha). An accuracy test was conducted by comparing the results of land-use classification from image processing with field validation data of 95 sample points. Based on the accuracy test, the results of land use classification using the maximum likelihood method are quite accurate, with the OA reaching 88% and KC value reaching 0.86.

Land use in the Sayung Sub-district is dominated by industrial and residential areas. According to the community and local officials, as a result of land subsidence and sea level rise, ponds have been damaged and no longer function; therefore, they are considered to be industrial areas, following the function of the surrounding area. In other sub-districts, in areas closer to the coast, the dominant classes were ponds and settlements. According to several studies, land subsidence has caused many changes in mangrove areas, settlements and rice fields to become ponds due to the increasing intensity of tidal surges each year[30][32].

Changes in land use further increase the risk of tidal flooding[33]. Land conversion in mangrove areas causes coastal areas to be prone to erosion. Rice fields used to be a pathway for water to enter the soil change function in ponds that are impermeable or cannot be passed by water. Residential areas that move and occupy the new area will result in land subsidence, such that the tidal inundation boundary becomes further inland.

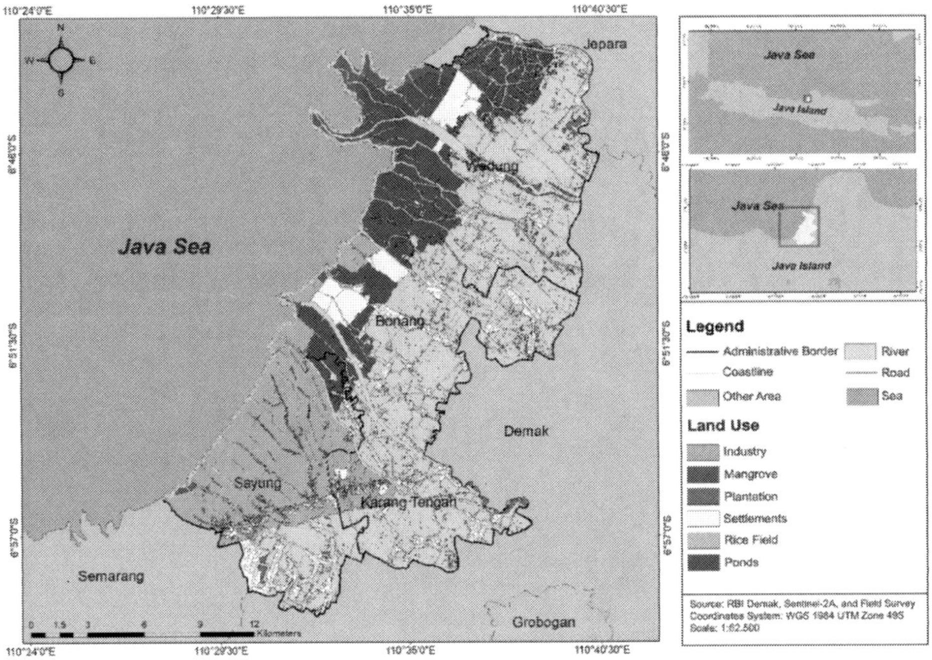

Fig. 3. Map of land use in Demak coastal area

3.3 Tides

Based on the research results, in the coastal waters of Demak Regency, the largest tide occurred on November 20, 2023 at 20.00 WIB with a height of 0.541 m, whereas the largest low tide occurred on November 22, 2023 at 05.00 WIB with a height of -0.535 meters. The Formzahl number was 1.502, and the difference between HHWL and MSL was 61.93 cm. The results of tidal data processing using the admiralty method are shown in Figure 4.

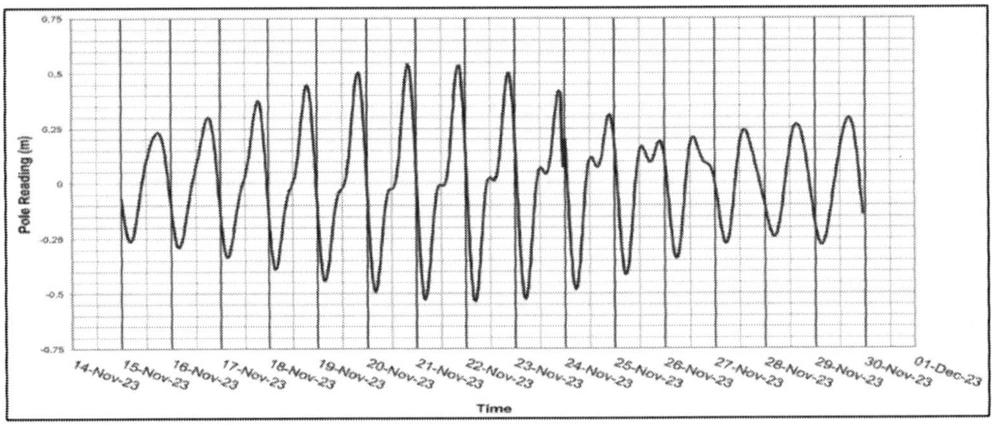

Fig. 4. Demak coastal tide chart in November 2023

The coastal waters of Demak Regency have a type of mixed tide leaning to a single day, where in one day there is one tide and ebb, but sometimes there are two tides and ebbs with different heights. This is reinforced by Sagala et al. (2024), who stated that the waters of

Demak Regency north of Java Island have a single daily tidal type[34]. High tides can cause tidal flooding because land elevation is lower than the tide. Tides in Demak waters in the west season tended to increase the intensity of tidal flooding compared to the east season. This is caused by the monsoon winds that blow from the South China Sea, carrying air masses that flow towards the Karimata Strait and Flores Sea. These winds also cross the South Coast of Kalimantan and the North Coast of Java Island, moving from west to east in accordance with the direction of the monsoon wind rotation. The influence of these monsoon winds causes higher tides[35].

3.4 Elevation

The coastal areas of the Demak Regency have elevations ranging from -3 m below sea level to 21 m above sea level, with an average of 1,893 m above sea level. Elevations were lower in areas closer to the coast owing to the influence of abrasion and land subsidence. Abrasion causes erosion of the land and reduces its elevation over time. Land subsidence can also cause a decrease in elevation in coastal areas, making them lower than areas farther from the coastline. The elevation of the study area is shown in Figure 5.

Elevation is one of the parameters that affects tidal flood inundation because the lower the elevation of an area, the higher the potential for tidal flooding[36]. Based on the research results, the coastal areas of the Demak Regency, especially those directly adjacent to the sea, have a high potential to be affected by tidal flooding because they have a lower elevation than the sea surface. In addition, the coastal areas of the Demak Regency have sloping contours, low elevation, and complex land use, making them easily affected by tides [37].

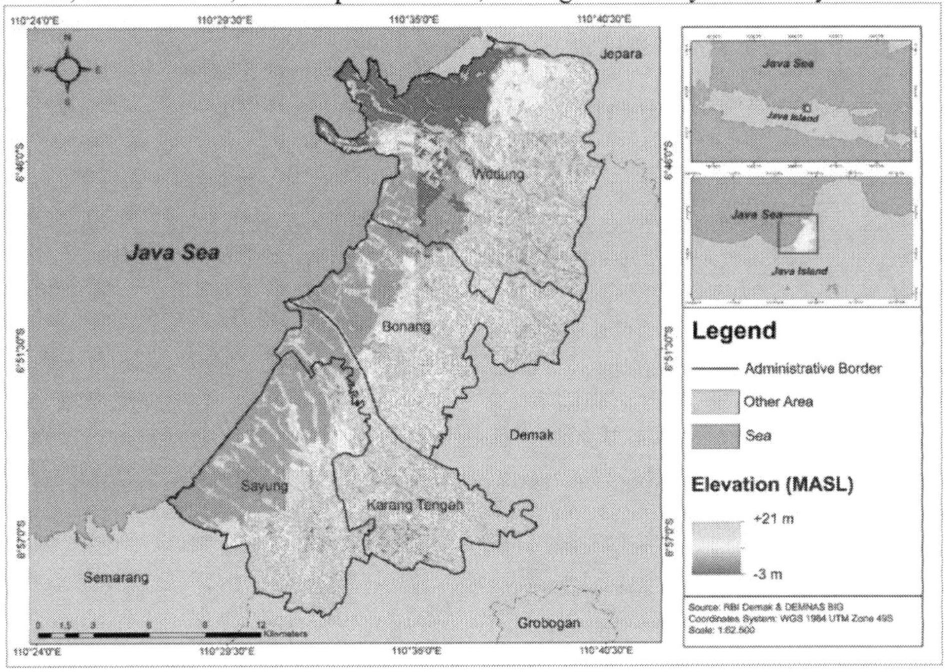

Fig. 5. Elevation of the study area

3.5 Rate of Sea Level Rise

The results of sea level anomaly data processing show that the rate of sea level rise in the coastal waters of the Demak Regency is 0.435 cm/year. This rate is higher than the global rate, which is 0.325 cm/year[38][39] and in line with the research of Lumban-Gaol et al. (2021) which states that Indonesian waters experience a sea level rise rate of 0.43 cm/year[40]. A high rate of sea-level rise can increase the risk of tidal flooding, especially during high tides. Land subsidence in this area also increases the risk of tidal flooding because every year, more land is below sea level. A graph of the rate of sea level rise is shown in Figure 6.

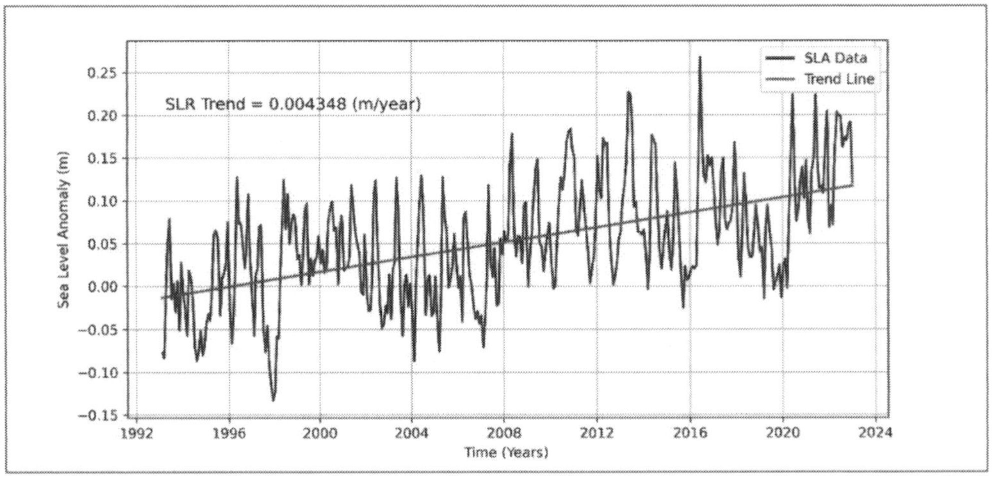

Fig. 6. Rate of sea level rise in coastal Demak

3.6 Rob Flood Inundation Height and Duration

The duration and height of the tidal flood inundation were obtained through interviews with communities in the Demak coastal area. Based on the results of interviews with residents of coastal Demak in 2024, it is known that the height of tidal flooding varies, ranging from the ankle of an adult to the calf of an adult, with a range of ± 20–30 cm. According to the local community, the worst tidal flooding occurred in the Sayung subdistrict. Tidal flooding significantly affects the social and economic lives of the coastal communities of Demak. Losses occur in the form of loss of goods, loss of income sources, damage to infrastructure, and casualties due to drifting or drowning. In addition, tidal floods disrupt transportation and the loading and unloading of goods and cause health and sanitation problems.

3.7 Wind

Seasonal wind changes greatly affect the current patterns in the coastal waters of the Demak Regency. During the western season, global current circulation is heavily influenced by changes in the monsoon winds worldwide. Winds moving from the South China Sea carry water masses, which then flow towards the Karimata Strait to the Flores Sea. These winds also cross the southern coast of Kalimantan and the northern coast of Java Island, moving from West to East. This causes current patterns in the coastal waters of the Demak Regency to be influenced by the transportation of these water masses, increasing the potential for tidal flooding[41]. The wind speed and direction measured in the coastal area of Demak Regency in November 2023 are displayed in the form of a wind rose, as shown in Figure 7. The wind rose aims to show the frequency of wind speed (m/s) and wind direction at the location and time of the study.

Winds in November 2023 on the coast of Demak Regency have a range of 0.5 m/s - 5.7 m/s with dominant wind speeds in the range of 0.5 m/s - 2.1 m/s and an average of 1.76 m/s. The dominant wind direction was from the northwest. Based on the Beaufort scale, the wind on the coast of the Demak Regency falls under the category of light breezes. According to research conducted by Fitrianti et al. (2018), light winds do not affect the distribution of tidal floods because they do not cause significant changes in tide heights, but moderate and strong winds can affect tidal floods[42]. The coastal area of Demak Regency has waters facing northwest; therefore, if there is moderate or strong wind during a tidal flood event, the tidal flood distribution will increase and exacerbate the tidal flood event.

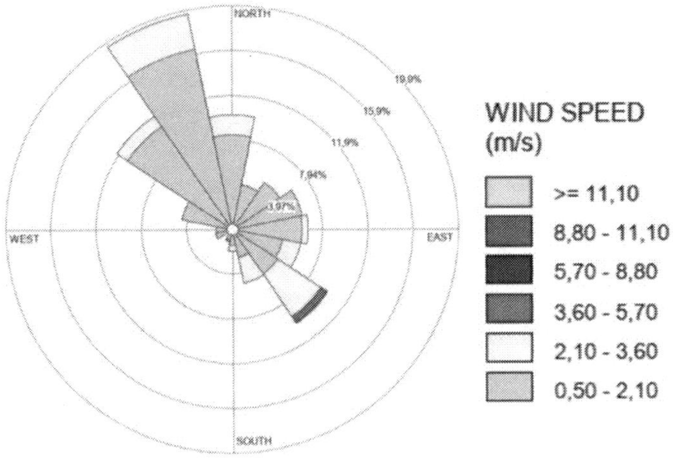

Fig. 7. Wind rose in November 2023 in coastal Demak

3.8 Rainfall

Decadal rainfall was calculated by dividing the total monthly rainfall into three periods of one month. According to the daily rainfall data, the beginning of the rainy season (AMH) occurred in mid-November. This refers to the BMKG standard; namely, if the rainfall value in three consecutive dashes is more than 50 mm, it can be said to be AMH[43]. Based on research conducted by Syafitri and Rochani (2021), the rainy season that lasts during November affects tidal flood inundation[44]. This is because rain causes soil to become saturated. Consequently, areas affected by tidal floods experience deeper inundation and take longer to recede.

High rainfall can also affect the expansion of tidal flood areas because it will increase the volume of water in the tidal inundation if it occurs at the same time as the tidal flood event, especially if there is tidal flooding around areas where there are many rivers, because it will cause river water to overflow inland[45]. A Dasarian histogram is shown in Figure 8.

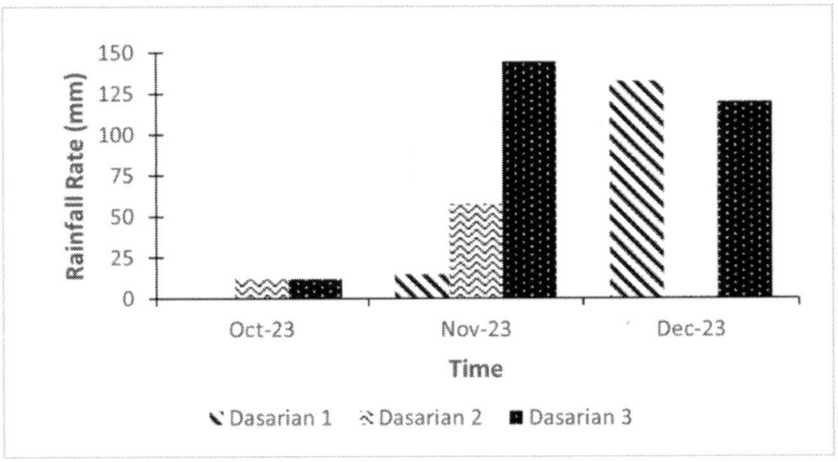

Fig. 8. Histogram of dasarian rainfall in coastal Semarang City

3.9 Distribution of Rob Flood in 2023 and its Prediction in 2031

The results of the spatial analysis of tidal flood inundation can be mapped onto the coastal land-use map of the Demak Regency, resulting in the tidal flood inundation distribution map presented in Figure 9. In 2023, all subdistricts bordering the sea in the Demak Regency experienced tidal flooding. The total inundation area reaches 8,954.6 ha or approximately 25.5% of the total area of the four sub-districts located on this coast. The most affected sub-district is Sayung and the least affected is Karang Tengah. Tidal flooding also inundates land use in this area. In the Bonang, Karang Tengah, and Wedung sub-districts, the most affected land use is ponds, causing significant losses to local communities, especially those that depend on the aquaculture sector.

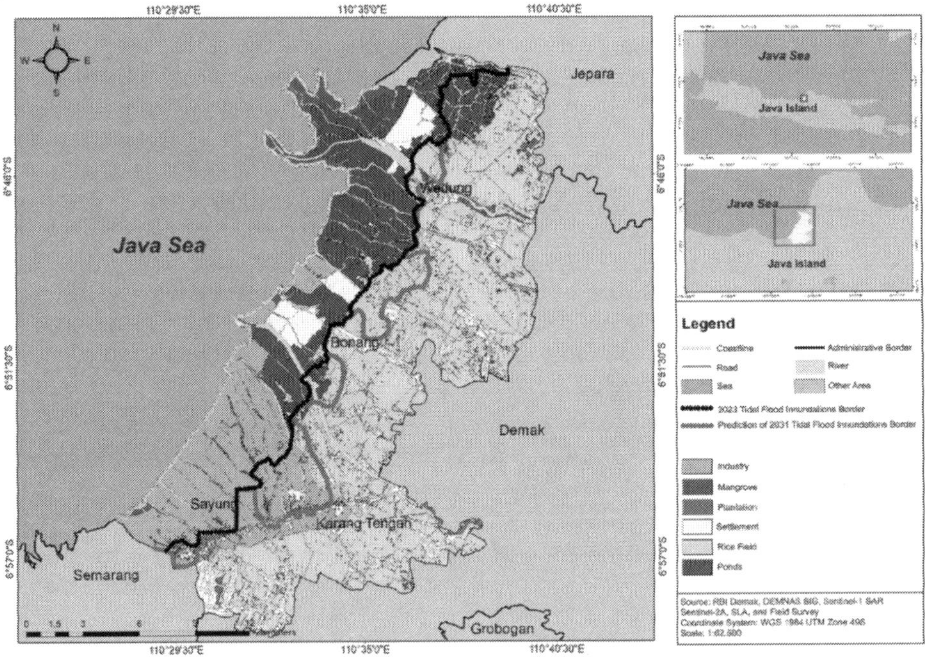

Fig. 9. Tidal flood inundation distribution map

In Bonang sub-district, affected are predicted to increase from 774.5 hectares (68.5%) in 2023 to 1,007.7 hectares (89.1%) in 2031. Settlements also increased from 527.2 hectares (27.4%) to 628.2 hectares (32.7%). Meanwhile, rice fields, which were initially only affected by 16 hectares (0.3%) in 2023, will increase significantly to 257.9 hectares (5%) in 2031. In Karang Tengah Sub-district, the affected industrial land increased from 195.3 hectares (18.5%) to 341.5 hectares (32.3%). Ponds also increased from 355.4 hectares (79.4%) to 407.5 hectares (91%). Settlements only experienced a slight increase, from 3.5 hectares (0.3%) to 5.3 hectares (0.5%).

In Sayung sub-district, the impact of tidal flood on industrial land increased significantly from 2,997.7 hectares (56.5%) to 3,765.9 hectares (71%). The number of affected ponds increased from 2,593 hectares (61.8%) to 2,978.4 hectares (70.9%). Affected paddy fields increased from 0.3 hectares (0.1%) to 34.7 hectares (2.2%). In the Wedung sub-district, the affected pond land increased from 2,593 hectares (61.8%) in 2023 to 2,978.4 hectares (70.9%) in 2031. Settlement land increased from 549.3 hectares (34.1%) to 567.5 hectares (35.2%). While paddy fields experienced a significant increase from 8.4 hectares (0.2%) to 202.8 hectares (3.7%).

Tidal flooding in the coastal Demak Regency is influenced by several factors, such as land subsidence, sea level rise, tides, and low area elevation. Land subsidence caused by groundwater exploitation and soil compaction (Novita et al. 2021), makes the already low coastal land of Demak Regency even lower. In line with previous research, wind direction, speed, and rainfall did not significantly affect the distribution of tidal flood inundation because tidal floods are not always accompanied by extreme wind and rainfall conditions[46][47].

Land subsidence lowers the land surface below sea level, making the area more vulnerable to inundation during high tides[48]. Sea-level rise leads to higher sea levels, making coastal areas more vulnerable to inundation during high tides or storms. This results in an increase in the frequency and intensity of tidal flooding, as higher seawater more easily overflows onto land[49].

Tidal flooding in the coastal areas of Demak Regency has a significant impact on land use. One of the affected areas is the mangrove forest, which plays a crucial role in mitigating tidal flooding owing to its strong roots and leaves that help reduce tidal surges[50]. Land conversion in mangrove areas has decreased protection against tidal flooding. Additionally, tidal flooding affects shrimp farming areas, disrupts production cycles, delays harvests, and damages farms with foreign materials, such as mud and sand[51]. Flooding also affects the industrial zones, settlements, plantations, and rice fields. In industrial areas, tidal flooding can disrupt operations and damage the factory infrastructure, thereby hindering production processes. In settlements, it damages homes and public facilities while also causing health issues such as skin diseases.

Flooding also disrupts planting and harvesting cycles in plantations and rice fields, forcing farmers to delay planting new crops and harvesting, directly impacting agricultural productivity, reducing crop yields, and affecting farmers' incomes. A comprehensive strategy, including the construction of flood barriers, improved drainage systems, and sustainable environmental management, is needed to address tidal flooding. Additionally, educating and empowering communities on climate adaptation and tidal flood risk mitigation are essential. The government should also collaborate with relevant stakeholders to implement mangrove reforestation as a natural barrier to erosion and flooding.

4 Conclusion

Spatial analysis of tidal flooding in the coastal areas of Demak Regency shows that in 2023, all sub-districts directly bordering the sea were affected, with a flood area of 8,954.6 ha, or approximately 25.5% of the total area of the four sub-districts. The Sayung Sub-district was the most affected, while the Karang Tengah Sub-district was the least impacted. Affected land uses included mangrove forests, shrimp ponds, settlements, industries, and rice fields. By 2031, the flood area is projected to increase to 11,394.5 ha, or approximately 32.5% of the total area of the four subdistricts, with the most severe impact still in Sayung and the least in Karang Tengah. The threat of tidal flooding has become more serious as even plantation areas, which are at higher elevations and farther from the sea, are now submerged. Factors contributing to tidal flooding in Demak include land subsidence, sea level rise, tidal surges, and the low elevation of the region. Tidal flooding has a significant impact on land use, including mangrove forests, shrimp ponds, industrial areas, settlements, plantations, and rice fields. The damage includes the disruption of industrial production, planting and harvesting cycles, destruction of infrastructure, and reduced income for local communities.

References

1. Y.T. Triana, Z. Hidayah, *Juvenil: Jurnal Ilmiah Kelautan dan Perikanan* **1**, 1 (2020)
2. [IPCC] International Panel for Climate Change. 2018, New York (US): Cambridge University Press.
3. X. Chen, X. Zhang, J.A. Church, C.S. Watson, M.A. King, D. Monselesan, B. Legresy, C. Harig, *Nature Climate Change* **7**, 7 (2017)
4. J.T. Fasullo, R.S. Nerem, *Altimeter-era emergence of the patterns of forced sea- level rise in climate models and implications for the future* in Proceedings of the National Academy of Sciences 18 Dec 2018, Toulouse, France (2018)
5. B. Fox-Kemper, H.T. Hewitt, C. Xiao, G. Aðalgeirsdóttir, S.S. Drijfhout, T.T. Edwards. 2021. *Ocean, cryosphere, and sea level change* In Climate Change, Cambridge (2021)
6. N.A. Azuga, *Jurnal Riset Kelautan Tropis* **3,** 2 (2021)
7. C. Quagliolo, P. Roebeling, F. Matos, A. Pezzoli, E. Comino, *Science of The Total Environment* **902** (2023)
8. R. Pelich, M. Chini, R. Hostache, P. Matgen, C. López-Martínez, *IEEE Geoscience and Remote Sensing Letters* **18**, 10 (2020)
9. E.S. Wulandari, R.A. Aziz, *Model prediksi dengan artificial neural network untuk kejadian banjir rob di wilayah pesisir Kota Bandar Lampung* in Prosiding Seminar Nasional Darmajaya 29 Aug 2022, Lampung, Indonesia (2022)
10. D. Kurniawan, S.Y. Prasetyo, C. Fibriani, *Indonesian Journal of Computing and Modeling* **4**, 1 (2021)
11. L.N. Istiqomah, L.M. Sabri, B. Sudarsono, *Jurnal Geodesi Undip* **9**, 2 (2020)
12. D. Wacano, A.A. Rif'an, E. Yuniastuti, R.W. Daulay, M.A. Marfai, *Seri Bunga Rampai Pengelolaan Lingkungan Zamrud Khatulistiwa* (Percetakan Kanisius, Yogyakarta 2013)
13. A. Asrofi, D.S. Hadmoko, *Jurnal Ketahanan Nasional* **23**, 2 (2017)
14. D. Nafisah, H. Setiyono, H. Hariyadi, Pemetaan sebaran genangan rob di pesisir bonang, Kabupaten Demak. *Journal of Oceanography* **6**, 3 (2017)
15. I.G. Darmawan, Z.K. Manurung, M. Nurul, W.P. Prihadita, K. Karyanto, *Jurnal Geosaintek.* **7**, 2 (2021)
16. E.D. Safitri, N. Hayati, F. Bioresita, *Jurnal Teknik ITS* **10**, 2 (2021)

17. B. Setiawan, G.D.P.N. Akbar, F.D. Suyeda, R. Permana, *Applicable Innovation of Engineering and Science Research (AVoER)* **14**, 1 (2022)
18. D.B. Susetyo, *Geodesy and Cartography* **49**, 4 (2023)
19. M. Prabandaru, *Jurnal Ilmiah Geomatika.* **2**, 1 (2022)
20. F.B. Simamora, B. Sasmito, Hani'ah, *Jurnal Geodesi Undip* **4** (2015)
21. O. Silitonga, D. Purnama, E. Nofriadiansyah, *Jurnal Enggano* **3**, 1 (2018)
22. O.S.R. Ongkosongo, Suyarso, Pasang Surut (Lembaga Ilmu Pengetahuan Indonesia, Jakarta, 1989)
23. Copernicus Climate Change Service, Sea level gridded data from satellite observations for the global ocean from 1993 to present (Copernicus Climate Change Service, Climate Data Store, 2018)
24. A. Ma'rufatin, A. Yananto, W.W. Pandoe, *Jurnal Teknologi Lingkungan* **25**, 1 (2024)
25. A. Suanda, D.M. Driptufany, D. Defwaldi, F. Fajrin, I. Armi, *Aerospace Engineering* **1**, 2 (2024)
26. B. Sasmito, *Elipsoida: Jurnal Geodesi dan Geomatika* **3**, 2 2020)
27. I. Gumilar, H.Z. Abidin, L.M. Hutasoit, D.M. Hakim, D.A. Sarsito, H. Andreas, T.P. Sidiq, *Indonesian Journal of Geospatial* **1**, 4 (2012)
28. B.D. Yuwono, H.Z. Abidin, M. Hilmi, *Analisa geospasial penyebab penurunan muka tanah di Kota Semarang* in Prosiding Sains Nasional dan Teknologi 5-11 Juli 2013 , Semarang, Indonesia (2013)
29. A.W. Syafitri, A. Rochani, *Jurnal Kajian Ruang* **1**, 1 (2021)
30. A.C. Kusuma, I. Irwani, S. Widada, *Journal of Marine Research* **2**, 3 (2013)
31. S. Munasikhah, *Journal of Indonesian History* **10**, 2 (2021)
32. M.S. Kuba, N.A. Rumata, C.A. Amal, *Journal of Green Complex Engineering* **1**, 2 (2024)
33. P.M. Sagala, R.K. Bhomia, D. Murdiyarso, *Regional Studies in Marine Science* **70** (2024)
34. S. Widada, B. Rochaddi, H. Endrawati, *Buletin Oseanografi Marina* **1**, 2 (2012)
35. A.D. Ramdhany, H.W. Wiranegara, M.N. Luru, *Jurnal Bhuwana* **1**, 2 (2021)
36. G. Bariroh, H.K. Surtikanti, *Applied Environmental Science* **1**, 2 (2024)
37. J. Lumban-Gaol, S. Vignudelli, I.W. Nurjaya, N.M.N. Natih, M.E. Sinurat, R.E. Arhatin, E.E. Kusumaningrum, *IOP Conf. Ser. Earth Environ. Sci.* **944** (2021)
38. N. Fitrianti, Y.D. Haryanto, E.V. Simamora, H.F. Bintari, A. Hartoko, S. Anggoro, M. Zainuri, *IOP Conference Series: Earth and Environmental Science* **162**, 1 (2018)
39. M.G. Novita, M. Helmi, R. Widiarati, H. Hariyadi, A. Wirasatriya, *Indonesian Journal of Oceanography* **3**, 3 (2021)
40. J. Jamalludin, K.I. Fatoni, T.M. Alam, W.S. Pranowo, *Jurnal Chart Datum* **2**, 2 (2016)
41. P.M. Kurniawan, P. Widodo, H.J. Saragih, P. Suwarno, E. Legowo, *G-Tech: Jurnal Teknologi Terapan* **7**, 2 (2023)
42. N.R. Adawiyah, T. Triyatno, *YASIN* **4**, 3, (2024)
43. A.S. Afifah, M.M. Sari, S. Suhardono, I.W. Suryawan, *Jurnal Serambi Engineering* **8**, 4 (2023)
44. R. Dewi, A.P. Ivo, S. Aninda, *Jurnal Rekayasa, Teknologi, dan Sains* **7**, 2 (2023)
45. R. Erlani, W.H. Nugrahandika, *Journal of Regional and Rural Development Planning (Jurnal Perencanaan Pembangunan Wilayah dan Perdesaan)* **3**, 1 (2019)

https://doi.org/10.1051/bioconf/202516805013

Geospatial modelling of coastal exposure index along the Madura Strait

Herlambang Aulia Rachman[1*] and *Zainul* Hidayah[1]

[1]Department of Marine and Fisheries, Faculty of Agriculture, University of Trunojoyo Madura

Abstract. The Madura Strait region is an area with high community activity and population density within East Java Province. This region is also known as the estuary of several major rivers, resulting in a relatively well-preserved mangrove ecosystem.This study aims to evaluate the impact of the mangrove ecosystem in the Madura Strait using the Integrated Valuation of Ecosystem Services and Tradeoffs (InVEST) software. The analysis focused on the Coastal Exposure Index (CEI) under the scenarios with the prcenses od mangrove as habitats. The CEI results categorized the coastline into four vulnerability classes: Low, Moderate, High, and Critical. The results of this study indicate that the CEI categories in the Madura Strait primarily fall under the Low category, covering 324.51 km (48.98% of the total coastline length). The Critical and High categories span 117.97 km and 30.71 km, respectively, accounting for 17.81% and 4.64%. The regions with the highest vulnerability are Sidoarjo, Surabaya, Pamekasan, and Sampang. The primary factors influencing vulnerability in these areas are wave exposure and the elevation of the region. It is also noted that mangrove ecosystems play a significant role, with coastlines that have dense mangrove cover exhibiting relatively lower vulnerability levels.

1 Introduction

The Madura Strait is a region in East Java Province characterized by significant economic activity, largely due to its location in a densely populated area. It is situated at the confluence of two major rivers on the island of Java, the Bengawan Solo and Brantas rivers. These rivers deposit large amounts of sediment in the estuarine area, creating ideal conditions for the growth of mangrove ecosystems. The abundant sedimentation supports the establishment and development of mangrove forests. As a result, the local coastal communities in the Madura Strait are heavily dependent on the mangrove ecosystems for a range of ecological and economic services [1].

Mangrove are an important ecosystem in coastal areas that provide vital ecosystem services for the coastal environment, such as being breeding grounds for fish, carbon sink, filtering waste and sediment from the land, as well as protecting the coastline [2–4]. As one of the tropical forest ecosystems, mangroves have a complex structure that serves as a protector of coastlines from various phenomena such as abrasion and rising sea levels. Results from several previous studies indicate that areas located behind the mangrove

* Corresponding author: herlambang.rachman@trunojoyo.ac.id

ecosystem experience very little damage when coastal abrasion or storm phenomena occur at sea [5,6].

Climate change has emerged as a critical global issue in recent decades. Based on previous research using Global Mangrove Watch (GMW) data, mangrove in East Java Provinces tend to show an expansion in area coverages. The Madura Strait Region, particularly in Sampang Regency is frequently affected by flooding causing by the seawater or tidal flooding [7]. For island nations such as Indonesia, rising sea levels represent a substantial and immediate threat to both the environment and human settlements. [8,9]. The North Coast of Java (commonly referred to as Pantura) is a region highly susceptible to the impacts of sea-level rise and related threats. Presently, the Java Sea is experiencing an average sea-level rise of approximately 3.4 mm per year, indicating significant vulnerability to coastal flooding and erosion. [10]. Several studies have highlighted the high coastal vulnerability of major cities along the North Coast of Java, including Semarang, Pekalongan, and Jakarta. However, there is a relative lack of research on coastal vulnerability in the eastern region of East Java Province, particularly regarding the role of mangrove ecosystems. This study aims to address this gap by employing a geospatial modelling approach to assess the Coastal Exposure Index (CEI), incorporating the influence of mangrove ecosystems in the Madura Strait area.

2 Data and Methods

2.1 Study Location

This study was carried out in the Madura Strait, East Java, Indonesia (Figure 1). The region, which separates the Java mainland from Madura Island, is characterized by extensive mangrove ecosystems. According to the 2021 National Mangrove Map, the total mangrove area in this region covers 7,688.554 hectares. Analysis of historical data indicates a positive trend in mangrove expansion over the past two decades, with an average increase of 164 hectares per year [1,11]. The Madura Strait is a semi-enclosed body of water, with its eastern portion being notably wider than the western part. The western side of the strait is dominated by the large river estuaries of Java Island, including the Brantas and Bengawan Solo rivers.

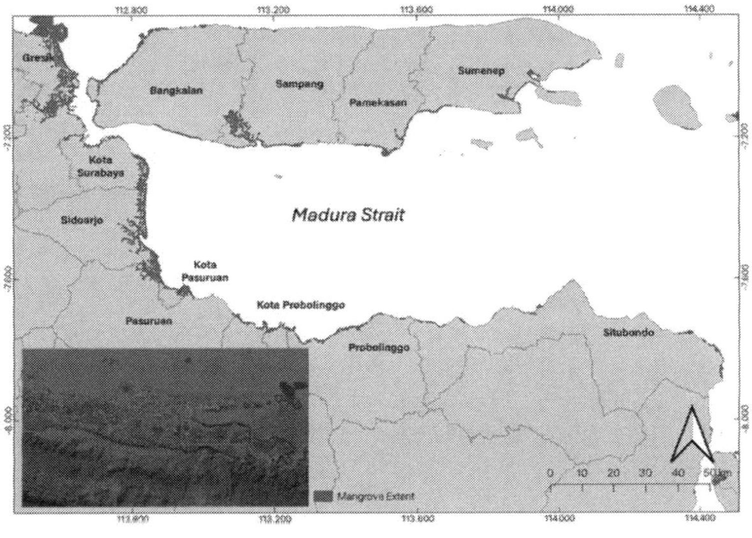

Fig. 1. Research Location of Madura Strait.

2.2 Data

This study utilizes several datasets as input for the model. Details of these datasets are presented in Table 1. The datasets will be processed using the Integrated Valuation of Ecosystem Services and Tradeoffs (InVEST) software.

Table 1. Dataset that used for calculate CEI using InVEST Software

Variable	Source
Digital Elevation Model	Reference Body (Initials followed by last name)
Continental shelf	Natural Capital Project, 2016
Landmass	Topographic Maps of Indonesia (RBI)
Sea Level Rise	Marine Copernicus Data
Natural Habitats	National Mangrove Maps [12]
Bathymetry	Geospatial Information Agency (BIG)
Coastal Geomorphology	Visual Interpretation from Sentinel-2 Imagery
Wind Dataset	Natural Capital Project, 2016

In this study, the Digital Elevation Model (DEM) and bathymetry data were obtained from the National DEM provided by Indonesia's Geospatial Information Agency (BIG), accessible through https://tanahair.indonesia.go.id/portal-web/. The DEM dataset integrates multiple data sources, including IFSAR, TERRASAR-X, and ALOS PALSAR, offering a fine spatial resolution of 0.27 arcseconds. Meanwhile, the National Bathymetry Data (BATNAS) is derived from gravity anomaly inversion based on altimetry data, supplemented with bathymetric survey measurements from BIG. The BATNAS dataset provides a spatial resolution of 6 arcseconds, referenced to the Mean Sea Level (MSL) datum.

In this study, physical data related to sea level rise and coastal geomorphology were derived from satellite imagery analysis. Sea level rise (SLR) was estimated using Sea Level Anomaly (SLA) data sourced from the Copernicus Marine Service (CMEMS), utilizing the GLORYS12V1 dataset. This dataset offers a monthly temporal resolution and a spatial resolution of 0.25 degrees, and can be accessed through the https://data.marine.copernicus.eu/product/GLOBAL_MULTIYEAR_PHY_001_030/description. The estimation of SLR was based on a linear trend analysis of the time series data spanning from 1993 to 2023. The equation for calculated the linear trend was using by calculating the slope for the linear regression. Next for the Coastal geomorphology was assessed through visual interpretation of Sentinel-2 satellite imagery, supported by high-resolution datasets such as Google Earth. The interpretation was conducted along the entire coastline of the study area, providing detailed insights into the geomorphological characteristics of the coastal zone.

Regarding coastal habitats, the data used in this research is the mangrove ecosystem coverage in the Madura Strait area, sourced from the National Mangrove Map by the Ministry of Environment and Forestry (MoEF) [12]. The dataset was generated through the integration of multi-satellite imagery, including Landsat, Sentinel, and SPOT, to develop a standardized dataset for mapping mangrove ecosystems across Indonesia.

2.3 Method

In this study, the Coastal Variability index is derived from the Coastal Exposure Index (CEI), calculated using the InVEST software. This software facilitates the valuation and analysis of ecosystem services, integrating spatial data. One of its key modules focuses on Coastal Vulnerability assessment. Several datasets, listed as input parameters (Table 1), are used to compute the CEI based on the following equation:

$$CEI = \left(\prod_{i=1}^{n} R_i\right)^{\frac{1}{n}}$$

Where R represents the ranking of each parameter calculated based on Table 2. The CEI formula is a modified approach previously used for calculating physical parameters in determining the Coastal Vulnerability Index (CVI).
The CEI formula was calculated by the value of Rank (R) of each parameters. This equation was modified approach based on previously used for calculating physical paraemeters in determining the Coastal Vulnerability Index (CVI) [13]. The notation n in the equation refers to the number of parameters used, which explained in Table 2. Where in this research we used 6 parameters (Table 2). The notation in the equation indicates that the values of R for each parameter are multiplied together and then raised to the power of 1/n, representing the number parameter used. The result is the CEI value with a maximum calculation of 5.

Table 2. Dataset that used for calculating CEI using InVEST Software

Variable	Very Low (1)	Low (2)	Medium (3)	High (4)	Very High (5)
Digital Elevation Model	81 – 100 percentile	61 to 80 percentile	41 – 60 percentile	21 – 40 percentile	0 – 20 percentile
Wave exposure	0 – 20 percentile	21 – 40 percentile	41 – 60 percentile	61 to 80 percentile	81 – 100 percentile
Sea Level Rise	0 – 20 percentile	21 – 40 percentile	41 – 60 percentile	61 to 80 percentile	81 – 100 percentile
Natural Habitats	Coral reef, mangrove, coastal forest	High dune, marsh	Low dune	Seagrass, kelp	No habitat
Coastal Geomorphol ogy	Rocky, high cliff, fjord, fiard, seawalls	Medium cliff, intended coast, bulkheads, small seawalls	Low cliff, glacial driff, alluvial plain, revertments, rip-rap walls	Cobble beach, estuary, lagoon, bluff	Barrier beach, sand beach, mud flat, delta
Surge potensial	0 – 20 percentile	21 – 40 percentile	41 – 60 percentile	61 to 80 percentile	81 – 100 percentile

The final CEI score was calculated using the weighted rankings of each parameter. This aggregated score was then classified into four categories based on quartile or Natural Break Point Method (25% percentile) distribution, resulting in the following classes: Critical, High, Moderate, and Low [14]. The subsequent analysis involved calculating the coastline length of each CEI class for each regency within the Madura Strait region. This process was performed using Quantum Geographic Information System (QGIS) and Python programming tools. To understand a complete description of each category can refer to Table 3 below.

Table 3. Description of each categories of CEI values

CEI Categories	Explanation
Critical	75 %> percentil of CEI value along Madura Strait. This categories means this location have the most exposure than other location based on CEI calculation. This location may have high exposure due and very vulnerable from the rising sea level, erotion, flooding, or any other phenomena.
High	50% – 75% percentile of CEI distribution. This categories means this location have the higher exposure than other averages CEI along the Madura Strait.
Moderate	25% - 50% percentile of CEI distribution. This categories means this location have the exposure slightly lower than averages along the location.
Low	<25% percentile of the CEI distribution. This categories means this location have the exposure is lower than other location. This could mean that the location is very safe due to its very low exposure values.

3 Result and Disscussion

The Coastal Exposure Index (CEI) for the Madura Strait coastline was assessed using a multi-parameter overlay analysis, incorporating data on Sea Level Rise (SLR), topographic relief, storm surge potential, wave action, and wind patterns. This overlay analysis produced CEI values, which were subsequently classified into four categories: Critical, High, Moderate, and Low. Overall, the CEI results provide an indicative measure of coastal vulnerability across the Madura Strait region. Figure 2 presents the weighted contribution of each input parameter used in the CEI calculation. Variations in the coastal color gradient illustrate the ranking of each parameter's influence in the assessment process. The findings reveal distinct value distributions across the parameters, highlighting their varying impacts in the vulnerability evaluation.

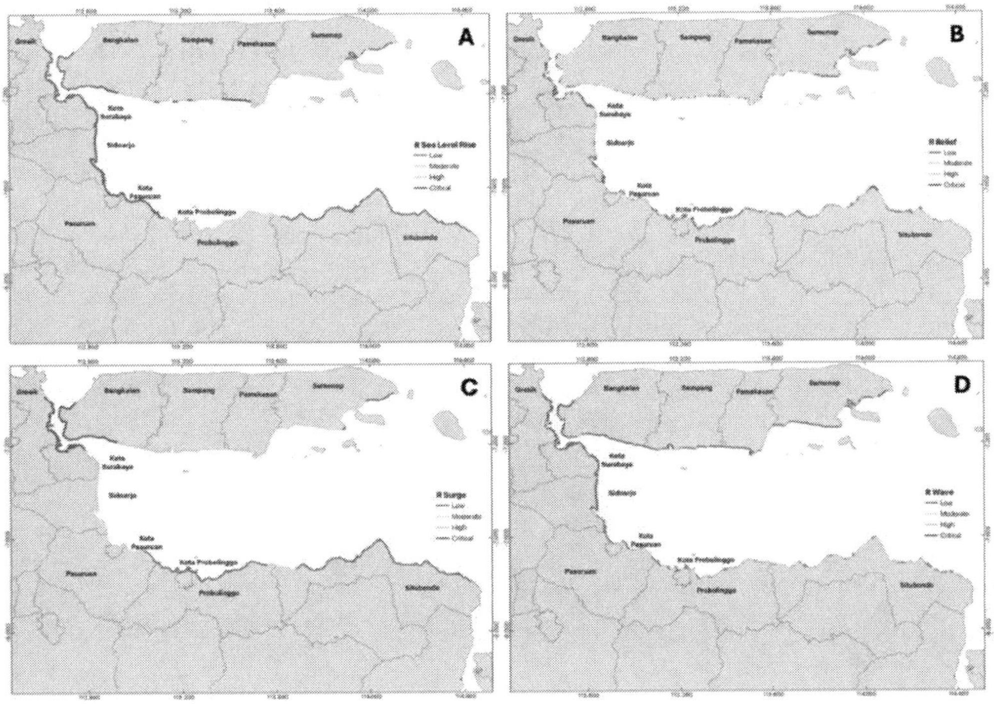

Fig. 2. Coastal Exposure Index (CEI) in Madura Strait. A) Sea Level Rise; B) Relief; C) Surge Potential; and D) Wave

The Sea Level Rise (SLR) parameter reveals that the areas with the highest risk (Critical) are located in Sidoarjo, Surabaya, and parts of Pamekasan (Figure 2A). The total coastline length classified as Critical is 318.17 km, which constitutes 42.61% of the overall coastline. The remaining categories each account for less than 20% of the total coastline in Madura Strait (Table 3). The overall rate of sea level rise in the Madura Strait, based on data analysis, is approximately 3.0 mm/year. This result aligns with previous studies, particularly in the Java Sea region, where the average sea level rise exceeds 3.2 mm/year [10]. Both regions are characterized by relatively low elevations around 0 – 5 m above sea level, rendering them highly vulnerable to the impacts of sea level rise. In contrast, areas categorized as Low and Moderate in terms of SLR are generally associated with gentler slopes.

Table 3. Dataset that use d for calculating CEI using InVEST Software

Class	Length of Rank per Parameters (km)				
	Wind	**Wave**	**Surge**	**SLR**	**Relief**
Critical	235.786	251.171	339.834	318.173	232.909
High	158.641	156.299	143.045	99.113	190.505
Moderate	188.541	186.157	118.039	157.984	178.223
Low	163.743	153.084	145.793	171.441	145.074
Total	**746.711**				

The relief parameter is an indicator derived from the slope gradient of the coastline. It is calculated using DEM (Digital Elevation Model) data, which is processed to determine slope values. The study area mainly consists of low slope gradients, leading to a large portion of the Relief parameter being classified as critical (Figure 2B). Several areas, including the coastal regions of Surabaya, Probolinggo, Situbondo, and parts of Sumenep, fall under the critical category. The coastline length identified as critical based on the Relief parameter is approximately 232.90 km (31.19%). The High category covers about 190.50 km (25.51%). These results suggest that the Madura Strait region, based on slope gradient, is highly vulnerable with significant exposure risks.

The Madura Strait is a semi-enclosed region where wind and wave exposure tend to be relatively moderate. Analysis indicates that the western part of the strait experiences higher levels of exposure, likely due to its direct connection with the open waters of the Java Sea. In contrast, the southern coast of Madura Island generally faces lower wave exposure compared to the southern coast of Java. This is because the southern part of Madura is directly influenced by wind and currents from the open waters of the Bali Sea, located to the east of the island. Meanwhile, the northern coast of Madura Strait is more protected from wind and wave action originating from the south. For a detailed overview of coastline lengths categorized by wind exposure levels, refer to Table 3.

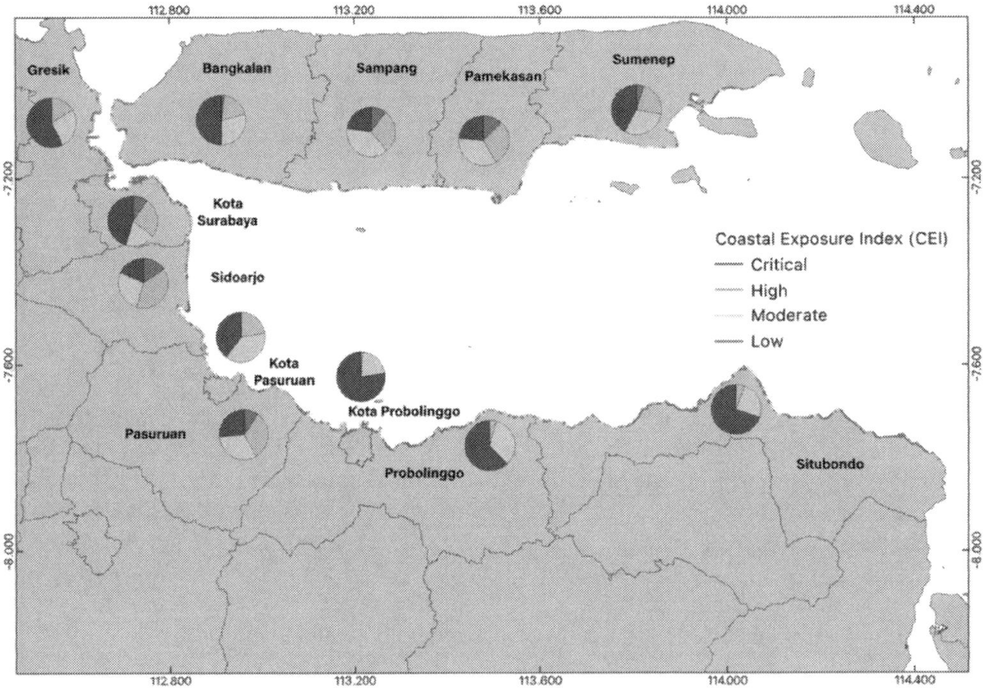

Fig. 3. Coastal Exposure Index (CEI) along the Madura Strait

Figure 3 presents the analysis of the Coastal Exposure Index (CEI) values along the Madura Strait coastline. The results indicate that the majority of the CEI categories in the Madura Strait fall within the Low category, covering 48.98% (324.51 km), followed by the Moderate category at 28.57% (189.27 km). The Critical category, however, only spans 30.71 km (4.64%). Generally, the Madura Strait is divided into two sections: the northern part (Southern of Madura Island) and the southern part (Northern of Java Island). The southern section, including the districts of Sidoarjo, Surabaya, and Pasuruan, predominantly exhibits Critical and High categories. Additionally, the northern districts of Sampang and Pamekasan also show relatively high proportions of Critical and High categories. In contrast, the eastern section, including the areas of Probolinggo and Situbondo, predominantly falls within the Low and Moderate categories.

A detailed analysis by district reveals that the longest stretch of the Critical category is found in Sidoarjo, covering 7.11 km, which constitutes 15.07% of its total coastline in Sidoarjo. This is followed by Surabaya and Pamekasan, with 6.36 km and 5.04 km, respectively, representing 9.9% and 12.6% of their total areas of each regency. In contrast, the longest stretches of the Low category are observed in Situbondo, covering 89.7 km (69.77%), and Sumenep, spanning 32.51 km (42.35%). A more detailed breakdown of the CEI category lengths for each district is provided in Table 4.

Table 4. Length of the CEI Class per Regency along the Madura Strait

Regency	CEI Class Length (km)			
	Critical	**High**	**Moderate**	**Low**
Bangkalan	1.172	12.076	18.267	31.394
Gresik		9.066	14.655	31.894
Kota Pasuruan		2.091	3.549	3.705
Kota Probolinggo		0.279	4.05	14.864
Kota Surabaya	6.361	16.242	12.424	29.19
Pamekasan	5.045	11.72	13.869	9.405
Pasuruan	2.837	9.17	9.136	7.57
Probolinggo		3.673	31.095	56.78
Sampang	4.145	11.436	15.469	9.22
Sidoarjo	7.113	17.458	12.453	8.26
Situbondo	0.287	7.261	31.324	89.719
Sumenep	3.758	17.503	22.987	32.517

The coastal area of the Madura Strait is a semi-enclosed region, offering significant protection from the open ocean. It serves as the estuary for several major rivers, leading to high sedimentation rates, especially in the western section [15]. This has led to healthy development of mangrove ecosystems in the region, with several areas demonstrating notable increases in mangrove coverage [1,16]. The results of this study also indicate that mangrove ecosystems play a significant role in reducing the Coastal Exposure Index (CEI) values in the Madura Strait region.

Sidoarjo and Surabaya districts exhibit the highest critical levels of the Coastal Exposure Index (CEI) among all assessed areas. Historically, these regions were part of the delta and estuarine system of the Brantas River Basin [17]. This leads to the region having lower and more gently sloping terrain. Furthermore, the impact of sea level rise in this area is likely to be more pronounced, resulting in higher vulnerability compared to the eastern side of the Madura Strait. Previous research also suggests that Surabaya City may face significant vulnerability if coastal urban planning is not effectively managed [18,19].

On the Madura Island side, the districts of Sampang and Pamekasan exhibit relatively high CEI levels. These regions are known to be prone to flooding during high tides or the rainy season [7,20]. Similar to Surabaya and Sidoarjo, both Sampang and Pamekasan have relatively low elevations, which contributes to their high vulnerability. Previous studies have shown that the elevation of a location is a critical parameter in determining the level of coastal vulnerability [21]. Areas with lower elevations tend to be more vulnerable compared to those with higher elevations.

The geospatial approach in this study aims to obtain a comprehensive assessment of coastal exposure conditions. The availability of Earth Observation data, whether from Remote Sensing Satellites, Modelling Approaches, or a combination or assimilation of several data sources, plays a critical role. Geospatial modelling in this study can provide an overview of the exposure conditions of an area based on the available data. However, this approach should ideally be supported by field verification, either for input data or for the outputs of the generated model. The results of this study show that geospatial modelling can provide a rapid assessment in calculating coastal vulnerability exposure values.

4 Conclusion

This study reveals that the majority of the Madura Strait region exhibits relatively low vulnerability levels. The areas with higher vulnerability include Sidoarjo, Surabaya, Sampang, and Pamekasan. The primary contributing factors to vulnerability in these regions are elevation and wave exposure. Furthermore, the presence of mangrove ecosystems plays a crucial role in mitigating coastal vulnerability. It is recommended that further research be conducted to better understand the extent to which mangrove ecosystems in the Madura Strait act as protective buffers for the coastal zone.

Acknowledgments

We would like to express our gratitude to the Institute of Research and Community Services for providing funding for this study through the Early Research Scheme, with contract number 087/UN46.4.1/PT.01.03/RISMAN/2024.

References

1. Z. Hidayah, Abd. R. As-syakur, and H. A. Rachman, Mar. Policy **163**, 106128 (2024).
2. M. A. Kusumaningtyas, T. L. Kepel, T. Solihuddin, A. A. Lubis, A. D. P. Putra, U. Sugiharto, R. N. A. Ati, H. L. Salim, E. Mustikasari, A. Heriati, A. Daulat, N. Sudirman, D. D. Suryono, and A. Rustam, Ecol. Res. **37**, 80 (2022).
3. T. Warningsih, K. Kusai, L. Bathara, Z. Zulkarnain, and D. Deviasari, IOP Conf. Ser. Earth Environ. Sci. **695**, 012036 (2021).
4. E. Sumarga, A. Sholihah, F. A. E. Srigati, S. Nabila, P. R. Azzahra, and N. P. Rabbani, Forests **14**, 1796 (2023).
5. N. Sudirman, H. L. Salim, A. Rustam, R. N. A. Ati, R. Hernina, Y. Marini, and D. D. Suryono, IOP Conf. Ser. Earth Environ. Sci. **1148**, 012041 (2023).
6. F. Fatimatuzzahroh, S. P. Hadi, and H. Purnaweni, E3S Web Conf. **31**, 08028 (2018).
7. L. Y. Irawan, M. R. Nadhifah, Sumarmi, A. K. Putra, and A. Tanjung, IOP Conf. Ser. Earth Environ. Sci. **485**, 012050 (2020).
8. R. T. Vinata, M. T. Kumala, and C. Yustisia Serfiyani, Mar. Policy **148**, 105443 (2023).
9. M. Zikra, Suntoyo, and Lukijanto, Procedia Earth Planet. Sci. **14**, 57 (2015).
10. M. W. Strassburg, B. D. Hamlington, R. R. Leben, P. Manurung, J. Lumban Gaol, B. Nababan, S. Vignudelli, and K.-Y. Kim, Clim. Past **11**, 743 (2015).
11. P. Bunting, A. Rosenqvist, R. M. Lucas, L.-M. Rebelo, L. Hilarides, N. Thomas, A. Hardy, T. Itoh, M. Shimada, and C. M. Finlayson, Remote Sens. **10**, 1669 (2018).

12. Ditjen PDASRH, *Peta Mangrove Nasional Tahun 2021* (Kementerian Lingkungan Hidup dan Kehutanan, 2021), p. 181.
13. V. Gornitz, Paleogeography Palaeoclimatol. Palaeoecol. **89**, 379 (1991).
14. B. Ai, Y. Tian, P. Wang, Y. Gan, F. Luo, and Q. Shi, Sustainability **14**, 6913 (2022).
15. V. D. Prasita, R. S. Bintoro, I. N. Permatasari, S. Widagdo, E. Sugianto, and N. Rosana, Indones. J. Geogr. **55**, (2023).
16. F. Sidik, D. Neil, and C. E. Lovelock, Mar. Pollut. Bull. **107**, 355 (2016).
17. S. Adi, I. Jänen, and T. C. Jennerjahn, Asian J. Water Environ. Pollut. **10**, 5 (2013).
18. I. H. Hariyanto and I. Gede Bhadra Pranaya Sardjitha, IOP Conf. Ser. Earth Environ. Sci. **1276**, 012064 (2023).
19. A. R. Okvitasari, A. R. Fatoni, A. Bahtiar, A. Hermanto, and M. F. N. Aulady, (n.d.).
20. L. Y. Irawan, Sumarmi, D. Panoto, I. H. Pradana, R. Faizal, M. M. R. Devy, and D. B. P. Putra, IOP Conf. Ser. Earth Environ. Sci. **1089**, 012013 (2022).
21. N. M. Noor and K. N. Abdul Maulud, J. Mar. Sci. Eng. **10**, 595 (2022).

AUTHOR INDEX